T0093976

Unmanned Aircraft Systems

Kimon P. Valavanis · Paul Y. Oh · Les A. Piegl

Unmanned Aircraft Systems

International Symposium on Unmanned Aerial Vehicles, UAV '08

Previously published in the Journal of Intelligent & Robotic Systems
Volume 54, Issues 1–3, 2009

 Springer

Kimon P. Valavanis
Department of Electrical and
Computer Engineering
School of Engineering and Computer Science
University of Denver
Denver, CO 80208
USA
kimon.valavanis@du.edu

Les Piegl
Department of Computer Science & Engineering
University of South Florida
4202 E. Fowler Ave.
Tampa FL 33620
USA
piegl@csee.usf.edu

Paul Oh
Applied Engineering Technology
Drexel University
3001 One Drexel Plaza
Market St., Philadelphia PA 19104
USA
paul@coe.drexel.edu

Library of Congress Control Number: 2008941675

ISBN- 978-1-4020-9136-0   e-ISBN- 978-1-4020-9137-7

Printed on acid-free paper.

Contents

UAS SIMULATION TESTBEDS AND FRAMEWORKS

Guest Editorial for the Special Volume On Unmanned Aircraft Systems (UAS)

Kimon P. Valavanis · Paul Oh · Les Piegl

Originally published in the Journal of Intelligent and Robotic Systems, Volume 54, Nos 1–3, 1–2.
© Springer Science + Business Media B.V. 2008

Dear colleagues,

This special volume includes reprints and enlarged versions of papers presented in the *International Symposium on Unmanned Aerial Vehicles*, which took place in Orlando FL, June 23–25.

The main objective of UAV'08 was to bring together different groups of qualified representatives from academia, industry, the private sector, government agencies like the Federal Aviation Administration, the Department of Homeland Security, the Department of Defense, the Armed Forces, funding agencies, state and local authorities to discuss the current state of unmanned aircraft systems (UAS) advances, the anticipated roadmap to their full utilization in military and civilian domains, but also present current obstacles, barriers, bottlenecks and limitations to flying autonomously in civilian space. Of paramount importance was to define needed steps to integrate UAS into the National Airspace System (NAS). Therefore, UAS risk analysis assessment, safety, airworthiness, definition of target levels of safety, desired fatality rates and certification issues were central to the Symposium objectives.

Symposium topics included, among others:

AS Airworthiness
UAS Risk Analysis
UAS Desired Levels of Safety
UAS Certification
UAS Operation
UAS See-and-avoid Systems
UAS Levels of Autonomy

K. P. Valavanis (✉) · P. Oh · L. Piegl
Department of Electrical and Computer Engineering,
School of Engineering and Computer Science, University of Denver,
Denver, CO 80208, USA
e-mail: kvalavan@du.edu, kimon.valavanis@du.edu

UAS Perspectives and their Integration in to the NAS
UAS On-board systems
UAS Fail-Safe Emergency Landing Systems
Micro Unmanned Vehicles
Fixed Wing and Rotorcraft UAS
UAS Range and Endurance
UAS Swarms
Multi-UAS coordination and cooperation
Regulations and Procedures

It is expected that this event will be an annual meeting, and as such, through this special volume, we invite everybody to visit http://www.uavconferences.com for details. The 2009 Symposium will be in Reno, NV, USA.

We want to thank all authors who contributed to this volume, the reviewers and the participants. Last, but not least, The Springer people who have been so professional, friendly and supportive of our recommendations. In alphabetical order, thank you Anneke, Joey, Gabriela and Nathalie. It has been a pleasure working with you.

We hope you enjoy the issue.

Development of an Unmanned Aerial Vehicle Piloting System with Integrated Motion Cueing for Training and Pilot Evaluation

James T. Hing · Paul Y. Oh

Originally published in the Journal of Intelligent and Robotic Systems, Volume 54, Nos 1–3, 3–19.
© Springer Science + Business Media B.V. 2008

Abstract UAV accidents have been steadily rising as demand and use of these vehicles increases. A critical examination of UAV accidents reveals that human error is a major cause. Advanced autonomous systems capable of eliminating the need for human piloting are still many years from implementation. There are also many potential applications of UAVs in near Earth environments that would require a human pilot's awareness and ability to adapt This suggests a need to improve the remote piloting of UAVs. This paper explores the use of motion platforms to augment pilot performance and the use of a simulator system to asses UAV pilot skill. The approach follows studies on human factors performance and cognitive loading. The resulting design serves as a test bed to study UAV pilot performance, create training programs, and ultimately a platform to decrease UAV accidents.

Keywords Unmanned aerial vehicle · Motion cueing · UAV safety · UAV accidents

1 Introduction

One documented civilian fatality has occurred due to a military UAV accident (non-US related) [1] and the number of near-mishaps has been steadily rising. In April 2006, a civilian version of the predator UAV crashed on the Arizona–Mexico border within a few hundred meters of a small town. In January 2006, a Los Angeles County Sheriff lost control of a UAV which then nose-dived into a neighborhood. In our own experiences over the past six years with UAVs, crashes are not uncommon. As Fig. 1 illustrates, UAV accidents are much more common than other aircraft and are increasing [2]. As such, the urgent and important issue is to design systems

J. T. Hing (✉) · P. Y. Oh
Drexel Autonomous Systems Laboratory (DASL),
Drexel University, Philadelphia, PA 19104, USA
e-mail: jth23@drexel.edu

P. Y. Oh
e-mail: paul@coe.drexel.edu

K. P. Valavanis et al. (eds.), *Unmanned Aircraft Systems.* DOI: 10.1007/978-1-4020-9136-0_2

Fig. 1 Comparison of
accident rates (data [2])

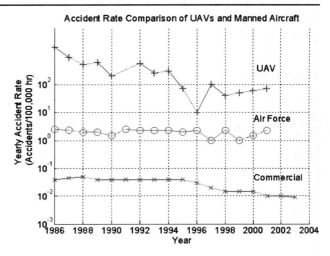

and protocols that can prevent UAV accidents, better train UAV operators, and
augment pilot performance. Accident reconstruction experts have observed that
UAV pilots often make unnecessarily high-risk maneuvers. Such maneuvers often
induce high stresses on the aircraft, accelerating wear-and-tear and even causing
crashes. Traditional pilots often fly by "feel", reacting to acceleration forces while
maneuvering the aircraft. When pilots perceive these forces as being too high, they
often ease off the controls to fly more smoothly. The authors believe that giving the
UAV pilot motion cues will enhance operator performance. By virtually immersing
the operator into the UAV cockpit, the pilot will react quicker with increased control
precision. This is supported by previous research conducted on the effectiveness of
motion cueing in flight simulators and trainers for pilots of manned aircraft, both
fixed wing and rotorcraft [3–5]. In this present study, a novel method for UAV
training, piloting, and accident evaluation is proposed. The aim is to have a system
that improves pilot control of the UAV and in turn decrease the potential for UAV
accidents. The setup will also allow for a better understanding of the cause of UAV
accidents associated with human error through recreation of accident scenarios and
evaluation of UAV pilot commands. This setup stems from discussions with cognitive
psychologists on a phenomenon called shared fate. The hypothesis explains that
because the ground operator does not share the same fate as the UAV flying in
the air, the operator often makes overly aggressive maneuvers that increase the
likelihood of crashes. During the experiments, motion cues will be given to the pilot
inside the cockpit of the motion platform based on the angular rates of the UAV. The
current goals of the experiments will be to assess the following questions in regards
to motion cueing:

1. What skills during UAV tasks are improved/degraded under various conditions?
2. To what degree does prior manned aircraft experience improve/degrade control
    of the UAV?
3. How does it affect a UAV pilot's decision making process and risk taking
    behaviors due to shared fate sensation?

This paper is part one of a three part development of a novel UAV flight training
setup that allows for pilot evaluation and can seamlessly transition pilots into a

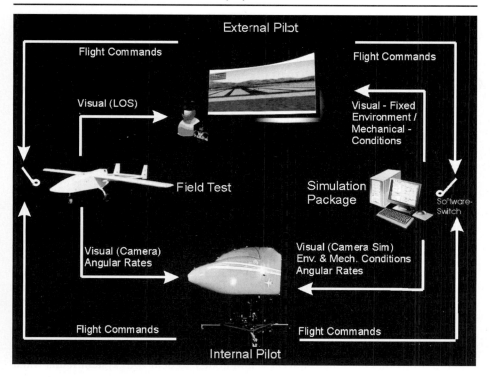

Fig. 2 Experimental setup for evaluating the effectiveness of motion cueing for UAV control. The benefit of this system is that pilots learn on the same system for simulation as they would use in the field

mission capable system. Part two will be the research to assess the effectiveness of the system and Part three will be the presentation of the complete trainer to mission ready system. As such, this paper presents the foundation of the UAV system which includes the software interface for training and the hardware interface for the mission capable system. Figure 2 shows the system and its general parts. This paper explores the use of motion platforms that give the UAV pilot increased awareness of the aircraft's state. The middle sections motivate this paper further by presenting studies on UAV accidents and how these aircraft are currently flown. It details the setup for simulation, training, human factor studies and accident assessment and presents the tele-operation setup for the real-world field tests. The final sections present and discuss experimental results, the conclusions and outlines future work.

2 UAV Operation and Accidents

While equipment failure has caused some of the accidents, human error has been found to be a significant causal factor in UAV mishaps and accidents [6, 7]. According to the Department of Defense, 70% of manned aircraft non-combat losses are attributed to human error, and a large percentage of the remaining losses have human error as a contributing factor [6]. Many believe the answer to this problem is full autonomy. However, with automation, it is difficult to anticipate all possible contingencies that can occur and to predict the response of the vehicle to all possible events. A more immediate impact can be made by modifying the way that a pilot is trained and how they currently control UAVs [8].

Many UAV accidents occur because of poor operator control. The current modes of operation for UAVs are: (1) external piloting (EP) which controls the vehicle by line of sight, similar to RC piloting; (2) internal piloting (IP) using a ground station and on board camera; and (3) autonomous flight. Some UAV systems are operated using a single mode, like the fully autonomous Global Hawk. Others are switched between modes like the Pioneer and Mako. The Pioneer used an EP for takeoff/landing and an IP during flight from a ground station. The current state of the art ground stations, like those for the Predator, contain static pilot and payload operator consoles. The pilot controls the aircraft with a joystick, rudder pedals and monitoring screens, one of which displays the view from the aircraft's nose.

The internal pilot is affected by many factors that degrade their performance such as limited field of view, delayed control response and feedback, and a lack of sensory cues from the aircraft [7]. These factors lead to a low situational awareness and decreased understanding of the state of the vehicle during operation. In turn this increases the chance of mishaps or accidents. Automating the flight tasks can have its draw backs as well. In a fully autonomous aircraft like the Global Hawk, [9] showed that because of the high levels of automation involved, operators do not closely monitor the automated mission-planning software. This results in both lowered levels of situational awareness and ability to deal with system faults when they occurred.

Human factors research has been conducted on UAV ground station piloting consoles leading to proposals on ways to improve pilot situational awareness. Improvements include new designs for head up displays [10], adding tactile and haptic feedback to the control stick [11, 12] and larger video displays [13]. To the author's knowledge, no research has been conducted in the use of motion cueing for control in UAV applications.

Potential applications of civilian UAVs such as search and rescue, fire suppression, law enforcement and many industrial applications, will take place in near-Earth environments. These are low altitude flying areas that are usually cluttered with obstacles. These new applications will result in an increased potential for mishaps. Current efforts to reduce this risk have been mostly focused on improving the autonomy of unmanned systems and thereby reducing human operator involvement. However, the state of the art of UAV avionics with sensor suites for obstacle avoidance and path planning is still not advanced enough for full autonomy in near-Earth environments like forests and urban landscapes. While the authors have shown that UAVs are capable of flying in near-Earth environments [14, 15], they also emphasized that autonomy is still an open challenge. This led the authors to focus less on developing autonomy and more on improving UAV operator control.

3 Simulation and Human Factor Studies

There are a few commercial UAV simulators available and the numbers continue to grow as the use of UAV's becomes more popular. Most of these simulators are developed to replicate the state of the art training and operation procedures for current military type UAVs. The simulation portion of our system is designed to train pilots to operate UAVs in dynamic environment conditions utilizing the motion feedback we provide them. The simulation setup also allows for reconstruction of UAV accident scenarios, to study in more detail of why the accident occurred, and allows for the placement of pilots back into the accident situation to train them on how to recover. The simulation utilizes the same motion platform and cockpit that would be used for the real world UAV flights so the transfer of the training skills to real world operation should be very close to 100%.

3.1 X-Plane and UAV Model

The training system utilizes the commercial flight simulator software known as X-Plane from Laminar Research. Using commercial software allows for much faster development time as many of the necessary items for simulation are already packaged in the software. X-Plane incorporates very accurate aerodynamic models into the program and allows for real time data to be sent into and out of the program. X-Plane has been used in the UAV research community as a visualization and validation tool for autonomous flight controllers [16]. In [16] they give a very detailed explanation of the inner workings of X-Plane and detail the data exchange through UDP. We are able to control almost every aspect of the program via two methods. The first method is an external interface running outside of the program created in a Visual Basic environment. The external program communicates with X-Plane through UDP. The second method is through the use of plug-ins developed using the X-Plane software development kit (SDK) Release 1.0.2 (freely available from http://www.xsquawkbox.net/xpsdk/). The X-Plane simulator was modified to fit this project's needs. Through the use of the author created plug ins, the simulator is capable of starting the UAV aircraft in any location, in any state, and under any condition for both an external pilot and an internal pilot. The plugin interface is shown on the right in Fig. 5. The benefit of the plugin is that the user can start the aircraft in any position and state in the environment which becomes beneficial when training landing, accident recovery and other in air skills. Another added benefit of the created plugin is that the user can also simulate a catapult launch by changing the position, orientation, and starting velocity of the vehicle. A few of the smaller UAVs are migrating toward catapult launches [17]. Utilizing X-Plane's modeling software, a UAV model was created that represents a real world UAV currently in military operation. The Mako as seen in Fig. 3 is a military drone developed by Navmar Applied Sciences Corporation. It is 130 lb, has a wingspan of 12.8 ft and is operated via an external pilot for takeoff and landings. The vehicle is under computer assisted autopilot during flight. For initial testing, this UAV platform was ideal as it could be validated by veteran Mako pilots in the author's local area. Other models of UAVs are currently available online such as the Predator A shown on the right in Fig. 3. The authors currently have a civilian Predator A pilot evaluating the accuracy of the model. The trainer is setup for the Mako such that an external pilot can train on flight tasks using an external view and RC control as in normal operation seen in Fig. 4. The system is then capable of switching to an internal view (simulated nose camera as seen in Fig. 4) at any moment to give control and send motion cues to a pilot inside of the motion platform.

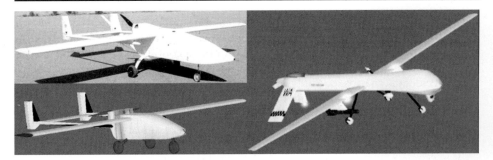

Fig. 3 *Top left* Mako UAV developed by NAVMAR Applied Sciences. *Bottom left* Mako UAV recreated in X-Plane. *Right* predator A model created by X-Plane online community

3.2 Human Factor Studies

Discussions with experienced UAV pilots of Mako and Predator A & B UAVs on current training operations and evaluation metrics for UAV pilots has helped establish a base from which to assess the effectiveness of the proposed motion integrated UAV training/control system.

The external pilot of the Mako and internal pilot of the Predator systems learn similar tasks and common flight maneuvers when training and operating the UAVs. These tasks include taking off, climbing and leveling off. While in the air, they conduct traffic pattern maneuvering such as a rectangular course and flight maneuvers such as Dutch rolls. On descent, they can conduct traffic pattern entry, go around procedures and landing approaches. These tasks are conducted during training and mission operations in various weather, day and night conditions. Each condition requires a different skill set and control technique. More advanced training includes control of the UAV during different types of system failure such as engine cutoff or camera malfunction. Spatial disorientation in UAVs as studied by [18] can effect both internal and external pilots causing mishaps. The simulator should be able to train pilots to experience and learn how to handle spatial disorientation without the financial risk of losing an aircraft to an accident.

Fig. 4 Simulator screen shots using the Mako UAV model. *Left* external pilot view point with telemetry data presented on screen. In the real world, this data is normally relayed to the pilot through a headset. *Right* internal view point with telemetry data presented. The view simulates a nose camera position on the aircraft and replicates the restricted field of view

Assessing the effectiveness of integrating motion cueing during piloting of a UAV will be conducted by having the motion platform provide cues for yaw, pitch and roll rates to the pilots during training tasks listed earlier. During simulation, the motion cues will be based on aircraft state information being fed out of the X-Plane simulation program. During field tests, the motion cues will be received wirelessly from the inertial measurement unit (IMU) onboard the aircraft. The proposed subjects will be groups of UAV internal pilots (Predator) with manned aircraft experience, UAV internal pilots without manned aircraft experience, and UAV external pilots without manned aircraft experience.

Results from these experiments will be based on quantitative analysis of the recorded flight paths and control inputs from the pilots. There will also be a survey given to assess pilot opinions of the motion integrated UAV training/control system. The work done by [19] offers a comprehensive study addressing the effects of conflicting motion cues during control of remotely piloted vehicles. The conflicting cues produced by a motion platform were representative of the motion felt by the pilot when operating a UAV from a moving position such as on a boat or another aircraft. Rather than conflicting cues, the authors of this paper will be studying the effects of relaying actual UAV motion to a pilot. We are also, in parallel, developing the hardware as mentioned earlier for field testing to validate the simulation. The authors feel that [19] is a good reference to follow for conducting the human factor tests for this study.

3.3 X-Plane and Motion Platform Interface

The left side of Fig. 5 shows the graphical user interface (GUI) designed by the authors to handle the communication between X-Plane and the motion platform ground station described in a later sections. The interface was created using Visual Basic 6 and communicates with X-Plane via UDP. The simulation interface was designed such that it sends/receives the same formatted data packet (via 802.11) to/from the motion platform ground station as an IMU would during real world flights. This allows for the same ground station to be used during simulation and field tests without any modifications. A button is programmed into the interface that allows either the attached RC controller command of the simulated UAV or the pilot inside the motion platform command at any desired moment. This would represent the external pilot control of the vehicle (RC controller) and the internal pilot control (from inside the motion platform) that would be typical of a mission setup. Currently the authors are sending angular rate data from X-Plane to the motion platform ground station and reading back into X-Plane the stick commands from the internal pilot inside the motion platform cockpit. Another powerful aspect of the program interface is that it allows the user to manipulate the data being sent out of and back into X-Plane. Noise can be easily added to the data, replicating real-world transmissions from the IMU. Time lag can also be added to data going into and out of X-plane which would represent real world data transmission delay. For example, Predator and other UAV pilots have seen delays on the order of seconds due to the long range operation of the vehicle and the use of satellite communication links [20]. Inexperienced pilots of the Predator have experienced pilot induced oscillations due to the time lag which has been the cause of some UAV mishaps.

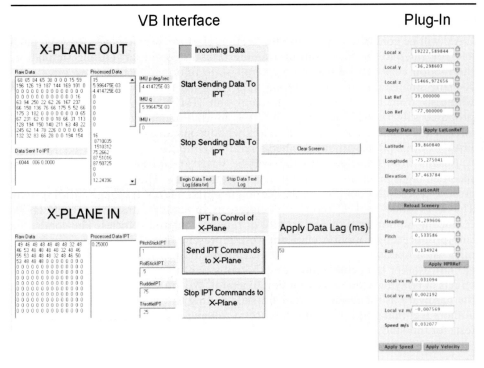

Fig. 5 *Left* graphical user interface for communication between X-Plane and IPT ground station. *Right* plugin interface running inside of X-Plane

4 Tele-operation Setup

The tele-operated system is made up of five major parts: (1) the motion platform, (2) the aerial platform, (3) the on board sensors including wireless communication, (4) the PC to remote control (RC) circuit and (5) the ground station.

4.1 Motion Platform

To relay the motion of the aircraft to the pilot during both simulation and field tests, the authors utilized a commercially available 4-*dof* flight simulator platform from Environmental Tectonics Corporation (ETC) shown in Fig. 6. ETC designs and manufactures a wide range of full-motion flight simulators for tactical fighters, general fixed-wing aircraft and helicopters. For initial development, a 4-*dof* integrated physiological trainer (IPT) system was employed because of its large workspace and fast accelerations. These are needed to replicate aircraft flight. The motion system capabilities are shown in Table 1. The cockpit is modified for specific aircrafts offering a high fidelity experience to the pilot. The visual display inside the motion platform can handle up to a 120° field of view. Basic output from the motion platform utilized in this work are the flight commands from the pilot in the form of encoder positions of the flight stick (pitch and roll), rudder pedals (yaw), and throttle.

Fig. 6 IPT 4-*dof* motion platform from ETC being wirelessly controlled with the MNAV

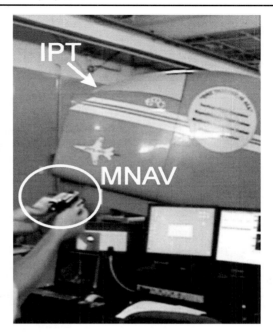

The motion platform generates the appropriate motion cues to the pilot based on the angular velocities that it receives from the ground station. Motion cues are brief movements in the direction of acceleration which give the sensation of constant motion to the pilot but are "washed out" before the motion platform exceeds its reachable workspace. Washout algorithms are commonly used by the motion platform community to return the platform to a neutral position at a rate below the threshold that humans can sense [21]. This allows the platform to simulate motions much greater than its reachable workspace. For the IPT motion platform in particular, angular rate data streaming from the MNAV is filtered and then pitch and roll rates are washed out. The yaw rate is fed straight through due to the continuous yaw capabilities of the IPT motion platform.

4.2 Aerial Platform

The authors are particularly interested in UAV rotorcraft because they are well suited to fulfill missions like medevac and cargo transport which demand hovering, pirouettes and precision positioning. For proof of concept, the immediate goal was

Table 1 Select ETC GYRO IPT II motion system capabilities

Degree of freedom	Displacement	Speed	Acceleration
Pitch	$\pm25°$	0.5–25°/s	0.5–50°/s^2
Roll	$\pm25°$	0.5–25°/s	0.5–50°/s^2
Continuous yaw	$\pm360°$ continuous	0.5–150°/s	0.5–15°/s^2

For complete specs please see ETC website

Fig. 7 The Sig Giant Kadet
model aircraft used as the
testing platform

to ensure a master-slave setup where the UAV's motions can be reproduced (in real-time) on a motion platform. To build system components, a fixed-wing UAV was used for initial demonstrations.

Rather than start with a Mako which costs on the order of thousands of dollars, the Sig Kadet offers a much cheaper, and quicker crash recovery solution for initial tests. With the Sig Kadet, the proper sensor suite and communication issues can be worked out before switching to an aircraft like the Mako shown in the earlier simulation section of this paper. The Sig Kadet shown in Fig. 7 is a very stable flight platform and is capable of carrying a sensor suite and camera system. It uses five servo motors controlled by pulse position modulated (PPM) signals to actuate the elevator, ailerons, rudder and throttle. With its 80 in. wingspan, it is comparable in size to the smaller back packable UAVs like the FQM-151 Pointer and the Raven [17].

4.3 On Board Sensors

On board the aircraft is a robotic vehicle sensor suite developed by Crossbow inertial systems. The MNAV100CA (MNAV) is a 6-*df* inertial measurement unit (IMU) measuring on board accelerations and angular rates at 50 Hz. It is also capable of measuring altitude, airspeed, GPS and heading. The MNAV is attached to the Stargate, also from Crossbow, which is an on board Linux single board computer. The Stargate is set to transmit the MNAV data at 20 Hz to the ground station via a wireless 802.11 link. As shown in Fig. 8, the MNAV and Stargate fit inside the cockpit of the Sig Kadet close to the aircraft's center of gravity.

On board video is streamed in real time to the ground station via a 2.4 GHz wireless transmission link. The transmitter is held under the belly of the Sig Kadet and the camera is located off the left wing of the aircraft. The current camera used has a 70° field of view and is capable of transmitting images at 30 FPS and 640 × 480 to a distance of 1.5 miles (AAR03-4/450 Camera from wirelessvideocameras.net). This is relatively low quality as compared with high definition camera systems but it is inexpensive, making it a decent choice for initial tests. Future tests will include much higher resolution cameras for a better visual for the pilots and a more strategic placement of the camera to replicate a pilot's on board view.

Fig. 8 MNAV and Stargate in the cockpit of the aircraft (top view)

MNAV+Stargate

4.4 PC to RC

Encoder positions of the flight stick, rudder pedals, and throttle inside the motion platform are transmitted via an Ethernet link to the ground station. The signals are then routed through a PC to RC circuit that converts the integer values of the encoders to pulse position modulated (PPM) signals. The PPM signals are sent through the buddy port of a 72 MHz RC transmitter which then transmits the signal to the RC receiver on board the aircraft. The PPM signals are routed to the appropriate servos to control the position of the ailerons, elevator, rudder, and throttle of the aircraft. The positions of the IFT flight controls are currently sent through the PC to RC link at a rate of 15 Hz.

4.5 Ground Station

The ground station used for the tele-operation system is a highly modified (by the authors) version of the MNAV Autopilot Ground station freely distributed on SourceForge.net. The modified ground station does three things. (1) It receives all the information being transmitted wirelessly from the MNAV and displays it to the user operating the ground station. (2) It acts as the communication hub between the aircraft and the motion platform. It relays the MNAV information via Ethernet link to the motion platform computers and sends the flight control positions of the motion platform to the PC to RC circuit via USB. (3) It continuously monitors the state of the communication link between the motion platform and the MNAV. If something fails it will put both the motion platform and aircraft (via the MNAV/Stargate) into a safe state. Determining if the ground station responds to an IMU or X-Plane data packets is set by assigning either the IP address of the IMU or the IP address of the simulator in the IPT ground station.

4.6 Field Tests

Current field tests have been conducted at a local RC flying field with the aircraft under full RC control. The field is approximately a half mile wide and a quarter mile deep. Avionics data such as angular velocity rates, accelerations and elevation was collected and recorded by the MNAV attached to the aircraft during flight. Video from the onboard camera was streamed wirelessly to the ground station and recorded. During each flight, the RC pilot conducted take off, figure eight patterns and landing with the Sig Kadet.

5 Initial Test Results and Discussion

As of writing this paper, the simulation portion was coming to completion and preparing for pilot testing and verification. In this section, the authors will present initial test results from the hardware control portion of the UAV system. In this prototyping stage, development was divided into three specific tasks that include: (1) motion platform control using the MNAV, (2) control of the aircraft servos using the IPT flight controls and (3) recording of actual flight data from the MNAV and replay on the IPT.

5.1 Motion Platform Control with MNAV

Aircraft angular rates are measured using the MNAV and this information is transmitted down to the ground station via a 20 Hz wireless link. Task A demonstrated the MNAV's ability to communicate with the ground station and the IPT. The MNAV was held in hand and commanded pitch, roll and yaw motion to the IPT by rotating the MNAV in the pitch, roll and yaw directions as seen in Fig. 6 (showing pitch).

Motions of the MNAV and IPT were recorded. Figure 9 shows a plot comparing MNAV and IPT data. The IPT is designed to replicate actual flight motions and therefore is not capable of recreating the very high angular rates commanded with the MNAV during the hand tests in the roll and pitch axis. The IPT handles this by decreasing the value of the rates to be within its bandwidth and it also filters out some of the noise associated with the MNAV sensor. Overall, the IPT tracked the motion being commanded by the MNAV fairly well. The IPT is limited by its reachable work space which is why the amplitude of the angular rates does not match at times.

Of considerable interest is the lag between the commanded angular rates and the response from the IPT motion platform, particularly with the yaw axis. This may be a limitation of the motion platform and is currently being assessed. Minimal lag is desired as significant differences between the motion cues from the IPT and visuals from the video feed will cause a quick onset of pilot vertigo.

5.2 Control of Aircraft Servos

Transmitting wirelessly at 15 Hz, no lag was observed between the instructor's flight box commands and the servo motor response. This is significant because it means

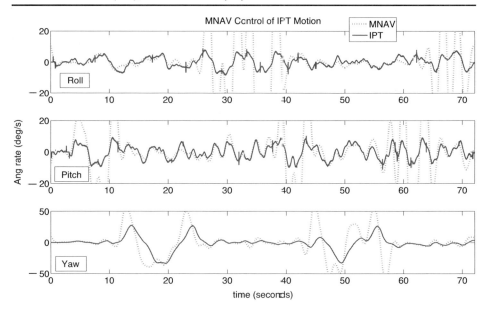

Fig. 9 Comparison of the angular rates during MNAV control of the IPT

that the pilot sitting inside the motion platform can control the aircraft through the RC link. This underscores fidelity; the aircraft will respond as if the pilot was inside its cockpit and flying the aircraft. This has only been tested during line of sight control. RC is limited in range and as stated earlier, satellite communication links for long range distances can introduce delays in data transfer. However the authors imagine near-Earth UAV applications will be conducted with groundstations near the operation site.

5.3 Record and Replay Real Flight Data

Task A demonstrated that the MNAV is able to transmit motion data to the IPT. During this task the MNAV was subjected to extreme rates and poses. Such extremes are not representative of actual aircraft angular rates but serve to demonstrate master-slave capability. To test the IPT's ability to respond to actual aircraft angular rates being sent from the MNAV, angular rate data was recorded directly from a field flight of the Sig Kadet. This data was replayed on the IPT along with on board flight video. The recorded video and flight data simulate the real time streaming information that would occur during a field tele-operation experiment. An example of the recorded angular rates from one of the field tests is shown in Fig. 10 and a still shot of the on board video recording is shown in Fig. 11.

Initial results showed errors in the angular rates between the observed motion and the recorded data. For example, the pitch rate (Fig. 10), while it is oscillating, rarely

Fig. 10 Filtered angular rates during actual aircraft flight

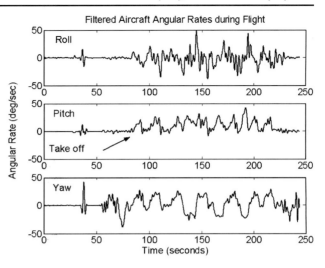

goes negative. This means that the sensor is measuring a positive pitch rate during most of the flight. Comparison of the rates with onboard aircraft video shows the error varying throughout the data so it is not a simple offset fix. This was consistently the case for multiple flights. The authors emphasize that this phenomenon was only seen during flights. Hand held motions always produced correct and expected angular rates. The recorded flight data was replayed on the IPT motion platform. This caused the IPT to travel and remain at its kinematic joint limits as was expected because of the aforementioned positive pitch rate.

The IMU was re-visited to output angular rates that reflect the bias correction made in the Kalman filter for the rate gyros [22]. A plot of the biases during a real flight is shown in Fig. 12. The resulting biases were very small and did little to fix the positive pitch rate phenomenon during flights. Alternative IMUs are thus

Fig. 11 Onboard camera view off of the left wing during flight

Fig. 12 Rate gyro biases during actual aircraft flight

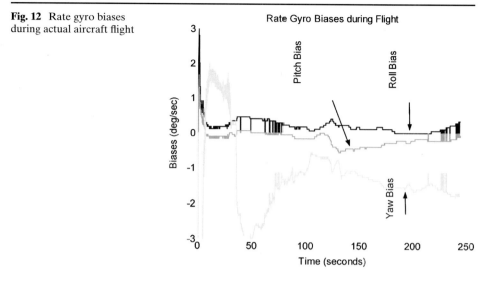

being explored at this prototyping stage. None the less, the integration of an IMU and motion platform was successfully developed. This underscores that the wireless communication interface and low-level avionics work as designed.

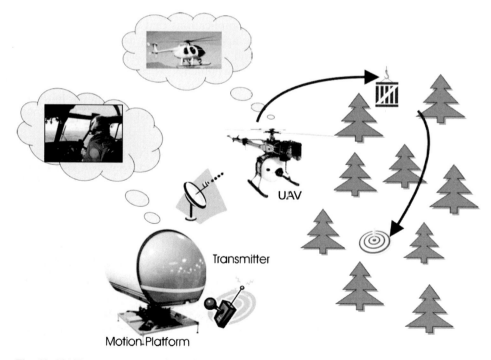

Fig. 13 UAV cargo transport in a cluttered environment using a radio link that slaves robotic helicopter motions to the motion platform. Through a "shared fate" sensation the pilot flies by "feeling" the UAV's response to maneuvers commanded by the pilot

6 Conclusion and Future Work

While the future of UAVs is promising, the lack of technical standards and fault tolerant systems are fundamental gaps preventing a vertical advance in UAV innovation, technology research, development and market growth. This paper has presented the development of the first steps toward a novel tele-operation paradigm that employs motion cueing to augment UAV operator performance and improve UAV flight training. This method has the potential to decrease the number of UAV accidents and increase the applicability of unmanned technology.

Leveraging this work, future development includes research to eliminate, reduce, or compensate for the motion lag in the motion platform. Also to be examined are additional cues like sight, touch and sound that may improve UAV control. Utilizing the system for accident reconstruction will also be assessed. The net effect is that from such understanding, one can analytically design systems to better control UAVs, train UAV pilots and help eliminate UAV accidents.

The shared fate and motion cueing will have tremendous benefit in near-Earth flying. Figure 13 depicts a notional mission involving cargo pickup and transport through a cluttered terrain to a target location. The motion platform can be used to implement a virtual "shared fate" infrastructure to command a robotic helicopter. The visuals from the helicopter's on board cameras would be transmitted to the motion platform cockpit. Added cues like audio, vibration, and motion would enable the pilot to perform precision maneuvers in cluttered environments like forests or urban structures. Future work demands the look at rotorcraft because their potential applications extend beyond the capabilities of current fixed wing UAVs. There are still a number of beneficial, life saving applications that are unachievable with current UAV methods. Among these are applications such as search and rescue and fire fighting. Even cargo transport is still very difficult to achieve autonomously in non-optimal conditions and cluttered environments. These tasks require quick, precise maneuvers and dynamic mission plans due to quickly changing environment conditions and close quarter terrain. To date these missions can only be flown by experienced, on board pilots, who still incur a great deal of risk.

Acknowledgements The authors would like to thank NAVMAR Applied Sciences for their support on the development of the UAV model and granting access to UAV pilots. The authors would also like to thank Brian DiCinti for his help with the construction of the Sig Kadet and piloting the aircraft. Acknowledgment also goes out to Canku Calargun, Caglar Unlu, and Alper Kus for their help interfacing the IPT motion platform with the MNAV. Finally the authors acknowledge Bill Mitchell, president of ETC, for his generosity in donating time on the IPT Motion platform, the supporting man power, and his overall support of this project.

References

1. Flight International: Belgians in Congo to probe fatal UAV incident. 10 October (2006)
2. Weibel, R.E., Hansman, R.J.: Safety considerations for operation of unmanned aerial vehicles in the national airspace system. Tech. Rep. ICAT-2005-1, MIT International Center for Air Transportation (2005)
3. Parrish, R.V., Houck, J.A., Martin, D.J., Jr.: Empirical comparison of a fixed-base and a moving-base simulation of a helicopter engaged in visually conducted slalom runs. NASA Tech. Rep. **D-8424**, 1–34 (1977)

4. Ricard, G.L., Parrish, R.V.: Pilot differences and motion cuing effects on simulated helicopter hover. Hum. Factors **26**(3), 249–256 (1984)
5. Wiegmann, D.A., Goh, J., O'Hare, D.: The role of situation assessment and flight experience in pilot's decisions to continue visual flight rules flight into adverse weather. Hum. Factors **44**(2), 189–197 (2001)
6. Rash, C.E., Leduc, P.A., Manning, S.D.: Human factors in U.S. military unmanned aerial vehicle accidents. Adv. Hum. Perform. Cognit. Eng. Res. **7**, 117–131 (2006)
7. Williams, K.W.: Human factors implications of unmanned aircraft accidents: flight-control problems. Adv. Hum. Perform. Cognit. Eng. Res. **7**, 105–116 (2006)
8. Schreiber, B.T., Lyon, D.R., Martin, E.L., Confer, H.A.: Impact of prior flight experience on learning predator UAV operator skills. Tech. rep., Air Force Research Laboratory Human Effectiveness Directorate Warfighter Training Research Division (2002)
9. Tvaryanas, A.P.: USAF UAV mishap epidemiology, 1997–2003. In: Human Factors of Uninhabited Aerial Vehicles First Annual Workshop Scottsdale, Az (2004)
10. Williams, K.W.: A summary of unmanned aircraft accident/incident data: human factors implications. Tech. Rep. DOT/FAA/AM-04/24, US Department of Transportation Federal Aviation Administration, Office of Aerospace Medicine (2004)
11. Calhoun, G., Draper, M.H., Ruff, H.A., Fontejon, J.V.: Utility of a tactile display for cueing faults. In: Proceedings of the Human Factors and Ergonomics Society 46th Annual Meeting, pp. 2144–2148 (2002)
12. Ruff, H.A., Draper, M.H., Poole, M., Repperger, D.: Haptic feedback as a supplemental method of altering UAV operators to the onset of turbulence. In: Proceedings of the IEA 2000/ HFES 2000 Congress, pp. 3.14–3.44 (2000)
13. Little, H.: Raytheon announces revolutionary new 'cockpit' for unmanned aircraft—an industry first. In: Raytheon Media Relations (2006)
14. Sevcik, K.W., Green, W.E., Oh, P.Y.: Exploring search-and-rescue in near-earth environments for aerial robots. In: IEEE International Conference on Advanced Intelligent Mechatronics Monterey, California, pp. 699–704 (2005)
15. Narli, V., Oh, P.Y.: Hardware-in-the-loop test rig to capture aerial robot and sensor suite performance metrics. In: IEEE International Conference on Intelligent Robots and Systems, p. 2006 (2006)
16. Ernst, D., Valavanis, K., Garcia, R., Craighead, J.: Unmanned vehicle controller design, evaluation and implementation: from matlab to printed circuit board. J. Intell. Robot. Syst. **49**, 85–108 (2007)
17. Defense, D.O.: Unmanned aircraft systems roadmap 2005–2030. Tech. rep., August (2005)
18. Self, B.P., Ercoline, W.R., Olson, W.A., Tvaryanas, A.: Spatial disorientation in unihabited aerial vehicles. In: Cook, N. (ed.) Human Factors of Remotely Operated Vehicles, vol. 7, pp. 133–146. Elsevier Ltd. (2006)
19. Reed, L.: Visual-proprioceptive cue conflicts in the control of remotely piloted vehicles. Tech. Rep. AFHRL-TR-77-57, Brooks Airforce Base, Air Force Human Resources Laboratory (1977)
20. Mouloua, M., Gilson, R., Daskarolis-Kring, E., Kring, J., Hancock, P.: Ergonomics of UAV/UCAV mission success: considerations for data link, control, and display issues. In: Human Factors and Ergonomics Soceity 45th Annual Meeting, pp. 144–148 (2001)
21. Nahon, M.A., Reid, L.D.: Simulator motion-drive algorithms: a designer's perspective. J. Guid. Control Dyn. **13**, 356–362 (1990)
22. Jang, J.S., Liccardo, D.: Automation of small UAVs using a low cost mems sensor and embedded computing platform. In: 25th Digital Avionics Systems Conference, pp. 1–9 (2006)

Networking Issues for Small Unmanned Aircraft Systems

Eric W. Frew · Timothy X. Brown

Originally published in the Journal of Intelligent and Robotic Systems, Volume 54, Nos 1–3, 21–37.
© Springer Science + Business Media B.V. 2008

Abstract This paper explores networking issues that arise as a result of the operational requirements of future applications of small unmanned aircraft systems. Small unmanned aircraft systems have the potential to create new applications and markets in civil domains, enable many disruptive technologies, and put considerable stress on air traffic control systems. The operational requirements lead to networking requirements that are mapped to three different conceptual axes that include network connectivity, data delivery, and service discovery. The location of small UAS networking requirements and limitations along these axes has implications on the networking architectures that should be deployed. The delay-tolerant mobile ad-hoc network architecture offers the best option in terms of flexibility, reliability, robustness, and performance compared to other possibilities. This network architecture also provides the opportunity to exploit controlled mobility to improve performance when the network becomes stressed or fractured.

Keywords Unmanned aircraft system · UAS · Airborne communication networks · Controlled mobility · Heterogeneous unmanned aircraft system · Mobile ad-hoc networking · Delay tolerant networking

1 Introduction

The proliferation of small unmanned aircraft systems (UAS) for military applications has led to rapid technological advancement and a large UAS-savvy workforce poised

E. W. Frew (✉)
Aerospace Engineering Sciences Department, University of Colorado,
Boulder, CO 80309, USA
e-mail: eric.frew@colorado.edu

T. X. Brown
Interdisciplinary Telecommunications Program Electrical
and Computer Engineering Department, University of Colorado, Boulder, CO 80309, USA
e-mail: timxb@colorado.edu

to propel unmanned aircraft into new areas and markets in civil domains. Small unmanned aircraft systems have already been fielded for missions such as law enforcement [29], wildfire management [34], pollutant studies [10], polar weather monitoring [11], and hurricane observation [26]. Proposed UAS span numerous more future civilian, commercial, and scientific applications. A recent study concluded that in 2017 the civil UAS market in the USA could reach $560 M out of a total (civil plus military) UAS market of approximately $5.0 B [32]. That study projects 1,500 civil UAS will be in service in 2017 and that approximately 85% of those will be small UAS.

As the number of fielded small UAS grows, networked communication will become an increasingly vital issue for small UAS development. The largest current barrier to the use of unmanned aircraft in the National Airspace System (NAS) of the USA is satisfaction of Federal Aviation Administration (FAA) regulations regarding safe flight operations and Air Traffic Control (ATC). In particular, the FAA requires almost all aircraft operating in the NAS to have a detect, sense, and avoid (DSA) capability [3] that provides an equivalent level of safety compared to manned aircraft [1, 33]. While onboard sensors are expected to be a component of future DSA solutions, communication to ATC and operator intervention will also be required, either from a regulatory or practical perspective. Thus, one of the primary concerns of the FAA regarding the ability of UAS to meet safety regulations without conflicting with existing systems is the availability and allocation of bandwidth and spectrum for communication, command, and control [2]. Although the particular regulations just mentioned refer to operation in the USA, similar concerns apply to the operation of small UAS anywhere.

Small unmanned aircraft (UA) are defined here to encompass the Micro, Mini, and Close Range categories defined in [5]. This classification means small UA have maximum takeoff weight less than or equal to 150 kg, maximum range of 30 km, and maximum altitude of 4,000 m mean sea level (MSL). The weight limit effectively means small UA can not carry the equivalent weight of a human operator. The altitude limit taken here means small UA cannot fly into Class A airspace (the airspace from 18,000 to 60,000 ft MSL where commercial aircraft fly). Although it may be possible for vehicles in this category to fly at higher altitudes, the regulatory issues are significantly more challenging and it is reasonable to assume most small UA will not fly in that airspace. In fact, most small UA would probably fly substantially closer to the ground. Likewise, the maximum range of 30 km represents typical operational limits on this class of aircraft and there can be notable exceptions [11]. Finally, note that a small UAS can be comprised of multiple heterogeneous small UA with highly varying capabilities.

Unlike larger unmanned aircraft, small UAS are in a unique regime where the ability to carry mitigating technology onboard is limited yet the potential for damage is high. Given the size and payload constraints of small UAS, these unmanned aircraft have limited onboard power, sensing, communication, and computation. Although the payload capacity of a small UAS is limiting, the kinetic energy stored in a 150 kg aircraft can cause significant damage to other aircraft, buildings, and people on the ground. Furthermore, the limited sizes of small UAS make them accessible to a wider audience (e.g. a variety of universities already have small UAS programs [4, 9, 16, 21, 28]) than larger systems and the percentage of small UAS deployed in the future will likely be high relative to larger unmanned aircraft systems [32]. The limited capabilities of small UAS lead to unique operational requirements compared to larger UA that can more easily assimilate into the existing ATC framework (e.g. larger UA can carry the same transponder equipment as manned aircraft).

This paper explores networking issues that arise as a result of the operational requirements of future applications of small unmanned aircraft systems. These requirements are derived from a set of representative application scenarios. The operational requirements then lead to networking requirements (e.g. throughput, which is the rate at which data can be sent over a communication link, and latency or delay) that greatly exceed those of current manned aircraft. Further, the networking requirements are mapped to three different conceptual axes that include network connectivity, data delivery, and service discovery. The location of small UAS networking requirements and limitations along these axes has implications on the networking architectures that should be deployed.

Of the existing possible network architectures for small UAS, only delay-tolerant mobile ad-hoc networking architectures will provide the needed communication for the large number of small aircraft expected to be deployed in the future. Since small UA are relatively cheap, future UAS will likely deploy multiple vehicles coordinated together. Many small UAS applications will require quick response times in areas where permanent supporting communication infrastructures will not exist. Furthermore, current approaches using powerful long-range or satellite communications are too big and expensive for small aircraft while smaller radios fundamentally limit the small UAS operational envelope in terms of range, altitude, and payload. The delay-tolerant mobile ad-hoc network architecture offers the best option in terms of flexibility, reliability, robustness, and performance compared to other possibilities. This network architecture also provides the opportunity to exploit controlled mobility to improve performance when the network becomes stressed or fractured.

2 Communication Requirements

2.1 Operational Requirements

This work is motivated by the Heterogeneous Unmanned Aircraft System (HUAS) developed at the University of Colorado as a platform to study airborne communication networks and multivehicle cooperative control (Fig. 1 shows the various small UA included in HUAS). Specific applications studied to date include the impact of mobility on airborne wireless communication using off the shelf IEEE 802.11b (WiFi) radios [7]; net-centric communication, command, and control of small UAS [16]; sensor data collection [22]; delay tolerant networking [6]; and a framework for controlled mobility that integrates direct, relay, and ferrying communication concepts [13].

As an example application consider a UAS to track a toxic plume. In this scenario a toxic plume has been released in an accident and the goal is to locate the plume extent and source [17]. To characterize the plume, multiple small UA fly while sensing the plume with onboard chemical sensors. Different sensors may be in different UA because it may not be possible or desirable for every small UA to carry every sensor. UA with the same chemical sensors onboard need to find each other to form gradient seeking pairs. Chemical gradients can be defined by sharing sensor

Fig. 1 The HUAS vehicle fleet includes (*clockwise from top left*) the CU Ares, the CU MUA, the Velocity XL, the MLB Bat 3, the CU ground control station, and the Hobico NextGen

data and the UAS can cooperatively track boundaries or follow the gradients to the source. UA can potentially move far away from their launching ground stations and each other. In this example, the UAS consists of potentially many heterogeneous UA. They need to fly freely over a large area and be able to dynamically form associations autonomously without relying on a centralized controller.

As a second example consider a UAS deployed to act as a communication network over a disaster area. Here, normal communication infrastructure has been damaged but various entities on the ground such as first responders, relief agencies, and local inhabitants require communication in order to organize a coordinated response. An unmanned aircraft system flying overhead can provide a meshed communication architecture that connects local devices, e.g. laptops with wireless networking or cell phones, with each other or back to the larger communication grid. Since communication demand will vary as the severity of the disaster is assessed and relief efforts are mounted, the UAS must be able to reposition itself in response. Since the actions of the ground units are in direct response to the emergency situation, the actions of the UAS must be dependent on them and not limit their efforts. Also, the UAS will likely operate in the vicinity of buildings and other manned aircraft so obstacle and collision avoidance will be critical.

The two scenarios described above share properties with many other potential small UAS applications and lead to communication requirements for the UAS itself. In particular, these communication needs can be broadly classified into platform safety, remote piloting, and payload management (Fig. 2). In general, the UAS will communicate with multiple external parties that could include ATC, the pilot, and payload operators who may be in widely separate locations.

2.1.1 Platform Safety

From a regulatory perspective, platform safety is the most critical component of an unmanned aircraft system. Like a manned aircraft, the pilot of the UAS must communicate with ATC in most controlled airspace [33]. This communication may be mediated by the UAS whereby the UAS communicates via conventional radio

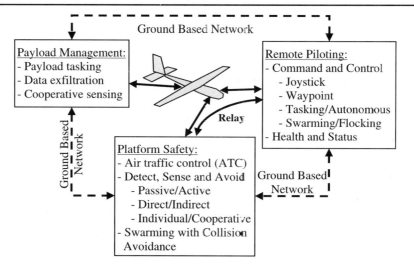

Fig. 2 Types of communication in an unmanned aircraft system

to the ATC and this communication is then backhauled to the pilot. This implies an inherent inefficiency. A single ATC radio channel is shared by all planes in an area. But, each UA requires a separate backhaul to its respective operator, and multiplies the communication requirements. Future non-voice approaches to managing aircraft are being contemplated [14]. In principle, ATC commands (e.g. to change altitude) could be acted upon directly by the UA without pilot intervention obviating the need for the inefficient backhaul. However, it is likely that a pilot will always be expected to be "on the loop" so that UAS operations are completely transparent to ATC. Future communication system analysis estimates the average ATC voice and data rates to be about 10 kbps per aircraft particularly in autonomous operations areas typical of UAS operations [14].

Other platform safety communication is related to detect, sense, and avoid requirements which generally require the UAS to have equivalent ability to avoid collisions as manned aircraft [1, 33]. This may require onboard radar (active sensing), backhaul of image data (passive sensing), transponders, or cooperative sharing of information between UA. The communication requirements here can depend significantly on the approach. The least communication demands are required when the aircraft uses active sensing, only reports potential collisions to the operator, and autonomously performs evasive maneuvers when collisions are imminent. The communication requirements here are negligible. More demanding systems send full visual situational awareness to the operator which can require 1 Mbps or more (e.g. Fig. 3).

The communication requirements for small UAS are simplified in many instances. Small UAS often fly in uncontrolled airspace. No transponders are required, nor is communication with ATC. Small UAS can operate directly in uncontrolled airspace without the need of an airport. They are often catapult or hand launched and can have parachute, net, or snag recovery systems. Small UAS generally fly over smaller regions than larger UAS. Small UAS are still subject to DSA requirements, however, for very short ranges, the pilot or spotters on the ground can provide the see and

Fig. 3 Situational awareness provide by imagery from onboard a small UA

avoid. For larger ranges active or passive techniques are required. The smaller platform size limits the ability to carry onboard autonomous DSA systems.

Small UAS participating in the example scenarios described in Section 2.1 will clearly be operating in environments with significant other air traffic so platform safety will be important. Manned aircraft for emergency response and from news agencies will surely operate in the environment where the UAS will be deployed. The UAS pilot will require significant communication with ATC to coordinate operation with these other aircraft. From the perspective of network or radio bandwidth and throughput, the requirements for this communication traffic are low since messages are limited in size (or length) and are sent sporadically. However, the safety critical nature of this traffic will require high reliability with low latency.

2.1.2 Remote Piloting

Remote piloting of the vehicle has requirements that vary with the type of flight control. On one extreme is direct joystick control of the aircraft. This requires low delay and high availability. At the other extreme, tasking commands are sent to the aircraft which are autonomously translated to flight paths (Fig. 4 shows the user interface for the Piccolo autopilot that allows for point and click commands [31]). Here delays can be longer and gaps in availability can be tolerated. The UA to

Fig. 4 Cloud cap technologies piccolo autopilot command center [31]

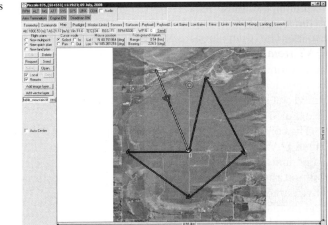

pilot link contains not only commands from the pilot to the UA but also essential health and status information from the aircraft back to the pilot. As examples, on the joystick end of the spectrum commercial digital radio control (R/C) links have data rates below 10 kbps and one-way delays below 100 ms are preferred. On the autonomous end of the spectrum, an Iridium satellite link is sufficient for waypoint flying of the Predator UAS. Iridium has 2.1 kbps throughput, delays of 1–7 s, and has gaps in connectivity with 96% average availability [25].

Small UAS are lower cost and are more likely to operate in cooperative groups. There is a strong push to enable one-to-many pilot-aircraft interactions for UAS [27]. This mode of operation would require increased amounts of UA autonomy with the pilot in charge of higher level mission planning and tasking. As such, UA must be capable of autonomous collision avoidance and therefore plane-to-plane communication becomes another significant communication component. Collision avoidance between two UA will also have low data rate and low latency requirements. However, the presence of multiple vehicles all performing plane-to-plane communication complicates the networking and introduces the need for bandwidth and congestion control. The possibility of varied capabilities and aircraft attrition also necessitates dynamic service discovery routines whereby system capabilities can be updated internally.

2.1.3 Payload Management

Communication with the payload can range from a few bits per second for simple sensor readings to megabits per second for high-quality images (Fig. 5). For instance, the Predator uses a 4.5 Mbps microwave link to communicate payload imagery when in line-of-site of the ground station [23]. The types of payload communication needed by small UAS can be highly varied. For example, depending on the type of chemical plume being tracked, real-time data assimilation may not be needed. In that case large amounts of data can be stored at intermediate nodes and transmitted opportunistically back to the end user. In contrast, if a toxic substance is released in an urban setting, source localization could take priority over all other requirements including DSA. Using multiple UA to provide information to multiple dispersed users also necessitates dynamic service discovery routines.

In summary, the communication requirements for UAS are modest for ATC communication and remote piloting while UAS can potentially require data rates in the megabits per second for payload management and DSA. It is this requirement for multiple connections, some of which are at high data rates, that distinguishes UAS from manned aircraft communications. There are also other considerations

Fig. 5 High quality payload imagery from the MLB Bat small UAS [24]

than data rates, latency, and availability. ATC, remote piloting, and other flight safety communication will likely be required to operate in protected spectrum that is not shared with payload and non-essential communication [14, 20].

2.2 Operational Networking Requirements

The communication needs can be explored along three axis (Fig. 6). The first is connectivity. In traditional networks, node connectivity is well defined; a physical wire connects two nodes. These links are designed to be reliable with rare transmission errors. Further, the links and nodes are stable with only rare failures. This yields a well defined network topology with clear notions of graph connectivity that is consistent across network nodes. In a small UAS, the links are less reliable wireless links with connectivity that ranges from good when nodes are close to poor for nodes that are further away. Even when connectivity is good packet error rates are high relative to wired standards. The transmission is broadcast and can reach multiple receivers so that connections are not simple graph edges. Further, broadcast transmissions interfere with each other so that the ability of two nodes to communicate depends on the other transmissions at the same time. As we add UA mobility, connectivity becomes dynamic and as UA speeds increase relative to the nominal communication range different UA may have inconsistent notions of connectivity.

The second axis is data delivery. In traditional networks, connectivity is well defined and stable so that data delivery is based on an end-to-end model. For instance, with TCP protocols the source and destination end points manage data delivery over a presumed reliable and low latency network. As these connectivity assumptions break down this model of delivery is not possible. As already noted, small UAS connectivity is unreliable and dynamic. Furthermore, small UAS may become spread out over a mission so that end-to-end connectivity simply does not exist for data delivery.

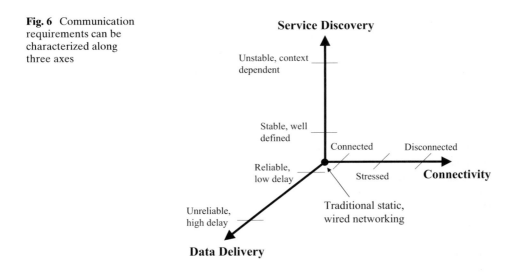

Fig. 6 Communication requirements can be characterized along three axes

The third axis is service discovery. In traditional networks, resources are stable and can be found through well defined procedures. To find a webpage, we use a URL (e.g. http://www.springer.com), this is translated by the DNS service into a network address, and we request the webpage from the server at that address. In contrast, with small UAS nodes, services, and users can come and go over the life of a mission and the resources sought may be ephemeral or context dependent. Each UA may have different capabilities (e.g. the chemical sensors in the plume tracking example) and onboard capabilities may be available for use by other UA (e.g. coordinating cameras for stereo imaging). As small UAS spread out these resources, even when they exist, may be difficult to discover. This concept of service discovery can be pushed down from aircraft to the subsystems aboard the aircraft. By connecting onboard subsystems via local subnets to the external meshed network, dispersed operators as well as other aircraft in the UAS can discover and communicate directly to different components of the aircraft avionics system.

3 Networking for Small UAS

The goal of small UAS networking is to address the communication needs given that small UAS differ from traditional networks along the connectivity, data delivery, and service discovery axes. To this end, this section describes the merits of different communication architectures and introduces the delay tolerant networking delivery mechanism. It also discusses how the mobility of the small UAS can be exploited to improve networking.

3.1 Communication Architectures

There are four basic communication architectures which can be used for small UAS applications: direct link, satellite, cellular, or mesh networking (Fig. 7). Each has advantages and disadvantages which we outline here. A direct link between the ground control station and each UA is the simplest architecture. It assumes connectivity is maintained over dedicated links to each UA and therefore data delivery is reliable with low latency. Since the ground station communicates to each UA, service discovery is easily managed by a centralized agent at the ground station. Unfortunately the direct architecture is not suited for dynamic environments and non-line-of-sight (NLOS) communication. Obstructions can block the signal, and at longer ranges the UA requires a high-power transmitter, a steerable antenna, or significant bandwidth in order to support high data rate downlinks. The amount of bandwidth scales with the number of UA so that many UAS may not operate simultaneously in the same area. Finally, plane-to-plane communication will be inefficiently routed through the ground control station in a star topology and not exploit direct communication between cooperative UA operating in the same area.

Satellite provides better coverage than a direct link to the ground control station. As a result, the UAS network can remain well connected, however this connectivity would still be provided by routing data through a centralized system. Data delivery is relatively poor using satellite. Lack of satellite bandwidth already limits existing UAS operations and will not scale with the increasing demand of 1,000s of small UAS operations in a region. For high data rate applications, a bulky steerable dish antenna mechanism unsuitable in size, weight and cost for small UAS is necessary.

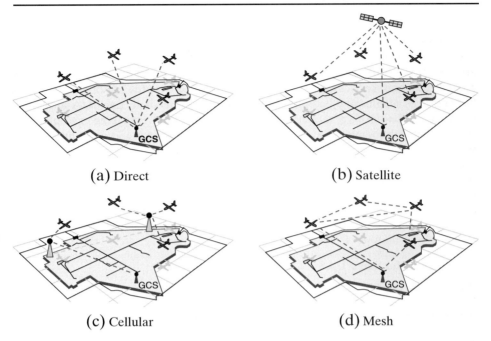

(a) Direct   (b) Satellite

(c) Cellular   (d) Mesh

Fig. 7 Four basic communication architectures for small UAS. The ground control station (GCS) represents the operator or end user

Further, the ground control station requires a connection to the satellite downlink network. The ground control station may have obstructed satellite views because of terrain or clutter. Finally, multiple UA operating in an area will suffer high delays if their communication is mediated by satellite.

Cellular refers to an infrastructure of downlink towers similar to the ubiquitous mobile telephone infrastructure. The cellular architecture has several advantages that can provide good levels of network connectivity and reliable data delivery. First, coverage can be extended over large areas via multiple base stations. UAS would hand-off between different base stations as needed during flight. Second, the multiple base stations provide a natural redundancy so that if one link is poor another link may perform better. Third, a limited bandwidth can be reused many times over a region and capacity increased as needed to meet demand. The reuse can grow by adding more base stations as the number of users grows. Fourth, the infrastructure can be shared by different UAS. Once installed, many UAS can each pay for the fraction of the infrastructure that they use. These advantages must be weighed against the cost. A typical mobile telephone base station is expensive for the tower, tower site, radio equipment, and associated networking infrastructure. Such a solution applies where the infrastructure investment can be amortized across frequent and regular UAS flights. Examples might include agricultural monitoring or border surveillance. Such architecture is not suited for applications like wildfire management or polar climatology where demand is transient. The existing mobile telephone infrastructure is not designed for air to ground communication. A single UA transmitter can blanket a large area with its signal degrading system performance. Therefore, small UAS operations may require a dedicated cellular infrastructure.

Meshing refers to a networking architecture where each node (i.e. a radio on a UA or ground node) can act as a relay to forward data. Communication between a UA and a ground control station can take place over several ŞhopsŤ through intermediate nodes. The shorter range simplifies the link requirements and bandwidth can be reused more frequently and thus more efficiently Plane-to-plane communication can be direct and also benefit from the mesh routing protocols that employ additional relays as needed to maintain communication. However, such meshing requires intermediate nodes to be present for such relaying to take place. Furthermore, nodes may be required to move specifically in order to support communication.

Mobile ad-hoc networking (MANET) is a particular example of a mesh architecture comprised of a self-configuring network of mobile routers which are free to move randomly throughout the environment. The MANET topology changes rapidly due to the motion and the nodes in the network rapidly self-organize in response. The MANET approach is promising for UAS applications where infrastructure is not available and multiple UA are operating cooperatively. Data relaying in MANETs reduces the connectivity requirements since source and destination nodes only need to be connected through the intermediate nodes. Due to the decrease of radio transmission power, and hence communication capacity, with separation distance [30], the presence of the intermediate relay nodes can actually improve data delivery performance over direct communication [18]. Further, since MANETs are design to self-heal, they can respond well to the dynamic network topologies that result from UA motion. Service discovery in MANETs is more important and more complex than the other architectures since the network tends to become fractured for periods of time and there is no centralized node to coordinate network activities.

Meshing can also leverage the other technologies described above. A direct, satellite, or cellular link to any node in a connected mesh enables communication with all the nodes providing additional redundancy in the communication. Meshing combined with mobility can extend range. For instance, as a group of UA move beyond the range of a direct link, some of the UA can be assigned to stay behind forming a chain of links back to the direct link. In an extreme form of this UA can fly back and forth to ferry data between nodes that have become widely separated. It is this flexibility, robustness, and added range that makes meshing an integral part of any small UAS operations.

3.2 Delay Tolerant Networking

Real-time communication is challenging for mobile ad-hoc networks on small UAS because of inherent variability in wireless connections combined with the fact that nodes may become spread out and sparsely connected. Connections that exist can be dynamic and intermittent, antenna patterns shift with aircraft maneuvering, sources of interference come and go, and low flying UAS can be separated by intervening terrain. In the extreme case of sparsely connected, moving nodes, some nodes might not be able to connect with any other node for a long time. In such environments, traditional end-to-end network protocols such as the ubiquitous TCP perform poorly and only delay-tolerant communication is feasible. So-called delay tolerant networks (DTN) are designed for these challenged environments [8, 15]. DTN provide a smooth spectrum of communication ability. Data is delivered quickly when end-to-end connections exist and as quickly as opportunities appear when intermittently connected. The DTN also supports data ferrying where, for instance, a ground sensor can deliver data to an overflying UA that then physically carries the data back to a network gateway to an observer.

The data flowing through the MANETs deployed on small UAS will have large variety of data types and quality of service requirements. For example, Voice over IP (VOIP) between first responders, UA control data, or critical process monitoring will require prioritized real-time flow. For a wireless ad-hoc network to be able to carry real-time data there needs to be a contemporaneous, reliable connection from sender to receiver. The more nodes that participate in the network in a given area, the more likely a path can be established. Other data types, such as email, non-critical messaging, or sensor data from long-term experiments, will not carry the same sense of urgency and consist of delay-tolerant traffic where only eventual, reliable reception is important. Continuous multi-hop links from the sender to receiver do not need to be maintained as long as a data packet is carried or forwarded to its destination in some reasonable amount of time. The DTN architecture provides seamless mechanisms for integrating these various requirements into a single network.

DTN are a current topic of research. An example DTN was implemented on the University of Colorado's HUAS for delivering sensor data to one or more external observers outside the UAS [22]. Sensors on UA or on the ground generate data and a DTN procedure delivers the data in stages through gateways to the observers. Each stage takes custody of the data and stores it in the network until a connection opportunity to the next stage arises. End nodes can operate with no knowledge of the DTN through proxy interfaces.

In addition to extending the capability of MANET architectures to sparse or fractured networks, the DTN concept is also important for maintaining system responsibility and accountability in the face of UAS failure. In traditional MANET architectures, data is not transmitted if an end-to-end connection is not present and therefore data will be lost if a link goes down due to failure. In contrast, the DTN protocols store data at intermediate nodes until it is safely transmitted to the source. Thus telemetry, health and status, and other data collected during periods when a node is disconnected from the network can still be collected. This includes information from moments before a failure which may be stored and can be collected for post analysis.

3.3 Exploiting Controlled Mobility

Unlike terrestrial networks which tend to be static and satellite communication systems which are in fixed orbits, meshed networks of small UA offer a unique opportunity to exploit controlled mobility. Even when a continuous link is established, environmental factors such as multipath, interference, or adversarial jamming, can degrade real-time performance relative to expected models. In these cases the mobility of the nodes themselves can be exploited to improve the network performance. In sparse networks, node mobility enables data ferrying, i.e. physically carrying data packets through the environment, between otherwise disconnected nodes. In the case of a connected network, node mobility enables adjustment of local network behavior in response to unmodeled disturbances.

Given the presence of node mobility in an ad-hoc network, transmission of data between a source and destination can take three forms (Fig. 8). *Direct*

Fig. 8 Three modes of maintaining a communication link between two static nodes A and B

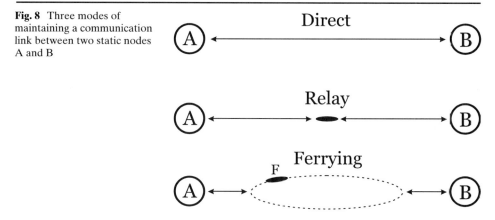

communication occurs when two nodes transmit data directly to one another. *Relaying* occurs when additional nodes are used to receive a transmission from a source and retransmit it to a destination. Finally, *data ferrying* occurs when a mobile node physically stores and carries data from one location to another. Each of these modes has a place in communication. For instance, at 5 km a low-power wireless direct link might support only a few 10's of kbps, while a ferry with a 50 megabyte buffer and velocity of 30 m/s can deliver data at a rate of over 1 Mbps. However, the direct link can deliver a packet in under a second while the ferry will require minutes. Thus, using mobility requires understanding the tradeoffs in delay and data rate. These ideas have been explored in [6] which defines the different operational regions and the tradeoffs in each.

To further illustrate the role of mobility in determining network performance, consider an example with two task nodes that wish to communicate with one another and additional mobile helper nodes in the environment that can act as relays between the task nodes. The two task nodes could represent a time-critical sensing system and a human-operated ground station or two relay nodes electronically leashed to convoys of ground vehicles. As the distance between the task nodes increases, one or more helper nodes are needed to relay data between them. While networking protocols controlling the flow of data between nodes have been extensively studied, the issue of how to best deploy or position these helper nodes is still open. Position-based solutions break down in the presence of noise sources and terrain that distort the radio propagation and power models from simple cases [12]. For example, Fig. 9 shows how a single noise source distorts the contours of end-to-end chain capacity as a function of relay node position for a simple 3-node network. For a multiple UA chain providing multi-hop communication, decentralized control laws can optimize chain capacity based only on measures of the local 3-node network perceived by each UA [12].

As the separation distance between the task nodes grows relative to the communication range of the helper nodes (i.e. the density scale decreases) positioning of the helper nodes becomes more difficult. Given relatively static node placement and knowledge of their position, deployment of the helper nodes becomes a resource allocation problem [18]. As the density scale decreases and the separation between nodes grows, it becomes difficult for helper nodes to build consensus on deployment.

Fig. 9 Contours of end-to-end chain capacity for a three-node network and the vector field showing the gradient of capacity. Interference distorts radio propagation from standard models and makes position-based solutions for relay placement suboptimal. **a** Contours and vector field with no noise; **b** Contours and vector field with a localized noise source

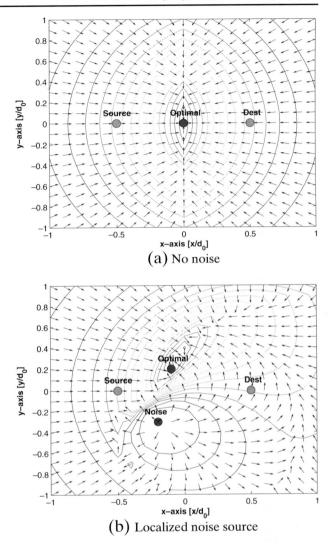

(a) No noise

(b) Localized noise source

Centralized strategies in which the helpers converge to build consensus trade advantages with distributed approaches that enable helper nodes to make decisions based only on local information. The best approach is not apparent. Furthermore, as the separation distance between nodes grows beyond a certain point it becomes impossible to establish a connected chain between them and the helper nodes must exploit their own mobility to ferry data back and forth [19]. While this seems to be a basic process, the fact that the nodes are not in a fully connected network at this point hinders consensus building among the helpers and the implementation of globally optimal behavior.

Although deployment of helper nodes is challenging in relatively static scenarios, the problem is further complicated when nodes (task nodes and helper nodes) are capable of moving quickly relative to their communication range (i.e. the dynamic scale increases). If we allow the task nodes the freedom to perform their primary task, the helper nodes must cooperatively distribute themselves through the environment to provide multi-hop communication. Motion of the helper nodes has a greater effect on the performance of the communication network and the problem shifts from resource allocation to continuous coordinated control.

This simple example scenario illustrates the effects of mobility on communication in delay tolerant networks. It is clear that the establishment and maintenance of networks of mobile nodes, especially given a subset of nodes dedicated to primary missions such as sensing, requires some form of distributed cooperative control. Furthermore, in order to exploit node mobility in a distributed control architecture, communication metrics such as quality of service requirements must be incorporated explicitly as objectives in the mobility control scheme.

4 Conclusion

This paper explored requirements for networking of small unmanned aircraft systems. The networked communication demands of unmanned aircraft systems in general, and small unmanned aircraft system in particular, are large compared to manned aircraft since telemetry, command and control, health and safety, and payload data must be sent from multiple aircraft to multiple dispersed users such as the unmanned aircraft system operator, air traffic control, and other aircraft in the vicinity.

The communication environment for small UAS was shown to deviate significantly form traditional networking assumptions. Connectivity is lower quality, less well defined, and dynamic. Data delivery can not depend on reliable, low-latency, and end-to-end connections. Network service discovery must operate in isolated networks to discover dynamically available resources.

In this environment, we described how mesh networking, supported by delay tolerant networking, and robust service discovery can operate. Mesh networking is designed to work in mobile environments and allows two UA to communicate by dynamically piecing together links to form a communication path. Delay tolerant networking is designed to work in sparse connectivity environments and inserts intermediate custody points so that communication can make progress in smaller stages over time as connections become available. Service discovery allows aircraft to advertise their capabilities and so others can find and use them.

The small UAS networking environment is challenging but provides some opportunities. A small UAS will likely have multiple aircraft. When needed some of these can be devoted to support communications, moving to locations to relay between disconnected groups or in extreme cases the aircraft can ferry data by physically carrying the data back and forth.

Small UAS provide unique challenges to networking. What is reported here is based on our experience implementing and testing networks on small UAS. Our research continues to address these challenges. These include methods for seamlessly and safely integrating flight critical command and control data with less critical payload date on a single communication network.

Acknowledgements The authors would like to thank the members of the University of Colorado AUGNet Research Group. This work was supported by the US Air Force under Grant FA9550-06-1-0205, by the Federal Aviation Administration under Grant 07-G-014, and by L-3 Communications.

References

1. Federal Aviation Administration: order 7610.4k. Special military operations (2004)
2. Federal Aviation Administration: meeting the challenge: unmanned aircraft systems. In: Federal Aviation Administration R&D Review, vol. 4 (2006)
3. Federal Aviation Administration: title 14 code of federal regulations (14 cfr) part 91 (2008)
4. Beard, R., McLain, T., Nelson, D., Kingston, D., Johanson, D.: Decentralized cooperative aerial surveillance using fixed-wing miniature UAVs. Proc. I.E.E.E. **94**(7), 1306–24 (2006)
5. van Blyenburgh, P.: Unmanned Aircraft Systems: The Global Perspective. UVS International, Paris, France (2007)
6. Brown, T.X., Henkel, D.: On controlled node mobility in delay-tolerant networks of unmanned aerial vehicles. In: Proc. of Intl Symposium on Advanced Radio Technologies, Boulder, CO (2006)
7. Brown, T.X., Argrow, B.M., Frew, E.W., Dixon, C., Henkel, D., Elston, J., Gates, H.: Experiments using small unmanned aircraft to augment a mobile ad hoc network. In: Bing, B. (ed.) Emerging Technologies in Wireless LANs: Theory, Design, and Deployment, chap. 28, pp. 123–145. Cambridge University Press (2007)
8. Cerf, V.G., Burleigh, S.C., Durst, R.C., Fall, K., Hooke, A.J., Scott, K.L., Torgerson, L., Weiss, H.S.: Delay-Tolerant Network Architecture. Internet Draft, IETF (2006)
9. Claus Christmann, H., Johnson, E.N.: Design and implementation of a self-configuring ad-hoc network for unmanned aerial systems. In: Collection of Technical Papers - 2007 AIAA InfoTech at Aerospace Conference, vol. 1, pp. 698–704. Rohnert Park, CA (2007)
10. Corrigan, C.E., Roberts, G., Ramana, M., Kim, D., Ramanathan, V.: Capturing vertical profiles of aerosols and black carbon over the Indian Ocean using autonomous unmanned aerial vehicles. Atmos. Chem. Phys. Discuss **7**, 11,429–11,463 (2007)
11. Curry, J.A., Maslanik, J., Holland, G., Pinto, J.: Applications of aerosondes in the arctic. Bull. Am. Meteorol. Soc. **85**(12), 1855–1861 (2004)
12. Dixon, C., Frew, E.W.: Decentralized extremum-seeking control of nonholonomi vehicles to form a communication chain. Lecture Notes in Computer Science, vol. 369. Springer-Verlag (2007)
13. Dixon, C., Henkel, D., Frew, E.W., Brown, T.X.: Phase transitions for controlled mobility in wireless ad hoc networks. In: AIAA Guidance, Navigation, and Control Conference, Keystone, CO (2006)
14. EUROCONTROL/FAA: Future Communications Study Operational Concepts and Requirements Team: communications operating concept and requirements (COCR) for the future radio system. Tech. Rep. 1.0 (2006)
15. Fall, K.: A delay-tolerant network architecture for challenged internets. In: SIGCOMM '01, pp. 27–34 (2003)
16. Frew, E.W., Dixon, C., Elston, J., Argrow, B., Brown, T.X.: Networked communication, command, and control of an unmanned aircraft system. AIAA Journal of Aerospace Computing, Information, and Communication **5**(4), 84–107 (2008)
17. Harvey, D.J., Lu, T.F., Keller, M.A.: Comparing insect-inspired chemical plume tracking algorithms using a mobile robot. IEEE Trans. Robot. **24**(2), 307–317 (2008)
18. Henkel, D., Brown, T.X.: Optimizing the use of relays for link establishment in wireless networks. In: Proc. IEEE Wireless Communications and Networking Conference (WCNC), Hong Kong (2008a)
19. Henkel, D., Brown, T.X.: Towards autonomous data ferry route design through reinforcement learningi. In: Autonomic and Opportunistic Communications Workshop (2008b)
20. Henriksen, S.J.: Estimation of future communications bandwidth requirements for unmanned aircraft systems operating in the national airspace system. In: AIAA InfoTech@Aerospace, vol. 3, pp. 2746–2754. Rohnert Park, CA (2007)
21. How, J., King, E., Kuwata, Y.: Flight demonstrations of cooperative control for uav teams. In: AIAA 3rd "Unmanned-Unlimited" Technical Conference, Workshop, and Exhibit, vol. 1, pp. 505–513. Chicago, IL (2004)
22. Jenkins, A., Henkel, D., Brown, T.X.: Sensor data collection through gateways in a highly mobile mesh network. In: IEEE Wireless Communications and Networking Conference, pp. 2786–2791. Hong Kong, China (2007)
23. Lamb, G.S., Stone, T.G.: Air combat command concept of operations for endurance unmanned aerial vehicles. Web page, http://www.fas.org/irp/doddir/usaf/conops_uav/ (1996)

24. MLB Company: MLB Company—The Bat. http://www.spyplanes.com/bat3.html (2008)
25. Mohammad, A.J., Frost, V., Zaghloul, S., Prescott, G. Braaten, D.: Multi-channel Iridium communication system for polar field experiments. In: International Geoscience and Remote Sensing Symposium (IGARSS), vol. 1, pp. 121–124. Anchorage, AK (2004)
26. NOAA: NOAA News Online: NOAA and partners conduct first successful unmanned aircraft hurricane observation by flying through Ophelia. http://www.noaanews.noaa.gov/stories2005/s2508.htm (2005)
27. Office of the Secratary of Defense: Unmanned Aircraft Systems Roadmap: 2005–2030 (2005)
28. Ryan, A., Xiao, X., Rathinam, S., Tisdale, J., Zennarc, M., Caveney, D., Sengupta, R., Hedrick, J.K.: A modular software infrastructure for distributed control of collaborating UAVs. In: AIAA Guidance, Navigation, and Control Conference, Keystone, CO (2006)
29. Sofge, E.: Houston cops test drone now in Iraq, operator says. Web page, http://www.popularmechanics.com/science/air_space/4234272.html (2008)
30. Taub, B., Schilling, D.L.: Principles of Communicatior Systems. McGraw-Hill, New York (1986)
31. Vaglienti, B., Hoag, R., Niculescu, M.: Piccolo systrem user's guide: software v2.0.4 with piccolo command center (pcc). http://www.cloudcaptech.com/resources_autopilots.shtm#downloads (2008)
32. Wagner, B.: Civilian market for unmanned aircraft struggles to take flight. In: National Defense Magazine (2007)
33. Weibel, R., Hansman, R.J.: Safety considerations for operation of unmanned aerial vehicles in the national airspace system. Tech. Rep. ICAT 2005-C1 (2006)
34. Zajkowski, T., Dunagan, S., Eilers, J.: Small UAS communications mission. In: Eleventh Biennial USDA Forest Service Remote Sensing Applications Conference, Salt Lake City, UT (2006)

UAVs Integration in the SWIM Based Architecture for ATM

Nicolás Peña · David Scarlatti · Aníbal Ollero

Originally published in the Journal of Intelligent and Robotic Systems, Volume 54, Nos 1–3, 39–59.
© Springer Science + Business Media B.V. 2008

Abstract The System Wide Information Management (SWIM) approach has been conceived to overcome the capacity and flexibility limitations of the current ATM systems. On the other hand the commercial applications of Unmanned Aerial Vehicles (UAVs) require the integration of these vehicles in the ATM. From this perspective, the unavoidable modernization of the ATM is seen as an opportunity to integrate the UAVs with the rest of the air traffic. This paper is devoted to study the feasibility and impact of the aggregation of UAVs on the future ATM supported by a SWIM inspired architecture. Departing from the existing technical documents that describe the fundamentals of SWIM we have explored the compatibility with a potential UAVs integration and also explored how the UAVs could help to improve the future ATM system. We will use the weather application as an example in both cases.

Keywords UAV · Air traffic management · System wide information management

1 Motivation and Objectives

The number of aircrafts in operation all around the world has grown steadily from the hundreds to the tens of thousands. As technology advanced, new systems and services were added to the Air Traffic Management (ATM) to improve safety and capacity, but due to the lack of a standard procedure to insert new functionalities in the ATM, these systems were designed independently, with very different interfaces

N. Peña (✉) · A. Ollero
University of Seville, Robotics, Vision and Control Group,
Avd. de los Descubrimientos s/n, 41092, Sevilla, Spain
e-mail: nicolas.grvc@gmail.com

D. Scarlatti
Boeing Research & Technology Europe – Cañada Real de las Merinas, 1-3,
Building 4-4th floor, 28042 Madrid, Spain

K. P. Valavanis et al. (eds.), *Unmanned Aircraft Systems*. DOI: 10.1007/978-1-4020-9136-0_4   39

and had to be hard wired to each other in a specific way for every combination that needed to interoperate. As the number of present systems grew, the cost of inserting a new one was always higher.

The result of this trend is that the current ATM is a rigidly configured, complex collection of independent systems interconnected by very different technologies from geographically dispersed facilities. Then, they are expensive to maintain, and their modifications are very costly and time consuming. Future capacity demands require the implementation of new network-enabled operational capabilities that are not feasible within the current ATM systems. In fact, the safety, capacity, efficiency and security requirements to meet the expected demand require the application of new flexible ATM architectures. A new approach to face this future demand is the so-called System Wide Information Management (SWIM) [1, 2]. This system enables shared information across existing disparate systems for network-enabled operations, and improves air traffic operations by integrating systems for optimal performance [2, 4].

The commercial applications of Unmanned Aerial Vehicles (UAVs) require the integration of these vehicles in the ATM [5]. Currently, the UAVs operate in a completely segregated aerial space due to the absence of protocols for their integration in the Air Traffic Management systems. From this perspective, the planned modernization of the ATM is seen as an opportunity to integrate the UAVs with the rest of the air traffic in a single aerial space. In fact the implementation of the SWIM concept makes easier the integration of the UAVs in the ATM than the architecture in use. Moreover, the standardization of interfaces and the involved network centric concepts involved in SWIM are additional benefits for the integration of UAVs.

This paper is devoted to study the impact of the aggregation of UAVs on the future ATM supported by a SWIM inspired architecture. There are some publications and on going projects on the UASs integration in the aerospace. These usually focus the aspects that need to be improved before this integration happens such as the autonomous sense and avoid [6] or the safety requirements [7]. In [8] a new architecture for this integration is proposed.

This paper studies the integration of the UAVs in the ATM from the point of view of the actual plans for the future SWIM inspired, ATM [2, 3]. The effect at the different layers of the planned ATM structure is discussed. The paper will also explore the possibility that arises from the integration of the UAVs in the ATM that can be achieved in several layers of the proposed architecture. For example, regarding the network layer, which is heavily stressed by the ever growing aircrafts density, it is possible to consider stratospheric UAVs providing in a dynamic way additional bandwidth in areas with such requirements. Another example, in the application layer, is the service provided by the weather application that could be improved by UAVs acquiring information from areas with higher uncertainty on weather conditions.

On the other hand, for a proper integration in SWIM, the software and hardware architectures of the UAVs should be adapted. In terms of the hardware on-board, the limited payload should be considered to prioritize (in terms of safety) the on-board equipment included in the UAV. For instance, the ACAS (Airborne Collision Avoidance System) system should be included and integrated in the software on-board to allow an automated response for collision avoidance in the same way that manned aircrafts do.

We have centered our study in two general aspects of this integration. One is at the application level, where the functionality of the different services as surveillance and weather is offered to the SWIM clients. The other one looks at the layers below the application one to try to evaluate if the approaches proposed for some of the inners of the future ATM, like the proposed client server model and the data models, would be a problem or an advantage for the integration of the UAVs. The paper contains a section regarding to each of the two aspects studied by the authors.

2 Application Layer

This section presents some considerations regarding the SWIM application layer that are useful to put in context some concepts that will be presented later on.

The integration of the UAVs and their associated infrastructure in SWIM applications will be studied from two different approaches:

- UAVs using SWIM applications during their operation, and
- UAVs providing services intended to improve the performance of some SWIM applications.

Regarding the first approach, a common application, such as the weather application, will be selected in order to describe how an UAV will use the weather services. Thus, the components interacting with the UAV will be considered and the differences with respect to a conventional aircraft at the application level (i.e. automated periodic weather reporting instead of on pilot demand), as well as the requirements for the UAVs (on-board probabilistic local weather model, permanent link with a broker, required bandwidth, etc) will be examined. Figure 1 shows the proposed architecture of the weather application in a SWIM enabled NAS (National Airspace System). The elements shown in the Figure will be used in the rest of this section in a simplified manner referring to them as weather data producers, weather data repositories, weather brokers and finally, weather application clients.

The second approach considers UAVs providing an improvement in the applications of SWIM and even new services for manned aircrafts, UAVs and all clients in general. Some examples could be: provide weather information from locations with high uncertainty, response in emergencies, serve as flying repositories of pseudo-static information such as mid and long-term weather information, etc.

In this section, the weather application will be taken as a case study in order to clarify some aspects related to the integration of the UAVs in SWIM.

2.1 UAVs Using SWIM Applications During their Operation

In the application layer, it is relevant to distinguish between UAVs with minor human intervention and UAVs with a remote human pilot or operator. In this section, the analysis will be focused on the first one due to the following reasons:

- When the UAV is teleoperated, the human operator can play the role of a conventional pilot in terms of using the SWIM applications (i.e. processing the messages from a SWIM weather application). Therefore, in this case and from

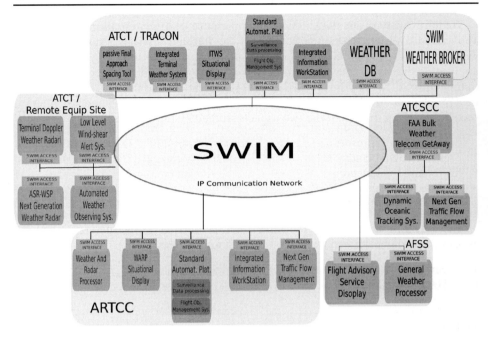

Fig. 1 The weather application elements adapted to the SWIM concepts

the point of view of the application layer, there is no relevant difference between an UAV and any other conventional aircraft. As it will be discussed later, in other layers several differences arise.

- Although nowadays almost all UAVs require in some extend the intervention of a human teleoperator, it is expected a transition towards full autonomy [9], allowing one operator to manage several UAVs. Furthermore, full autonomy would be even possible in some SWIM applications as weather remote sensing using stratospheric UAVs [10], reducing their operational cost.

Then, in the following, only autonomous UAVs will be considered. Therefore, in the weather application for example, problems related to autonomous weather messages processing and decision making should be addressed.

In the current ATM systems, after receiving one message, the human pilot has to "decode" and process it in order to make a decision taking into account other parameters such as the type of aircraft, payload, remaining fuel, etc. METAR (Meteorological Terminal Aviation Routine Weather Report) and TAF (Terminal Aerodrome Forecast) information are essential for flight planning and in-flight decisions. TAF messages are a very concise, coded 24-hour forecast for a specific airport that, opposed to a public weather forecast, only addresses weather elements critical to aviation as wind, visibility, weather and sky condition. A more recent type of message is the Transcribed Weather Enroute Broadcasts (TWEBs) which are composed by some Weather Forecast Offices (WFOs) and contain very similar information to a TAF but for a 50 mile wide corridor between two or three frequently connected airports.

As far as an autonomous UAV should perform this whole message processing by itself, the following requirements could be considered for the decisional level of an UAV:

- Estimate when a weather report is required. If the uncertainty about the weather in a given area of the flight plan is higher than a given threshold, the UAV should ask for a weather report in order to reduce the uncertainty of its local weather model. The weather reports requests will have a set of parameters such as the area of interest for example. Although a periodic weather reports request scheme could be adopted, the capability to autonomously estimate when a report is required will decrease the limited available data bandwidth.
- Process and decode the standard weather messages formats. Some examples of those formats are shown in Fig. 2.

In the proposed SWIM data models exposed in [3], several approaches are possible regarding the weather messages:

(a) Embedding the formats currently used into SWIM weather messages: The UAV should have to decode and process those formats.

There are several software projects which deal with METAR and TAF messages and even web sites providing access to some of these messages in a human readable format as [11].

In particular, the metaf2xml software [12] parses and decodes aviation routine weather reports and aerodrome forecasts (i.e. METAR and TAF messages) and stores the components in XML. They can then be converted to plain language, or other formats (see an example in Fig. 3). Similar software could be running on board the UAV, parsing the SWIM weather messages. After the parsing process, the information should be used to update the local weather model. A similar operation is performed by a software module of the FlightGear project [13] which updates a global weather model from parsed weather messages.

```
SBGL 091800Z 14008KT 9999 FEW020 BKN035 23/15 Q1013

SBGL 091550Z 091818 20015KT 8000 BKN020 PROB40 2024 4000 RA BKN015 BECMG 0002 24005KT

  BKN015 TEMPO 0210 4000 RA BR BKN010 BECMG 1113 20010KT 4000 RA BR TX22/19Z TN18/08Z

KJFK 091751Z 34013KT 10SM SCT038 21/11 A2952

KJFK 091425Z 091412 34012KT P6SM OVC025 TEMPO 1416 SCT025 FM1600 32013G18KT P6SM
```

```
  SCT025 BKN040 FM1800 31010KT P6SM SCT050 FM2200 28010KT P6SM SKC

RJTT 091830Z 34005KT CAVOK 14/09 Q1020 RMK A3012

RJTT 091500Z 091524 30004KT 9999 FEW020

RJTT 091500Z 100018 05005KT 9999 FEW030 BECMG 0003 14006KT BECMG 0912 30010KT
```

Fig. 2 Some METAR and TAF messages from Rio (SBGL), New York (KJFK), and Tokyo (RJTT)

msg: SBGL 091800Z 14008KT 9999 FEW020 BKN035 23/15 Q1013			
METAR	**METAR Report**		
SBGL	Airport-Id:	SBGL	
091800Z	Report time:	on the 9., 18:00 UTC	
14008KT	Wind:	from the SE (140°) at 14.8 km/h	8 kt = 9.2 mph = 4.1 m/s
9999	Visibility:	>=10 km	>=6.2 US-miles
FEW020 BKN035	ceiling:	at 3500 ft	1070 m
	Sky condition:	few clouds at 2000 ft	610 m
		broken clouds at 3500 ft	1070 m
23/15	Temperature:	23 °C	73.4 °F
	Dewpoint:	15 °C	59 °F
	relative humidity:	61%	
Q1013	Pressure:	1013 hPa	29.91 in. Hg

Fig. 3 Example of an automatic parsing and conversion to HTML of a METAR message

(b) Changing the formats currently used: One of the advantages of the proposed network-centric brokers based SWIM architecture is that allows fusing information from several sources to generate a response for a given request. The message with the response can have a new unified SWIM format designed to be easily decoded and processed. In fact, as the messages are generated by a broker merging several sources of information, the UAVs themselves only need to provide basic weather reports and a complex local weather model is not required.

The approach (a) would allow an easier and faster adaptation to SWIM (not for the UAVs integration in SWIM), but (b) takes full advantage of the SWIM architecture to provide better weather services.

- Autonomous decision making integrating the weather information with other parameters such as UAV dynamics, payload, remaining fuel, relevance of its mission, etc. Resulting plans should be directly reported with a proper data structure. In any case, changes in data structures reported periodically such the planned 4D trajectory, would also inform about the new local plan to other SWIM entities subscribed.

Regarding the UAV's autonomous decision functionality, it would be mandatory to periodically check that this mechanism is working properly. For example, the local plan or 4D trajectories reports could include the list of identifiers of weather messages used by the autonomous decision software. Furthermore, the decision making mechanism could be replicated in a ground facility in order to detect failures in the UAV operation (UAV malfunction), or eventually to protect against malicious intervention. In such a case, a human operator could take the control over the UAV through SWIM specialized channels that offer the CoS needed for teleoperation.

Finally, in order to illustrate an example of an UAV using the weather application over SWIM, the following storyboard based on Section 8 of [3] is presented. In a first stage (see Fig. 4), a weather database is updated with the messages sent from the weather data sources connected to SWIM. Following the publish/subscribe paradigm, a broker is acting as an intermediate element in the transaction, decoupling the data sources from the database itself.

In a second step, an UAV detects that its local weather model has a high level of uncertainty along its route and requests additional information. This query is managed by the nearest SWIM broker and sent to a weather data repository, where it is interpreted and processed (see Fig. 5). Another possible and more complex approach could involve a broker with the capability to process the query and translate it into a set of simple requests for different weather repositories. The two options differ in that the second one requires more intelligence in the brokers and less in the Data Repositories, taking the brokers a more than 'middle man' role. While the later can be seen as a more complex approach, it is more in line with the global SWIM philosophy of a really distributed, network centric system where data brokers for each service play a key role, isolating as much as possible the service clients from the inners of the backend elements of the service and the other way around.

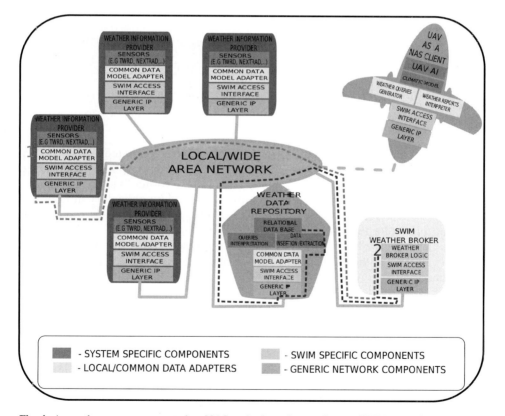

Fig. 4 A weather sensor connected to NAS and adapted according to SWIM reporting data. The information is sent to a database repository (or many) following the publish/subscribe paradigm

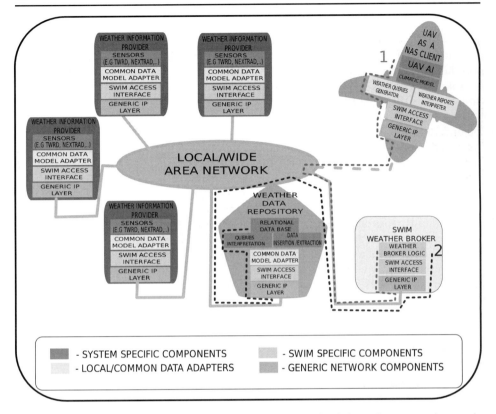

Fig. 5 When the UAV decision making core requires more weather information, a query is sent, via the closer broker, to a weather data repository

In the final stage (see Fig. 6), the required information is generated by the weather repository, formatted to be processed by the UAV and sent in a message to the broker, which just forwards it to the UAV. The message formatting task could have been also performed by the broker. Actually this would be obligatory in the ideal case of more intelligent brokers commented before. In any case, this would have been transparent for the rest of SWIM elements.

Once the response has arrived to the UAV the new weather information would be finally inserted in the local weather model satisfying the need that initially caused the request.

2.2 UAVs Improving the Performance of SWIM Applications

The adoption of the SWIM concept provides a unique opportunity to improve existing services and provide new ones. The current architecture has independent subsystems and services, and different physical connections for each type of information, whereas the SWIM architecture is network-centric and designed to grow and adapt to future demands. Scalability and connectivity are inherent features in SWIM and the cost of adding new services is reduced when comparing with the current architecture [14].

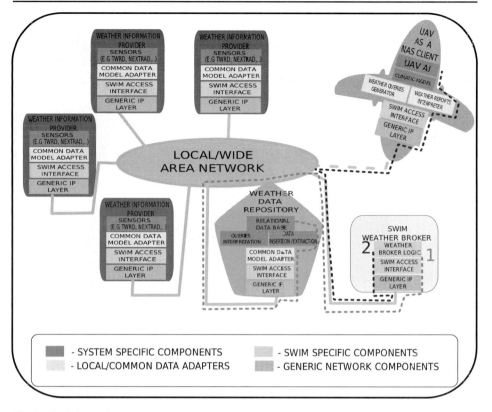

Fig. 6 The information requested is sent to the UAV

Taking into account the inherent scalability of SWIM and the emerging UAV technologies, it is expected to have UAVs supporting SWIM applications and even new UAV based applications. In the next subsections, several examples are provided assuming a flexible SWIM architecture with data fusion capabilities consistent with the COP concept [3]. Moreover, it is assumed that the brokers can manage "abstract" requests (not only forwarding raw data provided by sensors).

2.2.1 UAVs Acting as Weather Sensors

A team of stratospheric UAVs could be used to gather weather information [10] from areas with a weather estimation uncertainty higher than a given threshold. The autonomy of this type of UAVs is being improved by the recent developments in photovoltaic technology with the goal of achieving unlimited autonomy [15].

If it is expected to have traffic in a zone with high weather uncertainty, the weather servers could compute a list of waypoints to be visited for gathering weather data. Those servers would have access to the future traffic requesting this information to the corresponding brokers. Therefore, the waypoints could be prioritized depending on the expected routes in this zone.

The waypoints list could be sent to a stratospheric UAVs team that will autonomously allocate the waypoints among themselves using distributed algorithms trying to optimize some criteria, such as minimizing the total mission time or cost.

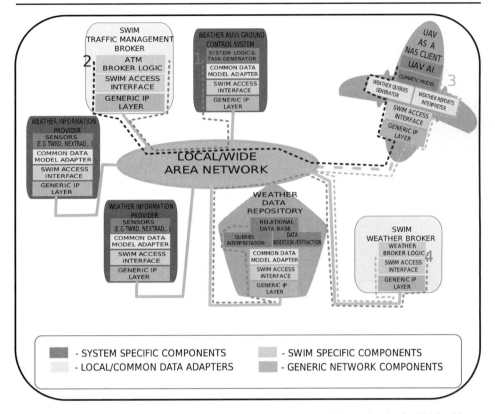

Fig. 7 UAV acting as a weather sensor. It is possible to insert this service in the NAS without changing the SWIM design

Figure 7 shows how the UAVs, acting as weather data sources, fit well in the proposed SWIM architecture. In this figure, four stages have been identified by different types of connecting lines in the information flow. In the first phase, composed by two stages, the UAV receives (via an ATM broker) a request consisting in a "goto" command. The UAV executes it and go to the specified location to gather weather information with the sensors on-board. In the second phase, the UAV reports weather data to the closest SWIM weather broker and then send the information to the weather repository.

2.2.2 Surveillance in Specific Locations and Emergency Response

Nowadays, UAVs are mainly used in surveillance missions taking aerial images from a given area. Those missions are mainly for military purposes [9], but some civil applications can also be found [16, 17]. In the future, those surveillance functionalities could be integrated in SWIM providing services such as autonomous surveying of a disaster area or assistance to identify aircrafts present in the surveillance system that are not responding to the radio.

There are several companies that provide satellite images of specific areas during a given period. Low orbit satellites can provide sequences of images during a period of time limited by their orbital speed. Those companies could adopt the use of UAVs to provide more flexible services at a lower cost. Their clients could even have access in real time to the images via a web browser in a computer connected to SWIM. A quote from [18]: "The progressive implementation of the SWIM principles in AIM (Aeronautical Information Management) is in fact AIM's evolution to IM, or Information Management that is fully SWIM based and which is the ultimate goal".

A fully SWIM compatible system could be easily interconnected with other network-centric platforms allowing to increase the number of services provided. Furthermore, as far as the amount of shared information and resources will be increased, the cost of implementing new services will decrease.

In order to illustrate an application that could exploit the benefits of interconnecting SWIM to other existing networks, the following situation will be considered, as a generalization of the application of UAVs for fire monitoring [19]. A fire station in Madrid receives a fire alarm from the sensors located in a given building. Then, an autonomous visual confirmation process starts sending a request to GIM (General Information Management) for images from the building area. GIM is part of a global network integrated by many networks including SWIM and therefore the request is finally routed to a SWIM broker. Several manned and unmanned aircrafts are flying over the city and the broker selects a proper one equipped with a camera on-board to establish a direct link between the camera and the fire station. The camera pan&tilt is pointed to the building area and the video streaming is received in the fire station allowing confirm/discard the fire (this streaming is allowed in the third broker model presented in [3]).

3 Middleware and the Network-Centric Nature

3.1 Introduction

In this section, the impact of the UAVs insertion in the SWIM middleware is studied. As far as SWIM is being designed to be very scalable and generic with the stated goal of easing the integration of new elements, the following analysis can be considered as an indicator that it is well designed for it, at least in the UAVs case.

In the current SWIM specification, it is described how piloted aircrafts (or any other data source/sink) can publish their information and subscribe to any authorized data channel. In this section it will be analyzed if those procedures would also allow the integration of UAVs in a transparent way for other SWIM users. For this purpose, the main SWIM architecture concepts are revised in the next subsection to check if they are general enough for this seamless integration. In the next subsections, required changes or additions for the UAVs integration that had been spotted during our analysis are listed and explained.

3.2 SWIM Main Concepts and UAVs

3.2.1 Network-centric Nature

One of the main ideas present in SWIM is to connect all the systems integrating ATM in a uniform manner with well defined common interfaces, all connected to a network with information flowing from the sources to the sinks sharing generic routing channels. This information can be also processed in any intermediate subsystem

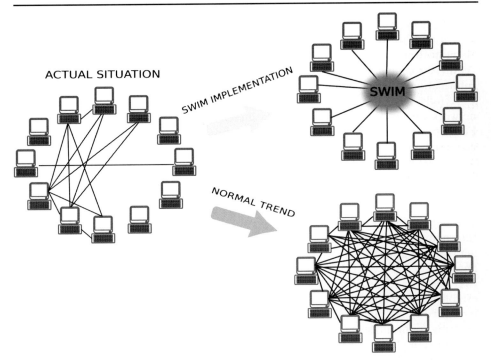

Fig. 8 SWIM network-centric approach versus point to point dedicated data link connections

demanding it. This concept represents a real revolution in the civil aviation as far as nowadays, ATM is mainly composed by many independent subsystems connected by dedicated channels in a very rigid way (see Fig. 8). Therefore, any change such as inserting a new subsystem has a significant associated cost.

On the other hand, in the recent research area of multi-UAV systems the architecture adopted has usually been network-oriented as it is the natural solution that leads to a flexible and cost effective interconnection among multiple systems. Regarding the physical communication layer, it is possible to find solutions with different channels for different types of information, i.e. high bandwidth analog channels for images transmission [20] or dedicated channels for teleoperation [21]. But the principal trend nowadays is to have one unique digital channel (or more if fault tolerance is a requirement) shared by different types of information. The progress in telecommunications technology is making this approach feasible.

Therefore, as the recent developments in multi-UAV systems follow a network-centric approach close to the SWIM architecture, from this point of view the introduction of the UAVs in SWIM is possible and even can be considered as natural.

3.2.2 The Publish/Subscribe Paradigm

Once a network-centric approach is adopted, the elements connected to the network can interchange information according to different models. Some of them are listed in Table 1, extracted from [3], which compares them with respect to the degree of decoupling between data producers and consumers.

Table 1 Different data distribution models compared w.r.t the degree of decoupling between information producers and consumers

Abstraction	Space decoupling?	Time decoupling?	Synchronization decoupling?
Message Passing	No	No	Publisher side
RPC/RMI	No	No	Publisher side
Async. RPC /RMI	No	No	YES
Notifications	No	No	YES
Shared Spaces	Yes	Yes	Publisher side
Message Queuing	Yes	Yes	Publisher side
Publish Subscribe	Yes	Yes	Yes

A high degree of decoupling leads to more robust solutions and to lower costs for the adaptation or insertion of new elements, which are properties required in SWIM. In general, the publish/subscribe paradigm allows a total decoupling between the sources and the sinks of information. Furthermore, new sources or sinks can be added or removed dynamically with a minor impact on the operation of the network. In fact, the publish/subscribe paradigm has been already adopted in other areas such as wireless sensor networks or multi-robot research [17, 22, 23], where the properties mentioned above are also relevant. Moreover, the performance of the network scales well against changes in the demand of a given type of information if dynamic replication techniques are applied.

In the publish/subscribe paradigm, the flow of information between sources and sinks is managed dynamically by one or several intermediate entities, usually called data brokers (see Fig. 9). The middleware is composed by these intermediate entities and their communication protocols. As a result, there is no direct communication between sources and sinks and if a new subsystem has to be added to the network, it is only necessary to develop its interface with the data brokers. Therefore, compatibility tests between this subsystem and other existing subsystems in the network are not necessary, decreasing the integration cost and time. In the design of SWIM, the publish/subscribe architecture was adopted from the beginning to allow an easy integration of new subsystems in the NAS, such as the UAVs themselves.

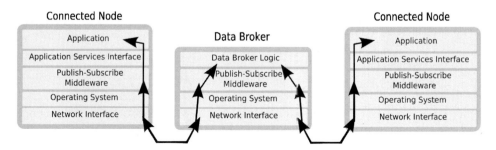

Fig. 9 Pure publish/subscribe model, with no direct interaction between clients

3.2.3 Architectures for the Brokers

There are several possible architectures for the data brokers, depending on the required services and capabilities for the whole communication system. Form the architectural point of view, the simplest one could be similar to the solution adopted in [17], where there is no independent element managing the information exchanges. The implementation of the middleware interface in the clients provides this function-ality in a distributed way as shown in Fig. 10. This architecture does not follow a pure publish/subscribe paradigm, but allows low latencies and it is a good solution when real time operation is required.

On the other hand, the most complex architecture could use different protocols to offer pure publish/subscribe functionality for services requiring low bandwidth and a direct connection oriented protocol for streaming data services.

There are three different brokers architectures proposed for SWIM in [3] and all the information related to the SWIM middleware design is consistent with the analysis provided in this paper. Table 2, extracted from the executive summary of [3], shows the main differences between the architectures proposed for the SWIM brokers. It should be pointed out that, due to the network centric nature of SWIM, more than one model can operate at the same time in the same network depending on the application and data types involved in the communication.

On the other hand, the complexity of the communication system, and hence the fault probability, would be increased, at least linearly, with the number of different broker models implemented. Therefore, this number should be kept as low as possible and the integration of the UAVs services should be adapted to the models proposed in Table 2, which are general enough:

- Model "Pub/Sub Broker": It follows the strict publish/subscribe paradigm, so the UAVs will only have to communicate with the brokers and the integration would be easier. The latency associated with this model makes it incompatible with some of the most common UAV applications nowadays, such as teleoperation. Furthermore, there are also services offered by the UAVs such as surveillance video streaming that generates a massive amount of information and the use of data brokers could represent a bottleneck, increasing the latency and decreasing the quality of service.

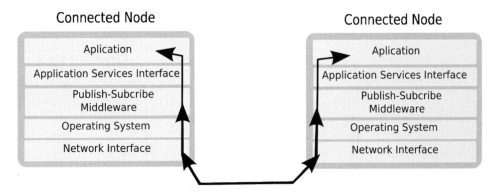

Fig. 10 Distributed version of the publish/subscribe paradigm

Table 2 Three different architectures proposed for the SWIM brokers

Broker Model	Description	Advantages	Disadvantages
Pub/Sub Broker	Implements the full set of functions needed to offer a pure Pub/sub middleware so that all operations between publishers and subscribers are completely decoupled in time, space and synchronization	Changes in publishers or subscribers completely transparent to each other; unified management of exchanged information	Possible extra latency in the process as data flows through the broker; This fact also means that the broker is a potential bottle neck limiting the ability to handle large streams of data
Lightweight Broker	Implements a subset of the functions needed to offer a pure Pub/sub middleware and implements the rest using traditional messaging mechanisms. This mixture means that not all operations between publishers and subscribers are completely decoupled	Supports implementation of several variants of Pub/Sub schemes	Publishers and Subscribers are not fully decoupled; predefined data channels need to be established
VC Broker	VC Broker is a superset of the functionalities offered by the pure Pub/Sub model. It can offer completely decoupled communication between the endpoints but also implements the primitives needed to establish a Virtual Circuit between to of them to achieve better latency or higher bandwidth	Broker approach is tailored to data type; Can offer different QoS for different data types	Extra complexity in information management functions such as monitoring; When connected by virtual circuits the Publishers and Subscribers are not fully decoupled

- Model "Lightweight Broker": This model is intended to simplify the adoption of SWIM by using more classical approaches that would allow reusing some of the current interfaces. But the integration of the UAVs would require to implement those interfaces as special cases that do not follow the publish/subscribe policy implemented in the brokers. This option can be cost effective in the short term, but it is not as uniform or powerful as the other two models.
- Model "VC Broker": In this solution, the broker offers a different kind of service depending on the data type. It works as the first model when it is enough to fulfill the service requirements. But when low latency or high bandwidth is required, the broker provides a virtual connection between the data producer and consumer to prevent the potential bottleneck due to centralized brokers. This model is general enough to fulfill all the requirements from the point of view of the UAVs integration. For most of the services, the UAV could use the

publish and subscribe pure model, as any other SWIM client. But for latency sensible applications such as teleoperation [24] or high bandwidth requirements as real time surveillance video transmission, virtual circuits between the UAV and the client application can be created by the broker dynamically.

3.2.4 Access Solutions Proposed for the SWIM Clients

Figure 11 shows three different models proposed to connect to SWIM. In the first model there is a SWIM interface software which is running on the hardware on-board and interacting with the brokers. This model is preferred for the UAVs integration as far as it is the most flexible solution in terms of the potential spectrum of services that can be offered and, additionally does not require increasing the payload. The downside of this solution is that the interface is not so decoupled from the UAV specific software and hardware as in the other options, and special tests could be required to check the implementation and performance of the SWIM interface in order to avoid security problems.

The second model is based on specific hardware to support the connection of currently NAS integrated subsystems to SWIM. Those subsystems have specific interfaces that require a hardware adaptation (see Section 8 in [3]). This model is not necessary for the UAVs as far as the hardware on board is quite flexible and updatable. In fact, the next generation of UAVs could be designed with SWIM hardware compatibility. Anyway, this model allows addressing the security issues much better than the others because the SWIM interface is based on a specific hardware that can be designed following "trusted computing" principles.

Finally, the third model is based on standard web browsers, whose services can be "upgraded" by new web 2.0 technologies in the near future. In the last years, web browsers have been used for teleoperation applications in robotics and even in some

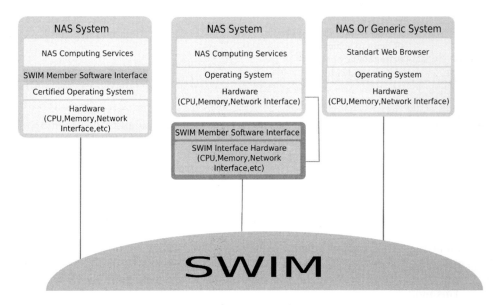

Fig. 11 Different models to connect to the SWIM network

UAVs control centers. Therefore, it is possible to have SWIM clients based on web browsers providing access to the UAV services and, in the opposite direction, the UAVs could also have access to information servers, such as map servers, through a web interface. In any case, limitations due to the latency and associated overhead involved in this model should be considered in order to select which interface to use for each service provided or required. The advantage of using this third model is that it is easier to develop and that the web browser/server protocols are well tested and designed with security in mind.

A combination of the first and third models (for non critical data queries) could be the best option.

3.2.5 UAVs Interfaces for Data and Services Access

The SWIM data model is described in the Appendix G of [3], where a detailed study of the different factors to be taken into account in the specification of the SWIM data structures is provided. Moreover, the impact of the data model on the flexibility and complexity of adding new subsystems is also presented.

Regarding the data model, the main design decisions made up to now and found in the used references can be summarized as follows:

- All the information should have a unified format, which has been called "SWIM Common Data Model" (see Fig. 12). Therefore, all the messages must be embedded in a standard data container, whose header should include standard fields independent of the type of message, such as its source, the data time stamp, etc. This mandatory header required in all the SWIM messages contains a set of fields which are usually referred as "SWIM Common Data Fields". During the development of SWIM it is expected to have sub-headers corresponding to sub-classes of data types leading to a tree of standard data structures. For every type of message, this tree will contain the list of mandatory fields and their associated

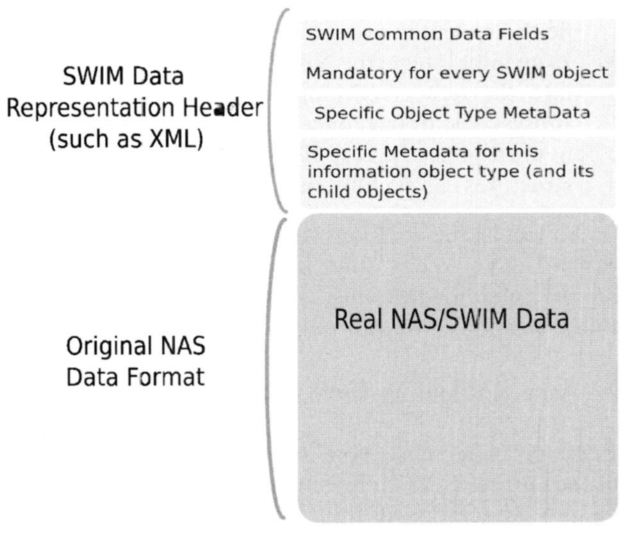

Fig. 12 Common data model diagram

Fig. 13 Example of a SWIM
common data format base
class, exposing examples of
possible fields of the Common
Data Fields

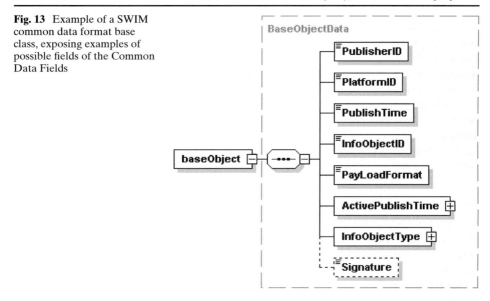

data structures. An example of a SWIM common data format base class is shown
in Fig. 13.

- The message structure should be embedded in the messages themselves, allowing
  a client to process a message without any previous knowledge about its internal
  data structure. This feature can be implemented using metalanguages or markup
  languages, such as XML (eXtensible Markup Language), and provides flexibility
  in the messages generation. For example, if a field is not applicable in a given
  message, this field is not included. Other advantage when using markup languages is the low probability for bad interpreted messages. When a new message
  is received, only the fields with known identifiers are processed whereas the rest
  of fields are ignored. This characteristic also makes easier the migration from an
  old type of message to a new one. During the transition both the old and the new
  fields are sent in order to keep offering the same services.
- In an initial stage, the NAS messages in their current format should be embedded
  into the new SWIM messages to allow an easier migration of the existing
  subsystems. In fact, once the original NAS message has been extracted from the
  new message, the rest of the process will remain the same. Later, those old data
  structures can be fully substituted for XML messages.

Almost all the decisions summarized above makes easier the integration in SWIM
of the UAVs, as any other kind of new NAS clients, due to the fact that they were
adopted with the flexibility criteria in mind. Of course, this statement does not apply
to the last point, which is only oriented to decrease the migration cost.

3.3 Aspects Requiring some Adaptation

In the previous subsection, it has been mentioned that in contrast to the rigid nature
of the current NAS architecture, the flexible network-centric design of SWIM allows
an easy integration of the UAVs. But there are several aspects that could require

further considerations due to the particular characteristics of the UAVs. It should be pointed out that the required adaptations presented in the next subsections do not come from the autonomous nature of the UAVs. In fact, the software on-board can make this autonomy characteristic nearly transparent for the rest of NAS subsystems.

In the following subsections, several aspects that could require some adaptation to tackle with some particular characteristics of the UAVs are presented. Moreover, possible solutions built on top of the basic functionality provided by SWIM are also depicted. Those solutions are based on elements that could be also added to the current SWIM specification in order to provide services requiring high flexibility.

3.3.1 Dynamic Brokers

The current UAV applications usually involve taking-off and landing from temporal or improvised locations. From the proposed architecture for the hierarchical connection and distribution of SWIM brokers [3], it seems that the idea is to have at least one broker in every airport that manages the insertion of new aircrafts into SWIM.

If the integration process of the UAVs implies to operate from conventional airports, their functionally will be drastically reduced. Therefore, an UAV taking-off from a given location needs a procedure to signal that it has started to operate and requires SWIM services (even if it is far away from an airport) such as receiving the "clear for take-off" message. The whole procedure requires a connection to a broker.

Nowadays, the ground control stations of the UAVs are continuously connected to them, and in some aspects, act as airports control towers (ATCT) for the UAVs. Then, it seems a natural solution to equip the ground control station with a broker that can link the UAV with other NAS elements. This solution involves dynamic brokers that are subscribing and unsubscribing to SWIM continuously from different locations.

3.3.2 High Bandwidth Channels on Demand

The teleoperation of the UAVs or some specific UAVs services such as surveillance could require the transmission of high bandwidth data in real-time during certain periods. Therefore, protocols to establish dedicated communication channels on demand are also required.

3.3.3 Collaborative UAVs Surveillance System

Regarding surveillance, due to the small dimensions and furtive nature of some UAVs, all the UAVs should be equipped with GPS and should continuously broadcast their positions at a given rate. This positional information could be merged with the rest of the surveillance related information (as the one provided by the primary and secondary radars) by the related data brokers. Given that the GPS infrastructure has been improved in recent years with the goal of making it more reliable and useful for the FAA this possibility should be easy to implement.

New elements designed to increase the usefulness of the GPS system for ATM applications are:

- The Wide Area Augmentation System (WAAS) [24]: Created by the Federal Aviation Administration (FAA) to augment GPS with additional signals for increasing the reliability, integrity, accuracy and availability of GPS for aviation users in the United States.

- The Local Area Augmentation System (LAAS): Created to allow GPS to be used for landing airplanes. LAAS is installed at individual airports and is effective over just a short range. This system should help the autonomous landing and take off of UAVs in normal airports.

3.3.4 Special Communication Technologies

As far as satellite links are intended to be used for global coverage in SWIM (Satellite enhanced CNS), the payload and budget limitations of some types of UAVs in terms of communication equipment on-board could be a limitation. In such cases, it could be the responsibility of the UAVs associated ground station to provide the communications link with its UAVs. As this is not the best, most general solution, it is the most straight forward one as the only link between the UAVs and the rest of the SWIM components would by its associated ground control station. This is coherent with the proposed figure of dynamic data brokers presented in Section 3.3.1.

4 Conclusions

The transition from the current rigidly configured ATM approach to the SWIM based ATM architecture will represent an important change in ATM concepts and procedures. This transition could represent a good opportunity to facilitate the introduction of the UAVs in non-segregated aerial spaces, which has been recognized as one of the main barriers for commercial UAV applications.

The UAV integration is examined in this paper at different layers of the ATM structure from global concepts, as the network centric nature and the pub-lish/subscribe paradigm, to the particular interfaces, broker and data models required to implement SWIM.

Furthermore, the UAVs integration could also help to improve basic ATM services, such as the weather information, and to offer new services such as on demand surveillance.

Then, it can be seen that the required extensions to include UAVs in the air traffic management of non-segregated aerial spaces are minimal and compatible with the proposed SWIM based ATM architecture.

References

1. Boeing Technology: Phantom works. Advanced Air Technology Management. http://www. boeing.com/phantom/ast/61605_08swim.pdf. Accessed 24 October 2007
2. SWIM Program Overview: http://www.swim.gov (redirects to www.faa.gov). Accessed 9 May 2008
3. System-Wide Information Management (SWIM) Architecture and Requirements. CNS-ATM Task 17 Final Report, ITT Industries, Advanced Engineering and Sciences Division, 26 March 2004
4. Jin, J., Gilbert, T., Henriksen, S., Hung, J.: ATO-P (ASD 100)/ITT SWIM Architecture Devel-opment. CNS-ATM, 29 April 2004

5. Koeners, G.J.M., De Vries, M.F.L., Goossens, A.A.H.E., Tadema, J., Theunissen, E.: Exploring network enabled airspace integration functions for a UAV mission management station. 25th Digital Avionics Systems Conference, 2006 IEEE/AIAA, Oct. 2006, pp. 1–11
6. Carbone, C., Ciniglio, U., Corraro, F., Luongo, S.: A novel 3D geometric algorithm for aircraft autonomous collision avoidance. In: 45th IEEE Conference on Decision and Control (CDC'06), pp. 1580–1585. San Diego, California, December 2006
7. UAV safety issues for civil operations (USICO), FP5 Programme Reference: G4RD-CT-2002-00635
8. Le Tallec, C., Joulia, A.: IFATS an innovative future air transport system concept. In: 4th Eurocontrol Innovative Research Workshop, December 2005
9. UAV Roadmap 2005–2030 – Office of the Secretary of Defense, August 2005
10. Everaerts, J., Lewyckyj, N., Fransaer, D.: Pegasus: design of a stratospheric long endurance UAV system for remote sensing. In: Proceedings of the XXth ISPRS Congress, July 2004
11. Weather Information Interface: http://aviationweather.gov/. Accessed 9 April 2007
12. metaf2xml: convert METAR and TAF messages to XML. Project web site: http://metaf2xml.sourceforge.net/. Accessed 25 May 2006
13. FlightGear Flight Simulator Project Homepage: http://www.flightgear.org/. Accessed 5 May 2008
14. Meserole, C.: Global communications, navigation, & surveillance systems program – progress and plans. In: 5th Integrated CNS Technologies Conference & Workshop, May 2005
15. Romeo, G., Frulla, G.: HELIPLAT: high altitude very-long endurance solar powered UAV for telecommunication and earth observation applications. Aeronaut. J. **108**(1084), 277–293 (2004)
16. Merino, L., Caballero, F., Martínez-de Dios, J.R., Ollero, A.: Cooperative fire detection using unmanned aerial vehicles. In: Proceedings of the 2005 IEEE, IEEE International Conference on Robotics and Automation, Barcelona (Spain), April 2005
17. Ollero, A., Maza, I.: Multiple Heterogeneous Unmanned Aerial Vehicles. Springer Tracts on Advanced Robotics. Springer, Berlin (2007)
18. Aeronautical Information Management Strategy, V4.0. EUROCONTROL, Brussels, Belgium March 2006
19. Merino, L., Caballero, F., Martínez-de Dios, J.R., Ferruz, J., Ollero, A.: A cooperative perception system for multiple UAVs: application to automatic detection of forest fires. J. Field Robot **23**(3), 165–184 (2006)
20. Beard, R.W., Kingston, D., Quigley, M., Snyder, D., Christiansen, R., Johnson, W., McLain, T., Goodrich, M.: Autonomous vehicle technologies for small fixed-wing UAVs. J. Aerosp. Comput. Inform. Commun. **2**(1), 92–108 (2005)
21. Alcázar, J., Cuesta, F., Ollero, A., Nogales, C., López-Pichaco, F.: Teleoperación de helicópteros para monitorización aérea en COMETS (in Spanish). XXIV Jornadas de Automática (JA 2003), León (Spain), 10–12 Septiembre 2003
22. Sørensen, C.F., Wu, M., Sivaharan, T., et al.: A context-aware middleware for applications in mobile AdHoc Environments. In: Proceedings of the 2nd Workshop on Middleware for Pervasive and Ad-hoc Computing Table of Contents. Toronto (2004)
23. Soetens, H., Koninckx, P.: The real-time motion control core of the Orocos project Bruyninckx, Robotics and Automation. In: Proceedings. ICRA '03. (2003)
24. Lam, T.M., Mulder, M., van Paassen, M.M.: Collision avoidance in UAV tele-operation with time delay, conference on systems, man and cybernetics. ISIC. IEEE International. Montreal, October 2007
25. Loh, R., Wullschleger, V., Elrod, B., Lage, M., Haas, F.: The U.S. wide-area augmentation system (WAAS). Journal Navigation **42**(3), 435–465 (1995)

A Survey of UAS Technologies for Command, Control, and Communication (C3)

Richard S. Stansbury · Manan A. Vyas ·
Timothy A. Wilson

Originally published in the Journal of Intelligent and Robotic Systems, Volume 54, Nos 1–3, 61–78.
© Springer Science + Business Media B.V. 2008

Abstract The integration of unmanned aircraft systems (UAS) into the National Airspace System (NAS) presents many challenges including airworthiness certification. As an alternative to the time consuming process of modifying the Federal Aviation Regulations (FARs), guidance materials may be generated that apply existing airworthiness regulations toward UAS. This paper discusses research to assist in the development of such guidance material. The results of a technology survey of command, control, and communication (C3) technologies for UAS are presented. Technologies supporting both line-of-sight and beyond line-of-sight UAS operations are examined. For each, data link technologies, flight control, and air traffic control (ATC) coordination are considered. Existing protocols and standards for UAS and aircraft communication technologies are discussed. Finally, future work toward developing the guidance material is discussed.

Keywords Command, control, and communication (C3) ·
Unmanned aircraft systems (UAS) · Certification

R. S. Stansbury (✉) · T. A. Wilson
Department of Computer and Software Engineering,
Embry Riddle Aeronautical University,
Daytona Beach, FL 32114, USA
e-mail: stansbur@erau.edu

T. A. Wilson
e-mail: wilsonti@erau.edu

M. A. Vyas
Department of Aerospace Engineering,
Embry Riddle Aeronautical University,
Daytona Beach, FL 32114, USA
e-mail: vyas85a@erau.edu

Abbreviations

ATC   air traffic control
BAMS   broad area maritime surveillance [24]
BLOS   radio frequency beyond line-of-sight
C2   command and control
C3   command, control, and communication
CoA   certificate of authorization
CFR   Code of Federal Regulations
DSA   detect, sense, and avoid
FAA   Federal Aviation Administration
FAR   Federal Aviation Regulations
HF   high frequency
ICAO   International Civil Aviation Organization [13]
IFR   Instrument Flight Rules
LOS   radio frequency line-of-sight
NAS   National Airspace System
NATO   North Atlantic Treaty Organization
STANAG   standardization agreement
SUAV   small UAV
TFR   temporary flight restriction
TUAV   tactical UAV
UA   unmanned aircraft
UAV   unmanned aerial vehicle
UAS   unmanned aircraft system
UCS   UAV Control System [22]
UHF   ultra high frequency
VHF   very high frequency
VSM   vehicle specific module [22]

1 Introduction

The integration of unmanned aircraft systems (UAS) into the US National Airspace System (NAS) is both a daunting and high priority task for the Federal Aviation Administration, manufacturers, and users. To modify the existing federal aviation regulations (FARs) under 14 CFR [1] to certify the airworthiness of a UAS, a tremendous effort is needed taking more time than the interested parties desire. An alternative approach is the development of guidance materials to interpret existing airworthiness standards upon manned aircraft systems that they may be applied to UAS.

The research presented in this paper assists the effort of this new approach. To interpret existing standards for UAS systems, a survey of existing UAS technologies must be performed, a system model must then be developed, and finally a regulatory gap analysis performed. This paper presents the technology survey of command, control, and communication (C3) systems on-board existing UAS.

1. Problem and Motivation

Before unmanned aircraft systems (UAS) are certified for general operation within the national airspace, airworthiness standards must be adopted. These standards may be derived through two possible approaches. The first approach would require the adoption of a new part to 14 CFR (also known as the Federal Aviation Regulations, or FARs). The time required to make such a change would be much greater than desired by the FAA, the US Military, UAS manufacturers, and UAS users.

The second option would be the provide guidance materials regarding the interpretation of the FARs with respect to the components of the unmanned aircraft system. This will likely be the approach taken. This research is meant to support its progress. A two part study is being conducted to analyze the technology associated with command, control, and communication of UAS, and from the technology survey determine issues and regulatory gaps that must be addressed.

The technology survey is essential to supporting the regulatory gap analysis. From an understanding of the technologies, it is then possible to examine the FARs and determine instances in which the existing regulation is not aligned with the state-of-the-art. These instances are the regulatory gaps that will be documented as the second phase of this project.

1.2 Approach

To assist the technology survey, a C3 system model shown in Fig. 1 was developed. The goal of this system model is to logically break the survey down into logical categories. For UAS operations, aircraft may operate within radio frequency line-of-sight, or beyond line-of-sight. Technologies and operating procedures related to command, control, and communication of UAS are divided into one of these two categories.

Fig. 1 C3 system model

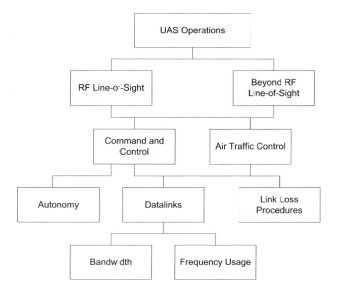

Under each category of RF LOS and BLOS, UAS technical issues may be divided into two categories: Command and Control (C2) and Air Traffic Control (ATC). For C2, the survey will explore technologies and issues necessary to safely support flight operations of the UAS from a remote pilot and/or control point-of-view. For ATC, technologies and issues related to the interaction of the aircraft or pilot-in-command with air traffic controllers while operating in the National Airspace System.

Under C2 and ATC, the various data links are examined including their respective frequency and data rates. The current link loss procedures are enumerated. Finally, for C2 only, the issue of autonomy, remote pilot versus autopilot, is examined for a variety of aircraft.

Communication related to detect, sense, and avoid (DSA) currently lies outside of the scope of this research. Such systems may consume additional command and control bandwidth, or require their own independent data links.

1.3 Paper Layout

Our system model divides our investigation of C3 technologies and issues between operation under RF line-of-sight and RF beyond line-of-sight conditions. The survey will discuss UAS operating under these conditions, the technological issues associated with command, control, and communication, and their interaction with ATC.

It is important to note that BLOS UAS do contain some LOS technologies. Figure 2 illustrates the overlap between these operating conditions and the class of UAS that can operate within these areas. The line-of-sight section shall include a discussion of all surveyed aircraft, but the beyond line-of-sight section shall only discuss medium and high-endurance UAS capable of operating beyond the RF line-of-sight of the pilot-in-command.

After discussing the surveyed UAS, security issues of C3 data links and flight controls are considered. Existing communication protocols, message sets, and standards are considered with respect to UAS C3.

For the language of this document, inline with current FAA documents, the term unmanned aircraft system will be used to describe the unmanned aircraft and its ground station. The term unmanned aerial vehicle, or its acronym UAV will be used only during discussion protocols in which the previous terminology is used.

Fig. 2 BLOS operations
subset of LOS operations

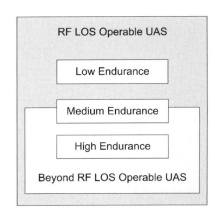

2 RF Line-of-Sight C3 Technologies and Operations

Line-of-sight operation may be divided between three classes of unmanned aircraft, which are low endurance, medium endurance, and high endurance. The first class operates almost entirely in line-of-sight. Surveyed low endurance aircraft include Advance Ceramic Research's Manta B [3], Advance Ceramic Research's Silver Fox [3], Meridian [10], Aerovironment's Raven [4], and Aerovironment's Dragon Eye [4]. The second and third class operates in both line-of-sight and beyond line-of-sight conditions. Surveyed medium endurance UA include Insitu's ScanEagle [15], Insitu's Georanger [15] and AAI Corp.'s Shadow [2], while the high endurance UA include General Atomics' Predator [7], General Atomics' Mariner [7], Northrup Grumman's Global Hawk [24], Northrup Grumman's BAMS [24], and Aerovironment's Global Observer [4]. Table 1 lists some examples of LOS C3 Technologies.

2.1 Data Links

The literature research revealed that LOS command and control data links commonly use a variety of frequencies from Very High Frequency (35 MHz) to C Band (6 GHz) [23]. It was observed from the technology survey that the most common LOS data link employed for current UAS is C Band. C Band uses low GHz frequencies for downlink, 3.7–4.2 GHz, and 5.9–6.4 for uplink [28]. C Band is strategically chosen for LOS C2 because the low GHz frequencies are less affected by extreme weather condition. For example, Mariner, Predator, and Predator's civilian versions such as Altair [5] use C Band for LOS C2.

Some small UA like ScanEagle and Georanger [15], Meridian [10], Shadow [2], Dragon [4], and Raven [4] use UHF for LOS command and control. It is not uncommon for these aircraft to utilize 72 MHz hand-held remote control similar or identical to those used by hobbiests.

Another option explored is Common Data Link (CDL) or Tactical CDL. CDL is a jam resistant spread spectrum digital microwave link only used by the military [8]. CDL is mostly used for BLOS operations; however, it can be used for LOS operations to ensure continuously safe and seamless communication when deployed in hostile territory. CDL will be discussed in later sections of this paper.

2.2 Flight Control Technologies and Operations

Aircraft autonomy varies dramatically amount unmanned aircraft. At one extreme, aircraft operate entirely by remote control. If the pilot is taken out of the loop, the aircraft would crash. At the other extreme, the aircraft may be controlled by an autopilot from takeoff to landing. The pilot-in-command stays outside of the C2 loop, but may intervene in the case of an emergency to override the autopilot.

Almost all modern UAS are equipped with both manual remote control and autopilot technologies. Flexibility to operate either way make a UAS more marketable as it fits the need of wide range of customers. From our study, a generalization could be made that the control technology for which the UAS operates under is highly specific to the mission characteristics and the ability of the aircraft to operate RF beyond line-of-sight.

Table 1 Line-of-sight communication for a sample of surveyed unmanned aircraft

Aircraft	Manufacturer	LOS Communication	Characteristics
Predator	General Atomics Aeronautical Systems	C-Band	Wing span: 20.1 m Length: 10.9 m Payload: 385.5 kg Max. altitude: 15,240 m Max endurance: 30 h
Global Hawk	Northrop Grumman Integrated Systems	CDL (137 Mbps, 274 Mbps); UHF SATCOM	Wing span: 39.9 m Length: 14.6 m Payload: 1,360.7 kg Max. altitude: 18,288 m Max endurance: 36 h
ScanEagle	Insitu Group	900 MHz of spread spectrum frequency hopping; UHF command/telemetry	Wing span: 3.1 m Length: 1.2 m Payload: 6.0 kg Max. altitude: 4,998 m Max endurance: 20 h
Meridian	University of Kansas	72MHz Futaba radio 16 km; 2.4GHz microband radio	Wing span: 8 m Length: 5.1 m Payload: 54.4 kg Max. altitude: 4,572 m Max endurance: 9 h
Desert Hawk	Lockheed Martin	Military 15 km data link	Wing span: 1.4 m Length: 0.9 m Payload: 3.1 kg Max. altitude: 152 m Max endurance: 1 h
Dragon Eye	AeroVironment	Military 10 km data link @ 9,600 baud	Wing span: 1.2 m Length: 0.9 m Payload: 2.3 kg Max. altitude: 152 m Max endurance: 1 h
Manta B	Advance Ceramic Research	Military band/ISM band radio modem; 24–32 km radio	Wing span: 2.7 m Length: 1.9 m Payload: 6.8 kg Max. altitude: 4,876 m Max endurance: 6 h

For LOS operations, a low endurance UAS typically uses remote control for part if not all of the flight. For takeoff and landing, a remote pilot will control the aircraft. Once airborne, the pilot may decide to fly the aircraft manually for the entirety of the flight path or allow the autopilot to perform simple waypoint navigation along a flight path. Some common UAS operating in such manner are Manta B, Meridian, Raven A and B, Dragon Eye, Silver Fox, and Shadow.

On the contrary, the high endurance UAS typically use autopilot for all LOS flight operations. The entire flight plan is programmed into the autopilot through a user interface at the ground control station. Once the mission begins, the aircraft will autonomously takeoff and follow the predefined path. The pilot remains out of the C2 loop, but monitors the flight operations for unusual situations, and, if need be, interrupts the autopilot and takes the manual remote C2. The Predator, Marnier,

ScanEagle, Georanger, Global Hawk, and BAMS are the example of UAS typically using autopilot technologies.

2.3 Link-Loss Procedures

The loss of a data link must be addressed by a link-loss procedure. It is important that the aircraft always operates in a predictable manner. From the survey, it was revealed that the most common link-loss procedure is for the aircraft to fly to a predefined location. Once at the predefined location, the UAS can either loiter until the link is restored, it can autonomously land, or it can be remotely piloted via secondary data link [21, 25, 27, 30]. In this section, specific examples are discussed.

The BAT III UA's link loss procedure involves a simple return home functionality, where it turns to the last known location of the ground control station and flies directly toward it [27]. A backup remote control radio at 72 MHz is available onboard the aircraft. Once within sufficient range to the base, a remote pilot will control the aircraft to land.

NASA and Boeing PhantomWorks X-36 follows a similar method of returning to base and loitering [30]. Rather than simply return to the base directly, the aircraft follows a pre-defined return path. To guide the UA to the path, steering points also exist. This proved troublesome as the X-36 would choose the nearest steering point, which in some cases may be behind the aircraft.

Similar to the X-36 return procedure, researchers at NASA Dryden have worked on a path planning algorithm for return-to-base and loss link operations that ensures the UA stays within its authorized flight zone [21]. Prior to departure, a return flight path was defined for use in the event of a loss of the control data link. Onboard autonomy will detect the loss of the link, and then maneuver the aircraft toward the return flight path.

Officials at Fort Carson have drafted a document for *Unmanned Aerial Vehicle Flight Regulations*. The military base includes two potential flight areas, one is restricted airspace, and the other is non-restricted airspace requiring either a Certificate of Authorization (CoA) or Temporary Flight Restriction (TFR) from the FAA [29]. They defined their classes of unmanned aircraft as Tactical UAV (TUAV) for operation beyond visual line-of-sight or over 1,000 ft; and Small UAV (SUAV) for flight under 1,000 ft or within visual line-of-sight. For the restricted airspace, if a TUAV loses link, it returns to a predefined grid location and loiters at 8,000 ft. If the SUAV loses link in the restricted airspace, it returns to the center of mass of the restricted airspace and lands. In both cases, necessary military authorities are contacted. When operating under CoA or TFR, the procedures modified in that FAA or other civilian authorities will be notified. If in either case the aircraft is likely to leave its restricted airspace, the flight will be terminated by some undisclosed means.

Finally, as disclosed from the NTSB report on the US Border Patrol Predator-B crash, the Predator-B will follow some predefined course back to base to loiter until communications are restored. In this case, as with others, the loss link procedure has on occasions been initiated for reasons besides an actual loss of data link [25]. In this case, operators attempted to use the link-loss procedure to restore the aircraft to a known working state.

2.4 ATC Communication and Coordination

ATC communications is one of the key factors allowing safe flight operations in NAS. It is utmost important that a UAS remains in contact with ATC during its entire mission even if traveling long distances from the ground control station. The literature research revealed that most ATC communications occur via a VHF transceiver. Special phone numbers were also setup in some instances to provide immediate contact with ATC in the event of a loss of data link.

In case of low endurance UAS for LOS operations, the pilot at ground control station is directly linked with ATC via VHF link. Redundant systems are employed for this purpose; a land based phone line is one of the options.

Our survey revealed very few details related to specific aircraft operations. Currently, unless operating under restricted military airspace, a Special Airworthiness Certificate or Certificate of Authorization (CoA) was required. To obtain such permission, procedures for interaction with ATC must be coordinated prior to flight. New interim regulations discussed later in this paper spell out more specific requirements for future UAS.

2.4.1 ATC Link Loss Procedures

ATC communication link loss procedures are handled quite differently from C2 link loss procedure. The primary objective in case of ATC link loss is to re-establish voice communication between the ground control station and the ATC facility. So in case of loss of direct link with ATC for LOS operation, a land based phone line is the only option currently used. Some unmanend aircraft are also equipped with multiple VHF transceivers that could be used to establish a ground control-aircraft-ATC voice communication link. Most link loss occurs due to weather so it is very likely that the later option is also unavailable, as all VHF voice communications will be affected in the same vicinity.

3 Beyond RF Line-of-Sight C3 Technologies and Operations

The survey of beyond line-of-sight UAS covers primarily high endurance UAS, but a few medium endurance UAS that operate beyond line-of-sight. In the first category, Predator [7], Marnier [7], Global Hawk [24], and BAMS [24] are surveyed. For the second category, Meridian [10], ScanEagle [15], and Georanger [15] are studied. Table 2 lists some examples of BLOS C3 Technologies aboard UAS.

3.1 Data Links

BLOS C2 data links range from Ultra High Frequency (300 MHz) to Ku Band (15 GHz) [23]. Ku Band SATCOM data links are widely used for BLOS C2 system. It has a frequency range from 11.7–12.7 GHz for downlink and 14–14.5 for uplink [28]. Ku Band is used by a bulk of high endurance UAS like Global Hawk, BAMS, Predator and its derivatives. INMARSAT SATCOM data links are also used by high endurance UAS including BAMS, Marnier and Global Hawk. It has frequency range from 1,626.5–1,660.5 MHz for uplink and 1,525–1,559 MHz for downlink [14]. L Band Iridium SATCOM data links are used by smaller, low or medium endurance,

Table 2 Beyond line-of-sight communication for a sample of surveyed unmanned aircraft

Aircraft	Manufacturer	BLOS Communication	Characteristics
Predator	General Atomics Aeronautical Systems	Ku-Band SATCOM	Wing span: 20.1 m Length: 10.9 m Payload: 385.5 kg Max. altitude: 15,240 m Max endurance: 30 h
Global Hawk	Northrop Grumman Integrated Systems	Ku-Band SATCOM: Inmarsat	Wing span: 39.9 m Length: 14.6 m Payload: 1,360.7 kg Max. altitude: 18,288 m Max endurance: 36 h
ScanEagle	Insitu Group	Iridium	Wing span: 3.1 m Length: 1.2 m Payload: 6.0 kg Max. altitude: 4,998 m Max endurance: 20 h
Meridian	University of Kansas	Iridium A3LA-D Modem 2.4 Kbits/s 1,616–1,626.5 MHz	Wing span: 8 m Length: 5.1 m Payload: 54.4 kg Max. altitude: 4,572 m Max endurance: 9 h
Desert Hawk	Lockheed Martin	No BLOS Operations Disclosed	Wing span: 1.4 m Length: 0.9 m Payload: 3.1 kg Max. altitude: 152 m Max endurance: 1 h
Dragon Eye	AeroVironment	No BLOS Operations Disclosed	Wing span: 1.2 m Length: 0.9 m Payload: 2.3 kg Max. altitude: 152 m Max endurance: 1 h
Manta B	Advance Ceramic Research	No BLOS Operations Disclosed	Wing span: 2.7 m Length: 1.9 m Payload: 6.8 kg Max. altitude: 4,876 m Max endurance: 6 h

research UAS. It has a frequency range from 390 MHz–1.55 GHz [17]. Georanger, a medium endurance unmanned aircraft, and Meridian, a research unmanned aircraft, use Iridium modems as part of the avionics communication package.

Investigating satellite communication providers, low earth orbiting (LEO) and geosynchronous earth orbiting (GEO) satellites represent two extremes. LEO satellites operate at an altitude of 1500 km. GEO satellites operate at an altitude of 35000 km. In [26], a constellation of 80 LEO satellites was compared with a six satellite GEO constellation with equivalent coverage area using Ka Band. The LEO constellation outperformed the GEO constellation with reduced latency, lower path losses, and reduced launch cost. A LEO satellite constellation does have higher operational costs. As satellites pass overhead, service may be temporarily disrupted as the communication is automatically handed-off to the next satellite. Examples of widely used LEO constellations include Iridium [17] and Globalstar [9].

One potential data link under consideration for BLOS communication is the military's existing Common Data Link [8]. It is unknown whether such a system would be applicable for civilian use or not, but it appears that CDL links are used on many of the larger UAS operated by the military;e.g. Predator-B, Global Hawk, etc. While no documents explicitly support this claim, the aircrafts' previously cited data sheets show identical specifications without explicitly stating that it is a CDL link. In addition to CDL, there is also the Tactical CDL, which also includes additional security.

Two technologies exist for CDL links. The first uses an I-band satcom link. The second data link uses Ku Band at 14.5–15.38 GHz in order to increase available bandwidth [8].

3.2 Flight Control Technologies and Operations

Satellite-based communications as are discussed in the last section are the primary means of beyond line-of-sight command and control communication with unmanned aircraft. Latency is a key issue encountered with SATCOM data links. At such latencies, remote piloting becomes less feasible. Autopilots are therefore required for control of most UAS under beyond RF line-of-sight operations.

The high endurance UAS uses autopilot for all flight operations. Entire flight plan is programmed in the autopilot GUI. The pilot remains out of the C2 loop but monitors the flight operations for unusual situations. A prime example of such a mission was the Altair UAS that flew a restricted flight plan imaging wildfires in western states. Altair is a NASA owned UA used for research purposes; a CoA was provided for each flight based on FAA approval of flight plans submitted by NASA [5]. Unfortunately, specific details of autopilot design and operations for the majority of high-endurance UAS were not disclosed.

The medium endurance UAS has the opportunity of using either LOS or BLOS technologies during the course of its flight. The research unmanned aircraft Merdian from the University of Kansas provided key insight into the transitions between LOS and BLOS and how operations change. Meridian is integrated with an autopilot for command and control [10], which allowed both remote piloting and waypoint navigation. Currently, a WePilot [31] autopilot is utilized (W. Burns, Personal communications, University of Kansas, 2008). When within close proximity, a 2.4 GHz LOS modem may be used and the pilot will receive real-time telemetry such that he may visualize the aircraft and control it similar to Instrument Flight Approach. Once the aircraft has exceeded the 2.4 GHz radio's range, the aircraft transitions to its Iridum data link. While using the Iridium due to latency and bandwidth, operation is limited to high-priority aircraft telemetry and waypoint navigation (W. Burns, Personal communications, University of Kansas, 2008) [10].

3.3 Lost-Link Procedures

From the survey, it was found that link loss procedures for BLOS operation in either medium endurance or high endurance unmanned aircraft are nearly identical to LOS operations. When the avionics detect a loss of a data link, a preplanned control sequence is executed. The aircraft either returns to base or emergency rendezvous point, taking a pre-determined flight path, or it loiters within a pre-determined airspace.

Altair [5] flew in NAS for Western States Fire Imaging Mission with CoA from FAA. During one of its mission, the Altair had modem malfunction, resulting in BLOS Ku Band C2 link loss. As a result, the aircraft switched to C Band and flew to pre-determined air space until the modem returned to normal functioning and the Ku Band link was established [5].

3.4 ATC Communication and Coordination

Since high endurance UAS may operate across multiple ATC regions, the current paradigm for ATC communication is to utilize the UAS as a communication relay. The relay allow a ground operator to remain in constant contact with the ATC of the UAS's current airspace. As part of a UAS's Certificate of Authorization, it is also a standard procedure to coordinate with all ATC sites along the UAS's path [5].

As discusses previously, the primary objective in case of ATC link loss is to re-establish the communication data link between ground control station and the ATC facility. For BLOS operations this is only possible by carrying multiple VHF transceivers and having redundant voice communication systems on board. This is necessary because as the aircraft travels through various ATC regions it contacts local ATC facility and the ground control station is connected to ATC via the aircraft as a relay.

For the Altair link loss, the FAA and ATC were provided with detailed flight plans, making sure that the ATC knew aircraft's location. Additionally, the missions were planned meticulously with respect to ATC coordination, such that all potential ATC facilities are notified. The mode of notification was not explicitly disclosed [5].

Using the unmanned aircraft as a relay between the ground control station and the overseeing ATC facility is not without several issues. The handoff of the aircraft between facilities presents an issue. For manned aircraft, as it transitions from one ATC cell to another, the onboard pilot dials the VHF radio to the appropriate ATC channel as instructed through the handoff procedure. For several existing CoAs and aircraft, the aircraft perform a rapid assent to uncontrolled airspace and maintain this altitude for the duration of the flight. As a result, interaction along a flight path involving multiple ATC facilities is not common, and proper procedures to handle operations within controlled airspace has not been formally developed. For UAS to operate within ATC controlled airspace in the NAS beyond line-of-sight, further protocols must be established regarding the handling of the handoffs, and setting of the new frequencies of the aircraft's ground-to-ATC relay.

Another potential issue of using UAS as a relay is the spectrum availability to handle the additional 25 KHz voice channels needed to support each unmanned aircraft (S. Heppe, Personal communications, Insitu, 2008). A proposed alternative would be the inclusion of a ground-based phone network that connects ground stations to the ATC facility for which the UA is operating under. These issues must be studied further.

4 Security Issues of UAS C3 Technology and Operations

Data link spoofing, hijacking, and jamming are major security issue facing UAS C2 and ATC communications. UAS are different than conventional aircraft from the point of "immediate control" of the aircraft. Pilot in "immediate control" means in an adverse event, the pilot can fly without putting aircraft or other aircraft in the immediate vicinity at risk of collision. In case of UAS, there is a medium between pilot at the ground control station and aircraft which is not the case with conventional aircraft. This medium being the data link, it is easily susceptible to threats mentioned previously. A hacker can create false UAS signals, jam the data link or even hi-jack the data link and take the control of UA. This issue must be addressed while picking the appropriate data link for future UAS C2 and ATC communication, as data links are vital to the safety and seamless functioning of the UAS.

In order to make C2 and ATC communication foolproof, security features can be built into the system. For example, one approach is for the aircraft to acknowledge or echo all commands it receives. This will ensure the pilot-in-command that all commands sent are received and acknowledged (S. Heppe, Personal communications, Insitu, 2008). Such an approach will also notify the pilot in command if the aircraft receives commands from an unauthorized entity. The military uses secured data links like CDL and Link 16 [20] with built-in validating functions. No such permanent solution is available for civilian market and the area must be explored.

5 Relevant Existing Protocols, Standards, and Regulations

5.1 Standardization Agreement 4586

NATO Standardization Agreement (STANAG) 4586 [22] was ratified by the NATO countries as an interoperability standard for unmanned aircraft systems. Now in its second edition, it was made available in April 20, 2004. One of the goals of STANAG 4586 is to achieve the highest defined level of interoperability of unmanned aircraft from NATO member nations. Several levels of interoperability are defined, and throughout the document, the relationships to these interoperability levels are indicated. It must be noted that the standard covers UAS command and control (C2) as well as payload communications.

STANAG 4586 is divided into two annexes. The first annex provides a glossary to support the second annex. The second annex provides an overview of the communication architecture, which is supported by three appendices. Appendix B1 discusses the data link interface. Appendix B2 discusses the command and control interface. More specifically B2 covers the military architecture that connects the ground control station with the military command hierarchy. Finally, Appendix B3 discusses the human and computer interfaces (HCI). For this study, the Annex B introduction and Appendix B1 are relevant.

Figure 3 presents the reference architecture for STANAG 4586. The UA is comprised in the UAV Air Component. Our interests lie in the Air Vehicle Element (AVE) as payload is not included as part of this technology survey. The AVE communicates with the UAV Surface Component (or ground station) through some data link. This Surface Component interacts with any external C4I infrastructure, any

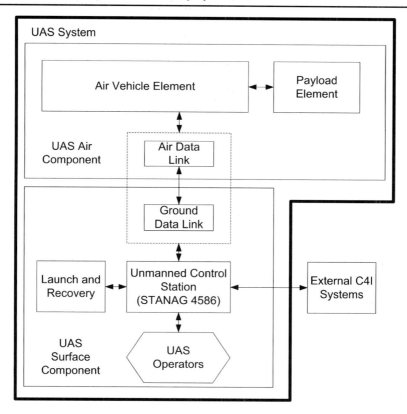

Fig. 3 STANAG 4586 system architecture [22]

ground operators controlling the aircraft, and the launch and recover system. The well defined interfaces between these components is presented as a block diagram in Fig. 4. Each interface is emboddied in a system module that operates as a wrapper to support previously incompatible technologies.

For this investigation, the Vehicle System Module (VSM) defines the interface between the control station and the aircraft. Its Data Link Interface (DLI) is studied further as it defines the message set for command and control to the aircraft and telemetry from the aircraft. Figure 5 illustrates the interaction between the aircraft and ground station via STANAG 4586 interfaces.

STANAG 4586 has strong potential as a standard for UAS command and control in the National Airspace System. It has already been adopted by NATO member countries as the current interoperability standard for military UAS. Unlike JAUS [19], discussed in the next section, which was developed to cover all unmanned systems (ground, aerial, surface, and underwater), STANAG 4586 is specifically written to support UAS and has grained broader acceptance in the UAS industry than JAUS.

From the certification point-of-view, the adoption of an interoperability standard could dramatically simplify the certification process. The natural division of UAS systems by STANAG 4586 supports the ability to certify these components inde-

Fig. 4 STANAG 4586
components and interfaces
[22]

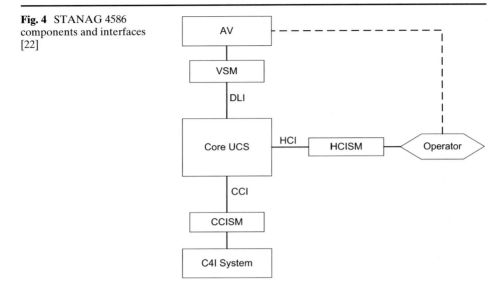

pendently. Unmanend aircraft can be certified as compliant independent of any
ground station as its implementation of the VSM must support the data link interface.
Likewise, the ground station may be certified by evaluating its interoperability with
unmanned aircraft and operator control interfaces. Unfortunately, the FAA has not
yet adopted modular certification.

There are some weaknesses of STANAG 4586. First, the standard includes messages that are specifically written to support communication with a UA's payload.
Since payload is not incorporated into the certification process, these messages much
be eliminated or modified. Fortunately, few messages incorporate both control and
payload aspects.

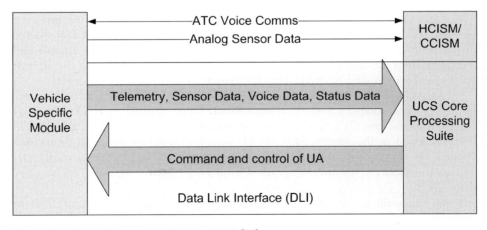

Fig. 5 STANAG 4586 ground station to aircraft [22]

5.2 Joint Architecture for Unmanned Systems

The Joint Architecture for Unmanned Systems (JAUS) [19] provides an alternate interoperability standard for unmanned aircraft. Unlike STANAG 4586, JAUS specifies interoperability of all unmanned systems and not just unmanned aircraft systems. Originally developed as a Department of Defense effort, JAUS is now being integrated into new standards by SAE through the AS-4 Technical Committee [16].

JAUS is an architecture defined for the research, development, and acquisition of unmanned systems. The documentation is divided into three volumes: Domain Model, Reference Architecture, and Document Control Plan [19]. The JAUS Working group [19] define in these volumes the architecture for representing unmanned systems, the communication protocols between these components, and the message set.

The technical constraints of JAUS are: platform independence, mission isolation, computer hardware independence, and technology independence [18]. The word System is used to define a logical grouping of subsystems. A subsystem is one or more unmanned system functions as a single localized entity within the framework of a system. A Node is defined as a distinct processing capability within a subsystem to control flow of message traffic. A component has a unique functionality capability for the unmanned system. Figure 6 presents a system topology using these terms.

Given the abstract definition of systems and subsystems and its applicability to unmanned systems other than aerial vehicles, JAUS does not seem to be sufficiently rigorous to meet the needs of regulators for defining a command and control protocol that may be certified for unmanned systems. As the architecture significantly changes toward new standards defined by AS-4, this concern may decrease, and warrants future consideration.

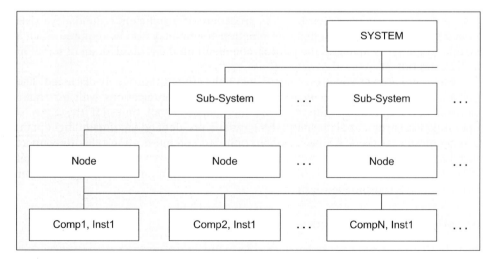

Fig. 6 Joint architecture for unmanned systems Topology [18]

5.3 International Civil Aviation Organization (ICAO) Annex 10

International Civil Aviation Organization is a United Nations agency responsible for adopting aviation standards and introducing as well as researching new practices for safety of international aviation. For the purpose of this technology survey, ICAO Annex 10 Aeronautical Telecommunications Volume I [12] and III [11] were researched. These standards apply to all aircraft and not only UAS. Volume III: Communication Systems discusses systems level requirements for various communication systems, namely aeronautical telecommunication networks, mobile satellite services, VHF air to ground digital data links, HF data links, etc., defining basic functionality, requirements and recommendations, and message protocols for developers.

5.4 FAA Interim Operational Approval Guidance 08-01

In March 2008, the FAA's Aviation Safety Unmanned Aircraft Program Office (AIR-160) with the cooperation of a number of other program offices at the FAA released interim guidelines for UAS operations in the NAS [6]. This document was produced to fill the need of some guidelines that may be shared between the FAA, UAS manufacturers, and UAS users until more permanent guidelines are put in place. The guidance may be used in order to assist in the distribution of Certificates of Authorization for public NAS users such as government agencies and Special Airworthiness Certificates for civilian NAS users. These regulations were meant to be flexible with periodic revisions.

The regulations do define some of the operational requirements of UAS relevant to Command, Control, and Communication. Since existing UAS detect, sense, and avoid (DSA) technologies are not safety certifiable at this time, a chase aircraft or ground observer is required during the operation of the UAS. The fact that some autonomy may exist for the unmanned aircraft is acknowledged and acceptable so long as a human-in-the-loop exists, or the system possesses a mechanism for immediate human intervention in its operation. Link-loss procedures must be defined for the aircraft such that it operates in a predictable manner. The guidance material does not suggest any particular link-loss procedure. If insufficient redundancy exists to maintain safe operation, a flight termination system may also be required. Such a system would be available to the pilot-in-command on the ground, or an observer in a chase aircraft.

Besides link loss procedures, control data links are not thoroughly discussed. The guidance material does specify requirements for communications with air traffic control. ATC communication must be maintained at all times. If the UAS is operating in instrument flight rules (IFR) under a pre-defined instrument flight path, it is required that the UAS possesses an onboard communication relay to connect its pilot-in-command (on the ground or in the chase aircraft) with the aircraft's local ATC. Lastly, under IFR, the UAS must possess at a minimum a mode C transponder (with a mode S transponder desired).

6 Conclusion and Future Work

From this research, an initial survey of technologies for command, control, and communication for unmanned systems was conducted. From this study, line-of-sight and beyond line-of-sight systems were compared. For each, the study addressed issues of communication technologies, command and control, lost link procedure, and ATC communications. This survey has assisted in gathering knowledge related to the technologies and issues associated with command, control, and communication.

The second phase of this project is a regulatory gap analysis. With a through understanding of current C3 technology, issues may now be enumerated with respect to FAA regulations, including, but not limited to, 14 CFR Part 21, 23, and 91 [1]. From these enumerated issues, the regulatory gaps will be analyzed. Regulatory gaps identify the areas of the existing regulations that are insufficient for the current state-of-the-art. Finally, the analysis will be shared with the FAA to hopefully assist in determining guidance material regarding airworthiness certification requirements for UAS users and manufacturers that wish for their aircraft to operate within the National Airspace System.

Acknowledgements This project was sponsored by the Federal Aviation Administration through the Air Transportation Center of Excellence for General Aviation, and was conducted by the members indicated. The Center of Excellence for General Aviation Research is comprised of the following universities: Embry-Riddle Aeronautical University, Florida A&M University, University of Alaska, University of North Dakota and Wichita State University. However, the Agency neither endorses nor rejects the findings of this research. The presentation of this information is made available in the interest of invoking technical community comment on the results and conclusions of the research.

Special thanks to Xiaogong Lee and Tong Vu of the FAA Research and Technology Development Branch—Airport and Aircraft Safety Team for their support.

References

1. 14 CFR: United States Code of federal regulations title 14 aeronautics and space. Online at: http://ecfr.gpoaccess.gov/cgi/t/text/text-idx?c=ecfr&tpl=ecfrbrowse/Title14/14tab_02.tpl (2008)
2. AAI Corp.: Unmanned aircraft systems. Online at http://www.aaicorp.com/New/UAS/index.htm (2008)
3. Advance Ceramic Research: Unmanned vehicle systems. Online at http://www.acrtucson.com/UAV/index.htm (2008)
4. AeroVironment: Unmanned aircraft systems. AeroVironment Inc. Online at http://www.avinc.com/UAS_products.asp (2008)
5. Ambrosia, V.G., Cobleigh, B., Jennison, C., Wegener, S.: Recent experiences with operating UAS in the NAS. In: AIAA Infotech Aerospace 2007 Conference and Exhibit. Rohnert Park, California (2007)
6. FAA: Interim operational approval guidance 08-01: unmanned aircraft systems operations in the national arspace system. Technical report, Federal Aviation Administration. Aviation Safety Unmanned Aircraft Program Office AIR-160 (2008)
7. General Atomics: Aircraft platforms. General Atomics Aeronautical Systems Inc. Online at http://www.ga-asi.com/products/index.php (2008)
8. Global Security: Common data link. Online at http://www.globalsecurity.org/intell/systems/cdl.htm (2008)
9. Globalstar: Globalstar, Inc. — Worldwide satellite voice and data products and services for customers around the globe'. Online at http://www.globalstar.com (2008)
10. Hale, R.D., Donovan, W.R., Ewin, M., Siegele, K., Jager R., Leong, E., Liu, W.B.: The meridian UAS: detailed design review. Technical Report TR-124, Center for Remote Sensing of Ice Sheets. The University of Kansas, Lawrence, Kansas (2007)
11. ICAO (ed.): Annex 10 communication systems, vol. 3 1st edn. International Civil Aviation Organization (1995)
12. ICAO (ed.): Annex 10 radio navigation aides, vol. 1, 6th edn. International Civil Aviation Organization (2006)

13. ICAO: Fourth meeting of the AFI CNS/ATM implementation co-ordination sub-group. International Civil Aviation Organization. Online at http://www.icao.int/icao/en/ro/wacaf/apirg/afi_cnsatm_4/WP08_eng.pdf (2008)
14. INMARSAT: Aeronautical services. Online at http://www.inmarsat.com/Services/Aeronautical/default.aspx?language=EN&textonly=False (2008)
15. Insitu: Insitu unmanned aircraft systems. Online at http://www.insitu.com/uas (2008)
16. International, S.: Fact sheet SAE technical committee as-4 unmanned systems. Online at: http://www.sae.org/servlets/works/committeeResources.do?resourceID=47220 (2008)
17. Iridium: Aviation equipment. Online at http://www.iridium.com/products/product.htm (2008)
18. JAUS: Joint architecture for unmanned systems: reference architecture version 3-3. Technical Report, JAUS Working Group. Online at: http://www.jauswg.org/baseline/refarch.html (2007)
19. JAUS: Joint architecture for unmanned systems. JAUS Working Group. Online at: http://www.jauswg.org/ (2008)
20. Martin, L.: Tactical data links—MIDS/JTIDS link 16, and variable message format—VMF. Lockheed Martin UK—Integrated Systems & Solutions. Online at: http://www.lm-isgs.co.uk/defence/datalinks/link_16.htm (2008)
21. McMinn, J.D., Jackson, E.B.: Autoreturn function for a remotely piloted vehicle. In: AIAA Guidance, Navigation, and Control Conference and Exhibit. Monterey, California (2002)
22. NATO: NATO standarization agreement 4586: standard interfaces of UAV control system (UCS) for NATO UAV interoperability. North Atlantic Treaty Organization (2007)
23. Neale, M., Schultz, M.J.: Current and future unmanned aircraft system control & communications datalinks. In: AIAA Infotech Aerospace Conference and Exhibit. Rohnert Park, California (2007)
24. Northrop Grumman: Unmanned systems. Northrop Grumman Integrated Systems. Online at http://www.is.northropgrumman.com/systems/systems_ums.html (2008)
25. NTSB: NTSB incident CHI06MA121—full narrative. National Transportation Safety Board. Online at http://www.ntsb.gov/ntsb/brief2.asp?ev_id=20060509X00531&ntsbno=CHI06MA121&akey=1 (2008)
26. Peters, R.A., Farrell, M.: Comparison of LEO and GEO satellite systems to provide broadband services. In: 21st International Communications Satellite Systems Conference and Exhibit (2003)
27. Ro, K., Oh, J.-S., Dong, L.: Lessons learned: application of small UAV for urban highway traffic monitoring. In: 45th AIAA Aerospace Sciences Meeting and Exhibit. Reno, Nevada (2007)
28. Tech-FAQ: What is C band?. Online at http://www.tech-faq.com/c-band.shtml (2008)
29. US Army: Unmanned aerial vehicle—flight regulations 95-23. Technical Report AR 95-23, The Army Headquarters. Fort Carson, Colorado (2005)
30. Walker, L.A.: Flight testing the X-36—the test pilot's perspective. Technical Report NASA Contractor Report no. 198058, NASA—Dryden Flight Research Center, Edwards, California (1997)
31. WeControl: WePilot 2000 technical brief. Online at: http://www.wecontrol.ch/pdf/wePilot2000Brief.pdf (2008)

Unmanned Aircraft Flights and Research at the United States Air Force Academy

Dean E. Bushey

Originally published in the Journal of Intelligent and Robotic Systems, Volume 54, Nos 1–3, 79–85.
© Springer Science + Business Media B.V. 2008

Abstract The United States Air Force Academy is actively involved in unmanned aircraft research across numerous departments involving many projects, aircraft, government agencies, and experimental programs. The importance of these research projects to the Academy, the faculty, the cadets, the Air Force, and to the defense of the nation cannot be understated. In an effort to be proactive in cooperating with recent concerns from the FAA about the growth and proliferation of UAS flights, the Air Force has implemented several new guidelines and requirements. Complying with these guidelines, directives, and regulations has been challenging to the researchers and research activities conducted at USAFA. Finding ways to incorporate these new guidelines effectively and efficiently is critical to research and participation in joint projects and exercises. This paper explores the nature of research at USAFA current restrictions imposed by the various regulations, the current process, short term solutions, and a long term vision for research into UAS at the Academy.

Keywords UAV · UAS · Unmanned aircraft · Research · Education · Air force · Flight · Airspace

1 Introduction

The United States Air Force Academy (USAFA) is actively engaged in several different areas of research involving Unmanned Aerial Systems (UAS), across many departments, supporting many different projects, agencies, and exercises. Several varieties of unmanned aircraft are flown at the academy, some of them commercially procured and modified for research, and some of them experimental one-of-a-kind

D. E. Bushey (✉)
Department of Computer Science, UAV Operations, US Air Force Academy,
Colorado Springs, CO 80840, USA
e-mail: dean.bushey@usafa.edu

K. P. Valavanis et al. (eds.), *Unmanned Aircraft Systems*. DOI: 10.1007/978-1-4020-9136-0_6

aircraft. Recent changes to regulations; both FAA and military have necessitated a review of operating procedures of UAS flights at the Air Force Academy. This change has had, and will continue to have a major impact on research, education, development of these new technologies. The current process to approve these flights has arisen from the growing concern for the rapid growth in numbers and types of UAS and remote-controlled aircraft of various sizes airborne, and the safe flight of UAS by the US Military and other government agencies. These regulations, intended for larger scale operational flights, are being adapted for use with UAS, even micro controlled UAS systems used in several research projects.

In this paper we explore the nature of the research projects at USAFA, the goals, regulations affecting UAS operations, agencies involved. In addition, we examine both short term and long term solutions to these hurdles.

Lt Gen Regni, Superintendent of the USAF Academy recently highlighted the importance of UAS research at the Academy in an address to USAFA personnel [3]. The explosive growth in research, funding, scope and attention in recent years is expected to continue for the foreseeable future. Along with the research component of the Air Force Academy, one of the primary missions is producing future Air Force leaders that understand the nature of current technologies, uses, limitations, challenges, and possible avenues for growth and development.

2 Background

The United States Air Force Academy is an undergraduate engineering school designed to educate, train, and develop Cadets to become future leaders in the Air Force. Its primary focus is on developing cadets of character and leadership. There is no designated graduate research program at USAFA. However, one of the goals of USAFA is to encourage cadet and faculty research by developing and promoting leading-edge research into critical areas of interest to the US Air Force and the Department of Defense. Cadets are encouraged to explore many facets of UAS operations, including flight characteristics, cockpit design, human computer interaction, data transmissions, tactical use of UAS, persistent UAS, real-time video relay, automated visual and feature recognition, etc. A list of current projects is provided in attachment 1.

The Academy also supports and encourages faculty research and publication. Several researchers work with various other universities, agencies, and government departments on research and development into UAS and UAS related technologies. Part of this research helps to support many different organizations in the Department of Defense, Department of Homeland Security, intelligence agencies, Department of Immigration, and several other federal and state agencies.

2.1 Restricting Flight Operations at USAFA

In May, 2007 all UAV/UAS flights at the Academy were halted in response to guidance from the DoD and FAA requiring all UAS to have a Certificate of Authorization (COA) and to meet specific prescribed guidelines to fly in the national airspace system. On 24 Sep 2007, the Deputy Secretary of Defense signed a Memorandum of Agreement (MOA) with the FAA allowing the DoD to conduct

UAS flights without a COA provided that the flights abide by certain restrictions [1]. Current standard FAA regulations are two-tiered regarding UAS flights, one for civilian use and one for government use. Government use, which is broadly defined as public-owned aircraft operated in support of government operations, requires a Certificate of Authorization (COA) to fly in unrestricted airspace. This COA outlines airspace, altitude, and other operating restrictions and is generally granted per airframe as follows:

–   Past experience has shown it takes approximately 6 months to receive a COA, but the FAA has agreed to a 60 day process for DoD requests (beginning 24 Sep 2007)
–   The COA requires a detailed listing of aircraft performance, operational limitations and procedures, location of operations, and personnel authorized to operate the UAS under the COA.

   ■   The USAFA has submitted paperwork to the FAA for a blanket COA that would authorize all UAS flights within the AFA airspace. While this is under review Air Force Special Operations Command (AFSOC) is the lead agency in charge of UAS operations.
   ■   UAS operations with a valid COA require that the pilot and observer hold a valid class III medical certificate, and that the pilot holds a valid pilot certification issued by the FAA (Private Pilot's License, etc.)

–   FAA/DoD Memorandum of Agreement, signed 24 Sep 2007 allows for flight of UAS over military reservations without a COA however,

   o   Specific DoD guidance applies to all flights conducted over military reservations, and a COA is still required for flights outside of this airspace.
   o   MOA delineates between micro UAS (smaller than 20 pounds) and other UASs

      ■   Micro UASs may operate within Class G airspace under 1,200 ft AGL over military bases provided the operations meet the following criteria:

         •   UAS remains within clear visual range of the pilot, or a certified observer in ready contact with the pilot
         •   UAS remains more than 5 miles from any civil use airport
         •   DoD must publish a Notice to Airmen (NOTAM) no later than 24 h prior to the flight (blanket NOTAMs can be used if justified)
         •   Pilots/Observers are qualified by the appropriate Military Department (AFSOC)

   o   AFSOC has been designated lead command for developing small UAS guidance for qualification and operation for all USAF UAS.

      ■   Draft guidance released by AFSOC (AFSOC 11-2) is geared towards operational qualification of UAS pilots on Systems of Record intended for use in combat situations in close coordination with manned aircraft [2]. These requirements are being modified to meet the unique demands of USAFA researchers to meet due to time, manpower and money constraints

- AFRL has begun developing separate guidance for research and development UAS, however AFSOC is the designated lead and must approve any AFRL-developed guidance prior to implementation
- Current AFSOC guidance requires all UAS to obtain certification through a Special Operations Application Request (SOAR) process. Not originally designed for certifying micro UAS, this process is being adapted to accommodate experimental research UAS designed and used at USAFA.

2.2 AFSOC Guidance

AFSOC has published guidelines to follow for UAS operators seeking approval for National Airspace Operations. The published checklist states, "on September 24, 2007, FAA and DoD representatives signed a Memorandum of Agreement concerning the operation of DoD Unmanned Aircraft Systems (UAS) in the National Airspace System. Prior to approving such operations certain criteria must be met. The following checklist outlines the necessary requirements for the operation of DoD UAS in Class G Airspace ONLY. A "Yes" answer is required to the following statements in order to process your request." (AFSOC checklist for UAS operations)

1. This is a DoD or DOD-contracted UAS certified by one of the military departments as airworthy to operate in accordance with applicable DoD and Military Department standards
2. The UAS pilots, operators and observers are trained, certified and medically qualified by the appropriate Military Department to fly in Class G airspace.
3. The Unmanned Aircraft (UA) weighs 20 pounds or less.
4. The UA operations will be contained in Class G airspace, below 1200' AGL.
5. The Class G airspace is located over a military base, reservation or land protected by purchase, lease or other restrictions.
6. The UA will remain within clear visual range of the pilot, or certified observer in ready contact with the pilot, to ensure separation from other aircraft.
7. The UA operations will remain more than 5 miles from any civil use airport or heliport.
8. The applicant verifies that this operation has been thoroughly coordinated and approved with a government official within the unit and that the applicant has been appointed as the unit requesting authority.

2.3 Restrictions Imposed

The net results of these regulations and the military MOA is UAS research operations at the Academy must be revised to meet the new safety related guidelines. Currently applications for Special Operations Airworthiness Release (SOAR) approval on several UAS are under review. Guidance from AFSOC has been clarified to a certain extent, but hurdles remain.

2.3.1 Airworthiness Process

The SOAR process, as defined in the Airworthiness Circular dated 24 August 2007, spells out the criteria used to certify UAS as airworthy. The guidance for this process comes from several underlying regulations, including:

Air Force Policy Directive (AFPD) 62-6, *USAF Aircraft Airworthiness Certification*
Air Force Instruction 90-901, *Operational Risk Management*
Air Force Pamphlet 90-902, *Operational Risk Management (ORM) Guidelines and Tools*
Air Force Instruction 11-202V1, *Aircrew Training*
Air Force Instruction 11-202 V3, *General Flight Rules*
Airworthiness Certification Circular No. 4, *Certification Basis*
Airworthiness Certification Circular No. 5, *Risk Evaluation and Acceptance*
MIL-HDBK-514, *Operational Safety, Suitability, and Effectiveness (OSS&E) for the Aeronautical Enterprise*
MIL-HDBK-516, *Airworthiness Certification Criteria*
MIL-STD-882, *Standard Practice for System Safety*

- SOAR process through Aeronautical Systems Center (ASC) at Wright-Patterson AFB, certification:

  o The SOAR process is rapidly evolving to meet these new types of UAS, however the time line for some research projects may be too inflexible for such a review.

    ■ USAFA-led flight safety reviews, perhaps through the aeronautics department, might be faster, more flexible and still meet the spirit and intent of the FAA and AFSOC requirements

- Pilot Qualification

  o Rated Pilots, either military or FAA certified, may act as POC of flights.
  o USAFA may design student pilot designations as long as FAA ground school requirements have been met.

- UAS flights in other airspace

  o Ft Carson offers a restricted airspace that is available for use. This would preclude the need for COAs and SOAR process review. This is an interim solution as it is not feasible to schedule range time, commute 1.5 h each way to and from the range during the already too full academic day for cadets.
  o USAFA's own class D—initial consideration was given to extended the Academy's class D airspace to include the Cadet Area and proposed UAS flight airspace. From discussions with the FAA this may be very difficult to achieve, however it remains a topic of great interest.

3 The Way Ahead

It is essential to come up with a viable solution that meets the needs of the FAA and DoD to maintain safety requirements and to adhere to regulations, while at the same time allowing needed research and development to be conducted at USAFA. Possible solutions include:

- Academy Certification of micro UAS—certify an agency at USAFA, possibly the aerospace engineering department, as a certifier of safety of flight and COA for these micro UAS
- Class D airspace extension—as mentioned above, this is a possibility that would allow USAFA airfield managers to allow UAS flights with prior coordination. Modification of existing airspace designations is being pursued, but this process could take time, if approved at all.

3.1 Short Term Solutions

Working within the framework of the current process, several short term steps can be taken to streamline the approval process for UAS research at USAFA:

- Streamline and expediting of the current approval process if possible
- USAFA local approval authority for flight readiness review process for cadet capstone projects.
- Verbal Approval/Waiver Authority, pending through review of applications

3.2 Long Term Solutions

Many long term solutions have been proposed to ensure that this valuable research proceeds:

- Establishment of a USAFA UAS Research Center that would work closely with all agencies involved
- Establishment of a cadet UAS flight center that would allow all cadets the opportunity to experience UAS flight during their sophomore or junior years.

4 Summary

UAV/UAS research at the USAFA has becoming an increasing integral part of the cadet education. New, tighter restrictions and safety considerations have necessitated a review of operating procedures and temporarily halted certain research flights. Several people, agencies, and departments are working to solve the problems; high-level attention has been given to this issue; and the current process is being ironed out to facilitate research needs.

Samples of UAS Research by USAFA Faculty and Cadets. Revised 5 Apr 2008

Subject	Department	Faculty Point of Contact
Fly Eye	DFB with U of Wyoming	Dr. Mike Wilcox
Atmospheric Research (Requested by USAF Weather Service)	DFP	Dr. Matthew Mcharg
Viper Aircraft (*V*ersatile *I*ntegrated *Pla*tform for *E*xperimental *R*esearch)	DFAN	Lt Col Carl Hawkins
KC-135 Redesign–design, build and fly scale model	DFAN	Dr. Steve Brandt
Fighter Sized Target Study (design, build and fly scale model)	DFAN	Maj Jeremy Agte
SCARF (collect radio frequency data using software radios)	DFCS	Capt James Lotspeich
Situational Awareness Tool	DFCS	Dr. Steve Hadfield
Summer Space Program	DFCS	Capt James Lotspeich
Robust/Reliable UAV Platforms	DFAN,DFCS	Dr. Steve Brandt/ Capt James Lotspeich
Black Dart Exercise Support	DFCS and DFEM	Lt Col Bushey
Black Dart *Red Team* UAS Project	DFAN/Sys Engr	Lt Col James Greer
Improved UAV Batteries	DFC	Dr. John Wilkes
Multiple UAVs for Persistent ISR	DFEC	Dr. Daniel Pack
Intelligent Sensors for Persistent Tactical ISR	DFEC	Dr. Daniel Pack
Heterogeneous Active Sensor Network for Efficient Search and Detection of IED/EFP Associated Activities	DFEC	Dr. Daniel Pack
SECAF UAV Project Evaluate/Improve Predator Operator Control Stations	DFBL, DFCS, DFEC, Sys Engr	Lt Col David Bell
Micro Air Vehicles	DFEM	Dr. Dan Jensen

References

1. DOD–FAA Memorandum, 20070924 OSD 14887-07—DoD–FAA MoA UAS Operations in the NAS, 24 Sep (2007)
2. Air Force Special Operations Command Interim Guidance 11-2, UAV Operations, 1 Oct (2007)
3. Lt Gen Regni: Speech to USAFA personnel, 7 Mar (2008)

Real-Time Participant Feedback from the Symposium for Civilian Applications of Unmanned Aircraft Systems

Brian Argrow · Elizabeth Weatherhead · Eric W. Frew

Originally published in the Journal of Intelligent and Robotic Systems, Volume 54, Nos 1–3, 87–103.
© Springer Science + Business Media B.V. 2008

Abstract The Symposium for Civilian Applications of Unmanned Aircraft Systems was held 1–3 October 2007 in Boulder, Colorado. The purpose of the meeting was to develop an integrated vision of future Unmanned Aircraft Systems with input from stakeholders in the government agencies, academia and industry. This paper discusses the motivation for the symposium, its organization, the outcome of focused presentations and discussions, and participant survey data from questions and statements that were collected in real time during the meeting. Some samples of these data are presented in graphical form and discussed. The complete set of survey data are included in the Appendix.

Keywords Unmannned aircraft system · UAS · Civil applications

1 Introduction

An investment of over 20 Billion dollars for the development of unmanned aircraft capabilities has resulted in U.S. leadership in the field and unprecedented capabilities that can now be successfully applied to civilian applications [11]. Unmanned Aircraft Systems (UAS) are highly developed based on decades of developments for modern

B. Argrow · E. W. Frew (✉)
Aerospace Engineering Sciences Department,
University of Colorado, Boulder, CO 80309, USA
e-mail: eric.frew@colorado.edu

B. Argrow
e-mail: brian.argrow@colorado.edu

E. Weatherhead
Cooperative Institute for Research in Environmental Sciences,
University of Colorado, Boulder, CO 80309, USA
e-mail: betsy.weatherhead@cires.colorado.edu

aircraft. At this stage, a broad range of aircraft exist from small and lightweight ones that often hold cameras or remote sensors to large airplanes, with wingspans of over 130 feet that can hold advanced scientific equipment [11]. All have one quality that is invaluable to civilian applications: they can go where it might be dangerous to send a manned aircraft and they generally provide persistence beyond the capabilities of manned aircraft.

Unmanned aircraft (UA) are proving valuable in initial attempts at civilian applications. UA have flown far over the Arctic Ocean, flying low enough to take detailed measurements of the Arctic ice that could not be taken with manned aircraft [8]. UAS are successfully being used to support homeland security, assisting in border patrol missions [6], keeping our agents both safe and effective in their duties. Recent flights into hurricanes brought back important measurements that could not have been gathered any other way [1, 3, 4]. Other civilian applications for which UA have already been fielded include law enforcement [12], wildfire management [13], and pollutant studies [7]. These first attempts to use UAS for civilian applications have brought tremendous results while maintaining safety in the air and on the ground.

Atmospheric scientists and engineers at the University of Colorado convened the first ever community gathering of those interested in civilian applications of UAS [2]. Entitled *The Symposium for Civilian Applications of Unmanned Aircraft Systems (CAUAS)*, this meeting was held 1-3 October 2007 in Boulder, Colorado (Fig. 1). Leaders from industry, government and academia came together for three days to discuss successes, future needs, and common goals. Representatives from over thirty five industries, a dozen universities, and five agencies (NASA, NOAA, DHS, DOE and Federal Aviation Administration (FAA)) gathered to discuss, across disciplines, the priorities for the next ten years. The CAUAS Steering Committee determined that the next decade will be critical for the technologies, regulations, and civilian

Fig. 1 The symposium for CAUAS was held 1–3 October 2007 in Boulder, Colorado

applications that will drive the design and development of future UAS. Fundamental to the future success of these efforts is the development of the technology and infrastructure to allow civilian applications to proceed safely. The development of these capabilities will allow the U.S. to maintain its technological leadership while successfully addressing important societal needs to protect the homeland, improve our understanding of the environment, and to respond to disasters as they occur.

2 Symposium Structure

The Symposium was structured to answer the questions: 1) What are current US civilian capabilities, 2) Where does the civilian UAS community want to be in 10 years, and 3) How do we get there? Each day of the three-day symposium was structured to address one of the primary questions:

Day 1: Current US civilian capabilities: Successes and failures

1. Background
2. Public and Commercial Applications
3. Scientific Applications
4. Industry, FAA, agency Vision

Day 2: The Public Decade: Where do we want to be in ten years?

1. Scientific Applications
2. Public and Commercial Applications
3. Visions for the integration of UAS capabilities into scientific and public goals
4. Industry, FAA, agency Vision

Day 3: How do we get there? Steps in the Public Decade

1. Existing paradigms and new alternatives
2. Applications-driven engineering challenges
3. Society and policy
4. Industry, FAA, and agency vision

On Day-1, Session 1 focused on how UAS technology has arrived at the current state of the art. The session started with presentations that summarized military applications and how the Department of Defense has been, and continues, to be the primary driver for UAS development and deployment. This was followed by a contrast of wartime and commercial applications of UAS, an overview of international UAS capabilities, and a view of the current state from the FAA perspective. Session 2 focused on recent public and commercial applications, both successes and failures, that included recent joint-agency (NASA, NOAA, USFS, FAA) missions to support firefighting efforts in the Western US, and DHS (Customs and Border Protection) operations along the US-Mexico border. Session 3 looked at current scientific applications ranging from sea-ice survey missions in the Arctic, to the measurement of various surface-to-atmosphere fluxes, to the integration of satellite-based remote sensing with UAS from observation platforms. Each day concluded with a Session 4 that provided a panel discussion where agency, academia, and industry representatives provided their assessment of what they heard during the previous three sessions.

The focus of Day-2 was the *Public Decade* where the term *public* follows the FAA definition that categorizes aircraft and aircraft operations as either *civil* or *public* [9]. Civil refers to commercial aircraft and operations. Public aircraft and operations are those sponsored by federal or state governments, that includes agencies and universities. In the first session, scientists and agency representatives presented their vision of science applications enabled by UAS technologies in 2018. In Session 2 presentations focused on visions for public and commercial applications such as search and rescue, border patrol, and communication networking. Session 3 focused on visions of integrated sensing systems that might be deployed for applications in 2018 such as in situ hurricane observations and fire monitoring. The day was concluded with a Session 4 panel discussion.

With the first two days addressing the state of the art and visions of the near future, the final Day-3 addressed the question of "How do we get there: What are the steps forward during the public decade?" This day was planned to focus on the technical engineering challenges, and also on the societal perception and regulations challenges. Presentations in Session-1 focused on alternatives to manned flight and regulatory constraints, who can use UAS, does their development makes sense from a business perspective, and a discussion of an UAS open-design paradigm. Session-2 presentations addressed applications-driven engineering challenges that include sensor development, payload integration, and command, control, communications, and computing challenges. The Session-3 presentations focused on the social and regulatory challenges with issues such as public perception and privacy, education of the public, and the challenge to US leadership with the global development and operations of UAS. Session 4 was conducted in the manner of the previous days, and the meeting was concluded by an address from the Symposium chairs.

3 Results of Real-Time Participant Feedback

A unique approach to understand the thoughts of the broad range of experts and measuring the collective opinions of the participants was employed for the Symposium. With over 150 people in attendance most days, hearing from each individual on a variety of subjects was not feasible. Clickers, donated from *i-clicker* (a Macmillan US Company, http://www.iclicker.com/), were used to poll the attendees on a variety of subjects. All the questions and responses are included in the Appendix and a few selected charts from the tabulated data are discussed below. Most often, statements, such as "UAS provide an irreplaceable resource for monitoring the environment" were presented and the audience could respond with one of five choices: Strongly Agree (SA), Agree (A), Neutral (N), Disagree (D), and Strongly Disagree (SD). For a small category of issues, the audience was asked to grade the community on specific issues, with the grades being A, B, C, D, and F, and in some cases the five responses corresponded to specific answers. The broad range of responses for the questions reflect the large diversity of the UAS civilian community who gathered in Boulder. The compiled results are broadly categorized as 1) participant demographics; 2) safety, regulations, and public perception; 3) applications, missions, and technologies; and 4) economics.

Figure 2 describes the demographics of the symposium participants and roughly approximates the targeted mix of participants sought by the CAUAS Steering

Committee invitations, with 36% from government agencies, 26% from academia, and 21% from industry. About half (45%) of the participants had more than 5 years of UAS experience.

Safety, regulatory issues, and public perception proved to be the primary focus of the discussions. There continues to be widespread misunderstanding of what is the current FAA policy for the operation of UAS in the United States National Airspace System (NAS). (Note that in March 2008, FAA published the most recent UAS policy revision [10]). Once this topic was raised and the current policies were discussed, it was clear that many continue to conduct research and commercial operations in violation of these policies and regulations. This revelation added to the urgency of the discussion for the CAUAS community to work more closely with FAA to abide by regulations and to pursue the legal requirements that the FAA representatives described for NAS operations. A specific recommendation was for

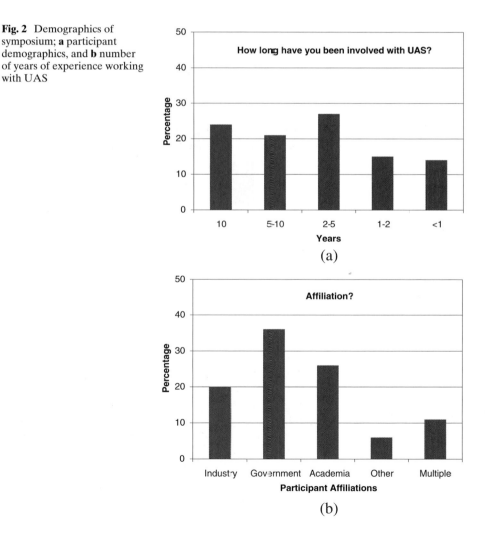

Fig. 2 Demographics of symposium; **a** participant demographics, and **b** number of years of experience working with UAS

UAS researchers to seek locations with official restricted airspace to conduct legal flights with a minimum of FAA regulatory constraints.

Figure 3 shows some of the response to specific i-clicker questions. The response in Fig. 3a indicates the optimism of the participants with over 60% of the participants agreeing and strongly agreeing with the prediction that UAS operations in the NAS will become routine in the next 10 years. However, this is countered with pessimism that regulatory policies can keep pace with technological developments that might help with the integration of UAS into the NAS, with more than 80% of the participants expecting regulatory policies will not keep pace.

In discussions of the most likely public applications to be pursued over the next 10 years and those that will prove to be most important, disaster response was selected as the most likely application. Figure 4a shows the consensus of the group that there is a compelling case to use UAS for public applications. Figure 4b displays the

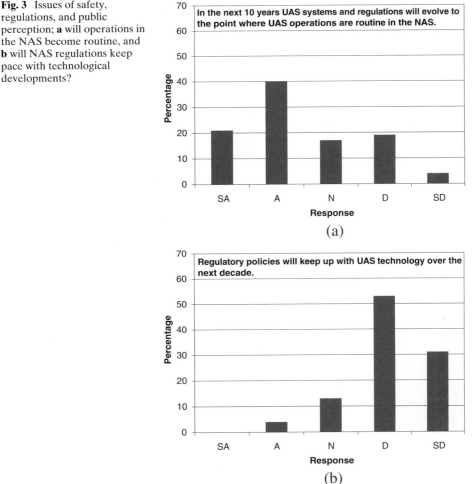

Fig. 3 Issues of safety, regulations, and public perception; **a** will operations in the NAS become routine, and **b** will NAS regulations keep pace with technological developments?

Fig. 4 Applications, missions, and technologies; **a** is there a compelling case for public UAS applications, and **b** will UAS become a key element in disaster response?

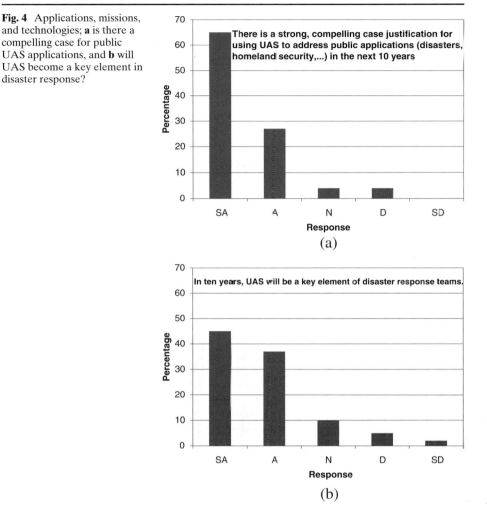

(a)

(b)

consensus that UAS will become an important and key element in disaster response teams during the next decade.

While Fig. 5a shows the general audience uncertainty that science applications will generate a significant UAS market, Fig. 5b shows strong agreement that market opportunities will be a primary driver if UAS are to be developed and applied in civilian applications. A participant pointed out that a current limitation is the lack of a civilian workforce trained to operate UAS. He suggested that this will begin to change with the return of U.S. service men and women who have operated UAS in theaters of conflict in Iraq and Afghanistan. Providing jobs for this workforce over the next decade might also have positive political payoffs to further enhance the economic future of UAS and their civilian applications.

Fig. 5 Applications, missions, and technologies; **a** is there a compelling case for public UAS applications, and **b** will UAS become a key element in disaster response?

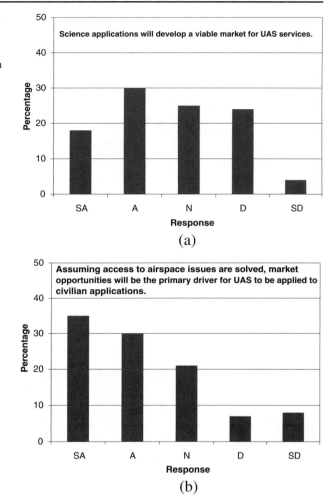

(a)

(b)

4 Outcome: UAS in the Public Decade

The CAUAS Steering Committee distilled the presentations and subsequent discussions into three civilian applications that are predicted to be of major importance during the Public Decade 2008–2018. These three applications are: 1) Disaster Response, 2) National Security, and 3) Climate Change. These applications were chosen, both because of their importance in which they were viewed in the participant feedback and because these are applications important to the agencies that will most likely have the financial support and the regulatory permission to carry them out.

For Disaster Response, UAS have already demonstrated their effectiveness for fire monitoring in six Western states. These preliminary flights helped fire-fighting efforts, directly saving costs, lives, and priority. UAS can also be used to survey disaster areas to provide immediate information for saving lives and property. Use of UAS after Hurricane Katrina could have expedited efforts to locate and rescue people stranded by floodwaters and dramatically helped in the disaster response.

UAS are becoming increasingly important for National Security applications focused upon protecting the homeland, helping Customs and Border Protection (CBP) streamline routine and event-driven operations along the countrys borders. Up to October 2007, CBP UAS helped personnel capture over 1,000 criminals crossing the Mexican border. Based on these successes, the Department of Homeland Security has significantly increased its investment in UAS.

UAS can fill an important gap not covered by existing systems for Climate Change observational applications. The observations would revolutionize the monitoring and scientific understanding of weather and climate. The Aerosonde, a small UAS manufactured and operated by AAI Corp., completed low-altitude penetration missions into both tropical cyclone Ophelia (2005) and Category-1 hurricane Noel (2007), providing real time observations of thermodynamic data and dangerous winds that in the future might be used in storm forecasting and storm track prediction [1, 3, 4]. UAS can also help measure carbon emissions providing necessary information for carbon trading.

The summary outcome of the meeting was condensed into an 8-page brochure entitled "CAUAS, Civilian Applications of Unmanned Aircraft Systems: Priorities for the Coming Decade [5]. An image of the cover page that captures the breadth of applications and technologies is shown in Fig. 6. The presentations

Fig. 6 Cover page for CAUAS brochure

Fig. 7 Participants response to the statement: "The UAS community dedicated to civilian applications has a clear vision for how to use UAS in the coming decade." This statement was presented at the start of the meeting (*Monday*) and repeated at the close of the meeting (*Wednesday*)

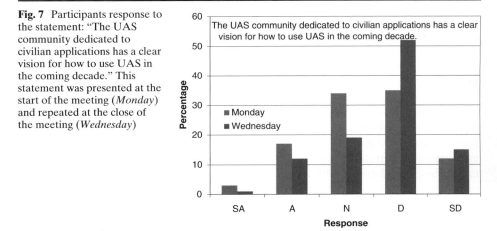

that formed the basis of this document can be found at the Symposium website http://cauas.colorado.edu.

To help gauge the impact of the meeting on the participants, the following i-clicker statement was asked on Monday, the first day of the symposium and again at the conclusion of the symposium on Wednesday: "The UAS community dedicated to civilian applications has a clear vision for how to use UAS in the coming decade". Figure 7 indicates that although a focus of the meeting was to discuss visions for the development and application of UAS technologies over the next 10 years, the participants indicated that they left with an even less clear vision of the future than before the meeting. There are at least two ways to interpret this result. First, it could mean that the presentation of a variety of visions for the future of UAS suggested the breadth of ideas of the possible. A second interpretation might be based on doubts that resulted from the clash of applications visions with the realities of regulatory requirements. In either case, the conclusion can be drawn that civilian community believes that there remains much uncertainty in the development of an integrated vision for the future of civilian applications of UAS.

5 Summary and Conclusions

CAUAS was the first symposium to bring together, in a single venue, the major stakeholders for the integration of UAS into the NAS for civilian applications, i.e., government agencies, industry, and academia. The meeting started with presentations that recounted the history of UAS and credited the current state-of-the-art to the development and maturation of military UAS. The remaining presentations focused on non-military, civilian applications and the need for conversion of military UAS for civilian applications, or for the development of UAS entirely for civilian applications. While technology has transfered between military and civilian uses, the view of the audience is that a separate profit-driven market must exist to support a UAS industry for civilian applications. Even if the number of such applications continues to increase, no substantial industry can currently exist unless UAS technologies can satisfy current FAA regulations developed for manned aircraft. This might soon be possible for larger UAS with UA that can carry equipment to emulate an onboard pilot and that can operate with the speed and performance of large manned aircraft. An alternative possibility is for new policies or regulations to be developed specifically for UAS integration of large UAS into the NAS. Either of these regulatory paths will continue to challenge or preclude the widespread integration of smaller UAS with performance requirements that require that they not operate like manned aircraft.

The CAUAS symposium was organized to focus on UAS applications for the *Public Decade*, 2008–2018; where "public" emphasizes the FAA distinction of civil and public aircraft. The CAUAS Steering Committee chose this as a reasonable time frame during which the government agencies might collaborate to take the lead and work with industry and academia to develop technologies, standards, and regulations to enable the integration of UAS into the NAS for civilian applications. If accomplished, this will expand opportunities and develop a significant market for the UAS industry. Of the applications presented and discussed, three rose to the level of a consistent organizing theme for to organize agency programs: Disaster Response, National Security, and Climate Change. These applications are summarized in the CAUAS summary document [2]. The purpose of this document is to lay out the case that can be made for civilian applications and to provide "policy relevant" data for those in positions to effect change.

The Symposium concluded with a general consensus that the gathered community had just begun to scratch the surface of the issues and obstacles to the integration of UAS into the NAS. Several recommended that CAUAS should reconvene for future symposia with specific emphasis on the civil and commercial applications and business cases, rather than the specific focus on public applications. At the time of this writing, the CAUAS organizers are considering a CAUAS-II meeting to be convened approximately two years after the original.

Acknowledgements The authors acknowledge the contributions of the CAUAS Steering Committee: Susan Avery, V. Ramanathan, Robin Murphy, Walt Oechel, Judy Curry, Doug Marshall, Sandy MacDonald, Steve Hipskind, Warren Wiscombe, Mike Kostelnik, Mike Francis, Raymond Kolibaba, Bob Curtin, Doug Davis, John Scull Walker, Steve Hottman, Mark Angier, and Rich Fagan. We also acknowledge the contributions of the CAUAS presenters whose presentations can be found at the Symposium website http://cauas.colorado.edu.

Appendix

Tables 1, 2, 3, and 4.

Table 1 CAUAS survey responses: demographics

	SA	A	N	D	SD
Affiliation: A) industry; B) government; C) academia; D) other; E) Multiple.	20	36	26	6	11
How long have you been involved with UAS? A) 10 years B) 5 to 10 years; C) 2 to 5 years D) 1–2 years; E) less than 1 year	24	21	27	15	14

Table 2 CAUAS survey responses: safety, regulatory issues, and public perception

	SA	A	N	D	SD
In the next 10 years UAS systems and regulations will evolve to the point where UAS operations are routine in the NAS.	21	40	17	19	4
DOD-FAA MOU should be expanded to include other agencies.	37	32	19	5	6
Regulatory policies will keep up with UAS technology over the next decade.	0	4	13	53	31
Civilian applications of UAS should be consistent and compliant with FAA rules and regulations.	52	30	10	6	2
Safety standards applied to UAS are currently more stringent than manned aircraft.	20	31	16	24	9
UAS are still seen as a threat to the UAS community.	20	32	23	16	9
The clarification of UAS regulations have negatively impacted our UAS plans.	31	20	34	9	7
The clarification of UAS regulations have negatively impacted our UAS plans.	27	28	27	11	8
Your institution or agency would supply real cost data of manned flights, both direct and indirect, to assist the UAS industry help benchmark its costing.	12	32	33	12	10
The general aviation community is concerned that more regulations and equipment requirements will be levied on them as a result of UAS in the NAS.	22	34	29	15	1
UAS are seen as a threat to public safety.	13	22	31	22	11
UAS civilian applications: safety from UAS accidents.	22	22	42	11	2
UAS civilian applications: threat to National security.	15	24	35	18	9
UAS civilian applications: leadership.	1	3	17	47	31
UAS civilian applications: coordination.	0	1	24	43	32
UAS civilian applications: collaboration.	0	5	31	42	23

Table 2 (continued)

	SA	A	N	D	SD
UAS civilian applications: redundancies.	8	18	38	33	5
UAS civilian applications: education.	1	6	29	40	20
All UAS are aircraft and the FAA has the requirement per Title 49 (law) to set safety standards within the US designated airspace.	43	40	6	8	4
Unrealistic fears exist about UAS safety, ignoring excellent safety records of UAS (including Iraq).	9	31	22	32	6
The first UAS accident in the US that results in major losses of civilian lives will be a major setback to the entire civilian applications UAS community.	61	28	4	6	1
The UAS community dedicated to civilian applications has a clear vision for how to use UAS in the coming decade (Monday).	3	17	34	35	12
The UAS community dedicated to civilian applications has a clear vision for how to use UAS in the coming decade (Wednesday).	1	12	19	52	15

Table 3 CAUAS survey responses: applications, missions, and technologies

	SA	A	N	D	SD
UAS have shown critical value to homeland security and environmental threats.	49	33	9	3	1
Number 1 barrier to civil UAS systems: A) technology; B) safety; C) business case; D) spectrum; E) equity between airspace users.	5	62	15	6	12
The number of possible applications and possible platforms has not been fully explored yet for civilian benefits.	62	32	5	0	1

Table 3 (continued)

	SA	A	N	D	SD
Efforts to develop UAS capabilities for civilian applications are currently scattered across agencies, universities and companies with little coordination such as might be gained from a dedicated or central facility focused on UAS.	32	35	21	6	5
There is a strong, compelling justification for UAS to address important environmental science questions in the next ten years.	69	23	6	3	0
Optimal science capabilities in ten years will require preferential development of: A) sensors; B) platforms; C) infrastructure; D) air space access; all of the above.	5	1	3	49	41
Ability to fill gaps in capability will be the primary driver for government public UAS success (missions that satellites, manned aircraft and remote sensors cannot do).	43	33	8	13	3
In ten years, UAS will be a key element of disaster response teams.	45	37	10	5	2
There are NOAA environmental collection missions that can only be done with small UAS, which will drive small UAS routine operations by A) 2009; B) 2012; C) 2017; D) 2022; E) Not foreseeable.	22	41	24	7	6
There are high altitude environmental missions that can only be accomplished with the range, persistence and endurance of a HALE UAS. That requirement will drive	22	36	25	13	4

Table 3 (continued)

	SA	A	N	D	SD
NOAA development, integration and routine operation of HALE systems by: A) 2010; B) 2012; C) 2017; D) 2022; E) Not foreseeable.					
There is a strong, compelling justification for using UAS to address public applications (disasters, homeland security), in the next ten years.	65	27	4	4	0
Given appropriate sensor and platform development, science questions can be addressed in the next next ten years which cannot be addressed effectively any other way.	45	37	12	5	1
The UAS community will be a primary driver for air traffic management over the next decade.	1	22	16	40	21
The UAS community must engage the manned aircraft community e.g. AOPA as a consensus partner for UAS NAS integration over the next decade.	51	33	10	7	0
Alternative paths to expedite UAS integration into the NAS. Which is likely to be the most effective over the next 10 years? A) Use the power of the industry B) Bring the technology forward (with data) C) Provide draft policy and guidance to FAA D) Press the faa to certify systems for commercial use E) Government agencies support FAA.	13	35	24	7	21
What would be the most likely application to push UAS development forward? A) Agriculture; B) Water Resources; C) Climate and Weather; D) Law Enforcement; E) Emergencies and Disasters.	5	1	29	20	44

Table 4 CAUAS survey responses: economics

	SA	A	N	D	SD
The economic investment in environmental science in the next 10 years should rise to be commensurate with the problems being addressed.	60	21	14	4	1
Science applications will develop a viable market for UAS services.	18	30	25	24	4
In ten years, cost benefit will be the primary driver for successful commercial civil UAS applications. (profit)	42	35	14	7	2
In ten years, endurance/ persistence will be the most attractive attribute for UAVs to have commercial value (dull, dirty, dangerous).	15	34	24	24	3
Assuming access to airspace issues are solved, market opportunities will be the primary driver for UAS to be fully applied to civilian applications.	35	30	21	7	8
The economic investment in public applications (disasters, homeland security), of UAS in the next ten years should rise to be commensurate with the problems being addressed.	42	33	14	8	2

References

1. UAS: Final report: First-ever successful UAS mission into a tropical storm Ophelia—2005). http://uas.noaa.gov/projects/demos/aerosonde/Ophelia_final.html (2005)
2. CAUAS: Civilian applications of unmanned aircraft systems (CAUAS) homepage. http://cauas.colorado.edu/ (2007)
3. NASA: NASA and NOAA fly unmanned aircraft into Hurricane Noel. http://www.nasa.gov/centers/wallops/news/story105.html (2007)
4. NOAA: Pilotless aircraft flies toward eye of hurricane for first time. http://www.noaanews.noaa.gov/stories2007/20071105_pilotlessaircraft.html (2007)
5. Argrow, B., Weatherhead, E., Avery, S.: Civilian applications of unammned aircraft systems: priorities for the coming decade. Meeting Summary. http://cauas.colorado.edu/ (2007)
6. Axtman, K.: U.S. border patrol's newest tool: a drone. http://www.usatoday.com/tech/news/techinnovations/2005-12-06-uav-border-patrol_x.htm (2005)
7. Corrigan, C.E., Roberts, G., Ramana, M., Kim, D., Ramanathan, V.: Capturing vertical profiles of aerosols and black carbon over the indian ocean using autonomous unmanned aerial vehicles. Atmos. Chem. Phys. Discuss **7**, 11,429–11,463 (2007)

8. Curry, J.A., Maslanik, J., Holland, G., Pinto, J.: Appl cations of aerosondes in the arctic. Bull. Am. Meteorol. Soc. **85**(12), 1855–1861 (2004)
9. Federal Aviation Administration: Government Aircra t Operations. Federal Aviation Administration, Washington, DC (1995)
10. Federal Aviation Administration: Interim Approval Guidance 08–01: Unmanned Aircraft Systems Operations in the U. S. National Airspace System. Federal Aviation Administration, Washington, DC (2008)
11. Office of the Secretary of Defense. Unmanned Aircraft Systems Roadmap: 2005–2030. Office of the Secretary of Defense, Washington, DC (2005)
12. Sofge, E.: Houston cops test drone now in Iraq, operator says. http://www.popular mechanics.com/science/air_space/4234272.html (2008)
13. Zajkowski, T., Dunagan, S., Eilers, J.: Small UAS communications mission. In: Eleventh Biennial USDA Forest Service Remote Sensing Applications Conference, Salt Lake City, 24–28 April 2006

Computer Vision Onboard UAVs for Civilian Tasks

Pascual Campoy · Juan F. Correa · Ivan Mondragón ·
Carol Martínez · Miguel Olivares · Luis Mejías ·
Jorge Artieda

Originally published in the Journal of Intelligent and Robotic Systems, Volume 54, Nos 1–3, 105–135.
© Springer Science + Business Media B.V. 2008

Abstract Computer vision is much more than a technique to sense and recover environmental information from an UAV. It should play a main role regarding UAVs' functionality because of the big amount of information that can be extracted, its possible uses and applications, and its natural connection to human driven tasks, taking into account that vision is our main interface to world understanding. Our current research's focus lays on the development of techniques that allow UAVs to maneuver in spaces using visual information as their main input source. This task involves the creation of techniques that allow an UAV to maneuver towards features of interest whenever a GPS signal is not reliable or sufficient, e.g. when signal dropouts occur (which usually happens in urban areas, when flying through terrestrial urban canyons or when operating on remote planetary bodies), or when tracking or inspecting visual targets—including moving ones—without knowing their exact UMT coordinates. This paper also investigates visual servoing control techniques that use velocity and position of suitable image features to compute the references for flight control. This paper aims to give a global view of the main aspects related to the research field of computer vision for UAVs, clustered in four main active research lines: visual servoing and control, stereo-based visual navigation, image processing algorithms for detection and tracking, and visual SLAM. Finally, the results of applying these techniques in several applications are presented and discussed: this

P. Campoy (✉) · J. F. Correa · I. Mondragón · C. Martínez · M. Olivares · J. Artieda
Computer Vision Group, Universidad Politécnica Madrid,
Jose Gutierrez Abascal 2, 28006 Madrid, Spain
e-mail: pascual.campoy@upm.es

L. Mejías
Australian Research Centre for Aerospace Automation (ARCAA),
School of Engineering Systems, Queensland University of Technology,
GPO Box 2434, Brisbane 4000, Australia

study will encompass power line inspection, mobile target tracking, stereo distance estimation, mapping and positioning.

Keywords UAV · Visual servoing · Image processing · Feature detection · Tracking · SLAM

1 Introduction

The vast infrastructure inspection industry frequently employs helicopter pilots and camera men who risk their lives in order to accomplish certain tasks, and taking into account that the way such tasks are done involves wasting large amounts of resources, the idea of developing an UAV—unmanned air vehicle—for such kind of tasks is certainly appealing and has become feasible nowadays. On the other hand, infrastructures such as oil pipelines, power lines or roads are usually imaged by helicopter pilots in order to monitor their performance or to detect faults, among other things. In contrast with those methods, UAVs appear as a cheap and suitable alternative in this field, given their flight capabilities and the possibility to integrate vision systems to enable them to perform otherwise human driven tasks or autonomous guiding and imaging.

Currently, some applications have been developed, among which we can find Valavanis' works on traffic monitoring [1], path planning for multiple UAV cooperation [2], and fire detection [3]. On the other hand, Ollero [4] has also made some works with multi-UAVs. There are, too, some other works with mini-UAVs and vision-based obstacle avoidance made by Oh [5] or by Serres [6]. Moreover, Piegl and Valanavis in [7] summarized the current status and future perspectives of the aforementioned vehicles. Applications where an UAV would manipulate its environment by picking and placing objects or by probing soil, among other things, can also be imagined and feasible in the future. In fact, there are plans to use rotorcraft for the exploration of planets like Mars [8, 9].

Additionally, aerial robotics might be a key research field in the future, providing small and medium sized UAVs as a cheap way of executing inspection functions, potentially revolutionizing the economics of this industry as a consequence. The goal of this research is to provide UAVs with the necessary technology to be visually guided by the extracted visual information. In this context, visual servoing techniques are applied in order to control the position of an UAV using the location of features in the image plane. Another alternative being explored is focused in the on-line reconstruction of the trajectory in the 3D space of moving targets (basically planes) to control the UAV's position [10].

Vision-based control has become interesting because machine vision directly detects a tracking error related to the target rather than indicating it via a coordinate system fixed to the earth. In order to achieve the aforemention detection, GPS is used to guide the UAV to the vicinity of the structure and line it up. Then, selected or extracted features in the image plane are tracked. Once features are detected and tracked, the system uses the image location of these features to generate image-based velocity references to the flight control.

In the following section briefly describe the different components that are needed to have an UAV ready to flight, and to test it for different applications. Section 3, explains with details the different approaches to extract useful information to achieve visual servoing in the image plane based on features and on appearance. Some improvements in 3D motion reconstruction are also pointed out. Section 4 describes visual control schemes employed to aim visual servoing, and the particular configuration of the control system assigned to close the visual control loop. Section 5 deals with the stereo configuration and theory to make motion and height estimation based on two views of a scene. Next, in Section 6 the simultaneous localization and Mapping problem based on visual information is addressed, with particular emphasis on images taken from an UAV. Section 7 shows experimental results of different applications, and Section 8, finally, deals with conclusions and future work.

2 System Overview

Several components are necessary to complete an operational platform equipped with a visual system to control UAVs. It is a multidisciplinary effort that encloses different disciplines like system modeling and control, data communication, trajectory planning, image processing, hardware architecture, software engineering, and some others. All this knowledge is traduced into an interconnected architecture of functional blocks. The Computer Vision Group at UPM has three fully operational platforms at its disposal, whereas two of them are gas powered Industrial Twim 52 c.c helicopters producing about 8 hp, which are equipped with an AFCS helicopter flight controller, a guidance system, a Mini Itx onboard vision computer, and an onboard 150 W generator. These helicopters are used for outdoors applications, as shown in Fig. 1, where one of the powered gas platforms performs an experimental autonomous flight. The third platform is a Rotomotion SR20 UAV with an electric motor of 1,300 W, 8A. It also has a Nano Itx onboard vision computer and WiFi ethernet for telemetry data. It is used on indoors and outdoors applications. In this section, a description of the main modules, their structure and some basic

Fig. 1 Aerial platform COLIBRI while is performing an experimental detection and tracking of external visual references

functionalities is provided. In general terms, the whole system can be divided into two components:

1. An onboard subsystem composed by:

   – Vision computer with the image processing algorithms and related image treatment programs.
   – Flight computer with Flight control software.
   – Cameras.
   – Communication interface with flight control and with ground subsystem.

2. A ground subsystem:

   – Ground computer for interaction with the onboard subsystem, and data analysis.
   – Communication interface.
   – Data storage.

Those components' division can be reorganized into subsystems, which are described below.

2.1 Flight Control Subsystem

Most complex physical systems' dynamics are nonlinear. Therefore, it is important to understand under which circumstances a linear modeling and control design will be adequate to address control challenges. In order to obtain a linear dynamic model, the hover state can be used as a point of work to approximate the helicopter dynamics by linear equations of motion. Using this approximation, linearization around this state gives a wide enough range of linearity to be useful for controlling purposes.

The control system is based on single-input single-output (SISO) proportional-integral-derivative (PID) feedback loops. Such a system has been tested to provide basic sufficient performance to accomplish position and velocity tracking near hover flight [11–13]. The advantage of this simple feedback architecture is that it can be implemented without a model of the vehicle dynamics (just kinematic), and all feedback gains can be turned on empirically in flight. The performance of this type of control reaches its limits when it is necessary to execute fast and extreme maneuvers. For a complete description of the control architecture, refer to [14–17].

The control system needs to be communicated with external processes (Fig. 2) in order to obtain references to close external loops (e.g. vision module, Kalman filter for state estimation, and trajectory planning). The communication is made through a high level layer that routes the messages to the specific process. The next subsection introduces the communication interface in detail.

2.2 Communication Interface

A client-server architecture has been implemented based on TCP/UDP messages, allowing embedded applications running on the computer onboard the autonomous helicopter to exchange data between them and with the processes running on the ground station. The exchange is made through a high level layer which routes the messages to the specific process. Switching and routing a message depends on the type of information received. For example, the layer can switch between position

Fig. 2 Control system interacting with external processes. Communication is made through a high level layer using specific messages routed for each process

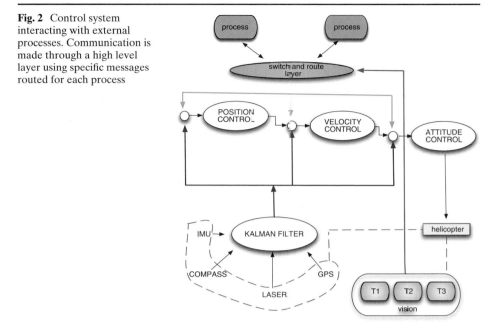

and velocity control depending on the messages received from an external process. The mechanism used for this purpose consists in defining data structures containing a field that uniquely identifies the type of information and the destination of a message. Some of the messages defined for flight control are: velocity control, position control, heading, attitude and helicopter state.

Figure 3, shows a case in which two processes are communicating through the switching layer. One process is sending commands to the flight control (red line), while the other one (blue line) is communicating with another process.

Fig. 3 Switching Layer. TCP/UDP messages are used to exchange data between flight controller and other process. Exchange is driven by a high level layer which routes the data to the specific process

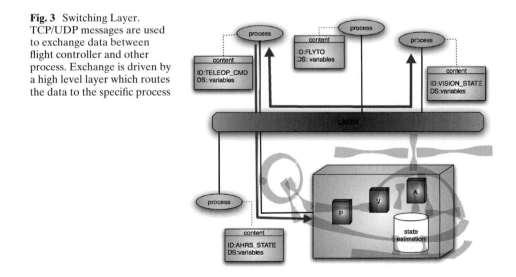

2.3 Visual Subsystem

The visual subsystem is a compound of a servo controlled Pan Tilt platform, an on-board computer and a variety of cameras and visual sensors, including analog/digital cameras (Firewire, USB, IP-LAN and other digital connections), with the capability of using configurations based on single, stereo cameras, arrays of synchronous multiple sensor heads and many other options. Additionally, the system allows the use of Gimbals' platforms and other kinds of sensors like IF/UV spectrum cameras or Range Finders. Communication is based on a long-range wireless interface which is used to send images for ground visualization of the onboard view and for visual algorithm supervision. Applications and approaches designed to perform visual tasks encompass optical flow, Hough transform, camera calibration, stereo vision to corner detection, visual servoing control implementation and Kalman filtering, among others.

Scene information obtained from image processing and analysis provides data related to the camera's coordinate system. This information is useful for purposes of automatic camera control, but not for the attitude and position control of the UAV. This issue is solved by fixating and aligning the camera's frame reference with the vehicle body-frame. Next section enumerates some basic algorithms of visual information extracted for controlling purposes.

3 Visual Tracking

The main interest of the computer vision group at UPM is to incorporate vision systems in UAVs in order to increase their navigation capabilities. Most of this effort is based on image processing algorithms and tracking techniques that have been implemented in UAVs and will be described below.

3.1 Image Processing

Image processing is used to find characteristics in the image that can be used to recognize an object or points of interest. This relevant information extracted from the image (called features) ranges from simple structures, such as points or edges, to more complex structures, such as objects. Such features will be used as reference for the visual flight control.

Most of the features used as reference are interest points, which are points in an image that have a well-defined position, can be robustly detected, and are usually found in any kind of images. Some of these points are corners formed by the intersection of two edges, and some others are points in the image whose context has rich information based on the intensity of the pixels. A detector used for this purpose is the Harris Corner detector [18]. It extracts a lot of corners very quickly based on the magnitude of the eigenvalues of the autocorrelation matrix. However, it is not enough to use this measure in order to guarantee the robustness of the corner, since the purpose of the features' extraction is to track them along an image sequence. This means that good features to track have to be selected in order to ensure the stability of the tracking process. The robustness of a corner extracted with the Harris detector can be measured by changing the size of the detection window, which is increased to test the stability of the position of the extracted corners. A measure of this variation is then calculated based on a maximum difference criteria. Besides, the magnitude of the eigenvalues is used to only keep features with eigenvalues higher than a minimum value. Combination of such criteria leads to the selection of the good features to track.

Another widely used algorithm is the SIFT (scale invariant feature transform) detector [19] of interest points, which are called keypoints in the SIFT framework. This detector was developed with the intention to use it for object recognition. Because of this, it extracts keypoints invariant to scale and rotation using the gaussian difference of the images in different scales to ensure invariance to scale. To achieve invariance to rotation, one or more orientations based on local image gradient directions are assigned to each keypoint. The result of all this process is a descriptor associated to the keypoint, which provides an efficient tool to represent an interest point, allowing an easy matching against a database of keypoints. The calculation of these features has a considerable computational cost, which can be assumed because of the robustness of the keypoint and the accuracy obtained when matching these features. However, the use of these features depends on the nature of the task: whether it needs to be done fast or accurate.

The use of other kind of features, such as edges, is another technique that can be applied on semi-structured environments. Since human constructions and objects are based on basic geometrical figures, the Hough transform [20] becomes a powerful technique to find them in the image. The simplest case of the algorithm is to find straight lines in an image that can be described with the equation $y = mx + b$. The main idea of the Hough transform is to consider the characteristics of the straight line not as image points x or y, but in terms of its parameters m and b. The procedure has more steps to re-parameterize into a space based on an angle and a distance, but what is important is that if a set of points form a straight line, they will produce sinusoids which cross at the parameters of that line. Thus, the problem of detecting collinear points can be converted to the problem of finding concurrent curves. To apply this concept just to points that might be on a line, some pre-processing algorithms are used to find edge features, such as the Canny edge detector [21] or the ones based on derivatives of the images obtained by a convolution of image intensities and a mask (Sobel [22], Prewitt). These methods have been used in order to find power lines and isolators in an inspection application [23].

3.2 Feature Tracking

The problem of tracking features can be solved with different approaches. The most popular algorithm to track features like corner features or interest points in consecutive images is the Lukas–Kanade algorithm [24]. It works under two premises: first, the intensity constancy in the vicinity of each pixel considered as a feature; secondly, the change in the position of the features between two consecutive frames must be minimum, so that the features are close enough to each other. Given these conditions to ensure the performance of the algorithm, it can be expressed in the following form: if we have a feature position $p_i = (x, y)$ in the image I_k, the objective of the tracker is to find the position of the same feature in the image I_{k-1} that fits the expression $p'_i = (x, y) + t$, where $t = (t_x, t_y)$. The t vector is known as the optical flow, and it is defined as the visual velocity that minimizes the residual function $e(t)$ defined as:

$$e(t) = \sum^{W}(I_k(p_i) - I_{k+1}(p_i + t))^2 w(W) \tag{1}$$

where $w(x)$ is a function to assign different weights to comparison window W. This equation can be solved for each tracked feature, but since it is expected that all features on physical objects move solidary, summation can be done over all features. The problem can be reformulated to make it possible to be solved in relation to all features in the form of a least squares' problem, having a closed form solution. In Section 3.3 more details are given. Whenever features are tracked from one frame to another in the image, the measure of the position is affected by noise. Hence, a Kalman filter can be used to reduce noise and to have a more smooth change in the position estimation of the features. This method is also desirable because it provides an estimation of the velocity of the pixel that is used as a reference to the velocity flight control of the UAV.

Another way to track features is based on the rich information given by the SIFT descriptor. The object is matched along the image sequence comparing the model template (the image from which the database of features is created) and the SIFT descriptor of the current image, using the nearest neighbor method. Given the high dimensionality of the keypoint descriptor (128), its matching performance is improved using the Kd-tree search algorithm with the Best Bin First search modification proposed by Lowe [25]. The advantage of this method lies in the robustness of the matching using the descriptor, and in the fact that this match does not depend on the relative position of the template and the current image. Once the matching is performed, a perspective transformation is calculated using the matched Keypoints, comparing the original template with the current image. Then, the RANSAC algorithm [26] is applied to obtain the best possible transformation, taking into consideration bad correspondences. This transformation includes the parameters for translation, rotation and scaling of the interest object, and is defined in Eqs. 2 and 3.

$$X_k = H X_0 \tag{2}$$

$$\begin{pmatrix} x_k \\ y_k \\ \lambda \end{pmatrix} = \begin{pmatrix} a & b & c \\ d & e & f \\ g & h & 1 \end{pmatrix} \begin{pmatrix} x_0 \\ y_0 \\ 1 \end{pmatrix} \tag{3}$$

where $(x_k, y_k, \lambda)^T$ is the homogeneous position of the matched keypoint against $(x_0, y_0, 1)^t$ position of the feature in the template image, and H is the homography transformation that relates the two features. Considering that every pair of matched keypoints gives us two equations, we need a minimum of four pairs of correctly matched keypoints to solve the system. Keeping in mind that not every match may be correct, the way to reject the outliers is to use the RANSAC algorithm to robustly estimate the transformation H. RANSAC achieves its goal by iteratively selecting a random subset of the original data points, testing it to obtain the model and then evaluating the model consensus, which is the total number of original data points that best fit the model. This procedure is then repeated a fixed number of times, each time producing either a model which is rejected because too few points are

Fig. 4 Experiments with planar objects in order to recover the full pose of the tracked object using SIFT. In the sub-figure **a** a template is chosen from the initial frame. In **b** the SIFT database is generated using the extracted keypoints. In **c** points are searched in a region twice the size of the template in the next image using the previous position as initial guess. **d** Subfigure shows the matching achieved by the tracking algorithm

classified as inliers or a model that better represents the transformation. If the total trials are reached, a good solution can not be obtained. This situation enforces the correspondences between points from one frame to another. Once a transformation is obtained, the pose of the tracked plane can be recovered using the information in the homography. Figure 4 shows an implementation of this method.

3.3 Appearance Based Tracking

Tracking based on appearance does not use features. On the other hand, it uses a patch of pixels that corresponds to the object that wants to be tracked. The method to track this patch of pixels is the same L–K algorithm. This patch is related to the next frame by a warping function that can be the optical flow or another model of motion. The problem can be formulated in this way: lets define X as the set of points that forms the template image $T(\mathbf{x})$, where $\mathbf{x} = (x, y)^T$ is a column vector with the coordinates in the image plane of the given pixel. The goal of the algorithm is to align the template $T(\mathbf{x})$ with the input image $I(\mathbf{x})$. Because $T(\mathbf{x})$ is a sub-image of $I(\mathbf{x})$, the algorithm will find the set of parameters $\mu = (\mu_1, \mu_2, ...\mu_n)$ for motion model function $W(\mathbf{x}; \mu)$, also called the warping function. The objective function of the algorithm to be minimized in order to align the template and the actual image is Eq. 4

$$\sum_{\forall \mathbf{x} \in X} (I(W(\mathbf{x}; \mu) - T(\mathbf{x}))^2 \tag{4}$$

Since the minimization process has to be made with respect to μ, and there is no lineal relation between the pixel position and its intensity value, the Lukas–Kanade algorithm assumes a known initial value for the parameters μ and finds increments of the parameters $\delta\mu$. Hence, the expression to be minimized is:

$$\sum_{\forall \mathbf{x} \in X} (I(W(\mathbf{x}; \mu + \delta\mu) - T(\mathbf{x}))^2 \tag{5}$$

and the parameter actualization in every iteration is $\mu = \mu + \delta\mu$. In order to solve Eq. 5 efficiently, the objective function is linearized using a Taylor Series expansion employing only the first order terms. The parameter to be minimized is $\delta\mu$. Afterwards, the function to be minimized looks like Eq. 6 and can be solved like a "least squares problem" with Eq. 7.

$$\sum_{\forall \mathbf{x} \in X} \left(I(W(\mathbf{x}; \mu) + \nabla I \frac{\partial W}{\partial \mu} \delta\mu - T(\mathbf{x}) \right)^2 \tag{6}$$

$$\delta\mu = H^{-1} \sum_{\forall \mathbf{x} \in X} \left(\nabla I \frac{\partial W}{\partial \mu} \right)^T (T(\mathbf{x}) - I(W(\mathbf{x}; \mu))) \tag{7}$$

where H is the Hessian Matrix approximation,

$$H = \sum_{\forall \mathbf{x} \in X} \left(\nabla I \frac{\partial W}{\partial \mu} \right)^T \left(\nabla I \frac{\partial W}{\partial \mu} \right) \tag{8}$$

More details about this formulation can be found in [10] and [27], where some modifications are introduced in order to make the minimization process more efficient, by inverting the roles of the template and changing the parameter update rule from an additive form to a compositional function. This is the so called ICA (Inverse Compositional Algorithm), first proposed in [27]. These modifications where introduced to avoid the cost of computing the gradient of the images, the Jacobian of the Warping function in every step and the inversion of the Hessian Matrix that assumes the most computational cost of the algorithm.

Besides the performance improvements that can be done to the algorithm, it is important to explore the possible motion models that can be applied to warp the patch of tracked pixels into the $T(\mathbf{x})$ space, because this defines the degrees of freedom of the tracking and constrains the possibility to correctly follow the region of interest. Table 1 summarizes some of the warping functions used and the degrees of freedom. Less degrees of freedom make the minimization process more stable and accurate, but less information can be extracted from the motion of the object. If a perspective transformation is applied as the warping function, and if the selected patch corresponds to a plane in the world, then 3D pose of the plane can be

Table 1 Warping functions summary

Name	Rule	D.O.F
Optical flow	$(x, y) + (t_x, t_y)$	2
Scale+translation	$(1 + s)((x, y) + (t_x, t_y))$	3
Scale+rotation+ translation	$(1 + s)(R_{2x2}(x, y)^T + (t_x, t_y)^T)$	4
Affine	$\begin{pmatrix} 1 + \mu_1 & \mu_3 & \mu_5 \\ \mu_2 & 1 + \mu_4 & \mu_6 \end{pmatrix}$	6
Perspective	$\begin{pmatrix} \mu_1 & \mu_2 & \mu_3 \\ \mu_2 & \mu_5 & \mu_6 \\ \mu_7 & \mu_8 & 1 \end{pmatrix}$	8

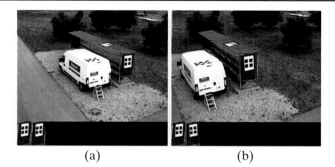

(a)   (b)

Fig. 5 Experiments using appearance based tracking we~e conducted to track a template in the scene. **a** Is the initial frame of the image sequence. Image region is manually selected and tracked along image sequence, using a scale + translation model (see Table 1). **b** Shows the tracked template 50 frames later from image (**a**). *Sub-images in the bottom of each figure* represent the initial template selected and the warped patch transformed into the template coordinate system

reconstructed from the obtained parameters. Figure 5 shows some tests carried out using a translation+scale motion model.

4 Visual Flight Control

4.1 Control Scheme

The flight control system is composed of three control loops arranged in a cascade formation, allowing it to perform tasks in different levels depending on the workspace of the task. The first control loop is in charge of the attitude of the helicopter. It interacts directly over the servomotors that define the four basic variables: cyclic/collective pitch of the principal rotor, cyclic/collective pitch of the tale rotor, longitudinal cyclic pitch, and the latitudinal cyclic pitch. The kinematic and dynamic models of the helicopter relate those variables with the six degrees of motion that this kind of vehicle can have in the cartesian space. As mentioned above in Section 2.1, the hover state can be used as a point of work to approximate the helicopter's dynamics by linear equations of motion. Using this approximation, linearization around this state gives a wide enough range of linearity that is useful for control purposes. For this reason, this control is formed of decoupled PID controllers for each of the control variables described above.

The second controller is a velocity-based control responsible of generating the references for the attitude control. It is implemented using a PI configuration. The controller reads the state of the vehicle from the state estimator and gives references to the next level, but only to make lateral and longitudinal displacements. The third controller (position based control) is at the higher level of the system, and is designed to receive GPS coordinates.The control scheme allows different modes of operation, one of which is to take the helicopter to a desired position (position control). Once the UAV is hovering, the velocity based control is capable of receiving references to keep the UAV aligned with a selected target, and it leaves the stability of the aircraft to the most internal loop in charge of the attitude. Figure 6 shows the structure of the

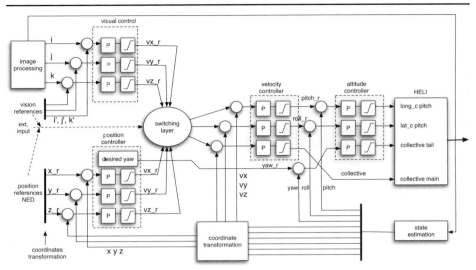

Fig. 6 Schematic flight control system. The inner velocity control loop is made of three cascade decoupled PID controllers. The outer position control loop can be externally switched between the visual based controller and the GPS based position controller. The former can be based on direct feature visual control or alternatively on visual estimated world positioning

flight control system with more details, and the communication interface described in Section 2.2, that is the key to integrate the visual reference as an external loop. Next subsection describes how this has been achieved.

4.2 Visual References Integration

The first step to design the control task in the image coordinates is to define the camera's model and the dynamics of a feature in the image, in order to construct a control law that properly represents the behavior of the task. Figure 7 shows the basic PinHole model of the camera, where $P^c(x, y, z)$ is a point in the camera coordinates system, and $p^c(i, j)^T$ denotes the projection of that point in the image plane π. Velocity of the camera can be represented with the vector $V = (v_x^c, v_y^c, v_z^c)^T$, while vector $\omega = (w_x^c, w_y^c, w_z^c)^T$ depicts the angular velocity. Considering that objects in the scene don't move, the relative velocity of a point in the world related to the camera's optical center can be expressed in this form:

$$\dot{P^c} = -\left(V + \omega \times P^c\right) \tag{9}$$

Using the well known Eq. 10 based on the camera calibration matrix that expresses the relationship between a point in the camera's coordinate system and its projection in the image plane, deriving Eq. 10 with respect to time, and replacing Eq. 9, it is possible to obtain a new Eq. 11 that describes a differential relation between the

Fig. 7 PinHole camera model to describe the dynamic model, where $P(x, y, z)$ is a point in the camera coordinates system, $p(i, j)^T$ represents the projection of that point in the image plane π and the vector $\omega = (w_x, w_y, w_z)^T$ is the angular velocity

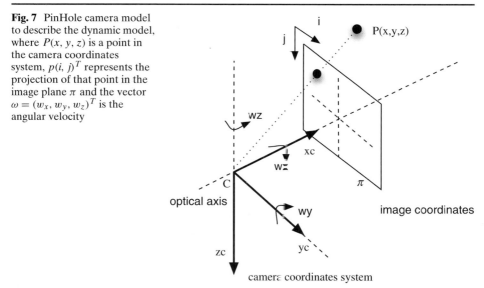

velocity of the projection of a point in the image and the velocity vector of the camera V and ω.

$$p^c = \mathbf{K}P^c \tag{10}$$

$$\dot{p}^c = -\mathbf{K}\left(V + \omega \times P^c\right) \tag{11}$$

Since the visual servoing task is designed to only make lateral and longitudinal displacements, and the camera is fixed looking to the front, it is possible to assume that the angular velocity is despicable because of the short range of motion of the pitch angles and the velocity constraint imposed to the system. Hence, Eq. 11 is reduced to this expression:

$$\dot{p}^c = \begin{bmatrix} \dfrac{di}{dt} \\ \dfrac{dj}{dt} \end{bmatrix} = - \begin{bmatrix} \dfrac{f}{x^c} & 0 \\ 0 & \dfrac{f}{x^c} \end{bmatrix} \begin{bmatrix} v_x^c \\ v_z^c \end{bmatrix} \tag{12}$$

This expression permits the introduction of the references described in Section 3 as a single measure, using the center of mass of the features or the patch tracked by the image processing algorithm, and using the velocity control module of the Flight Control System described above in this section.

5 Stereo Vision

This section shows a system to estimate the altitude and motion of an aerial vehicle using a stereo visual system. The system first detects and tracks interest points in the scene. The depth of the plane that contains the features is calculated matching features between left and right images and then using the disparity principle. Motion is recovered tracking pixels from one frame to the next one, finding its visual displacement and resolving camera rotation and translation by a least-square method [28].

5.1 Height Estimation

Height Estimation is performed on a stereo system using a first step to detect features in the environment with any of the technique mentioned in Section 3. This procedure is performed in each and every one of the stereo images.

As a second step, a correlation procedure is applied in order to find the correspondences between the two sets of features from the right and left images. Double check is performed by checking right against left, and then comparing left with right. The correlation stage is based on the ZNNC—zero mean normalized cross correlation—which offers good robustness against light and environmental changes [29].

Once the correspondence has been solved, considering an error tolerance, given that the correspondence is not perfect, and thanks to the fact that all pixels belong to the same plane, the stereo disparity principle is used to find the distance to the plane that contains the features. Disparity is inversely proportional to scene depth multiplied by the focal length (f) and baseline (b). The depth is computed using the expression for Z shown in Fig. 8.

Figure 9 shows the algorithm used to estimate the distance from the stereo system to the plane. In the helicopter, the stereo system is used in two positions. In the first one, the stereo system is looking down, perpendicular to ground, so that the estimated distance corresponds to the UAV altitude. In the second configuration, the stereo system is looking forward, and by so doing the estimated distance corresponds to the distance between the UAV and an object or feature.

Fig. 8 Stereo Disparity for aligned cameras with all pixel in the same plane. Stereo disparity principle is used to find the distance to the plane that contains the features

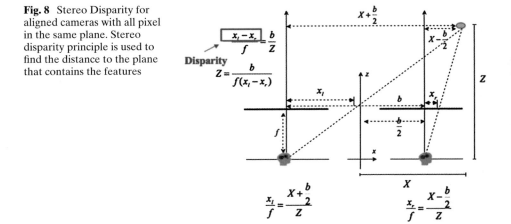

Fig. 9 Height estimation using the Harris Corner detector and ZNNC. Height is obtained employing the stereo disparity principle

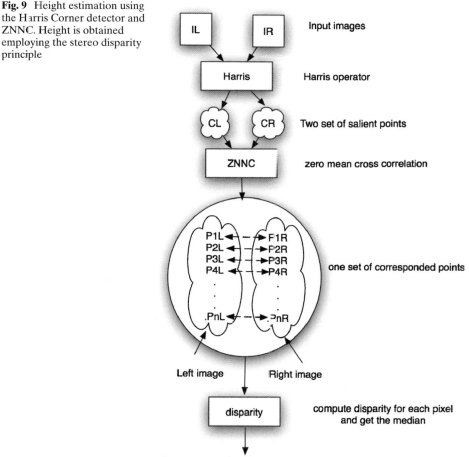

IL

IR

Input images

Harris

Harris operator

CL

CR

Two set of salient points

ZNNC

zero mean cross correlation

P1L ◄ — ▶ F1R
P2L ◄ — ▶ P2R
P3L ◄ — ▶ P3R
P4L ◄ — ▶ P4R
.
.
.
.PnL ◄ — ▶ PnR

one set of corresponded points

Left image   Right image

disparity

compute disparity for each pixel and get the median

direct computation of height applying the formula

5.2 Motion Estimation

Motion estimation is performed using at a first stage the same technique used for feature correspondence between left and right corners: the zero mean normalized cross-correlation (ZNNC). Correlation is performed within a certain pixel distance from each other keeping those points in a correlation coefficient higher than 0.85. The motion problem estimation is done aligning two sets of points whose correspondence is known, and finding the rotation matrix and translation vector, i.e, 3D transformation matrix T that minimizes the mean-squares' objective function $\min_{R,t} \sum_N \| TP_{k-1} - P_k \|^2$. Problem can be solved using Iterative Closes Point (ICP) registration and motion parameter estimation using SVD. Assuming there are two sets of points which are called data and model: $P = \{p_i\}_1^{N_p}$ and $M = \{m_i\}_1^{N_m}$ respectively with $N_p \neq N_m$, whose correspondence is known. The problem is how to compute the rotation (R) and translation (t) producing the best possible alignment of P and M by relation them with the equation $M = RP + t$. Lets define the closest point in the model to a data point p as $cp(p) = \arg \min_{m \in M} \| m - p \|$. Then, the ICP step goes like this:

1. Compute the subset of closest points (CP) , $y = \{m \in M \mid p \in P : m = cp(p)\}$
2. Compute the least-squares estimate of motion bringing P onto y:
   $(R, t) = \arg \min_{R,t} \sum_{i=1}^{N_p} \| y_i - Rp_i - t \|^2$
3. Apply motion to the data points, $P \leftarrow RP + t$
4. If the stopping criterion is satisfied, exit; else goto 1.

Calculating the rotation and the translation matrix using SVD can be summarized as follows: first, the rotation matrix is calculated using the centroid of the set of points. Centroid is calculated as $y_{c_i} = y_i - \bar{y}$ and $p_{c_i} = p_i - \bar{p}$, where $\bar{y} = \frac{1}{N_p} \sum_{N_p} cp(p_i)$ and $\bar{p} = \frac{1}{N_p} \sum_{N_p} p_i$. Then, rotation is found minimizing $\min_R \sum_{N_p} \| y_{c_i} - Rp_{c_i} \|^2$. This equation is minimized when trace (RK) is maximized with $K = \sum_{N_p} y_{c_i} p_{c_i}^T$. Matrix K is calculated using SVD as $K = VDU^T$. Thus, the optimal rotation matrix that maximizes the trace is $R = VU^T$. The optimal translation that aligns the centroid is $t = \bar{y} - P\bar{p}$.

Section 7.3, shows tests and applications' development using the stereo system and algorithms explained in this section.

6 Airborne Visual SLAM

This section presents the implementation of an aerial visual SLAM algorithm with monocular information. No prior information of the scene is needed for the proposed formulation. In this approach, no extra absolute or relative information, GPS or odometry are used. The SLAM algorithm is based on the features or corners' matching process using SURF features [30] or on the Harris Corner detector [18]. First, the formulation of the problem will be described. Then, the details of the Kalman filter will be explained and, finally, this section will end with the description of this approach's particularities.

6.1 Formulation of the Problem

The problem is formulated using state variables to describe and model the system. The state of the system is described by the vector:

$$X = [\mathbf{x}, \mathbf{s}_1, \mathbf{s}_2, \mathbf{s}_3, ...] \tag{13}$$

where $\mathbf{x}$ denotes the state of the camera and $\mathbf{s}_i$ represents the state of each feature. Camera state has 12 variables. The First six variables represent the position of the vehicle in iteration k and in the previous iteration. The Next six variables, vector $[p, q, r]$, represent the rotation at iteration k and k-1. Rotation is expressed using

Rodrigues' notation, which expresses a rotation around a vector with the direction of $\omega = [p, q, r]$ of an angle $\theta = \sqrt{p^2 + q^2 + r^2}$. The rotation matrix is calculated from this representation using

$$e^{\tilde{\omega}\theta} = I + \tilde{\omega}\sin(\theta) + \tilde{\omega}^2(1 - \cos(\theta)) \tag{14}$$

where I is the 3×3 identity matrix and $\tilde{\omega}$ denotes the antisymmetric matrix with entries

$$\tilde{\omega} = \begin{bmatrix} 0 & -r & q \\ r & 0 & -p \\ -q & p & 0 \end{bmatrix} \tag{15}$$

Therefore the state of the camera, not including the features, is composed by the following 12 variables,

$$\mathbf{x} = [x_k, x_{k-1}, y_k, y_{k-1}, z_k, z_{k-1}, p_k, p_{k-1}, q_k, q_{k-1}, r_k, r_{k-1}] \tag{16}$$

Another implementation of monocular SLAM uses quaternion to express the rotation [31]. The use of Rodrigues' notation, instead of quaternion, allows the reduction of the problem's dimension by only using three variables to represent the rotation.

Each feature is represented as a vector $[s_i]$ of dimension 6 using the inverse depth parametrization proposed by Javier Civera in [31]. This parametrization uses six parameters to define the position of a feature in a 3-Dimensional space. Each feature is defined by the position of a point (x_0, y_0, z_0) where the camera first saw the feature, the direction of a line based on that point and the inverse distance from the point to the feature along the line. This reference system allows the initialization of the features without any prior knowledge about the scene. This is important in exterior scenes where features with very different depths can coexist.

$$s_i = [x_0, y_0, z_0, \theta, \phi, \rho] \tag{17}$$

6.2 Prediction and Correction Stages

Extended Kalman filter (EKF) is used to implement the main algorithm loop, which has two stages: prediction and correction. In the prediction stage, uncertainty is propagated using the movement model. The correction stage uses real and predicted measurements to compute a correction to the prediction stage. Both stages need a precise description of the stochastic variables involved in the system.

There are mainly two approaches to implement this filter: extended Kalman filter and particle filter (FastSLAM). Both filters use the same formulation of the problem but have different approaches to the solution. The advantages of the Kalman filter are the direct estimation of the covariance matrix and the fact that it is a closed mathematical solution.

Its disadvantages are the increase of computational requirements as the number of features increase, the need of the model's linearization and the assumption of gaussian noise. On the other hand, particle filters can deal with non-linear, non-gaussian models, but the solution they provide depends on an initial random set of particles which can differ in each execution. Prediction stage is formulated using linear equations

$$\hat{X}_{k+1} = A \cdot X_k + B \cdot U_k$$
$$\hat{P}_{k+1} = A \cdot P_k \cdot A^T + Q \tag{18}$$

Where A is the transition matrix, B is the control matrix and Q is the model's covariance. Camera movement is modeled using a constant velocity model. Accelerations are included in a random noise component. For a variable n, which represents any of the position components (x, y, z) or the rotation components (p, q, r), we have:

$$n_{k+1} = n_k + v_k \cdot \Delta t \tag{19}$$

Where v_k is the derivative of n. We can estimate v_k as the differences in position,

$$n_{k+1} = n_k + \left(\frac{n_k - n_{k-1}}{\Delta t} \right) \Delta t = 2n_k - x_{n-1} \tag{20}$$

Feature movement is considered constant and therefore is modeled by an identity matrix. Now, full state model can be constructed

$$
\begin{bmatrix} x_{k+1} \\ x_k \\ y_{k+1} \\ y_k \\ z_{k+1} \\ z_k \\ r_{k+1} \\ r_k \\ p_{k+1} \\ p_k \\ q_{k+1} \\ q_k \\ s_{1,k+1} \\ \cdots \end{bmatrix}
=
\begin{bmatrix}
2 & -1 \\
1 & 0 \\
& & 2 & -1 \\
& & 1 & 0 \\
& & & & 2 & -1 \\
& & & & 1 & 0 \\
& & & & & & 2 & -1 \\
& & & & & & 1 & 0 \\
& & & & & & & & 2 & -1 \\
& & & & & & & & 1 & 0 \\
& & & & & & & & & & 2 & -1 \\
& & & & & & & & & & 1 & 0 \\
& & & & & & & & & & & & \mathbf{I} \\
& & & & & & & & & & & & & \ddots
\end{bmatrix}
\begin{bmatrix} x_k \\ x_{k-1} \\ y_k \\ y_{k-1} \\ z_k \\ z_{k-1} \\ r_k \\ r_{k-1} \\ p_k \\ p_{k-1} \\ q_k \\ q_{k-1} \\ s_{1,k} \\ \cdots \end{bmatrix}
\tag{21}
$$

Correction stage uses a non-linear measurement model. This model is the pin-hole camera model. The formulation of the Extended Kalman Filter in this scenario is

$$K_k = \hat{P}_k \cdot J^T \left(J \cdot P \cdot J^T + R \right)^{-1}$$
$$X_k = \hat{X}_k + K_k \cdot \left(Z_k - H\left(\hat{X}_k \right) \right)$$
$$P_k = \hat{P}_k - K_k \cdot J \cdot \hat{P}_k \tag{22}$$

where Z_k is the measurement vector, $H(X)$ is the non-linear camera model, J is the jacobian of the camera model and K_k is the Kalman gain.

The movement of the system is modeled as a solid with constant motion. Acceleration is considered a perturbation to the movement. A pin-hole camera model is used as a measurement model.

$$\begin{bmatrix} \lambda u \\ \lambda v \\ \lambda \end{bmatrix} = \begin{bmatrix} f & 0 & 0 \\ 0 & f & 0 \\ 0 & 0 & 1 \end{bmatrix} \cdot [R|T] \cdot \begin{bmatrix} x_w \\ y_w \\ z_w \\ 1 \end{bmatrix} \tag{23}$$

where u and v are the projected feature's central coordinates and λ is a scale factor. Distortion is considered using a four parameter model (k1, k2, k3, k4)

$$r^2 = u^2 + v^2$$

$$C_{dist} = 1 + k_0 r^2 + k_1 r^4$$

$$x_d = u \cdot C_{dist} + k_2 (2u \cdot v) + k_3 (r^2 + 2u^2)$$

$$y_d = v \cdot C_{dist} + k_2 (r^2 + 2v^2) + k_3 (2u \cdot v) \tag{24}$$

State error covariance matrix is initialized in a two-part process. First, elements related to the position and orientation of the camera, x, are initialized as zero or as a diagonal matrix with very small values. This represents that the position is known, at the first instant, with very low uncertainty. The initialization of the values related to the features, s_i, must be done for each feature seen for first time. This initialization is done using the results from [31]:

$$\mathbf{P}_{k|k}^{new} = J \begin{bmatrix} \mathbf{P}_{k|k} & & \\ & \mathbf{R}_i & \\ & & \sigma_\rho^2 \end{bmatrix} J^T \tag{25}$$

where

$$\mathbf{J} = \begin{bmatrix} I & & 0 & 0 \\ \frac{\partial \mathbf{s}}{\partial \mathbf{xyz}} & \frac{\partial \mathbf{s}}{\partial \mathbf{pqr}} & 0 & 0 & \cdots & \frac{\partial \mathbf{s}}{\partial x_d, y_d} & \frac{\partial \mathbf{s}}{\partial \rho_0} \end{bmatrix} \tag{26}$$

$$\frac{\partial \mathbf{s}}{\partial \mathbf{xyz}} = \begin{bmatrix} 1 & 0 & 0 \\ 0 & 1 & 0 \\ 0 & 0 & 1 \\ 0 & 0 & 0 \\ 0 & 0 & 0 \\ 0 & 0 & 0 \end{bmatrix} ; \frac{\partial \mathbf{s}}{\partial \mathbf{pqr}} = \begin{bmatrix} 0 & 0 & 0 \\ 0 & 0 & 0 \\ 0 & 0 & 0 \\ \frac{\partial\theta}{\partial p} & \frac{\partial\theta}{\partial q} & \frac{\partial\theta}{\partial r} \\ \frac{\partial\phi}{\partial p} & \frac{\partial\phi}{\partial q} & \frac{\partial\phi}{\partial r} \\ 0 & 0 & 0 \end{bmatrix} ; \frac{\partial \mathbf{s}}{\partial x_d, y_d} = \begin{bmatrix} 0 & 0 \\ 0 & 0 \\ 0 & 0 \\ \frac{\partial\theta}{\partial x_d} & \frac{\partial\theta}{\partial y_d} \\ \frac{\partial\phi}{\partial x_d} & \frac{\partial\phi}{\partial y_d} \\ 0 & 0 \end{bmatrix} ; \frac{\partial \mathbf{s}}{\partial \rho_0} = \begin{bmatrix} 0 \\ 0 \\ 0 \\ 0 \\ 0 \\ 1 \end{bmatrix} \tag{27}$$

Taking into account that a robust feature tracking and detection is a key element in the system, a Mahalanobis' test is used in order to improve the robustness of feature matching. The filter is implemented using Mahalanobis' distance between the predicted feature measurement and the real measurement. Mahalanobis' distance weighs Euclidean distance with the covariance matrix. This distance is the input to a χ^2 test which rejects false matches.

$$(Z - J \cdot X)^t \cdot C^{-1} (Z - J \cdot X) > \chi_n^2 \tag{28}$$

where

$$C = H \cdot P \cdot H^T + R \tag{29}$$

Finally, it should be noted that the reconstruction scale is an unobservable system state. This problem is dealt with using inverse depth parametrization [32], which avoids the use of initialization features of known 3D positions. This permits the use of the algorithm in any video sequence. Without these initialization features, the problem becomes dimensionless. The scale of the system can be recovered using the distance between two points or the position between the camera and one point. Computational cost is dependant on the number of features in the scene, and so an increase in the scene's complexity affects processing time in a negative way. Robust feature selection and matching are very important to the stability of the filter and a correct mapping. Experiments carried out successfully were made offline on sequences taken from the UAV.

7 Experimental Application and Tests

7.1 Visual Tracking Experiments

Tracking algorithms are fundamental to close the vision control loop in order to give an UAV the capability to follow objects. Hence, it is important to ensure the reliability of the tracker. Some experiments were conducted on images taken on test flights. Such experiments, where interest points were extracted with the Harris algorithm and tracked with the Lukas–Kanade algorithm, have proven to be fast enough so as to close the control loop at 17 Hz. However, if there are too many features selected to represent an object, the algorithm's speed slows down because of the calculation of the image derivatives.

SIFT features are very robust and rely on the advantage that the matching process does not depend on the proximity of two consecutive frames. On the other hand, the computational cost of the extraction is expensive. For that reason, they are suitable for visual servoing only if the displacements of the helicopter are forced to be very slow in order to avoid instabilities when closing the loop.

Tracking based on appearance proves to be very fast and reliable for acquired sequences at frame rates above 25 fps. This procedure is very sensitive to abrupt changes in the position of the tracked patch as long as the number of parameters of the motion model is higher than 3. This can be solved using stacks of trackers, each of which must have a different warping function that provides an estimation of the parameter to the next level of the stack. Simple warping functions give an estimation of more complex parameters. In the case of a simple tracker the translation-only warping function is the most stable one. Figure 10a shows the evolution of the parameters in a sequence of 1,000 images, and Fig. 10b the SSD error between the template image and the warped patch for each image.

7.2 Visual Servoing Experiments

The basic idea of visual servoing is to control the position of the helicopter based on an error in the image, or in a characteristic extracted from the image. If the

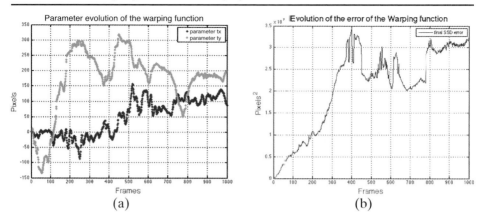

(a)   (b)

Fig. 10 Evolution of the translation parameter during the tracking process of a patch along 1,000 frames (**a**). **b** shows the SSD error of warped patch with respect to the template

control error is in the image plane, the measure of the error is a vector (in pixels) that represents the distance from the image's center to the feature's position. Figure 11 shows the basic idea of the error and the 2D visual servoing. In this sense, there are two ways to use this error in different contexts. One approach is to track features that are static in the scene. In this case, the control tries to move the UAV to align the feature's position with the image's center by moving the helicopter in the space.

Vision-based references are translated into helicopter displacements based on the tracked features. Velocity references are used to control the UAV, so that when the feature to track changes—as happens, for example, when another window of a building is chosen—velocity references change in order to align the UAV with the window.

The displacement of the helicopter when it tries to align with the feature being tracked is displayed in Fig. 12a. Vertical and lateral displacements of the helicopter are the consequence of the visual references generated from the vertical and horizontal positions of the window in the image. Figure 12b shows the displacement of the helicopter when the window above displayed was tracked, and Fig. 13 shows the velocity references when another window is chosen.

Fig. 11 Error measure in 2D visual servoing consists in the estimation of the distance of the reference point to the image's center

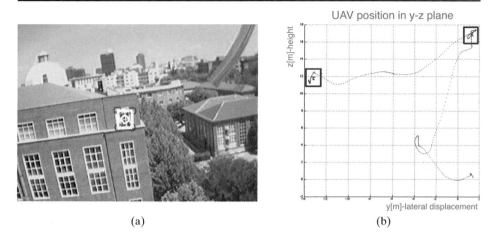

(a)   (b)

Fig. 12 Window being tracked during a visual servoing task (**a**), in which the UAV's vertical and lateral displacements are controlled by the visual control loop in order to fix the window in the center of the image, while the approaching movement is controlled by the GPS position controller. **b** Shows UAV vertical and lateral positions during the visual controlled flight. After taking off, the UAV moves to two positions (*marked with the red rectangles*) in order to consecutively track two external visual references that consist of two different windows

Another possible scenario is to keep the UAV hovering and to track moving objects in the scene. Experiments have been conducted successfully in order to proof variation of the method with good results. Control of the camera's Pan-Tilt Platform using 2D image servoing tries to keep a moving object in the image's center. In this case, position references are used instead of velocity in order to control the camera's pan and tilt positions. Figure 14 shows a car carrying a poster being tracked by moving the camera's platform.

Fig. 13 Velocity references change when a new feature is selected, in this case when another window is selected as shown in Fig. 12. Visual control takes the feature to the image center

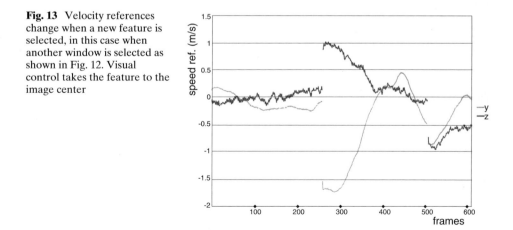

Fig. 14 Tracking of moving object. Servoing is perform on the pan-tilt platform. Notice that velocity in the cartesian coordinates is 0.0 (each component is printed on the image) since the UAV is hovering. Tracking is performed using corner features as explained in Section 3.2

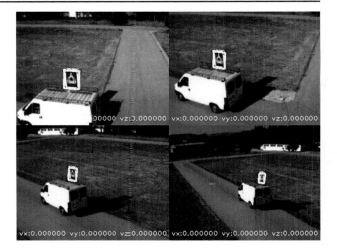

7.3 Height and Motion Estimation Using a Stereo System

Stereo tests are made using a Firewire stereo system camera onboard the UAV. In these experiments, the helicopter is commanded to fly autonomously following a given trajectory while the onboard stereo vision algorithm is running. The experiments find the correlation between the stereo visual estimation and the onboard helicopter state given by its sensor suite. Figure 15 shows the results of one flight trial in which the longitudinal displacement (X), lateral displacement (Y), altitude (H) and relative orientation are estimated. Altitude is computed negative since the helicopter's body frame is used as a reference system. Each estimation is correlated with its similar value taken from the onboard helicopter state, which uses an EKF to fuse onboard sensors. Table 2 shows the error analysis based on the mean square error of the visual estimation and the helicopter's state. Four measures of the mean squared error are used: the error vision-GPS Northting (MSE_N^V), the error vision-GPS Easting (MSE_E^V), the error vision-yaw (MSE_ψ^V) and the error vision-altitude (MSE_H^V).

7.4 Power Lines Inspection

Besides visual servoing and image tracking applications, other experiments have been conducted to achieve object recognition in inspection tasks. Major contributions and successful tests were obtained in power lines' inspection. The objective of the application developed at the computer vision group is to identify powered lines and electrical isolators. The methodology that has been employed is based on the Hough transform and on Corner detectors that find lines in the image that are associated with the catenary curve formed by the hanging wire. Interest points are used to locate the isolator. Once both components are detected in the image, tracking can be initiated to make close up shots with the appropriate resolution needed for expert inspection and detection of failures. Figure 16 shows images of the UAV

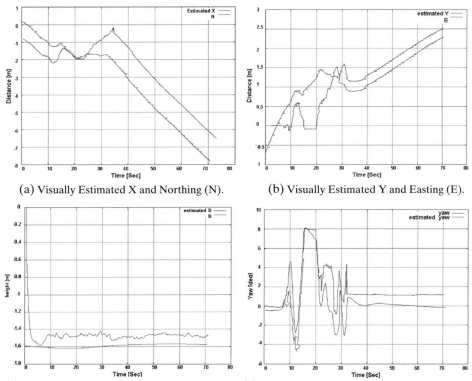

(a) Visually Estimated X and Northing (N).   (b) Visually Estimated Y and Easting (E).

(c) Visually Estimated H and helicopter altitude.   (d) Visually Estimated Yaw and helicopter Yaw.

Fig. 15 Results using a stereo system. Four parameters are estimated for this experiment: the longitudinal displacement (X) (**a**), the lateral displacement (Y) (**b**), altitude (H) (**c**) and relative orientation (yaw) (**d**)

approaching a power line while in the sub-image the onboard camera displays the detection of the line and the isolator.

Stereo System has also been used to estimate the UAV distance and altitude with respect to power lines. In these tests, the line is detected using the Hough Transform. If the camera's angles, stereo system calibration and disparity are known, it is possible to determine the position of the helicopter relative to the power line. Some tests using the Stereo system onboard the helicopter were carried out to obtain the distance to the power line from the helicopter. The power Line is detected using Hough transform in both images. In this test, the helicopter was initially 2 m below the power line. Afterwards, it rises to be at the same altitude of the cable and then

Table 2 Error analysis for the helicopter's experimental trials	Exp.	Test
	MSE_N^V **m**	1.0910
	MSE_E^V **m**	0.4712
	MSE_ψ^V **deg**	1.7363
	MSE_H^V **m**	0.1729

Fig. 16 Power line and
Isolator detection using the
UAV vision system

it returns to its original position. Figure 17 shows the distance and height estimated
from the UAV to the power line during this test. Additional tests can be seen on the
Colibri Project's Web Page [33].

7.5 Mapping and Positioning using Visual SLAM

The SLAM algorithm explained in Section 6 is used in a series of image sequences
of trajectories around a 3D scene that were performed flying in autonomous mode
navigation based on way points and desired heading values. The scene is composed
of many objects, including a grandstand, a van and many other elements, and also
of a series of marks feasible for features and corners' detection. For each flight test,
a 30 fps image sequence of the scene was obtained, associating the UAV attitude
information for each one. That includes the GPS position, IMU data (Heading, body
frame angles and displacement velocities) and the helicopter's position, estimated by
the Kalman filter on the local plane with reference to takeoff point. Figure 18 shows
a reconstruction of one flight around one scene test.

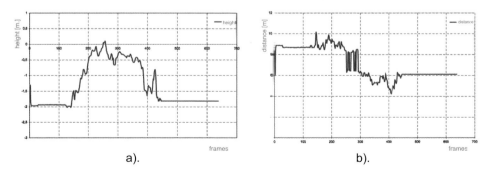

a).   b).

Fig. 17 Distance and height estimation to the power lines using a stereo system onboard the UAV

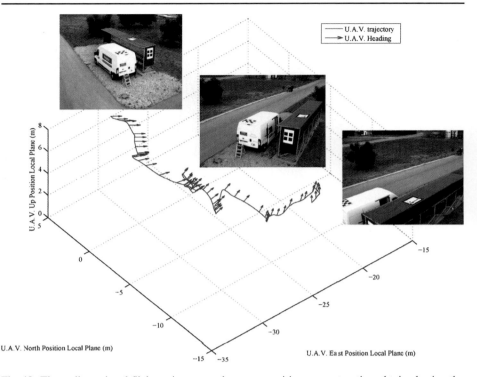

Fig. 18 Three-dimensional flight trajectory and camera position reconstruction obtained using the flightlog data. The *blue line* depicts the translational movement and the *red arrows* represent the heading direction of the camera (pitch and yaw angles). *Superimposed images* show the different perspectives obtained during the flight sequence around the semi-structured scene

Results for tests using a tracking algorithm for scene elements are shown on Fig. 19a. Reconstructed features are shown as crosses. In the figure, some reference planes were added by hand in order to make interpretation easier. Figure 19b shows an image from the sequence used in this test.

Results show that the reconstruction has a coherent structure but that the scale of the reconstruction is function of the initialization values. The scale can be recovered using the distance between two points or the positions of one point and the camera.

The camera movement relative to the first image is compared with the real flight trajectory. For this, the (x, y, z) axis on the camera plane are rotated so that they are coincident with the world reference plane used by the UAV. The heading or yaw angles (ψ) and the Pitch angle (θ) of the helicopter, in the first image of the SLAM sequence, define the rotational matrix used to align the camera and UAV frames.

The displacement values obtained using SLAM are rotated and then scaled to be compared with the real UAV trajectory. Figure 20 shows the UAV and SLAM trajectories and the medium square error (MSE) between real flight and SLAM displacement for each axe. The trajectory adjusts better to the real flight as the features reduce their uncertainty, because the more images are processed, more measurements refine features estimation.

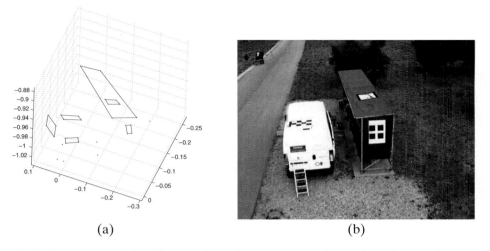

(a) (b)

Fig. 19 Scene reconstruction. The *upper figure* shows reconstructed points from the scene shown in the *lower figure*. *Points* are linked manually with lines to ease the interpretation of the figure

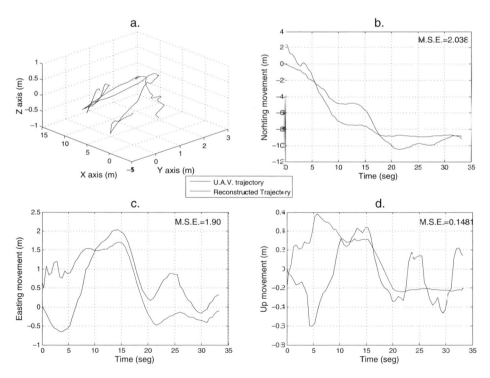

Fig. 20 SLAM reconstructed trajectory vs. UAV trajectory. **a** Three-dimensional flight, **b** north axe in meters, **d** east axe in meters, **c** altitude in meters. The reconstructed trajectory adjusts best to the real flight as soon as more images are processed and the uncertainty of the features is thus reduced

8 Conclusions

This paper dealt with the researches, results and discussion of the use of several techniques of computer vision onboard an UAV. These computer vision techniques are not merely used for acquiring environmental visual information that can be used afterwards by off-line processing. That's why the paper also shows how computer vision can play an important role on-line during the flight itself in order to acquire the adequate sequences necessary to actively track targets (fixed or moving ones) and to guide and control flight trajectories.

Image processing algorithms are very important, and are often designed to detect and track objects along the sequences, whether key points are extracted by the algorithm itself or are externally determined visual targets. Successful, wide spread algorithms onboard an UAV have test bed challenges and thus provide a source of inspiration for their constant improvement and for achieving their better robustness. Some of those test bed challenges are the non-structured and changing light conditions, the highly vibrating and quick and sharp movements, and on-line requirements when necessary.

Some improvements have been presented and tested in the following two types of image processing algorithms: feature tracking and appearance-based tracking, due to the above mentioned characteristics. When using the SIFT key point detector, the algorithm reduces and classifies the key points for achieving a more robust and quick tracking as stated in Section 3. When tracking a whole visual target, an ICA based algorithm is used in a multi-scale hierarchical architecture that makes it robust for scaling. In both type of algorithms, a Kalman filter has been implemented in order to improve the consistence of the features and targets' movements within the image plane, a feat that is particularly relevant in quick changing sequences, as stated in Section 3.3.

The filtered outputs of the image processing algorithms are the visual measurements of the external references that, when compared to their desired position, are introduced in a decoupled position control structure that generates the velocity references in order to control the position of the UAV according to those external visual references. Depending on the type of information extracted by the image processing algorithms (i.e. bi-dimensional translation, rotation, 3D measurements, among others), the UAV's position and orientation control can be a mix of visual based control for some UAV coordinates and GPS based control for some others. A Kalman filter can also be computed in future developments to produce unified UAV estimation and control based on visual, GPS, and inertial information.

This paper also shows that it is possible to obtain robust and coherent results using Visual SLAM for 3D mapping and positioning in vague structured outdoor scenes from a mini UAV. The SLAM algorithm has been implemented using only visual information without considering any odometric or GPS information. Nonetheless, this information has been later used in order to compare and evaluate the obtained results. The state of the system comprises a 12 variable array (position, orientation and their rates), where the inverse depth parametrization has been used in order to avoid the initialization of the distances to the detected visual features, that otherwise becomes a drawback when using SLAM outdoors in unknown environments. The rest of the state array is made up of the tracked features, being ten the minimum allowed number. The prediction stage in EKF has been modeled considering constant velocity for both the position-orientation coordinates and the feature movements in the image plane. The correlation stage in the EKF uses a non-linear camera model that includes a pin-hole distortion model for the sake of more accurate results. Within the implemented SLAM algorithm the Mahalanobis' distance is used to discharge far away matched pairs that can otherwise distort the results.

Based on the results of our work, we conclude that the UAV field has reached an important stage of maturity in which the possibility of using UAVs in civilian applications is now imaginable and in some cases attainable. We have experimentally demonstrated several capabilities that an autonomous helicopter can have by using visual information such as navigation, trajectory planning and visual servoing. The successful implementation of all these algorithms confirms the necessity of dotting UAVs with additional functionalities when tasks like outdoor structures' inspection and object tracking are required.

Our current work is aimed at increasing these capabilities using different visual information sources like catadioptric systems and multiple view systems, and extending them to 3D image-based visual servoing, where the position and orientation of the object will be used to visually conduct the helicopter. The challenge is to achieve real-time image processing and tracking algorithms to reduce the uncertainty of the measure. The field of computer vision for UAVs can be considered as a promising area for investing further research for the benefit of the autonomy and applicability of this type of aerial platforms, considering that reliability and safety have become major research issues of our community.

Acknowledgements   The work reported in this paper is the product of several research stages at the Computer Vision Group Universidad Politécnica de Madrid. The authors would like to thank Jorge León for supporting the flight trials and the I.A. Institute—CSIC for collaborating in the flights' consecution. We would also like to thank Enrique Muñoz Corral and Luis Baumela for helping us to understand and put into practice algorithms to track planar objects. This work has been sponsored by the Spanish Science and Technology Ministry under grants CICYT DPI2004-06624, CICYT DPI2000-1561-C02-02 and MICYT DPI2007-66156, and by the Comunidad Autónoma de Madrid under grant SLAM visual 3D.

References

1. Puri, A., Valavanis, K.P., Kontitsis, M.: Statistical profile generation for traffic monitoring using real-time UAV based video data. In: Control and Automation, 2007. MED '07. Mediterranean Conference on, MED, pp. 1–6 (2007)
2. Nikolos, I.K., Tsourveloudis, N.C., Valavanis, K.P.: Evolutionary algorithm based path planning for multiple UAV cooperation. In: Advances in Unmanned Aerial Vehicles, Intelligent Systems, Control and Automation: Science and Engineering, pp. 309–340. Springer, The Netherlands (2007)
3. Nikolos, I.K., Tsourveloudis, N.C., Valavanis, K.P.: A UAV vision system for airborne surveillance. In: Robotics and Automation, 2004. Proceedings. ICRA '04. 2004 IEEE International Conference on, pp. 77–83. New Orleans, LA, USA (2004), May
4. Nikolos, I.K., Tsourveloudis, N.C., Valavanis, K.P.: Multi-UAV experiments: application to forest fires. In: Multiple Heterogeneous Unmanned Aerial Vehicles, Springer Tracts in Advanced Robotics, pp. 207–228. Springer, Berlin (2007)
5. Green, W., Oh, P.Y.: The integration of a multimodal mav and biomimetic sensing for autonomous flights in near-earth environments. In: Advances in Unmanned Aerial Vehicles, Intelligent Systems, Control and Automation: Science and Engineering, pp. 407–430. Springer, The Netherlands (2007)

6. Belloni, G., Feroli, M., Ficola, A., Pagnottelli, S., Valigi, P.: Obstacle and terrain avoidance for miniature aerial vehicles. In: Advances in Unmanned Aerial Vehicles, Intelligent Systems, Control and Automation: Science and Engineering, pp. 213–244. Springer, The Netherlands (2007)
7. Dalamagkidis, K., Valavanis, K.P., Piegl, L.A.: Current status and future perspectives for unmanned aircraft system operations in the US. In: Journal of Intelligent and Robotic Systems, pp. 313–329. Springer, The Netherlands (2007)
8. Long, L.N., Corfeld, K.J., Strawn, R.C.: Computational analysis of a prototype martian rotorcraft experiment. In: 20th AIAA Applied Aerodynamics Conference, number AIAA Paper 2002–2815, Saint Louis, MO, USA. Ames Research Center, June–October 22 (2001)
9. Yavrucuk, I., Kanan, S., Kahn, A.D.: Gtmars—flight controls and computer architecture. Technical report, School of Aerospace Engineering, Georgia Institute of Technology, Atlanta (2000)
10. Buenaposada, J.M., Munoz, E., Baumela, L.: Tracking a planar patch by additive image registration. In: Proc. of International Workshop, VLBV 2003, vol. 2849 of LNCS, pp. 50–57 (2003)
11. Miller, R., Mettler, B., Amidi, O.: Carnegie mellon university's 1997 international aerial robotics competition entry. In: International Aerial Robotics Competition (1997)
12. Montgomery, J.F.: The usc autonomous flying vehicle (afv) project: Year 2000 status. Technical Report IRIS-00-390, Institute for Robotics and Intelligent Systems Technical Report, Los Angeles, CA, 90089-0273 (2000)
13. Saripalli, S., Montgomery, J.F., Sukhatme, G.S.: Visually-guided landing of an unmanned aerial vehicle. IEEE Trans. Robot Autom. **19**(3), 371–381, June (2003)
14. Mejias, L.: Control visual de un vehiculo aereo autonomo usando detección y seguimiento de características en espacios exteriores. PhD thesis, Escuela Técnica Superior de Ingenieros Industriales. Universidad Politécnica de Madrid, Spain, December (2006)
15. Mejias, L., Saripalli, S., Campoy, P., Sukhatme, G.: Visual servoing approach for tracking features in urban areas using an autonomous helicopter. In: Proceedings of IEEE International Conference on Robotics and Automation, pp. 2503–2508, Orlando, FL, May (2006)
16. Mejias, L., Saripalli, S., Sukhatme, G., Campoy, P.: Detection and tracking of external features in a urban environment using an autonomous helicopter. In: Proceedings of IEEE International Conference on Robotics and Automation, pp. 3983–3988, May (2005)
17. Mejias, L., Saripalli, S., Campoy, P., Sukhatme, G.: Visual servoing of an autonomous helicopter in urban areas using feature tracking. J. Field Robot. **23**(3–4), 185–199, April (2006)
18. Harris, C.G., Stephens, M.: A combined corner and edge detection. In: Proceedings of the 4th Alvey Vision Conference, pp. 147–151 (1988)
19. Lowe, D.G.: Distintive image features from scale-invariant keypoints. Int. J. Computer Vision **60**(2), 91–110 (2004)
20. Duda, R.O., Hart, P.E.: Use of the hough transformation to detect lines and curves in pictures. Commun. ACM **15**(1), 11–15 (1972)
21. Canny, J.: A computational approach to edge detection. IEEE Trans. Pattern Anal. Machine Intel. **8**(6), 679–698, November (1986)
22. Feldman, G., Sobel, I.: A 3 × 3 isotropic gradient operator for image processing. Presented at a talk at the Stanford Artificial Project (1968)
23. Mejías, L., Mondragón, I., Correa, J.F., Campoy, P.: Colibri: Vision-guided helicopter for surveillance and visual inspection. In: Video Proceedings of IEEE International Conference on Robotics and Automation, Rome, Italy, April (2007)
24. Lucas, B.D., Kanade, T.: An iterative image registration technique with an application to stereo vision. In: Proc. of the 7th IJCAI, pp. 674–679, Vancouver, Canada (1981)
25. Beis, J.S., Lowe, D.G.: Shape indexing using approximate nearest-neighbour search in high-dimensional spaces. In: CVPR '97: Proceedings of the 1997 Conference on Computer Vision and Pattern Recognition (CVPR '97), p. 1000. IEEE Computer Society, Washington, DC, USA (1997)
26. Fischer, M.A., Bolles, R.C.: Random sample concensus: a paradigm for model fitting with applications to image analysis and automated cartography. Commun. ACM **24**(6), 381–395 (1981)
27. Baker, S., Matthews, I.: Lucas-kanade 20 years on: A unifying framework: Part 1. Technical Report CMU-RI-TR-02-16, Robotics Institute, Carnegie Mellon University, Pittsburgh, PA, July (2002)
28. Mejias, L., Campoy, P., Mondragon, I., Doherty, P.: Stereo visual system for autonomous air vehicle navigation. In: 6th IFAC Symposium on Intelligent Autonomous Vehicles (IAV 07), Toulouse, France, September (2007)

29. Martin, J., Crowley, J.: Experimental comparison of correlation techniques. Technical report, IMAG-LIFIA, 46 Av. Félix Viallet 38031 Grenoble, France (1995)
30. Bay, H., Tuytelaars, T., Van Gool, L.: SURF: Speeded Up Robust Features. In: Proceedings of the Ninth European Conference on Computer Vision, May (2006)
31. Montiel, J.M.M., Civera, J., Davison, A.J.: Unified inverse depth parametrization for monocular slam. In: Robotics: Science and Systems (2006)
32. Civera, J., Davison, A.J., Montiel, J.M.M.: Dimensionless monocular slam. In: IbPRIA, pp. 412–419 (2007)
33. COLIBRI. Universidad Politécnica de Madrid. Computer Vision Group. COLIBRI Project. http://www.disam.upm.es/colibri (2005)

Vision-Based Odometry and SLAM for Medium and High Altitude Flying UAVs

F. Caballero · L. Merino · J. Ferruz · A. Ollero

Originally published in the Journal of Intelligent and Robotic Systems, Volume 54, Nos 1–3, 137–161.
© Springer Science + Business Media B.V. 2008

Abstract This paper proposes vision-based techniques for localizing an unmanned aerial vehicle (UAV) by means of an on-board camera. Only natural landmarks provided by a feature tracking algorithm will be considered, without the help of visual beacons or landmarks with known positions. First, it is described a monocular visual odometer which could be used as a backup system when the accuracy of GPS is reduced to critical levels. Homography-based techniques are used to compute the UAV relative translation and rotation by means of the images gathered by an onboard camera. The analysis of the problem takes into account the stochastic nature of the estimation and practical implementation issues. The visual odometer is then integrated into a simultaneous localization and mapping (SLAM) scheme in order to reduce the impact of cumulative errors in odometry-based position estimation

This work is partially supported by the AWARE project (IST-2006-33579) funded by the European Commission, and the AEROSENS project (DPI-2005-02293) funded by the Spanish Government.

F. Caballero (✉) · J. Ferruz · A. Ollero
University of Seville, Seville, Spain
e-mail: caba@cartuja.us.es

J. Ferruz
e-mail: ferruz@cartuja.us.es

A. Ollero
e-mail: aollero@cartuja.us.es

L. Merino
Pablo de Olavide University, Seville, Spain
e-mail: lmercab@upo.es

K. P. Valavanis et al. (eds.), *Unmanned Aircraft Systems*. DOI: 10.1007/978-1-4020-9136-0_9   137

approaches. Novel prediction and landmark initialization for SLAM in UAVs are presented. The paper is supported by an extensive experimental work where the proposed algorithms have been tested and validated using real UAVs.

Keywords Visual odometry · Homography · Unmanned aerial vehicles · Simultaneous localization and mapping · Computer vision

1 Introduction

Outdoor robotics applications in natural environments sometimes require different accessibility capabilities than the capabilities provided by existing ground robotic vehicles. In fact, in spite of the progress in the development of unmanned ground vehicles along the last 20 years, navigating in unstructured natural environments still poses significant challenges. The existing ground vehicles have inherent limitations to reach the desired locations in many applications. The characteristics of the terrain and the presence of obstacles, together with the requirement of fast response, may represent a major drawback to the use of any ground locomotion system. Thus, in many cases, the use unmanned aerial vehicles (UAVs) is the only effective way to reach the target to get information or to deploy instrumentation.

In the last ten years UAVs have improved their autonomy both in energy and information processing. Significant achievements have been obtained in autonomous positioning and tracking. These improvements are based on modern satellite-based position technologies, inertial navigation systems, communication and control technologies, and image processing. Furthermore, new sensing and processing capabilities have been implemented on-board the UAVs. Thus, today we can consider some UAVs as intelligent robotic systems integrating perception, learning, real-time control, situation assessment, reasoning, decision-making and planning capabilities for evolving and operating in complex environments.

In most cases, UAVs use the global position system (GPS) to determine their position. As pointed out in the Volpe Report [42], the accuracy of this estimation directly depends on the number of satellites used to compute the position and the quality of the signals received by the device; radio effects like multi-path propagation could cause the degradation in the estimation. In addition, radio frequency interferences with coexisting devices or jamming could make the position estimation unfeasible.

These problems are well known in robotics. Thus, odometry is commonly used in terrestrial robots as a backup positioning system or in sensor data fusion approaches. This local estimation allows temporally managing GPS faults or degradations. However, the lack of odometry systems in most aerial vehicles can lead to catastrophic consequences under GPS errors; incoherent control actions could be commanded to the UAV, leading to crash and the loss of valuable hardware. Moreover, if full autonomy in GPS-less environments is considered, then the problem of simultaneous localization and mapping (SLAM) should be addressed.

If small UAVs are considered, their low payload represents a hard restriction on the variety of devices to be used for odometry. Sensors like 3D or 2D laser scanners are too heavy and have an important dependence to the UAV distance to the ground. Although there exist small devices for depth sensing, their range is usually shorter than 15 m. Stereo vision systems have been successfully applied to low/medium size UAVs due to its low weight and versatility [4, 9, 18], but the rigid distance between the two cameras limits the useful altitude range.

Monocular vision seems to offer a good solution in terms of weight, accuracy and scalability. This paper proposes a monocular visual odometer and vision-based localization methods to act as backup systems when the accuracy of GPS is reduced to critical levels. The objective is the development of computer vision techniques for the computation of the relative translation and rotation, and for the localization of the vehicle based on the images gathered by a camera on-board the UAV. The analysis of the problem takes into account the stochastic nature of the estimation and practical implementation issues.

The paper is structured as follows. First, related work in vision based localization for UAVs is detailed. Then, a visual odometer based on frame-to-frame homographies is described, together with a robust method for homography computation. Later, the homography-based odometry is included in a SLAM scheme in order to overcome the error accumulation present in odometric approaches. The proposed SLAM approach uses the information provided by the odometer as main prediction hypothesis and for landmark initialization. Finally, conclusions and lessons learned are described.

1.1 Related Work

One of the first researches on vision applied to UAV position estimation starts in the nineties at the Carnegie-Mellon University (CMU). In [1], it is described a vision-based odometer that allowed to lock the UAV to ground objects and sense relative helicopter position and velocity in real time by means of stereo vision. The same visual tracking techniques, combined with inertial sensors, were applied to autonomous take off, following a prescribed trajectory and landing. The CMU autonomous helicopter also demonstrated autonomous tracking capabilities of moving objects by using only on-board specialized hardware.

The topic of vision-based autonomous landing of airborne systems has been actively researched [30]. In the early nineties, Dickmanns and Schell [13] presented some results of the possible use of vision for landing an airplane. Systems based on artificial beacons and structured light are presented [44, 45]. The BEAR project at Berkeley is a good example of vision systems for autonomous landing of UAVs. In this project, vision-based pose estimation relative to a planar landing target and vision-based landing of an aerial vehicle on a moving deck have been researched [36, 40]. A technique based on multiple view geometry is used to compute the real motion of one UAV with respect to a planar landing target. An artificial target allows to establish quick matches and to solve the scale problem.

Computer vision has also been proposed for safe landing. Thus, in [15], a strategy and an algorithm relying on image processing to search the ground for a safe landing spot is presented. Vision-based techniques for landing on a artificial helipad of known shape are also presented in [34, 35], where the case of landing on a slow moving helipad is considered. In [37], the landing strategies of bees are used to devise a vision system based on optical flow for UAVs.

Corke et. al [9] have analyzed the use of stereo vision for height estimation in small size helicopters. In Georgia Tech, vision-based aided navigation for UAVs has been considered. Thus, in [43] the authors present an Extended Kalman Filter approach that combines GPS measurements with image features obtained from a known artificial target for helicopter position estimation.

In a previous work [5], the authors present a visual odometer for aerial vehicles using monocular image sequences, but no error estimation is provided by the algorithm, and the approach is limited to planar scenes. In [6], it is shown how a mosaic can be used in aerial vehicles to partially correct the drift associated to odometric approaches. This technique is extended in [7] with a minimization process that allows to improve the spatial consistency of the online built mosaic. Recently, in [8] the authors propose a visual odometer to compensate GPS failures. Image matching with geo-referenced aerial imagery is proposed to compensate the drift associated to odometry.

Although vision-based SLAM has been widely used in ground robots and has demonstrated its feasibility for consistent perception of the environment and position of the robot, only a few applications have been implemented on UAVs. The researches carried out in the LAAS laboratory in France and the Centre for Autonomous Systems in Australia can be highlighted. The first of them has developed an stereo vision system designed for the KARMA blimp [18, 21], where interest point matching and Kalman filtering techniques are used for simultaneous localization and mapping with very good results. However, this approach is not suitable for helicopters, as the baseline of the stereo rig that can be carried is small, and therefore it limits the height at which the UAV can fly. UAV simultaneous localisation and map building with vision using a delta fixed-wing platform is also presented in [19]. Artificial landmarks of known size are used in order to simplify the landmark identification problem. The known size of the landmarks allows to use the cameras as a passive range/bearing/elevation sensor. Preliminary work on the use of vision-based bearing-only SLAM in UAVs is presented in [23]. In [22], vision and IMU are combined for UAV SLAM employing an Unscented Kalman Filter. The feature initialization assumes a flat terrain model, similarly to the present approach. Results in simulation are shown in the paper. In [25], an architecture for multi-vehicle SLAM is studied for its use with UAVs. The paper deals with the issues of data association and communication, and some simulation results are presented.

Visual servoing approaches has been also proposed for direct control of UAVs. The use of an omnidirectional camera for helicopter control has been presented in [17]. The camera is used to maintain the helicopter in the centroid of a set of artificial targets. The processed images are directly used to command the helicopter. The paper shows the feasibility of the procedure, but no actual control is tested. Omnidirectional vision is also used in [12] to estimate the attitude of an UAV. The method detects the horizon line by means of image processing and computes the attitude from its apparent motion. In the work of [27], vision is used to track features of buildings. Image features and GPS measurements are combined together to keep the UAV aligned with the selected features. Control design and stability analysis of image-based controllers for aerial robots are presented in [26]. In [32] recent work on vision-based control of a fixed wing aircraft is presented.

2 Homography-Based Visual Odometry for UAVs

Image homographies will be a basic tool for estimating the motion that an UAV undergoes by using monocular image sequences. A homography can be defined as an invertible application of the projective space $\mathbf{P}^2$ into $\mathbf{P}^2$ that applies lines into lines. Some basic properties of the homographies are the following:

- Any homography can be represented as a linear and invertible transformation in homogeneous coordinates:

$$\begin{bmatrix} \tilde{u} \\ \tilde{v} \\ k \end{bmatrix} = \underbrace{\begin{bmatrix} h_{11} & h_{12} & h_{13} \\ h_{21} & h_{22} & h_{23} \\ h_{31} & h_{32} & h_{33} \end{bmatrix}}_{\mathbf{H}} \begin{bmatrix} u \\ v \\ 1 \end{bmatrix} \tag{1}$$

Inversely, any transformation of this nature can be considered as a homography.
- Given the homogeneous nature of the homography $\mathbf{H}$, it can be multiplied by an arbitrary constant $k \neq 0$ and represent the same transformation. This means that the matrix $\mathbf{H}$ is constrained by eight independent parameters and a scale factor.

Given two views of a scene, the homography model represents the exact transformation of the pixels on the image plane if both views are related by a pure rotation, or if the viewed points lie on a plane. When a UAV flies at relatively high altitude, it is a usual assumption to model the scene as pseudo-planar. The paper will propose a method to extend the applicability of the homography model to non-planar scenes (computing the homography related to a dominant plane on the scene) in order to be able to perform motion estimation at medium or even low UAV altitude.

2.1 Robust Homography Estimation

The algorithm for homography computation is based on a point features matching algorithm, and has been tested and validated with thousands of images captured by different UAVs flying at different altitudes, from 15 to 150 m. This algorithm (including the feature matching approach) was briefly described in [29]. It basically consists of a point-feature tracker that obtains matches between images, and a combination of least median of squares and M-Estimator for outlier rejection and accurate homography estimation from these matches.

However, there are two factors that may reduce the applicability of the technique, mainly when the UAV flies at altitudes of the same order of other elements on the ground (buildings, trees, etc):

- Depending on the frame-rate and the vehicle motion, the overlap between images in the sequence is sometimes reduced. This generates a non-uniform distribution of the features along the images.
- In 3D scenes, the parallax effect will increase, and the planarity assumption will not hold. The result is a dramatic growth of the outliers and even the divergence of the M-Estimator.

They produce different problems when computing the homography. If the matches are not uniformly distributed over the images, an ill-posed system of equations for homography computation will be generated, and there may exist multiple solutions. On the other hand, if the parallax effect is significant, there may exist multiple planes (whose transformation should be described by multiple homographies); the algorithm should try to filter out all features but those lying on the dominant plane of the scene (the ground plane).

In the proposed solution, the first problem is addressed through a *hierarchy* of homographic models (see Fig. 1), in which the complexity of the model to be fitted is decreased whenever the system of equations is ill-constrained, while the second is tackled through outlier rejection techniques.

Therefore, depending on the quality of the available data, the constraints used to compute the homography are different; thus, the accuracy changes as well. An estimation of this accuracy will be given by the covariance matrix of the computed parameters.

A complete homography has 8 *df* (as it is defined up to a scale factor). The degrees of freedom can be reduced by fixing some of the parameters of the 3×3 matrix. The models used are the defined by Hartley in [16]: Euclidean, Affine and Complete Homographic models, which have 4, 6 and 8 *df* respectively (see Fig. 1). The percentage of successful matches obtained by the point tracker is used to have an estimation about the level of the hierarchy where the homography computation should start. These percentage thresholds were obtained empirically by processing hundreds of aerial images. Each level involves the following different steps:

-   Complete homography. Least median of squares (LMedS) is used for outlier rejection and a M-Estimator to compute the final result. This model is used if more than the 65% of the matches are successfully tracked.
-   Affine homography. If the percentage of success in the tracking step is between 40% and 65%, then the LMedS is not used, given the reduction in the number of matches. A relaxed M-Estimator (soft penalization) is carried out to compute the model.

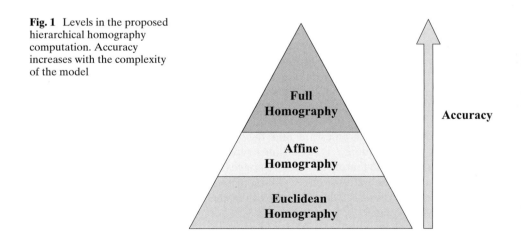

Fig. 1 Levels in the proposed hierarchical homography computation. Accuracy increases with the complexity of the model

–   Euclidean homography. If the percentage is below 40%, the set of data is too
    noisy and small to apply non-linear minimizations. The model is computed using
    least-squares.

In addition, it is necessary a rule to know when the current level is ill-posed
and the algorithm has to decrease the model complexity. The M-Estimator used
in the complete and affine computations is used for this purpose. It is considered
that the M-Estimator diverge if it reaches the maximum number of iterations and,
hence, the level in the hierarchy has to be changed to the next one.

2.2 Geometry of Two Views of the Same Plane

The odometer will extract the camera motion from the image motion modeled by the
estimated homography between two consecutive views. If we consider the position
and orientation of two cameras in the world coordinate frame, as shown in Fig. 2, it
can be seen that the two projections $\mathbf{m_1} \in \mathbb{R}^2$ and $\mathbf{m_2} \in \mathbb{R}^2$ of a fixed point $\mathbf{P} \in \mathbb{R}^3$
belonging to a plane Π are related by:

$$\tilde{\mathbf{m}}_2 = \underbrace{\mathbf{A}_2 \mathbf{R}_{12} \left(\mathbf{I} - \frac{\mathbf{t}_2 \mathbf{n}_1^T}{d_1} \right) \mathbf{A}_1^{-1}}_{\mathbf{H}_{12}} \tilde{\mathbf{m}}_1 \tag{2}$$

where $\mathbf{R}_{12}$ is the rotation matrix that transforms a vector expressed in the coordinate
frame of camera one into the coordinate frame of camera two, $\mathbf{t}_2$ is the translation of
camera two with respect to camera one expressed in the coordinate frame of camera
one, the Euclidean distance from the camera one to the plane Π is d_1 and the normal
of the plane Π (in the first camera coordinate frame) is given by the unitary 3-D
vector $\mathbf{n}_1$ (see Fig. 2).

Fig. 2 Geometry of two views
of the same plane

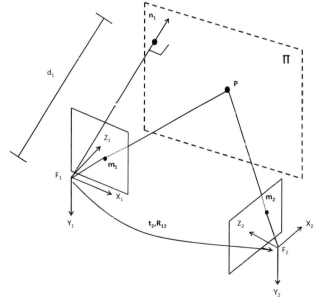

For this particular case, the transformation between the features $\mathbf{m}_1$ and $\mathbf{m}_2$ is a plane-to-plane homography, so $\tilde{\mathbf{m}}_2 = \mathbf{H}_{12}\tilde{\mathbf{m}}_1$. This homography is completely defined by the calibration matrices $\mathbf{A}_1$ and $\mathbf{A}_2$, the relative position of the cameras and the structure of the scene (the normal and distance of the plane). The problem can be reformulated as a single camera whose position and orientation change through time. In this case the calibration matrix is the same for both views, so $\mathbf{A}_1 = \mathbf{A}_2$.

Then, for the calibrated case, the relative position (rotation and translation) between the cameras and the plane normal can be obtained if the homography that relates two views of the same plane is known, for instance by obtaining a set of matches between the images, as described in the previous section. Moreover, it will be shown how to obtain an estimation of the covariance matrix for all these parameters.

2.3 Motion Estimation from Homographies

A solution based on the singular value decomposition (SVD) of the homography will be used. Consider a single camera that moves through time, the homography $\mathbf{H}_{12}$ that relates the first and the second view of the same planar scene and the camera calibration matrix $\mathbf{A}_1$. According to Eq. 2, the *calibrated homography* is defined as:

$$\mathbf{H}_{12}^c = \mathbf{A}_1^{-1}\mathbf{H}_{12}\mathbf{A}_1 = \mathbf{R}_{12}\left(\mathbf{I} - \frac{\mathbf{t}_2\mathbf{n}_1^T}{d_1}\right) \tag{3}$$

The elements can be extracted from the singular value decomposition (SVD) of the homography $\mathbf{H}_{12}^c = \mathbf{U}\mathbf{D}\mathbf{V}^T$, where $\mathbf{D} = diag(\lambda_1, \lambda_2, \lambda_3)$ stores the singular values. Once $\mathbf{U}$, $\mathbf{V}$ and $\mathbf{D}$ have been conveniently ordered such us $\lambda_1 > \lambda_2 > \lambda_3$, the singular values can be used to distinguish three types of movements carried out by the camera [39]:

- The three singular values of $\mathbf{H}_{12}^c$ are equal, so $\lambda_1 = \lambda_2 = \lambda_3$. It occurs when the motion consist of rotation around an axis through the origin only, i.e., $\mathbf{t}_2 = \mathbf{0}$. The rotation matrix is unique, but there is not sufficient information to estimate the plane normal $\mathbf{n}_1$.
- The multiplicity of the singular values of $\mathbf{H}_{12}^c$ is two, for example $\lambda_1 = \lambda_2 \neq \lambda_3$. Then, the solution for motion and geometrical parameters is unique up to a common scale factor for the translation parameters. In this case, the camera translation is parallel to the normal plane.
- The three singular values of $\mathbf{H}_{12}^c$ are different, i.e., $\lambda_1 \neq \lambda_2 \neq \lambda_3$. In this case two possible solutions for rotation, translation and plane normal exist and can be computed.

The presence on noise in both feature tracking and homography estimation always leads to different singular values for $\mathbf{H}_{12}^c$ and the third of the previous cases becomes the dominant in real conditions. Rotation, translation and normal to the plane is then given by the following expressions [39]:

$$\mathbf{R}_2 = \mathbf{U} \begin{bmatrix} \alpha & 0 & \beta \\ 0 & 1 & 0 \\ -s\beta & 0 & s\alpha \end{bmatrix} \mathbf{V}^T$$

$$\mathbf{t}_2 = \frac{1}{w} \left(-\beta \mathbf{u}_1 + \left(\frac{\lambda_3}{\lambda_2} - s\alpha \right) \mathbf{u}_3 \right)$$

$$\mathbf{n}_1 = w(\delta \mathbf{v}_1 + \mathbf{v}_3) \tag{4}$$

where:

$$\delta = \pm \sqrt{\frac{\lambda_1^2 - \lambda_2^2}{\lambda_2^2 - \lambda_3^2}}$$

$$\alpha = \frac{\lambda_1 + s\lambda_3 \delta^2}{\lambda_2 (1 + \delta^2)}$$

$$\beta = \pm \sqrt{1 - \alpha^2}$$

$$s = det(\mathbf{U})det(\mathbf{V})$$

and ω is a scale factor. We set that scale factor so that $\|\mathbf{n}_1\| = 1$. Each solution must accomplish that $sgn(\beta) = -sgn(\delta)$. For this case, Triggs algorithm [38] allows a systematic and robust estimation. This method has been implemented and tested in the experiments presented in this paper with very good results.

From Eq. 3 it can be seen that (as $\|\mathbf{n}_1\| = 1$) only the product $\frac{\|\mathbf{t}_2\|}{d_1}$ can be recovered. The scale can be solved, then, if the distance d_1 of camera 1 to the reference plane is known. If the reference plane is the ground plane, as it would be the case in the experiments, a barometric sensor or height sensor can be used to estimate this initial distance. Also, a range sensor can be used. In this paper, we will consider that this height is estimated for the first frame by one of these methods.

2.4 Correct Solution Disambiguation

Apart from the scale factor, two possible solutions $\{\mathbf{R}_2^1, \mathbf{t}_2^1, \mathbf{n}_1^1\}$ and $\{\mathbf{R}_2^2, \mathbf{t}_2^2, \mathbf{n}_1^2\}$ will be obtained. Given a third view and its homography with respect to the first frame $\mathbf{H}_{13}$, it is possible to recover an unique solution, as the estimated normal of the reference plane in the first camera coordinate frame, $\mathbf{n}_1$, should be the same.

A method to detect the correct solution is proposed. If a sequence of images is used, the set of possible normals is represented by:

$$S_n = \{ \mathbf{n}_{12}^1, \mathbf{n}_{12}^2, \mathbf{n}_{13}^1, \mathbf{n}_{13}^2, \mathbf{n}_{14}^1, \mathbf{n}_{14}^2, \ldots \} \tag{5}$$

where the superindex denotes the two possible normal solutions and the subindex $1j$ denotes the normal $\mathbf{n}_1$ estimated using image j in the sequence.

If $\mathbf{n}_{12}^1$ and $\mathbf{n}_{12}^2$ were correct, there would be two set of solutions, S_{n^1} and S_{n^2}. The uniqueness of the normal leads to the following constraints:

$$\left\| \mathbf{n}_{12}^1 - \mathbf{n}_{1j}^i \right\| \leq \epsilon_1 \; \forall \mathbf{n}_{1j}^i \in S_{n^1} \tag{6}$$

$$\left\| \mathbf{n}_{12}^2 - \mathbf{n}_{1j}^i \right\| \leq \epsilon_2 \; \forall \mathbf{n}_{j}^i \in S_{n^2} \tag{7}$$

where ϵ_1 and ϵ_2 are the minimal values that guarantee an unique solution for Eqs. 6 and 7 respectively. The pairs $\{S_{n^1}, \epsilon_1\}$ and $\{S_{n^2}, \epsilon_2\}$ are computed separately by means of the following iterative algorithm:

1. The distance among $\mathbf{n}_{12}^i$ and the rest of normals of S_n is computed.
2. ϵ_i is set to an initial value.
3. For the current value ϵ_i, check if there exist an unique solution.
4. If no solution is found, increase the value of ϵ_i and try again with the step 3. If multiple solutions were found decrease ϵ_i and try again with step 3. If an unique solution was found, then finish.

The algorithm is applied to $i = 1$ and $i = 2$ and the correct solution is then chosen between both options as the one that achieves the minimum ϵ.

2.5 An Estimation of the Uncertainties

An important issue with odometric measurements is to obtain a correct estimation of the associated drift. The idea is to estimate the uncertainties on the estimated rotation, translation and plane normal from the covariance matrix associated to the homography, which can be computed from the estimated errors on the point matches [6].

The proposed method computes the Jacobian of the complete process to obtain a first order approximation of rotation, translation and plane normal error covariance matrix. Once the calibrated homography has been decomposed into its singular values, the computation of the camera motion is straightforward, so this section will focus in the computation of the Jacobian associated to the singular value decomposition process.

Thus, given the SVD decomposition of the calibrated homography $\mathbf{H}_{12}^c$:

$$\mathbf{H}_{12}^c = \begin{bmatrix} h_{11} & h_{12} & h_{13} \\ h_{21} & h_{22} & h_{23} \\ h_{31} & h_{32} & h_{33} \end{bmatrix} = \mathbf{U}\mathbf{D}\mathbf{V}^T = \sum_{i=1}^{3} \left(\lambda_i \mathbf{u}_i \mathbf{v}_i^T \right) \tag{8}$$

The goal is to compute $\frac{\partial \mathbf{U}}{\partial h_{ij}}$, $\frac{\partial \mathbf{V}}{\partial h_{ij}}$ and $\frac{\partial \mathbf{D}}{\partial h_{ij}}$ for all h_{ij} in $\mathbf{H}_{12}^c$. This Jacobian can be easily computed through the robust method proposed by Papadopoulo and Lourakis in [31].

Taking the derivative of Eq. 8 with respect to h_{ij} yields the following expression:

$$\frac{\partial \mathbf{H}_{12}^c}{\partial h_{ij}} = \frac{\partial \mathbf{U}}{\partial h_{ij}} \mathbf{D}\mathbf{V}^T + \mathbf{U} \frac{\partial \mathbf{D}}{\partial h_{ij}} \mathbf{V}^T + \mathbf{U}\mathbf{D} \frac{\partial \mathbf{V}^T}{\partial h_{ij}} \tag{9}$$

Clearly, $\forall (k, l) \neq (i, j)$, $\frac{\partial h_{kl}}{\partial h_{ij}} = 0$ while $\frac{\partial h_{ij}}{\partial h_{ij}} = 1$. Since $\mathbf{U}$ is an orthogonal matrix:

$$\mathbf{U}^T \mathbf{U} = \mathbf{I} \Rightarrow \frac{\partial \mathbf{U}^T}{\partial h_{ij}} \mathbf{U} + \mathbf{U}^T \frac{\partial \mathbf{U}}{\partial h_{ij}} = \Omega_U^{ij\,T} + \Omega_U^{ij} = 0 \tag{10}$$

where Ω_U^{ij} is defined by

$$\Omega_U^{ij} = \mathbf{U}^T \frac{\partial \mathbf{U}}{\partial h_{ij}} \tag{11}$$

It is clear that Ω_U^{ij} is an antisymmetric matrix. Similarly, an antisymmetric matrix Ω_V^{ij} can be defined for $\mathbf{V}$ as:

$$\Omega_V^{ij} = \frac{\partial \mathbf{V}^T}{\partial h_{ij}} \mathbf{V} \tag{12}$$

By multiplying Eq. 9 by $\mathbf{U}^T$ and $\mathbf{V}$ from left and right respectively, and using Eqs. 11 and 12, the following relation is obtained:

$$\mathbf{U}^T \frac{\partial \mathbf{H}_{12}^u}{\partial h_{ij}} \mathbf{V} = \Omega_U^{ij} \mathbf{D} + \frac{\partial \mathbf{D}}{\partial h_{\cdot j}} + \mathbf{D}\Omega_V^{ij} \tag{13}$$

Since Ω_U^{ij} and Ω_V^{ij} are antisymmetric matrices, all their diagonal elements are equal to zero. Recalling that $\mathbf{D}$ is a diagonal matrix, it is easy to see that the diagonal elements of $\Omega_U^{ij}\mathbf{D}$ and $\mathbf{D}\Omega_V^{ij}$ are also zero. Thus:

$$\frac{\partial \lambda_k}{\partial h_{ij}} = u_{ik} v_{jk} \tag{14}$$

Taking into account the antisymmetric property, the elements of the matrices Ω_U^{ij} and Ω_V^{ij} can be computed by solving a set of 2×2 linear systems, which are derived from the off-diagonal elements of the matrices in Eq. 13:

$$\left. \begin{array}{r} d_l \Omega_{U\,kl}^{ij} + d_k \Omega_{V\,kl}^{ij} = u_{ik} v_{jl} \\ d_k \Omega_{U\,kl}^{ij} + d_l \Omega_{V\,kl}^{ij} = -u_{il} v_{jk} \end{array} \right\} \tag{15}$$

where the index ranges are $k = 1 \ldots 3$ and $l = i + 1 \ldots 2$. Note that, since the d_k are positive numbers, this system has a unique solution provided that $d_k \neq d_l$. Assuming for now that $\forall (k, l)$, $d_k \neq d_l$, the 3 parameters defining the non-zero elements of Ω_U^{ij} and Ω_V^{ij} can be easily recovered by solving the 3 corresponding 2×2 linear systems.

Once Ω_U^{ij} and Ω_V^{ij} have been computed, the partial derivatives are obtained as follows:

$$\frac{\partial \mathbf{U}}{\partial h_{ij}} = \mathbf{U}\Omega_U^{ij} \tag{16}$$

$$\frac{\partial \mathbf{V}}{\partial h_{ij}} = -\mathbf{V}\Omega_V^{ij} \tag{17}$$

Taking into account the Eqs. 14, 16 and 17 and the covariance matrix corresponding to the homography it is possible to compute the covariance matrix associated to $\mathbf{U}$, $\mathbf{V}$ and $\mathbf{D}$. Further details and demonstrations can be found in [31]. Finally, the

Jacobians of the equations used to extract the rotation, translation and normal, given by Eq. 4, are easily computed and combined with these covariances to estimate the final motion covariances.

2.6 Experimental Results

This section shows some experimental results in which the homography-based visual odometer is applied to monocular image sequences gathered by real UAVs.

The first experiment was conducted with the HERO helicopter (see Fig. 3). HERO is an aerial robotic platform designed for research on UAV control, navigation and perception. It has been developed by the "Robotics, Vision and Control Research Group" at the University of Seville during the CROMAT project, funded by the Spanish Government. HERO is equipped with accurate sensors to measure position and orientation, cameras and a PC-104 to allow processing on board. A DSP is used as data acquisition system and low level controller (position and orientation); the PC-104 runs the rest of tasks such as perception, communications or navigation. All the data gathered by the DSP are exported to the PC-104 through a serial line and published for the rest of the processes.

All the sensor data have been logged together with the images in order to avoid inconsistency among different sensor data. The position is estimated with a Novatel DGPS with 2 cm accuracy and updated at 5 Hz, while an inertial measurement unit (IMU) provides the orientation at 50 Hz, with accuracy of 0.5 degrees. In the experiment, the camera was oriented forty-five degrees with respect to the helicopter horizontal.

The visual odometer algorithm (feature tracking, robust homography computation and homography decomposition) has been programmed in C++ code and runs at 10 Hz with 320 × 240 images. The experiment image sequence is composed by 650 samples, or approximately 65 s of flight. A sharp movement is made around sample 400.

The DGPS measurements are used to validate the results. Along the flight, good GPS coverage was available at all time. It is important to notice that the

Fig. 3 HERO helicopter

odometry is computed taking into account the estimated translation and rotation, so it accumulates both errors. The estimated position by using the visual odometer is shown in Fig. 4. The figure presents the DGPS position estimation and the errors associated to the odometer. It can be seen how the errors grow with the image sample index. The errors corresponding to each estimation are added to the previous ones

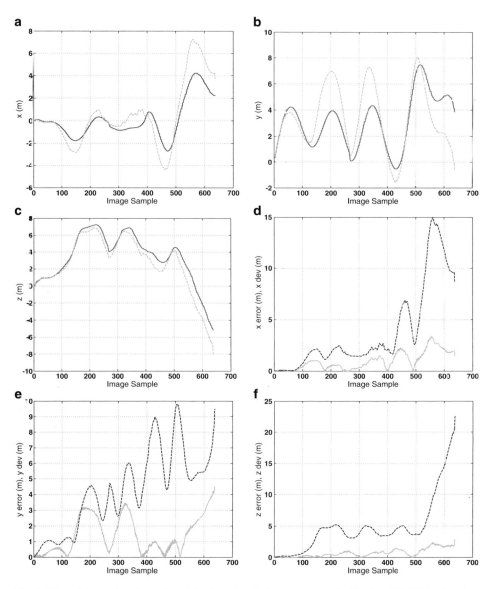

Fig. 4 *Up* position estimation using vision based technique (*green dashed line*) and DGPS estimation (*red solid line*). *Down* error of the vision based odometry (*green solid line*) and estimated standard deviation (*blue dashed line*)

and make the position estimation diverge through time. Moreover, it can be seen how the estimation of the standard deviation is coherent with the evolution of the error (which is very important for further steps).

Figure 5 shows the evolution of the estimated orientation by using the odometer and the on-board IMU. The orientation has been represented in the classic

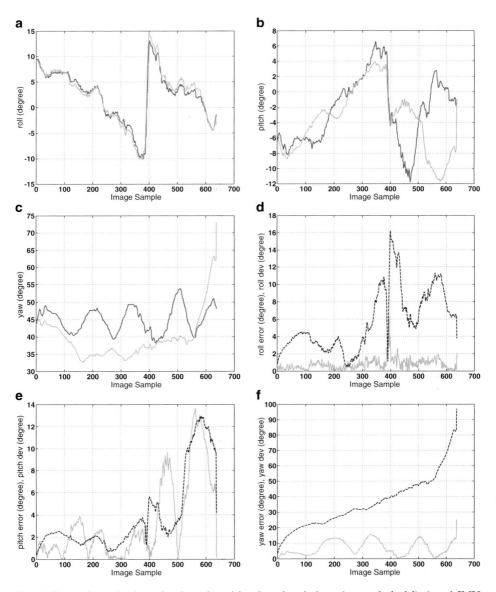

Fig. 5 *Top* estimated orientation by using vision based technique (*green dashed line*) and IMU estimation (*red solid line*). The orientation is represented in roll/pitch/yaw. *Bottom* errors in the vision based estimation (*green solid line*) and estimated standard deviation (*blue dashed line*)

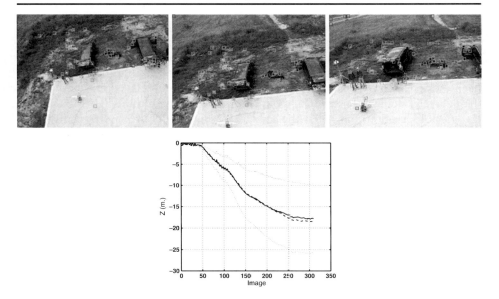

Fig. 6 Three images of the landing sequence and the estimated height computed by the visual odometer (*solid*) and DGPS (*dashed*). The average frame rate is 7 Hz

roll/pitch/yaw convention (Euler XYZ). It can be seen that the errors in the estimated orientation are small except for the pitch angle. The standard deviation is in general overall consistent.

Results have been also obtained with data gathered during an autonomous landing[1] by the autonomous helicopter Marvin, developed by the Technical University of Berlin [33]. Figure 6 shows three frames of the landing sequence with the obtained matches. It should be pointed out that there are no artificial landmarks for the matching process. Also, in this experiment, the concrete landing platform lacks of structure, which can pose difficulties for the matching procedure. Moreover, along the descent, the pan and tilt unit was moving the camera. Figure 6 shows the estimated translation compared with DGPS, along with the estimated errors. The results are very accurate, although the technique tends to overestimate the uncertainty.

Thus, the experimental results show that the visual odometer can be used to estimate the motion of the UAV; moreover, the estimated errors are consistent. It is important to highlight that all experiments were carried out by using natural landmarks automatically selected by the feature tracking algorithm, without the help of visual beacons.

3 Application of Homography-Based Odometry to the SLAM Problem

A SLAM-based technique is proposed to compensate the accumulative error intrinsic to odometry and to solve the localization problem. SLAM employing monocular

---

[1]The autonomous landing was done based on DGPS and ultrasonic sensors.

imagery is a particular case of the SLAM problem, called bearing-only SLAM or boSLAM, in which bearing only sensors are used, a camera in this case. boSLAM is a partially observable problem [41], as the depth of the landmarks cannot be directly estimated. This entails a difficult landmark initialization problem which has been tackled with two basic approaches: delayed and un-delayed initialization. In the delayed initialization case, landmarks are not included in the SLAM system in the first observation, but when the angular baseline in between observations has grown large enough to ensure a good triangulation. This method has the advantage of using well conditioned landmarks, but the SLAM system cannot take advantage of the landmark until its localization is well conditioned. Several approaches have been proposed in this area such as [10] where a Particle Filter is used to initialize the landmark depth, or [11], where non-linear bundle adjustment over a set of observations is used to initialize the landmarks.

On the other hand, un-delayed approaches introduce the landmark in the SLAM system with the first observation, but some considerations have to be taken into account due to the fact that the landmarks are usually bad conditioned in depth, and then divergence problems may appear in the SLAM filter. Most existing approaches are based on multiple hypotheses, as in [20], where a Gaussian Mixture is used for landmark initialization in a Kalman Filter. Recent research [28] proposes the inverse depth parametrization in a single-hypothesis approach for landmark initialization.

The technique presented in this Section is based on a classical EKF that simultaneously estimates the pose of the robot (6 *df*) and a map of point features, as in [2, 3, 14, 24]. The main contributions is a new undelayed feature initialization that takes advantage of the scene normal plane estimation computed in the Homography-based odometry. Indeed, the technique cannot be considered as boSLAM because information from a range sensor is used, combined with the normal vector to the plane, to initialize the landmark depth.

The use of the estimated rotation and translation provided by the odometer as the main motion hypothesis in the prediction stage of the EKF is another contribution made by this approach. Complex non-linear models are normally used to estimate vehicle dynamics, due to the lack of odometers in UAVs. This leads to poor prediction hypotheses, in terms of accuracy, and then a significant reduction of the filter efficiency. In [19] a solution based on merging model-based estimation and inertial measurements from local sensors (IMUs) is proposed, resulting in an accuracy growth. The integration of the IMU is also considered here in order to improve the position estimation. Next paragraphs describe the structure and implementation of this filter.

3.1 The State Vector

The robot pose $\mathbf{p}_t$ is composed by the position and orientation of the vehicle at time t in the World Frame (see Section 3.4), so:

$$\mathbf{p}_t = [\mathbf{t}_t, \mathbf{q}_t]^T = [x, y, z, q_x, q_y, q_z, q_w]^T \tag{18}$$

where $\mathbf{t}_t$ expresses the position at time t of the UAV in the world coordinate frame, and $\mathbf{q}_t$ is the unitary quaternion that aligns the robot to the world reference frame at time t. Using quaternions increases (in one) the number of parameters for the orientation with respect to Euler angles, but simplifies the algebra and hence, the error propagation. However, the quaternion normalization has to be taken into account after the prediction and update stages.

Landmarks will be represented by their 3D cartesian position in the World Frame $\mathbf{y}_n$. Thus, the state vector $\mathbf{x}_t$ is composed by the robot pose $\mathbf{p_t}$ and the set of current landmarks $\{\mathbf{y}_1, ..., \mathbf{y}_n\}$ so:

$$\mathbf{x}_t = \left[\mathbf{p}_t^T, \mathbf{y}_1^T, ..., \mathbf{y}_r^-\right]^T \tag{19}$$

3.2 Prediction Stage

Given the pose at time $t - 1$, the odometer provides the translation with respect to the previous position (expressed in the $t - 1$ frame) and the rotation that transforms the previous orientation into the new one (expressed in the t frame). Taking into account the quaternions algebra, the state vector at time t can be computed as:

$$\mathbf{t}_t = \mathbf{t}_{t-1} + \mathbf{q}_{t-1} \otimes \mathbf{t}_u \otimes \mathbf{q}_{t-1}^{-1} \tag{20}$$

$$\mathbf{q}_t = \mathbf{q}_{u}^{-1} \otimes \mathbf{q}_{t-} \tag{21}$$

where $\mathbf{t}_u$ and $\mathbf{q}_u$ represent the estimated translation and rotation from the odometer, and $\otimes$ denotes quaternion multiplication. Notice that prediction does not affect the landmark position because they are assumed to be motionless.

Computing the odometry requires to carry out the image processing between consecutive images detailed in Section 2: feature tracking, homography estimation and, finally, odometry. The estimated translation and rotation covariance matrices are used to compute the process noise covariance matrix.

3.3 Updating Stage

From the whole set of features provided by the feature tracking algorithm used in the prediction stage, a small subset is selected to act as landmarks. The features associated to the landmarks are taken apart and not used for the homography estimation in order to eliminate correlations among prediction and updating. Thus, the number of landmarks must be a compromise between the performance of the EKF and the performance of the homography estimation (and thus, the odometry estimation). In addition, the computational requirements of the full SLAM approach has to be considered.

Experimental results allowed the authors to properly tune the number of landmarks and features used in the approach. A set of one hundred features are tracked from one image to another, and a subset of ten/fifteen well distributed and stable features are used as landmarks. Therefore, for each new image, the new position of this subset of features will be given by the feature tracking algorithm; this information will be used as measurement at time t, $\mathbf{z}_t$.

If the prediction stage was correct, the projection of each landmark into the camera would fit with the estimated position of the feature given by the tracking algorithm. If the landmark $\mathbf{y}_n$ corresponds to the image feature $\mathbf{m}_n = [u, v]$, following the camera projection model (Fig. 7):

$$\tilde{\mathbf{m}}_n = \mathbf{A}\left(\mathbf{q}_t^{-1} \otimes (\mathbf{y}_n - \mathbf{t}_t) \otimes \mathbf{q}_t\right) \tag{22}$$

Fig. 7 Projection of a landmark into the camera. The landmark is represented by a *black dot*, the translation of the camera focal (F) in the world frame (W) is $\mathbf{t}_t$, the back-projection of the feature $\mathbf{m}_n$ is $\tilde{\mathbf{m}}_n$ and the position of the landmark in the world frame is $\mathbf{y}_n$

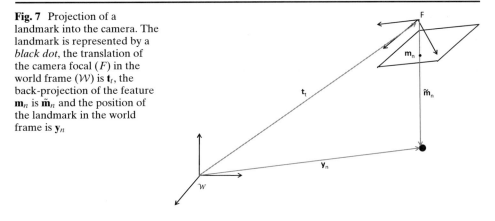

where $\mathbf{A}$ is the camera calibration matrix and $\tilde{\mathbf{m}}_n = [\tilde{u}, \tilde{v}, h]$, so the feature position is computed as $\mathbf{m}_n = [\tilde{u}/h, \tilde{v}/h]$. This measurement equation is applied to all the features correctly tracked from the previous image to the current one. The data association problem is solved by means of the feature matching algorithm.

In order to bound the computational cost needed for the SLAM approach, landmarks are not stored indefinitely in the EKF filter. Instead, they are maintained for a short period of time in the filter just to avoid transient occlusions, later they are automatically marginalized out from the filter and a new feature, provided by the tracker, initialized. If the corresponding landmarks are well conditioned, the measurement equation constraints the current position and orientation of the UAV.

3.4 Filter and Landmarks Initialization

The filter state vector will be initialized to a given position and orientation. This information can be provided by external devices such as GPS and IMU, and the process covariance matrix to the corresponding error information. The position can be also initialized to zero, so the first position is assumed as the origin and the corresponding covariances are zero too. This initial position defines the World Reference Frame where the landmarks and UAV pose are expressed.

In the following, a more sophisticated method for landmark initialization is proposed. When a new image feature is selected for being a landmark in the filter, it is necessary to compute its real position in the World frame. Due to the bearing only nature of the camera, the back-projection of the feature is given by a ray defined by the camera focal point and the image of the landmark. The proposed technique takes advantage of knowing the normal to the scene plane and the distance from the UAV to the ground at a given time. With this information the ground can be locally approximated by a plane and the landmark position as the intersection of the back-projection ray with this plane, as shown in Fig. 8.

If the World frame is aligned with the camera frame, the back-projection of the feature $\mathbf{m}_n = [u, v]$ will be the ray $\mathbf{r}$ defined by:

$$\mathbf{r} : \mathbf{A}^{-1}\tilde{\mathbf{m}}_n \tag{23}$$

Fig. 8 Landmark
initialization representation

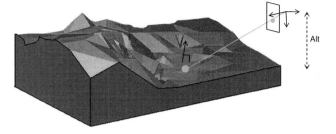

where $\mathbf{A}$ is the camera calibration matrix and $\tilde{\mathbf{m}}_n = [h\mathbf{m}_n, h]$. In addition, the odometer provides an estimation of the normal to the scene plane at time t denoted as $\mathbf{n}_t$. Given the distance to the plane d_t, the plane Π is defined as:

$$\Pi : d_t - \mathbf{n}_t^T \begin{bmatrix} x \\ y \\ z \end{bmatrix} \tag{24}$$

Then, the landmark position will be computed as the intersection of the ray $\mathbf{r}$ with the plane Π. If Eqs. 23 and 24 are merged, the value of λ can be easily computed as:

$$h = \left(\mathbf{n}_t^T \mathbf{A}^{-1} \tilde{\mathbf{m}}_n\right)^{-1} d_t \tag{25}$$

and the landmark can be computed as:

$$\mathbf{y}_n = \left(\mathbf{n}_t^T \mathbf{A}^{-1} \tilde{\mathbf{m}}_n\right)^{-1} d_t \mathbf{A}^{-1} \tilde{\mathbf{m}}_n \tag{26}$$

But this landmark is expressed in the camera coordinate frame. The UAV current position $\mathbf{d}_t$ and orientation $\mathbf{q}_t$ are finally used to express the landmark in the World frame:

$$\mathbf{y}_n = \mathbf{t}_t + \mathbf{q}_t \otimes \left(\left(\mathbf{n}_t^T \mathbf{A}^{-1} \tilde{\mathbf{m}}_n\right)^{-1} d_t \mathbf{A}^{-1} \tilde{\mathbf{m}}_n\right) \otimes \mathbf{q}_t^{-1} \tag{27}$$

There is a strong dependence of this approach on the planarity of the scene. The more planar the scene is, the better the plane approximation, leading to smaller noise in the plane normal estimation, and thus, to a better initialization.

Nevertheless, the back-projection procedure is still non-linear, and therefore, the Gaussian approximation for the errors has to be carefully considered. If the relative orientation of the ray $\mathbf{r}$ associated to a feature is near parallel with respect to the plane, the errors on the estimation can be high and a Gaussian distribution will not approximate the error shape adequately. Then, only those landmarks for which the relative orientation of the ray and the plane is higher than 30 degrees will be considered in the initialization process.

3.5 Experimental Results on Homography-Based SLAM

To test the proposed approach, experiments with the HERO helicopter were carried out. The image sequence was gathered at 15 m of altitude with respect to the ground and with the camera pointed 45 degrees with respect to the helicopter horizontal.

It is important to remark that no close-loop was carried out during the experiment, although there are some loops present in the UAV trajectory, this subject is out of the scope of this research work. Therefore, the result can be improved if a reliable data association algorithm is used for detecting and associating landmarks in the filter. The complete size of the trajectory is about 90 m long.

IMU information is used to express the results in the same frame than DGPS measurements. The results of the experiment are shown in Fig. 9, where the estimation

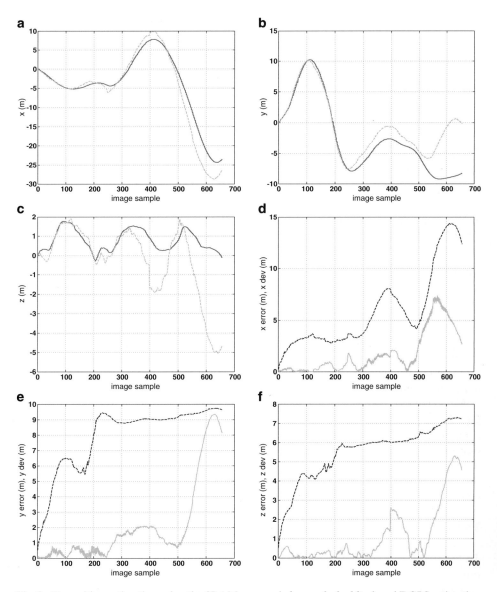

Fig. 9 *Up* position estimation using the SLAM approach (*green dashed line*) and DGPS estimation (bi). *Down* error of the SLAM approach (*green solid line*) and estimated standard deviation (*blue dashed line*)

Fig. 10 XY position estimation using the SLAM approach (*green dashed line*) and DGPS estimation (*red solid line*)

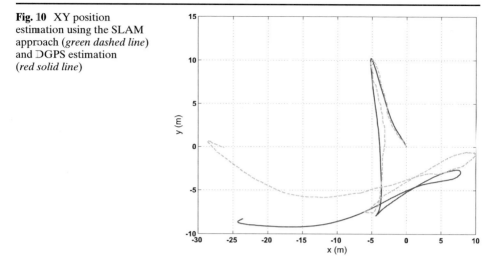

in each axis and the errors (with respect to DGPS outputs) are plotted. It can be seen how the uncertainty estimation is coherent with the measured errors. However, the position slowly diverges through time due to the absence of large loop closing. The instant orientation is not plotted because it is inherently taken into account in the computation of position. More details are shown in Fig. 10, where the XY DGPS trajectory is plotted together with the XY estimation.

3.6 Experimental Results Including an Inertial Measurement Unit

The errors shown in Fig. 9 are partially generated by a drift in the estimation of the UAV orientation. If the measurements of an inertial measurement unit (IMU) are incorporated into the SLAM approach, the errors introduced by the orientation estimation can be reset, and then the localization could be improved.

Fig. 11 XY position estimation using the SLAM approach with IMU corrections (*green dashed line*) and DGPS estimation (*red solid line*)

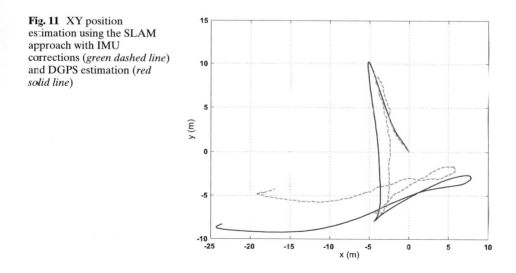

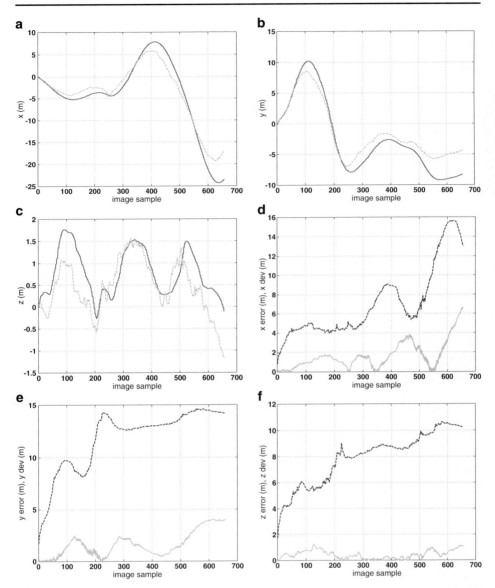

Fig. 12 *Up* position estimation using the SLAM approach with IMU corrections (*green dashed line*) and DGPS estimation (*red solid line*). *Down* error of the SLAM approach with IMU corrections (*green solid line*) and estimated standard deviation (*blue dashed line*)

The proposed SLAM approach can be easily adapted to include the IMU information by integrating its measurement in the prediction stage of the EKF. The IMU provides the complete orientation, so there is no error integration and it is bounded by the accuracy of the device.

This approach has been tested with the same data set. The XY estimation is plotted in Fig. 11. Figure 12 shows the estimation compared to the DGPS measurement. It can be seen that the errors in Z and Y are significantly smaller while in X are slightly smaller with respect to the approach without considering the IMU.

4 Conclusions

The paper presents contributions to the vision-based navigation of aerial vehicles. It is proposed a visual odometry system for UAVs based on monocular imagery. Homographic models and homography decomposition are used to extract the real camera motion and the normal vector to the scene plane. A range sensor is used to obtain the scale factor of the motion. The paper shows the feasibility of the approach through experimental results with real UAVs.

An important aspect of the proposed odometry approach is the use of natural landmarks instead of beacons or visual references with known positions. A general-purpose feature tracking is used for this purpose. Although natural landmarks increase the applicability of the proposed techniques, they also increase the complexity of the problem to be solved. In fact, outlier rejection and robust homography estimation are required.

The paper also proposes a localization technique based on monocular vision. An Extended Kalman filter-based SLAM is successfully used to compute the localization and mapping. Two basic contributions to SLAM with UAVs are proposed. First, the use of a vision based odometry as main motion hypothesis for the prediction stage of the Kalman filter and, second, a new landmark initialization technique that exploits the benefits of estimating the normal to the scene plane. Both techniques are implemented and validated with a real UAV.

Although no large loops are closed in the experiments, the estimated position and covariance are coherent, so the result could be improved if a reliable data association algorithm is used for detecting and associating landmarks in the filter.

Future developments will consider different features with better invariance characteristics in order to close loops. It should be pointed out that the method can be applied to piece-wise planar scenes, like in urban scenarios

Acknowledgements The authors would like to thank the Technical University of Berlin for providing images to validate the homography-based odometry approach. In addition, the authors thank the support of Victor Vega, Fran Real and Ivan Maza during the experiments with HERO helicopter.

References

1. Amidi, O., Kanade, T., Fujita, K.: A visual odometer for autonomous helicopter fight. In: Proceedings of the Fifth International Conference on Intelligent Autonomous Systems (IAS-5), June 1998
2. Betge-Brezetz, S., Hebert, P., Chatila, R., Devy, M.: Uncertain map making in natural environments. In: Proceedings of the IEEE International Conference on Robotics and Automation, vol. 2, pp. 1048–1053, April 1996
3. Betke, M., Gurvits, L.: Mobile robot localization using landmarks. IEEE Trans. Robot. Autom. **13**, 251–263 (1997)

4. Byrne, J., Cosgrove, M., Mehra, R.: Stereo based obstacle detection for an unmanned air vehicle. In: Proceedings 2006 IEEE International Conference on Robotics and Automation, pp. 2830–2835, May 2006
5. Caballero, F., Merino, L., Ferruz, J., Ollero, A.: A visual odometer without 3D reconstruction for aerial vehicles. applications to building inspection. In: Proceedings of the International Conference on Robotics and Automation, pp. 4684–4689. IEEE, April 2005
6. Caballero, F., Merino, L., Ferruz, J., Ollero, A.: Improving vision-based planar motion estimation for unmanned aerial vehicles through online mosaicing. In: Proceedings of the International Conference on Robotics and Automation, pp. 2860–2865. IEEE, May 2006
7. Caballero, F., Merino, L., Ferruz, J., Ollero, A.: Homography based Kalman filter for mosaic building. applications to UAV position estimation. In: IEEE International Conference on Robotics and Automation, pp. 2004–2009, April 2007
8. Conte, G., Doherty, P.: An integrated UAV navigation system based on aerial image matching. In: Proceedings of the IEEE Aerospace Conference, pp. 1–10 (2008)
9. Corke, P.I., Sikka, P., Roberts, J.M.: Height estimation for an autonomous helicopter. In: Proceedings of ISER, pp. 101–110 (2000)
10. Davison, A.: Real-time simultaneous localisation and mapping with a single camera. In: IEEE International Conference on Computer Vision, pp. 1403–1410, October 2003
11. Deans, M., Hebert, M.: Experimental comparison of techniques for localization and mapping using a bearings only sensor. In: Proceedings of the Seventh International Symposium on Experimental Robotics, December 2000
12. Demonceaux, C., Vasseur, P., Pegard, C.: Omnidirectional vision on UAV for attitude computation. In: Proceedings 2006 IEEE International Conference on Robotics and Automation, pp. 2842–2847, May 2006
13. Dickmanns, E.D., Schell, F.R.: Autonomous landing of airplanes using dynamic machine vision. In: Proc. of the IEEE Workshop Applications of Computer Vision, pp. 172–179, December 1992
14. Feder, H.J.S., Leonard, J.J., Smith, C.M.: Adaptive mobile robot navigation and mapping. Int. J. Rob. Res. **18**(7), 650–668 (1999) July
15. Garcia-Pardo, P.J., Sukhatme, G.S., Montgomery, J.F.: Towards vision-based safe landing for an autonomous helicopter. Robot. Auton. Syst. **38**(1), 19–29 (2001)
16. Hartley, R.I., Zisserman, A.: Multiple View Geometry in Computer Vision, 2nd edn. Cambridge University Press (2004)
17. Hrabar, S., Sukhatme, G.S.: Omnidirectional vision for an autonomous helicopter. In: Proceedings of the International Conference on Robotics and Automation, vol. 1, pp. 558–563 (2003)
18. Hygounenc, E., Jung, I.-K., Soueres, P., Lacroix, S.: The autonomous blimp project of LAAS-CNRS: achievements in flight control and terrain mapping. Int. J. Rob. Res. **23**(4-5), 473–511 (2004)
19. Kim, J., Sukkarieh, S.: Autonomous airborne navigation in unknown terrain environments. IEEE Trans. Aerosp. Electron. Syst. **40**(3), 1031–1045 (2004) July
20. Kwok, N.M., Dissanayake, G.: An efficient multiple hypothesis filter for bearing-only SLAM. In: Proceedings of the 2004 IEEE/RSJ International Conference on Intelligent Robots and Systems, vol. 1, pp. 736–741, October 2004
21. Lacroix, S., Jung, I.K., Soueres, P., Hygounenc, E., Berry, J.P.: The autonomous blimp project of LAAS/CNRS - current status and research challenges. In: Proceeding of the International Conference on Intelligent Robots and Systems, IROS, Workshop WS6 Aerial Robotics, pp. 35–42. IEEE/RSJ (2002)
22. Langedaan, J., Rock, S.: Passive GPS-free navigation of small UAVs. In: Proceedings of the IEEE Aerospace Conference, pp. 1–9 (2005)
23. Lemaire, T., Lacroix, S., Solà, J.: A practical 3D bearing only SLAM algorithm. In: Proceedings of the IEEE/RSJ International Conference on Intelligent Robots and Systems, pp. 2449–2454 (2005)
24. Leonard, J.J., Durrant-Whyte, H.F.: Simultaneous map building and localization for an autonomous mobile robot. In: Proceedings of the IEEE/RSJ International Workshop on Intelligent Robots and Systems, pp. 1442–1447, November 1991
25. Ling, L., Ridley, M., Kim, J.-H., Nettleton, E., Sukkarieh, S.: Six DoF decentralised SLAM. In: Proceedings of the Australasian Conference on Robotics and Automation (2003)
26. Mahony, R., Hamel, T.: Image-based visual servo control of aerial robotic systems using linear image features. IEEE Trans. Robot. **21**(2), 227–239 (2005)
27. Mejías, L., Saripalli, S., Campoy, P., Sukhatme, G.S.: Visual servoing of an autonomous helicopter in urban areas using feature tracking. J. Field. Robot. **23**(3–4), 185–199 (2006)

28. Montiel, J., Civera J, Davison, A.: Unified inverse depth parametrization for monocular SLAM. In: Robotics: Science and Systems, August 2006
29. Ollero, A., Ferruz, J., Caballero, F., Hurtado, S., Merino, L.: Motion compensation and object detection for autonomous helicopter visual navigation in the COMETS system. In: Proceedings of the International Conference on Robotics and Automation, ICRA, pp. 19–24. IEEE (2004)
30. Ollero, A., Merino, L.: Control and perception techniques for aerial robotics. Annu. Rev. Control, Elsevier (Francia), **28**, 167–178 (2004)
31. Papadopoulo, T., Lourakis, M.I.A.: Estimating the jacobian of the singular value decomposition: theory and applications. In: Proceedings of the 2000 European Conference on Computer Vision, vol. 1, pp. 554–570 (2000)
32. Proctor, A.A., Johnson, E.N., Apker, T.B.: Vision-only control and guidance for aircraft. J. Field. Robot. **23**(10), 863–890 (2006)
33. Remuss, V., Musial, M., Hommel, G.: Marvin - an autonomous flying robot-bases on mass market. In: International Conference on Intelligent Robots and Systems, IROS. Proceedings of the Workshop WS6 Aerial Robotics, pp. 23–28. IEEE/RSJ (2002)
34. Saripalli, S., Montgomery, J.F., Sukhatme, G.S.: Visually guided landing of an unmanned aerial vehicle. IEEE Trans. Robot. Autom. **19**(3), 371–380 (2003) June
35. Saripalli, S., Sukhatme, G.S.: Landing on a mobile target using an autonomous helicopter. In: Proceedings of the International Conference on Field and Service Robotics, FSR, July 2003
36. Shakernia, O., Vidal, R., Sharp, C., Ma, Y., Sastry, S.: Multiple view motion estimation and control for landing an aerial vehicle. In: Proceedings of the International Conference on Robotics and Automation, ICRA, vol. 3, pp. 2793–2798. IEEE. May 2002
37. Srinivasan, M.V., Zhang, S.W., Garrant, M.A.: Landing strategies in honeybees, and applications to UAVs. In: Springer Tracts in Advanced Robotics, pp. 373–384. Springer-Verlag, Berlin (2003)
38. Triggs, B.: Autocalibration from planar scenes. In: Proceedings of the 5th European Conference on Computer Vision, ECCV, vol. 1, pp. 89–105. Springer-Verlag, London, UK (1998)
39. Tsai, R.Y., Huang, T.S., Zhu, W.-L.: Estimating three-dimensional motion parameters of a rigid planar patch, ii: singular value decomposition. IEEE Trans. Acoust. Speech Signal Process. **30**(4), 525–534 (1982) August
40. Vidal, R., Sastry, S., Kim, J., Shakernia, O., Shim, D.: The Berkeley aerial robot project (BEAR). In: Proceeding of the International Conference on Intelligent Robots and Systems, IROS, pp. 1–10. IEEE/RSJ (2002)
41. Vidal-Calleja, T., Bryson, M., Sukkarieh, S., Sanfeliu, A., Andrade-Cetto, J.: On the observability of bearing-only SLAM. In: Proceedings of the 2007 IEEE International Conference on Robotics and Automation, pp. 1050–4729, April 2007
42. Volpe, J.A.: Vulnerability assessment of the transportation infrastructure relying on the global positioning system. Technical report, Office of the Assistant Secretary for Transportation Policy, August (2001)
43. Wu, A.D., Johnson, E.N., Proctor, A.A.: Vision-aided inertial navigation for flight control. In: Proc. of AIAA Guidance, Navigation, and Control Conference and Exhibit (2005)
44. Yakimenko, O.A., Kaminer, I.I., Lentz, W.J., Ghyzel, P.A.: Unmanned aircraft navigation for shipboard landing using infrared vision. IEEE Trans. Aerosp. Electron. Syst. **38**(4), 1181–1200 (2002) October
45. Zhang, Z., Hintz, K.J.: Evolving neural networks for video attitude and hight sensor. In: Proc. of the SPIE International Symposium on Aerospace/Defense Sensing and Control, vol. 2484, pp. 383–393 (1995) April

Real-time Implementation and Validation of a New Hierarchical Path Planning Scheme of UAVs via Hardware-in-the-Loop Simulation

Dongwon Jung · Jayant Ratti · Panagiotis Tsiotras

Originally published in the Journal of Intelligent and Robotic Systems, Volume 54, Nos 1–3, 163–181.
© Springer Science + Business Media B.V. 2008

Abstract We present a real-time hardware-in-the-loop simulation environment for the validation of a new hierarchical path planning and control algorithm for a small fixed-wing unmanned aerial vehicle (UAV). The complete control algorithm is validated through on-board, real-time implementation on a small autopilot having limited computational resources. We present two distinct real-time software frameworks for implementing the overall control architecture, including path planning, path smoothing, and path following. We emphasize, in particular, the use of a real-time kernel, which is shown to be an effective and robust way to accomplish real-time operation of small UAVs under non-trivial scenarios. By seamless integration of the whole control hierarchy using the real-time kernel, we demonstrate the soundness of the approach. The UAV equipped with a small autopilot, despite its limited computational resources, manages to accomplish sophisticated unsupervised navigation to the target, while autonomously avoiding obstacles.

Keywords Path planning and control · Hardware-in-the-loop simulation (HILS) · UAV

1 Introduction

Autonomous, unmanned ground, sea, and air vehicles have become indispensable both in the civilian and military sectors. Current military operations, in particular,

D. Jung (✉) · J. Ratti · P. Tsiotras
Georgia Institute of Technology, Atlanta, GA 30332-0150, USA
e-mail: dongwon.jung@gatech.edu

J. Ratti
e-mail: jayantratti@gatech.edu

P. Tsiotras
e-mail: tsiotras@gatech.edu

depend on a diverse fleet of unmanned (primarily aerial) vehicles that provide constant and persistent monitoring, surveillance, communications, and–in some cases–even weapon delivery. This trend will continue, as new paradigms for their use are being proposed by military planners. Unmanned vehicles are also used extensively in civilian applications, such as law enforcement, humanitarian missions, natural disaster relief efforts, etc.

During the past decade, in particular, there has been an explosion of research related to the control of small unmanned aerial vehicles (UAVs). The major part of this work has been conducted in academia [3–5, 10, 13–15, 23–25], since these platforms offer an excellent avenue for students to be involved in the design and testing of sophisticated navigation and guidance algorithms [17].

The operation of small-scale UAVs brings about new challenges that are absent in their large-scale counterparts. For instance, *autonomous* operation of small-scale UAVs requires both trajectory design (planning) and trajectory tracking (control) tasks to be completely automated. Given the short response time scales of these vehicles, these are challenging tasks using existing route optimizers. On-board, real-time path planning is especially challenging for small UAVs, which may not have the on-board computational capabilities (CPU and memory) to implement some of the sophisticated path planning algorithms proposed in the literature. In fact, the effect of limited computational resources on the control design of real-time, embedded systems has only recently received some attention in the literature [1, 29]. The problem is exacerbated when a low-cost micro-controller is utilized as an embedded control computer.

Autonomous path planning and control for small UAVs imposes severe restrictions on control algorithm development, stemming from the limitations imposed by the on-board hardware and the requirement for real-time implementation. In order to overcome these limitations it is imperative to develop computationally efficient algorithms that make use of the on-board computational resources wisely.

Due to the stringent operational requirements and the hardware restrictions imposed on the small UAVs, a complete solution to fully automated, unsupervised, path planning and control of UAVs remains a difficult undertaking. Hierarchical structures have been successfully applied in many cases in order to deal with the issue of complexity. In such hierarchical structures the entire control problem is subdivided into a set of smaller sub-control tasks (see Fig. 1). This allows for a more straightforward design of the control algorithms for each modular control task. It also leads to simple and effective implementation in practice [2, 27, 28].

In this paper, a complete solution to the hierarchical path planning and control algorithm, recently developed by the authors in Refs. [16, 21, 30], is experimentally validated on a small-size UAV autopilot. The control hierarchy consists of path planning, path smoothing, and path following tasks. Each stage provides the necessary commands to the next control stage in order to accomplish the goal of the mission, specified at the top level. The execution of the entire control algorithm is demonstrated through a realistic hardware-in-the-loop (HIL) simulation environment. All control algorithms are coded on a micro-controller running a real-time kernel, which schedules each task efficiently, by taking full advantage of the provided kernel services. We describe the practical issues associated with the implementation of the proposed control algorithm, while taking into consideration the actual hardware limitations.

We emphasize the use of a real-time kernel for implementing the overall control architecture. A real-time operating system provides the user with great flexibility in building complex real-time applications [11], owing to the ease in programming, error-free coding, and execution robustness. We note in passing that currently there exist many real-time kernels employed for real-time operation of UAVs. They differ in the kernel size, memory requirements, kernel services, etc. Some of these real-time kernels can be adopted for small micro-controller processor [6, 12]. An open source real-time Linux is used for flight testing for UAVs [8]. In this work we have used the MicroC/OS-II [26], which is ideal for the small microcontroller of our autopilot.

2 Hierarchical Path Planning and Control Algorithm

In this section, we briefly describe a hierarchical path planning and control algorithm, which has been recently developed by the authors, and which takes into account the limited computational resources of the on-board autopilot.

Figure 1 shows the overall control hierarchy. It consists of path planning, path smoothing, path following, and the low level autopilot functions. At the top level of the control hierarchy, a wavelet-based, multi-resolution path planning algorithm [21, 30] is employed to compute an optimal path from the current position of

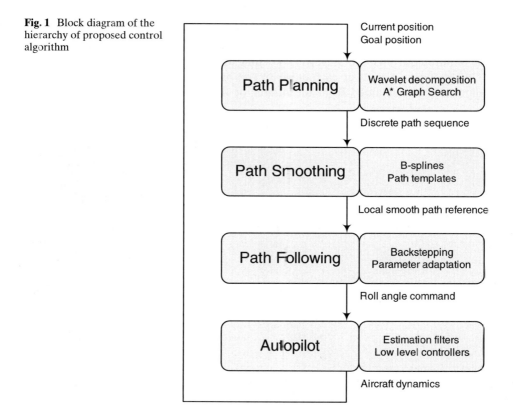

Fig. 1 Block diagram of the hierarchy of proposed control algorithm

the vehicle to the goal. The path planning algorithm utilizes a multiresolution decomposition of the environment, such that a coarser resolution is used far away from the agent, whereas fine resolution is used in the vicinity of the agent. The result is a topological graph of the environment of known a priori complexity. The algorithm then computes a path with the highest accuracy at the current location of the vehicle, where is needed most. Figure 2 illustrates this idea. In conjunction with the adjacency relationship derived directly from the wavelet coefficients [21], a discrete cell sequence (i.e., channel) to the goal destination is generated by invoking the $\mathcal{A}^*$ graph search algorithm [7, 9].

The discrete path sequence is subsequently utilized by the on-line path smoothing layer to generate a smooth reference path, which incorporates path templates comprised of a set of B-spline curves [22]. The path templates are obtained from an off-line optimization step, so that the resulting path is ensured to stay inside the prescribed cell channel. The on-line implementation of the path smoothing algorithm finds the corresponding path segments over a finite planning horizon with respect to the current position of the agent, and stitches them together, while preserving the smoothness of the composite curve.

After a local smooth reference path is obtained, a nonlinear path following control algorithm [20] is applied to asymptotically follow the reference path constructed by the path smoothing step. Assuming that the air speed and the altitude of the UAV are constant, a kinematic model is utilized to design a control law to command heading rate. Subsequently, a roll command to follow the desired heading rate is computed by taking into account the inaccurate system time constant. Finally, an autopilot with on-board sensors that provides feedback control for the attitude angles, air speed, and altitude, implements the low-level inner loops for command following to attain the required roll angle steering, while keeping the altitude and the air speed constant.

As shown in Fig. 1, at each stage of the hierarchy, the corresponding control commands are obtained from the output of the previous stage, given the initial

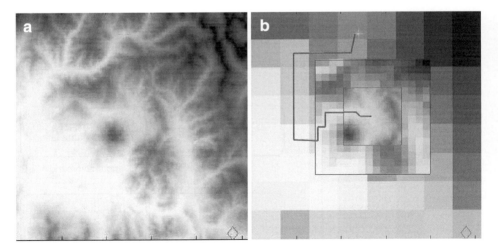

Fig. 2 Demonstration of multi-resolution decomposition of the environment (**a**) using square cells induced by the use of Haar wavelets (**b**). The current location of the agent (vehicle) is at the center of the red square (high-resolution region). The dynamics are included in the high-resolution region

environment information (e.g., a two dimensional elevation map). With the goal position specified by the user, this hierarchical control algorithm allows the vehicle to accomplish its mission of reaching the goal destination, while avoiding obstacles.

3 Experimental Test-Bed

3.1 Hardware Description

A UAV platform based on the airframe of an off-the-shelf R/C model airplane has been developed to implement the hierarchical path planning and control algorithm described above. The development of the hardware and software was done completely in-house. The on-board autopilot is equipped with a micro-controller, sensors and actuators, and communication devices that allow full functionality for autonomous control. The on-board sensors include angular rate sensors for all three axes, accelerometers along all three axes, a three-axis magnetic compass, a GPS sensor, and absolute and differential pressure sensors. An 8-bit micro-controller (Rabbit RCM-3400 running at 30 MHz with 512 KB RAM and 512 KB Flash ROM) is the core of the autopilot. The Rabbit RCM-3400 is a low-end micro-controller with limited computational throughput (as low as 7 μs for floating-point multiplication and 20 μs for computing a square root) compared to a generic high performance 32 bit micro-processor. This micro-controller provides data acquisition, data processing, and manages the communication with the ground station. It also runs the estimation algorithms for attitude and absolute position and the low-level control loops for the attitude angles, air speed, and altitude control. A detailed description of the UAV platform and the autopilot can be found in Refs. [17, 18].

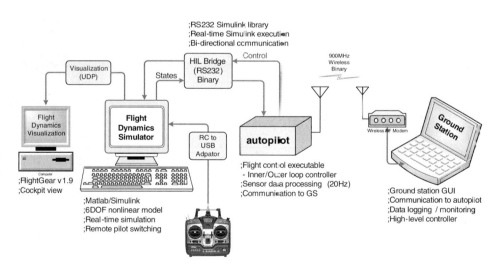

Fig. 3 High fidelity hardware-in-the-loop simulation (HILS) environment for rapid testing of the path planning and control algorithm

3.2 Hardware-in-the-Loop Simulation Environment

A realistic hardware-in-the-loop simulation (HILS) environment has been developed to validate the UAV autopilot hardware and software operations utilizing Matlab® and Simulink®. A full 6-DOF nonlinear aircraft model is used in conjunction with a linear approximation of the aerodynamic forces and moments, along with gravitational (WGS-84) and magnetic field models for the Earth. Detailed models of the sensors and actuators have also been incorporated. Four independent computers are used in the HILS, as illustrated in Fig. 3. A 6-DOF simulator, a flight visualization computer, the autopilot micro-controller, and the ground station computer console are involved in the HIL simulation. Further details about the HILS set-up can be found in Ref. [19].

4 Real-Time Software Environment

The software architecture of the on-board autopilot is shown in Fig. 4. It is comprised of several blocks, called *tasks*, which are allotted throughout different functioning layers such as the application level, the low level control, the data processing level, and the hardware level. The tasks at the hardware level, or hardware services, interact with the actual hardware devices to collect data from the sensors, communicate with the ground station, and issue commands to the DC servo motors. The

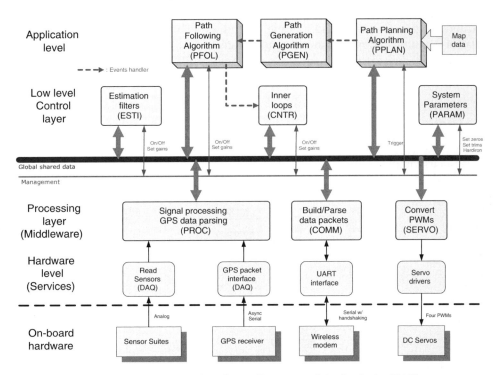

Fig. 4 Software architecture of the on-board autopilot system of the fixed-wing UAV

middleware tasks on top of the hardware services provide the abstraction of the inbound and outbound data, by supplying the processed data on a globally shared data bus or by extracting data from the global bus to the corresponding hardware services. Utilizing the processed data on the globally shared data bus, the lower level control layer achieves the basic control functions such as estimation of the attitude angles, estimation of the absolute position, and implementation of the inner loop PID controllers. Finally, three application tasks, which correspond to path planning, path generation, and path following, are incorporated to implement the hierarchical control architecture described in Section 2. The hierarchical control structure dictates all application tasks, in the sense that completion of the upper level task (event triggered) initiates the execution of a lower level task (event processed). This is shown by red dashed arrows in Fig. 4 each representing an event signal. In Fig. 4, besides exchanging the data via the global shared data bus, each task is managed by a global management bus, used for triggering execution of tasks, initializing/modifying system parameters, etc.

The task management, also called task scheduling, is the most crucial component of a real-time system. It seamlessly integrates the multiple "tasks" in this real-time software application. In practice however, a processor can only execute one instruction at a time; thus multitasking scheduling is necessitated for embedded control system implementations where several tasks need to be executed while meeting real-time constraints. Using multitasking, more than one task, such as control algorithm implementation, hardware device interfaces and so on, can appear to be executed in parallel. However, the tasks need to be prioritized based on their importance in the software flow structure so that the multitasking kernel correctly times their order of operation, while limiting any deadlocks or priority inversions.

4.1 Cooperative Scheduling Method: Initial Design

For the initial implementation, we developed a near real-time control software environment that is based predominately on the idea of cooperative scheduling. Cooperative scheduling is better explained by a large main loop containing small fragments of codes (tasks). Each task is configured to voluntarily relinquish the CPU when it is waiting, allowing other tasks to execute. This way, one big loop can execute several tasks in parallel, while no single task is busy waiting.

Like most real-time control problems, we let the main loop begin while waiting for a trigger signal from a timer, as shown by the red arrows in Fig. 5. In accordance with the software framework of Fig. 4, we classify the tasks into three groups: *routine tasks, application tasks,* and *non-periodic tasks*. The routine tasks are critical tasks required for the UAV to perform minimum automatic control. In our case these consist of the tasks of reading analog/GPS sensors (DAQ), signal processing (PROC), estimation (ESTI), inner loop control (CNTR), and servo driving (SERVO). The sampling period T_s is carefully chosen to ensure the completion of the routine tasks and allow the minimum sampling period to capture the fastest dynamics of the system. In order to attain real-time scheduling over all other tasks besides the routine tasks, a sampling period of $T_s = 50$ ms, or a sampling rate of 20 Hz, was used. On the other hand, some of the application tasks require substantial computation time and resources, as they deal with the more complicated, high level computational algorithms such as path planning (PPLAN), path generation (PGEN), and path

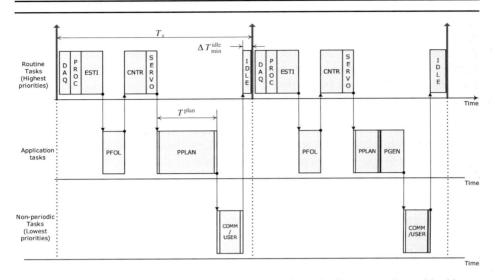

Fig. 5 A real-time scheduling method combining cooperative and naive preemptive multitasking

following (**PFOL**). In particular, the path planning algorithm in Ref. [21] turns out to have a total computation time greater than the chosen sampling period. As a result, and in order to meet the real-time constraints, we fragmentized the execution of the computationally intensive task, **PPLAN** into several slices of code execution, each with a finite execution time T^{plan}. The finite execution time is selected a priori by taking into account both T_s and the (estimated) total execution time of the routine tasks. The objective is to maximize the CPU usage to complete the task **PPLAN** as soon as possible, while meeting the criterion for real-time operation. Finally, non-periodic tasks such as communication (**COMM**) and user application (**USER**) are executed whenever the CPU becomes available, ensuring a minimum idling time of duration $\Delta T^{\text{idle}}_{\min}$ to allow the CPU to wait for other triggering signals.

Figure 6 shows a pseudo-code implementation of the cooperative scheduling scheme. Each `costate` implements the cooperative scheduling, while the `slice` statement implements the naive preemptive scheduling, which preempts the CPU over the finite execution window T^{plan}.

4.2 Preemptive Scheduling Method: Final Design

Given the a priori knowledge of the required tasks to be executed, in conjunction with an approximate knowledge of the total execution time, the use of *costate* blocks was shown to be an effective implementation of cooperative scheduling in the previous section. However, it is often a painstaking job for a programmer to meticulously synchronize and schedule all tasks in the application, many of which may have unpredictable execution time. Alternatively, it is possible to design a cooperative scheduler using conservative timing estimates for the corresponding tasks in a manner similar to that of Section 4.1. However, such an approach will result in poor performance in terms of the overall completion time. With a conservative estimate of execution times for the routine tasks, the portion allotted for the

Fig. 6 Pseudo-code implementation of the combined cooperative/preemptive scheduling scheme for the hierarchical path planning and control algorithm

```
main() {
    while (1) {
        costate {
            Wait_for_timer (T_s);
            Task DAQ;
            Task PROC;
            Task ESTI;
            if (EVENT(PFOL)) Task CNTR;
            Task SERVO;
        }
        costate {
            if (EVENT(PGEN)) Task PFOL;
        }
        costate {
            if (EVENT(PPLAN)) Task PGEN;
        }
        costate {
            Task COMM;
            Task PARAM;
            Task USER;
        }
        if (Δ T^idle > Δ T^plan) {
            slice (Δ T^plan, Task PPLAN);
        }
    }
}
```

execution of the computationally expensive tasks remains fixed regardless whether the CPU remains idle for the rest of the sampling period. This implies that the CPU does not make full use of its capacity, thus delaying the execution of the overall tasks by a noticeable amount of time. The throughput of the computationally intensive tasks may be improved by employing a *preemptive multitasking scheduler* [26]. Since the kernel has full access to CPU timing, it can allot the CPU resources to the lower level tasks whenever the higher level tasks relinquish their control. This effectively minimizes the CPU idle time and reduces the task completion time. In the next section we present an alternative framework for implementing the hierarchical path planning and control algorithm shown in Section 2 using a preemptive real-time kernel, namely, MicroC/OS-II.

The MicroC/OS-II is known to be a highly portable, low on memory, scalable, preemptive, real-time operating system for small microcontrollers [26]. Besides being a preemptive task scheduler which can manage up to 64 tasks, the MicroC/OS-II also provides general kernel services such as semaphores, including mutual exclusion semaphores, event flags, message mailboxes, etc. These services are especially helpful for a programmer to build a complex real-time software system and integrate tasks seamlessly. Its use also simplifies the software structure by utilizing a state flow concept. The MicroC/OS-II allows small on-chip code size of the real-time kernel. The code size of the kernel is no more than approximately 5 to 10 kBytes [26], adding a relatively small overhead (around 5.26%) to the current total code size of 190 kBytes for our application.

4.2.1 Real-time software architecture

Real-time software programming begins with creating a list of tasks. In this work we emphasize the real-time implementation of the path planning and control algorithm using HILS. This requires new tasks to deal with the additional HILS communication. The simulator transmits the emulated sensor data to the micro-controller via serial communication. Hence, the sensor/GPS reading task (**DAQ**) is substituted with the sensor data reading task (**HILS_Rx**), which continuously checks a serial buffer for incoming communication packets. Similarly, the servo driving task (**SERVO**) is replaced by the command writing task (**HILS_Tx**), which sends back PWM servo commands to the simulator. On the other hand, the communication task **COMM** is subdivided into three different tasks according to their respective roles, such as a downlink task for data logging (**COMM_Tx**), an uplink task for user command (**COMM_Rx**), and a user command parsing task (**COMM_Proc**). In addition, we create a path management (**PMAN**) task which coordinates the execution of the path planning and control algorithm, thus directly communicating with **PPLAN**, **PGEN**, and **PFOL**, respectively. Finally, a run-time statistics checking task (**STAT**) is created in order to obtain run-time statistics of the program such as CPU usage and the execution time of each task. These can be used to benchmark the performance of the real-time kernel. Table 1 lists all tasks created in the real-time kernel.

The MicroC/OS-II manages up to 64 distinct tasks, the priorities of which must be uniquely assigned. Starting from zero, increasing numbers impose lower priorities to be assigned to corresponding tasks. In particular, because the top and bottom ends of the priority list are reserved for internal kernel use, application tasks are required to have priorities other than a priority level in this protected range. Following an empirical convention of priority assignment, we assign the critical tasks with high priorities because they usually involve direct hardware interface. In order to minimize degradation of the overall performance of the system, the hardware

Table 1 List of tasks created by the real-time kernel

ID	Alias	Description	Priority
1	HILS_Tx	Sending back servo commands to the simulator	11
2	HILS_Rx	Reading sensor/GPS packets from the simulator	12
3	COMM_Rx	Uplink for user command from the ground station	13
4	COMM_Proc	Parsing the user command	14
5	ESTI_Atti	Attitude estimation task	15
6	ESTI_Nav	Absolute position estimation task	16
7	CNTR	Inner loop control task	17
8	PFOL	Nonlinear path following control task	18
9	COMM_Tx	Downlink to the ground station	19
10	PGEN	Path generation task using B-spline templates	20
11	PMAN	Control coordination task	21
13	PPLAN	Multiresolution path planning task	23
12	STAT	Obtaining run-time statistics	22

related tasks may need proper synchronization with the hardware, hence demanding immediate attention. It follows that routine tasks that are required for the UAV to perform minimum automatic control such as ESTI_Atti, ESTI_Nav, and CNTR are given lower priorities. Finally, application-specific tasks such as PFOL, PGEN, and PPLAN are given even lower priorities. This implies that these tasks can be activated whenever the highest priority tasks relinquish the CPU. Table 1 shows the assigned priority for each task. Note that the task COMM_Tx is assigned with a lower priority because this task is less critical to the autonomous operation of the UAV.

Having the required tasks created, we proceed to design a real-time software framework by establishing the relationships between the tasks using the available kernel services: A semaphore is utilized to control access to a globally shared object in order to prevent it from being shared indiscriminately by several different tasks. Event flags are used when a task needs to be synchronized with the occurrence of multiple events or relevant tasks. For inter-task communication, a mailbox is employed to exchange a message in order to convey information between tasks.

Figure 7 illustrates the overall real-time software architecture for the autopilot. In the diagram two binary semaphores are utilized for two different objects corresponding to the wireless modem and a reference path curve, respectively. Any task that requires getting access to those objects needs to be blocked (by semaphore pending) until the corresponding semaphore is either non-zero or is released (by semaphore posting). Consequently, only one task has exclusive access to the objects at a time, which allows data compatibility among different tasks. The event flags are posted by the triggering tasks and are consumed by the pending tasks, allowing synchronization of two consecutive tasks. Note that an event from the HILS_Rx triggers a chain of routine tasks for processing raw sensor data (ESTI_Atti, ESTI_Nav) and control implementation (CNTR). A global data storage is used to hold all significant variables that can be referenced by any task, while the global flags declaration block contains a number of event flag groups used for synchronization of tasks. Each mailbox can hold a byte-length message which is posted by a sender task with information indicated next to the data flow arrow symbol. Each task receiving the message will empty the mailbox and wait for another message to be posted. These mailboxes are employed to pass the results of one task to another task. It should be noted that when the task PMAN triggers the task PPLAN, the results of subsequent tasks are transmitted via mailboxes in the following order: (PPLAN $\to$ PGEN $\to$ PMAN $\to$ PFOL).

4.2.2 Benefits of using a real-time kernel

Robustness: The real-time kernel provides many error handling capabilities during deadlock situations. We have been able to resolve all possible deadlocks using the timing features of the Semaphore-Pend or Flag-Pend operations. The kernel provides time-out signals in the semaphore and flag calls with appropriate generated errors. These are used to handle the unexpected latency or deadlock in the scheduling operation.

Flexibility and ease of maintenance: The entire architecture for the autopilot software has been designed, while keeping in mind the object-oriented requirements for an applications engineer. The real-time kernel provides an easy and natural way to achieve this goal. The architecture has been designed to keep the code flexible enough to allow adding higher level tasks, like the ones required to process

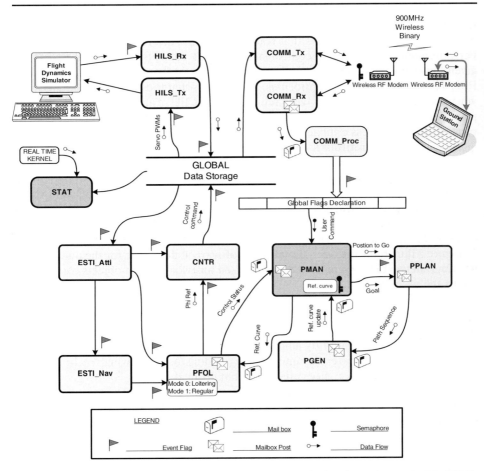

Fig. 7 The complete real-time software architecture for the path planning and control of the UAV

and execute the multi-resolution wavelet path planning algorithm. All this can be achieved without engrossing into the system level intricacies of handling and programming a microcontroller/microprocessor. The flexibility of the architecture also makes it extremely efficient to debug faults in low-level, mid-level or high-level tasks, without having to re-code/interfere with the other tasks.

5 Hardware-in-the-Loop Simulation Results

In this section we present the results of the hierarchical path control algorithm using a small micro-controller in a real-time HILS environment. The details of the implementation are discussed in the sequel.

5.1 Simulation Scenario

The environment is a square area containing actual elevation data of a US state, of dimension 128×128 units, which corresponds to 9.6×9.6 km. Taking into account the available memory of the micro-controller, we choose the range and granularity of the fine and coarse resolution levels. Cells at the fine resolution have dimensions 150×150 m, which is larger than the minimum turning radius of the UAV. The minimum turn radius is approximately calculated for the vehicle flying at a constant speed of $V_T = 20$ m/s with a bounded roll angle of $|\phi| \leq 30°$, resulting in a minimum turn radius of $R_{\min} \approx 70$ m.

The objective of the UAV is to generate and track a path from the initial position to the final position while circumventing all obstacles above a certain elevation threshold. Figure 8 illustrates the detailed implementation of the proposed path planning and control algorithm. Initially, the UAV is loitering around the initial position p_0 until a local path segment from p_0 to p_a is computed (Step A,B). Subsequently, the path following controller is engaged to follow the path (Step C,D). At step D, the UAV replans to compute a new path from the intermediate location

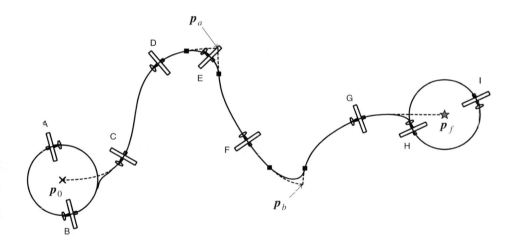

Step	Task description
A	Initially, the UAV is loitering around the initial position with the circle radius R_l
B	Calculate the first path segment from p_0 to p_a
C	Break away from the loitering circle, start to follow the first path segment
D	Calculate the second segment from p_a to p_b, and a transient path
E	UAV is on the transient path
F	Calculate the third path segment, and a transient path
G	UAV is approaching the goal position, no path is calculated
H	The goal is reached, end of the path control, get on the loitering circle
I	UAV is loitering around the goal position p_f

Fig. 8 Illustration of the real-time implementation of the proposed hierarchical path planning and control algorithm

Fig. 9 HILS results of the hierarchical path planning and control implementation. The *plots on the right* show the close-up view of the simulation. At each instant, the channel where the smooth path segment from the corresponding path template has to stay, is drawn by polygonal lines. The actual path followed by the UAV is drawn on top of the reference path. **a** $t = 64.5$ s. **b** $t = 333.0$ s. **c** $t = 429.0$ s. **d** $t = 591.5$ s

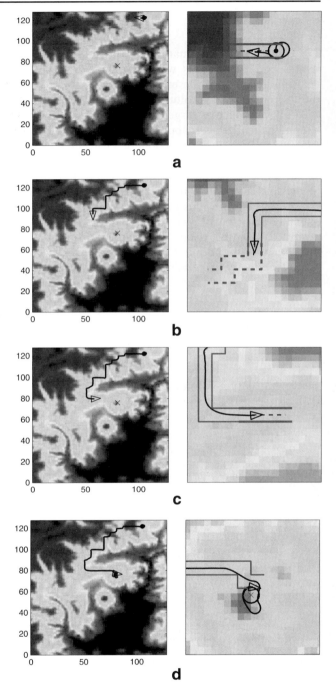

p_a to the goal, resulting in the second local path segments from p_a to p_b. The first and second path segments are stitched by a transient B-spline path assuring the continuity condition at each intersection point (marked by black squares). This process iterates

until the final position p_f is reached (Step H), when the UAV engages to loitering around the goal location.

5.2 Simulation Results

Figure 9 shows the simulation results of the hierarchical path planning and control implementation. Specifically, the plots on the right show the close-up view of the simulation. The channels are drawn by solid polygonal lines. The actual trajectory followed by the UAV is also shown. The UAV smoothly follows the reference path while avoiding all possible obstacles outside these channels. Finally, the UAV reaches the final destination in a collision-free manner, as seen in Fig. 9d.

5.3 Real-Time Kernel Run-Time Statistics

In order to evaluate the performance of the real-time software framework, we have used several distinct metrics that are available within the real-time kernel.

The amount of time for which the CPU is utilized by the kernel and task execution can be retrieved from the CPU usage metric, which is the percentage of the duty cycle of the CPU, provided by the real-time kernel at every second. When high CPU demanding tasks such as PPLAN and PGEN get activated, a higher percentage of CPU usage implies quicker completion of these tasks. Consequently, a higher percentage of CPU usage implies higher performance and efficiency. Figure 10a shows this metric during the simulation. It is noticed that for most of the time the CPU usage is about 40%, however, when a new path segment needs to be calculated (see Step D and F in Fig. 8), it goes up to nearly 100%, until the tasks PPLAN and PGEN are completed and can provide a new reference curve.

Context switches are performed when the real-time scheduler needs to switch from the current task to a different task. Each context switch is comprised of storing the current CPU register, restoring the CPU register from the stack of the new task, and resuming execution of the new task's code. Subsequently, this operation adds overhead to the existing application, which should be kept at a minimum. The number of context switches per second is provided by the real-time kernel and helps keep track of the system overhead. Although the performance of the real-time kernel should not be justified by this number, this metric can be interpreted as an indication of the health status of the real-time software. In other words, increased overhead may cause the scheduler to delay switching and make tasks be activated out of their proper timing due to this latency. If this happens, because the number of context switches represents the system overhead, hence it reveals the health status of the system. Figure 10b shows the number of context switches per second during our experiments. Given the sampling period of $T_s = 50$ ms, we notice that the average number of context switches per each sampling period is approximately seven. This indicates that the overhead is relatively small, taking into account the number of routine tasks that should be completed during each sampling period.

Individual task execution time is measured during run-time, by referencing a hook function OSTaskSwHook() when a context switching is performed. This function calculates the elapsed CPU time of the preempted task and updates a task-specific data structure with this information. Subsequently, when a statistics hook function is

Fig. 10 Real-time kernel
run-time statistics: CPU usage
(**a**) and number of context
switches per second (**b**)

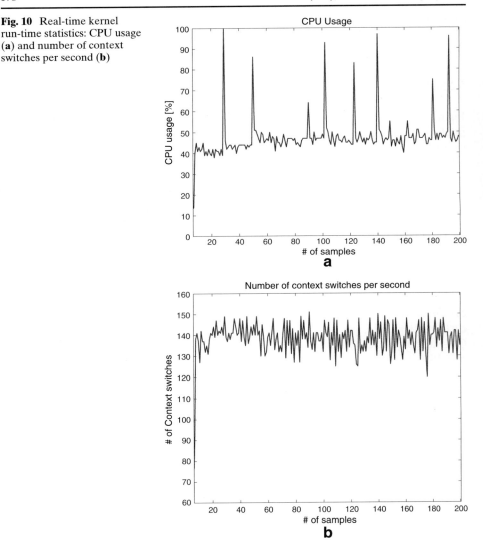

called by the real-time kernel, each task execution time $\text{TET}_i(k)$ is added together to get the total execution time of all tasks. On the other hand, the CPU ratio of different tasks, i.e., the percentage of time actually consumed by each task, is computed as follows,

$$\pi_i(k) = \frac{\text{TET}_i(k)}{\sum_{i=1}^{N} \text{TET}_i(k)} \times 100 \quad [\%], \tag{1}$$

where $\pi_i(k)$ denotes a percentage of time consumed by ith task at kth time. Figure 11a shows the run-time CPU ratio. Different colors are used to distinguish between different tasks, as depicted by the colorbar on the right side of each figure.

Fig. 11 Real-time kernel
run-time statistics: CPU ratio
over different tasks (**a**) vs.
Mean tasks execution time (**b**)

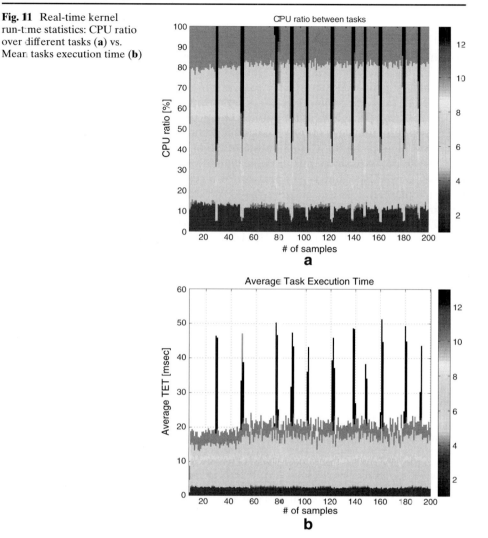

Finally, the execution time of each task over the given sampling period $T_s = 50$ ms is calculated as follows,

$$\overline{\mathrm{TET}}_i(k) = \frac{\mathrm{TET}_i(k)}{(t_k - t_{k-1})/T_s} \quad [\mathrm{ms}], \tag{2}$$

where t_k denotes the actual time of the kth statistical sample. Equation 2 basically calculates the mean value of the execution time of the ith task over one sampling period. This metric is updated at every statistical sample. As shown in Fig. 11b, this metric helps us check if the total sum of execution times of different tasks exceeds the given sampling period T_s.

6 Conclusions

We have implemented a hierarchical path planning and control algorithm of a small UAV on the actual hardware platform. Using a high fidelity HIL simulation environment, the proposed hierarchical path planning and control algorithm has been validated through the on-line, real-time implementation on a small autopilot. By integrating the control algorithms for path planning, path smoothing, and path following under two real-time/near real-time implementations, it has been demonstrated that the UAV equipped with a small autopilot having limited computational resources manages to accomplish the mission objective of reaching the goal destination, while avoiding obstacles. Emphasizing the use of a real-time kernel, we have discussed the implementation issues utilizing the kernel services, which allows flexibility of task management, robustness of code execution, etc. Provided with quantitative metrics, we have shown that the real-time kernel successfully accomplishes real-time operation of a small UAV for a non-trivial, realistic, path planning scenario.

Acknowledgements Partial support for this work has been provided by NSF award no. CMS-0510259 and NASA award no. NNX08AB94A.

References

1. Balluchi, A., Berardi, L., Di Benedetto, M., Ferrari, A., Girasole, G., Sangiovanni-Vincentelli, A.L.: Integrated control-implementation design. In: Proceedings of the 41st IEEE Conference on Decision and Control, pp. 1337–1342. IEEE, Las Vegas, December 2002
2. Beard, R.W., McLain, T.W., Goodrich, M., Anderson, E.P.: Coordinated target assignment and intercept for unmanned air vehicles. IEEE Trans. Robot. Autom. **18**, 911–922 (2002)
3. Bellingham, J., Richards, A., How, J.: Receding horizon control of autonomous aerial vehicles. In: Proceedings of the American Control Conference, pp. 3741–3745, Anchorage, May 2002
4. Bortoff, S.A.: Path planning for UAVs. In: Proceedings of the American Control Conference, pp. 364–368, Chicago, June 2000
5. Chandler, P., Pachter, M.: Research issues in autonomous control of tactical UAVs. In: Proceedings of the American Control Conference, Philadelphia, June 1998
6. Christophersen, H.B., Pickell, R.W., Neidhoefer, J.C., Koller, A.A., Kannan, S.K., Johnson, E.N.: A compact guidance, navigation, and control system for unmanned aerial vehicles. J. Aerosp. Comput. Inf. Commun. **3**(5), 187–213 (2006)
7. Gelperin, D.: On the optimality of A*. Artif. Intell. **8**(1), 69–76 (1977)
8. Hall, C.: On board flight computers for flight testing small uninhabited aerial vehicles. In: The 2002 45th Midwest Symposium on Circuits and Systems, vol. 2, pp. II-139–II-143. IEEE, August 2002
9. Hart, P., Nilsson, N., Rafael, B.: A formal basis for the heuristic determination of minimum cost paths. IEEE Trans. Sys. Sci. Cybern. **4**, 100–107 (1968)
10. Jang, J., Tomlin, C.: Autopilot design for the Stanford DragonFly UAV: validation through hardware-in-the-loop simulation. In: AIAA Guidance, Navigation, and Control Conference, AIAA 2001-4179, Montreal, August 2001
11. Jang, J., Tomlin, C.: Design and implementation of a low cost, hierarchical and modular avionics architecture for the DranfonFly UAVs. In: AIAA Guidance, Navigation, and Control Conference and Exhibit, AIAA 2002-4465, Monterey, August 2002
12. Jang, J.S., Liccardo, D.: Small UAV automation using MEMS. IEEE Aerosp. Electron. Syst. Mag. **22**(5), 30–34 (2007)
13. Jang, J.S., Tomlin, C.: Longitudinal stability augmentation system design for the DragonFly UAV using a single GPS receiver. In: AIAA Guidance, Navigation, and Control Conference, AIAA 2003-5592, Austin, August 2003

14. Jang, J.S., Tomlin, C.J.: Design and implementation of a low cost, hierarchical and modular avionics architecture for the dragonfly UAVs. In: AIAA Guidance, Navigation, and Control Conference and Exhibit, Monterey, August 2002

15. Johnson, E.N., Proctor, A.A., Ha, J., Tannenbaum, A.R.: Development and test of highly autonomous unmanned aerial vehicles. J. Aerosp. Comput. Inf. Commun. **1**, 486–500 (2004)

16. Jung, D.: Hierarchical path planning and control of a small fixed-wing UAV: theory and experimental validation. Ph.D. thesis, Georgia Institute of Technology, Atlanta, GA (2007)

17. Jung, D., Levy, E.J., Zhou, D., Fink, R., Moshe, J., Earl, A., Tsiotras, P.: Design and development of a low-cost test-bed for undergraduate education in uavs. In: Proceedings of the 44th IEEE Conference on Decision and Control, pp. 2739–2744. IEEE, Seville, December 2005

18. Jung, D., Tsiotras, P.: Inertial attitude and position reference system development for a small uav. In: AIAA Infotech at Aerospace, AIAA 07-2768, Rohnert Park, May 2007

19. Jung, D., Tsiotras, P.: Modelling and hardware-in-the-loop simulation for a small unmanned aerial vehicle. In: AIAA Infotech at Aerospace, AIAA 07-2763, Rohnert Park, May 2007

20. Jung, D., Tsiotras, P.: Bank-to-turn control for a small UAV using backstepping and parameter adaptation. In: International Federation of Automatic Control (IFAC) World Congress, Seoul, July 2008

21. Jung, D., Tsiotras, P.: Multiresolution on-line path planning for small unmanned aerial vehicles. In: American Control Conference, Seattle, June 2008

22. Jung, D., Tsiotras, P.: On-line path generation for small unmanned aerial vehicles using B-spline path templates. In: AIAA Guidance, Navigation and Control Conference, AIAA 2008-7135, Honolulu, August 2008

23. Kaminer, I., Yakimenko, O., Dobrokhodov, V., Lim, B.A.: Development and flight testing of GNC algorithms using a rapid flight test prototyping system. In: AIAA Guidance, Navigation, and Control Conference and Exhibit, Monterey, August 2002

24. Kingston, D., Beard, R.W., McLain, T., Larsen, M., Ren, W.: Autonomous vehicle technologies for small fixed wing UAVs. In: 2nd AIAA Unmanned Unlimited Conference and Workshop and Exhibit, Chicago 2003

25. Kingston, D.B., Beard, R.W.: Real-time attitude and position estimation for small UAVs using low-cost sensors. In: AIAA 3rd Unmanned Unlimited Technical Conference, Workshop and Exhibit, Chicago, September 2004

26. Labrosse, J.J.: MicroC/OS-II - The Real-Time Kernel 2nd edn. CMPBooks, San Francisco (2002)

27. McLain, T., Chandler, P., Pachter, M.: A decomposition strategy for optimal coordination of unmanned air vehicles. In: Proceedings of the American Control Conference, pp. 369–373, Chicago, June 2000

28. McLain, T.W., Beard, R.W.: Coordination variables, coordination functions, and cooperative timing missions. J. Guid. Control Dyn. **28**(1), 150–161 (2005)

29. Palopoli, L., Pinello, C., Sangiovanni-Vincentelli, A.L., Elghaoui, L., Bicchi, A.: Synthesis of robust control systems under resource constraints. In: HSCC '02: Proceedings of the 5th International Workshop on Hybrid Systems: Computation and Control, pp. 337–350. Stanford, March 2002

30. Tsiotras, P., Bakolas, E.: A hierarchical on-line path planning scheme using wavelets. In: Proceedings of the European Control Conference, pp. 2306–2812, Kos, July 2007

Comparison of RBF and SHL Neural Network Based Adaptive Control

**Ryan T. Anderson · Girish Chowdhary ·
Eric N. Johnson**

Originally published in the Journal of Intelligent and Robotic Systems, Volume 54, Nos 1–3, 183–199.
© Springer Science + Business Media B.V. 2008

Abstract Modern unmanned aerial vehicles (UAVs) are required to perform complex maneuvers while operating in increasingly uncertain environments. To meet these demands and model the system dynamics with a high degree of precision, a control system design known as neural network based model reference adaptive control (MRAC) is employed. There are currently two neural network architectures used by industry and academia as the adaptive element for MRAC; the radial basis function and single hidden layer neural network. While mathematical derivations can identify differences between the two neural networks, there have been no comparative analyses conducted on the performance characteristics for the flight controller to justify the selection of one neural network over the other. While the architecture of both neural networks contain similarities, there are several key distinctions which exhibit a noticeable impact on the control system's overall performance. In this paper, a detailed comparison of the performance characteristics between both neural network based adaptive control approaches has been conducted in an application highly relevant to UAVs. The results and conclusions drawn from this paper will provide engineers with tangible justification for the selection of the better neural network adaptive element and thus a controller with better performance characteristics.

Keywords Neural network · SHL · RBF · MRAC · Adaptive control · Comparison

R. T. Anderson (✉) · G. Chowdhary · E. N. Johnson
Department of Aerospace Engineering, Georgia Institute of Technology,
270 Ferst Drive, Atlanta GA 30332-0150, USA
e-mail: RAnderson@gatech.edu

G. Chowdhary
e-mail: Girish.Chowdhary@gatech.edu

E. N. Johnson
e-mail: Eric.Johnson@ae.gatech.edu

1 Introduction

Both the development and demand of unmanned aerial vehicle (UAV) technologies have considerably expanded over the previous decade. The creation of new micro-air vehicles, unmanned combat air vehicles, and long range surveillance aircraft has pushed the envelope of aviation, which was once confined by the physical limits of the human operator. Modern UAVs are required to perform complex maneuvers while operating in increasingly uncertain environment. As modern UAVs undergo increasingly complex operations, they demand control systems capable of accurately modeling the system dynamics with a high degree of precision.

One of the more recent approaches to control system design involves the use of neural networks for model reference adaptive control (MRAC) [1, 8–10]. This approach has been the focus of considerable research and can be seen in a wide range of recent projects, such as; the development of the F-36 tailless fighter aircraft (RESTORE) [2, 3] and Georgia Tech's Autonomous Helicopter Test-Bed (GTMAX) [4–6, 13]. A key advantage of neural network based MRAC is the neural network's capacity to model any continuous function over a compact domain to an arbitrary degree of accuracy [7, 8, 10]. Theoretically, the neural network can model all bounded and piecewise continuous nonlinear dynamics present in the UAV as well as all uncertainties in the plant, provided the appropriate control law and NN structure has been selected, and the neurons are properly tuned.

In both industry and academia, the neural network selected for adaptive control have traditionally been either a radial basis function (RBF) or single hidden layer (SHL) neural network; however, there is currently no justification for the selection of one architecture over the other. While mathematical theory can illustrate the differences between both neural networks; there have been no actual side by side comparisons conducted when the neural networks are applied to an actual controller. In this paper, a detailed comparison between both adaptive control architectures has been performed in an application that is highly relevant to the development of new UAV systems.

The performance characteristics of a longitudinal flight controller have been analyzed when both the SHL and RBF neural architectures are used in a model reference adaptive controller. The findings illustrate a number of key comparisons between both neural networks when applied to MRAC; such as the computational complexity, robustness in tuning parameters, and the variations in tracking error as various types of model uncertainties are added. As the intent of this paper is to draw a comparison between the two MRAC approaches, the plant dynamics have been selected to be both simple and relevant to practical application. The plant model used in our comparison is a simplified fixed wing platform that contains trigonometric nonlinearity. A sinusoidal function was included in the plant model, which serves to mimic the platform's phugoid mode without significantly increasing the plant complexity. To perform a thorough comparison between the RBF and SHL adaptive element characteristics, analyses have been conducted for a wide range of system dynamics. In our paper, a second order actuator has been added to the system, which introduces non-linearities to the system as dictated by the analysis while allowing the plant dynamics to remain unchanged. The actuator provides a comparison between both neural network adaptive controllers for a number of nonlinearities commonly found in actual control systems; which include time delay, viscosity effects, unmodelled coupling, and trigonometric nonlinearities.

This paper presents a detailed comparison of the performance characteristics between the RBF and HL neural networks when applied to MRAC for a system of varying complexity. The results of our findings will provide engineers with tangible justification for selecting one neural network architecture over another. The proper selection of the neural network architecture will provide an overall improvement to the performance of the controller, which will provide a boost to the effectiveness of MRAC in the latest UAV systems.

2 Adaptive Control Architecture

The RBF and SHL neural network characteristics have been analyzed through their implementation on a model reference adaptive controller. The neural network will serve as the adaptive element responsible for approximating the inversion error resulting from unmodelled higher order dynamics in the aircraft's plant model. The MRAC is a proven control architecture that has been successfully used in several flight tests as the control system for the GTMAX autonomous helicopter [4–6, 13] as well as older aircraft such as the F-15 [10]. When the adaptive element is correctly used, the MRAC will effectively provide precise and accurate responses to commanded changes that are both sudden and large. The MRAC structure in this paper, aside from the adaptive element, will remain constant to provide an impartial comparison between the RBF and SHL neural networks; only the dynamics in the actuator block will be modified to introduce uncertainties into the system.

The control architecture for the MRAC in this paper is based on the controller implemented by Johnson [1]. The key components include: a reference model, approximate dynamic inversion, P-D compensator, and the adaptive element. Figure 1

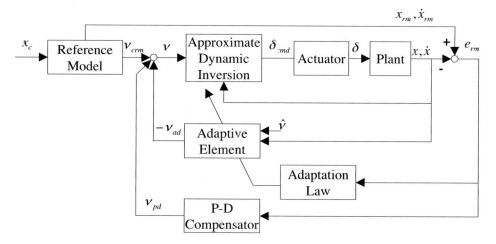

Fig. 1 Model reference adaptive controller using approximate model inversion

illustrates the controller architecture, which is capable of modeling an unstable and non-linear plant as well as account for higher order modeling uncertainties resulting from the actuator dynamics. The plant dynamics for this system are:

$$\ddot{x} = f(x, \dot{x}, \delta) \tag{1}$$

Where; $x \in \Re^2$ and $\dot{x} \in \Re^2$. Commanded input signals are represented as $x_c, \dot{x}_c$ and are fed into the reference model, where the reference model output signals are represented as $x_{rm}, \dot{x}_{rm}$ and $\ddot{x}_{rm}$. Therefore, the system model dynamics are represented as a function of the reference model output:

$$\ddot{x}_{rm} = v_{crm}(x_{rm}, \dot{x}_{rm}, x_c, \dot{x}_c) \tag{2}$$

Where v_{crm} can also be expressed as:

$$v_{crm} = f_{rm}(x_{rm}, \dot{x}_{rm}, x_c, \dot{x}_c) \tag{3}$$

The control signal that enters the approximate dynamic inversion is commonly referred to as the pseudo-control function and it is a combination of the reference model output (v_{crm}), a linear compensator (v_{pd}), and the neural network adaptive element (v_{ad}). Such that the pseudo-control is mathematically presented as:

$$v = v_{crm} + v_{pd} - v_{ad} \tag{4}$$

Ideally, the pseudo-control signal will exactly equal the system dynamics of the plant model. However, the controller deals with a non-linear system, which makes it impractical to perform an inversion for the true value. Therefore, the control signal that is generated, δ_{cmd}, will be an approximate inversion of v, represented as:

$$v = \hat{f}(x, \dot{x}, \delta_{cmd}) \tag{5}$$

The δ_{cmd} control signal will be modified by the unknown actuator dynamics, resulting in a modified control signal δ. The results in the modeling error between the pseudo control and the plant dynamics is represented as:

$$\ddot{x} = v + \Delta(x, \dot{x}, \delta) \tag{6}$$

Where the difference between the actual system model and the approximate system model is represented as:

$$\Delta(x, \dot{x}, \delta) = f(x, \dot{x}, \delta) - \hat{f}(x, \dot{x}, \delta) \tag{7}$$

And the reference model tracking model error (e) is defined as:

$$e = \begin{bmatrix} x_{rm} - x \\ \dot{x}_{rm} - \dot{x} \end{bmatrix} \tag{8}$$

While the model tracking error dynamics are found by differentiating the reference model tracking model error (e), which results in:

$$\Delta(x, \dot{x}, \delta) = f(x, \dot{x}, \delta) - \hat{f}(x, \dot{x}, \delta) \tag{9}$$

Where;

$$A = \begin{bmatrix} 0 & I \\ -K_p & -K_d \end{bmatrix}, B = \begin{bmatrix} 0 \\ I \end{bmatrix} \tag{10}$$

Provided $K_p > 0 \in \mathfrak{R}^{2\times2}$ and $K_d > 0 \in \mathfrak{R}^{2\times2}$. The model error, $\Delta(x, \dot{x}, \delta)$, will be approximated and cancelled by the adaptive element v_{ad} and can be written as:

$$\Delta(x, \dot{x}, \delta) = f(x, \dot{x}, \delta) - \hat{f}(x, \dot{x}, \delta) \tag{11}$$

Such that,

$$v_{ad} = \Delta(x, \dot{x}, \mathcal{E}) \tag{12}$$

The linear compensator used for this controller is a Proportional-Derivative compensator, which is represented as:

$$v_{pd} = \begin{bmatrix} K_p & K_d \end{bmatrix} e \tag{13}$$

The adaptive signal v_{ad} attempts to cancel the model error Δ by adapting to the model error using neural network based error parameterization. If the adaptation is exact, the dynamics of the system are reduced to that of a linear system in e. The error tracking equations that satisfy the Lyapunov equation is:

$$A^T P + PA + Q = 0 \tag{14}$$

Where $P \in \mathfrak{R}^{2n\times2n}$, $A \in \mathfrak{R}^{2\times2}$ and $Q \in \mathfrak{R}^{2\times2}$. Such that the P and Q matrices are positive definite, hence the linear system generated by the adaptive element, $\dot{x} = Ax$, is globally asymptotically stable. Therefore, the MRAC is guaranteed to be stable if a perfect parameterization of the model error can be achieved using neural networks.

3 Single Hidden Layer Neural Network

The SHL perceptron neural network is a universal approximator, because of its ability to approximate smooth non-linear functions to an arbitrary degree of accuracy. The SHL neural network contains a simple neural network architecture, which consists of an input signal that passes through a layer of neurons each containing a unique sigmoidal function (Fig. 2).

The input-output map of the SHL NN used in the adaptive element of the flight controller is represented as:

$$v_{adk} = b_w \theta_{w,k} + \sum_{j=1}^{n2} w_{j,k} \sigma_j \left(b_v \theta_{v,j} + \sum_{i=1}^{n1} v_{i,j} \bar{x}_i \right) \tag{15}$$

Where, $j = 1, \cdots, n_3$, b_w is the outer layer bias, θ_w is the kth threshold, b_v is the inner layer bias, and θ_v is the jth threshold. The sigmoid function, σ_j, contained in each neuron is represented as:

$$\theta_j(z_j) = \left(\frac{1}{1 + e^{-a_j z_j}} \right) \tag{16}$$

Where $j = 1, ..., n_3$, a is the activation potential and z is a function of the inner weight multiplied by the input signal. The activation potential will have a unique value for each hidden-layer neuron, which ensures proper functionality. The

Fig. 2 SHL neural network
architecture

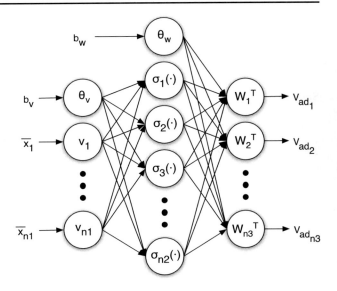

equation for the input output map can also be represen ted in a matrix form,
such that:

$$v_{adk}(W, V, \bar{x}) = W^T \sigma(V^T \bar{x}) \in \Re^{n_3 \times 1} \tag{17}$$

Provided, the following definitions are used:

$$\bar{x} = \begin{bmatrix} b_v \\ x_{in} \end{bmatrix} = \begin{bmatrix} b_v \\ \bar{x}_1 \\ \bar{x}_2 \\ \vdots \\ \bar{x}_{n1} \end{bmatrix} \in \Re^{(n_1+1) \times 1} \tag{18}$$

$$z = V^T \bar{x} = \begin{bmatrix} b_v \\ \bar{z}_1 \\ \bar{z}_2 \\ \vdots \\ \bar{z}_{n2} \end{bmatrix} \in \Re^{n_2 \times 1} \tag{19}$$

$$V = \begin{bmatrix} \theta_{v,1} & \cdots & \theta_{v,n_2} \\ v_{1,1} & \cdots & v_{1,n_2} \\ \vdots & \ddots & \vdots \\ v_{n_1,1} & \cdots & v_{n_1,n_2} \end{bmatrix} \in \Re^{(n_1+1) \times n_2} \tag{20}$$

$$W = \begin{bmatrix} \theta_{w,1} & \cdots & \theta_{w,n_2} \\ w_{1,1} & \cdots & w_{1,n_2} \\ \vdots & \ddots & \vdots \\ w_{n_1,1} & \cdots & w_{n_1,n_2} \end{bmatrix} \in \Re^{(n_1+1) \times n_2} \tag{21}$$

$$\sigma(z) = \begin{bmatrix} b_w \\ \sigma z_1 \\ \sigma z_2 \\ \vdots \\ \sigma z_4 \end{bmatrix} \in \mathfrak{R}^{\cdot n_2+1) \times 1} \tag{22}$$

At each input signal, there will be an ideal set of weights W and V such that the error between the non-linear signal and the neural network adaptation is guaranteed to be less than ε. The value where ε is bounded by $\bar{\varepsilon}$ is represented as:

$$\bar{\varepsilon} = \frac{sup}{\bar{x} \in D} ||W^T \sigma(V^T \bar{x}) - \Delta(\bar{x})|| \tag{23}$$

Where, $\Delta(\bar{x})$ is the error function being approximated and D is the compact set which contains all values for $\bar{x}$. As the number of neurons used in the hidden layer increase, the value of will theoretically become arbitrarily small. The following online adaptation laws for the weights W and V are based on the Lyapunov stability equation and are guaranteed the ultimate boundedness all input signals [1, 7].

$$\dot{W} = -\left[\left(\sigma - \sigma' V^T \bar{x} \right) + \kappa ||e|| W \right] \Gamma_W \tag{24}$$

$$\dot{V} = -\Gamma_V \left[\bar{x} \left(r^T W^T \sigma' \right) + \kappa ||e|| V \right] \tag{25}$$

Such that Γ_W and Γ_V are the global weight training weight and $r = e^T PB$.

4 Radial Basis Function Neural Network

The RBF neural network is another type of universal approximator. The RBF neural network architecture, Fig. 3, contains a vector of input signals that become modified as they are fed through a layer of neurons, each containing unique Gaussian functions.

The structure of the RBF neural network is nearly identical to the SHL neural network structure however there are several key distinctions. First, both neural networks use different functions within the hidden layer node; the SHL uses the sigmoid function while the RBF uses the Gaussian function. Second, the RBF does not have an inner weight matrix and instead distributes the input signals equally to each hidden layer node [8, 10]. Third, the Gaussian function operates within a locally bounded region [10] and must rely on nonlinear parameterization of the center location and width values to guarantee proper function approximation over a large domain [8].

This is accomplished through the use of adaptive laws based on the Lyapunov stability analysis, which allow the width and center values of each neuron in the neural network to recursively update with each time derivative. The adaptive element for a flight controller implementing a RBF neural network can be represented as:

$$v_{ad} = \hat{M}^T \hat{\phi} \tag{26}$$

Fig. 3 RBF neural network structure

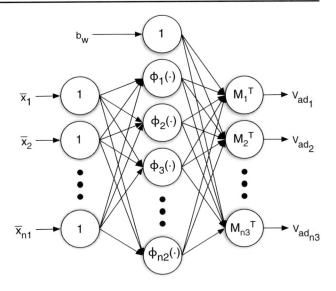

Where, $\hat{M}$ is a matrix containing the neuron weights and $\hat{\phi}$ is a column containing the neurons used in RBF. Each value of $\hat{\phi}$ is a unique Gaussian function, which will be represented as:

$$\hat{\phi}\,(\bar{x}, \xi, \eta) = exp\left(\frac{||\bar{x} - \xi||^2}{\eta^2}\right) \tag{27}$$

Where, $\bar{x}$ is the signal input, ξ is the neuron center position, and η is the neuron width. At each time increment, the values for the Gaussian functions, neuron weights, widths, and center locations are updated to ensure the neurons remain properly parameterized and continue to approximate the model error. Nardi [8] shows that the following adaptive laws for RBF neural networks in a MRAC architecture guarantee the ultimate boundedness of all signals, subject to certain initial tuning parameters:

$$\dot{\hat{\xi}} = G\left[2\left(\varsigma\,\hat{M}^T\hat{\phi}_\xi\right)^T - \lambda_\xi\left(\hat{\xi} - \xi_0\right)\right] \tag{28}$$

$$\dot{\hat{\eta}} = L\left[2\left(\varsigma\,\hat{M}^T\hat{\phi}_\eta\right)^T - \lambda_\eta\left(\hat{\eta} - \eta_0\right)\right] \tag{29}$$

$$\dot{\hat{M}} = F\left[2\left(\hat{\phi} - \hat{\phi}_\xi\hat{\xi} - \hat{\phi}_\eta\hat{\eta}\right)\varsigma - \lambda_M\left(\hat{M} - M_0\right)\right] \tag{30}$$

For the above functions: $\dot{\hat{\xi}}$ is the instantaneous time-rate change for the neurons' center positions, $\dot{\hat{\eta}}$ is the instantaneous time-rate change for the neurons' width, and $\dot{\hat{M}}$ is the instantaneous time rate of change for the neurons' weight values. The symbols G, L, and F are the center tuning value width tuning value, and the weight tuning value, respectively. The function ζ is defined as:

$$\zeta = e^T P b \tag{31}$$

With e being the approximation error between the controller and the reference model and P the positive definite solution for the Lyapunov equation in Section 2.

5 Flight Simulation

The comparison of both neural network adaptive elements has been drawn through the use of a simplified longitudinal plant model, which contains only sinusoidal nonlinearities. This simplified model is ideal for our comparison, because it excludes unwanted nonlinearities while retaining enough complexity to decently imitate the longitudinal plant model for an actual fixed wing UAV. As the intent of this paper is to compare the RBF and SHL adaptive elements across a spectrum of common nonlinear dynamics present in all aerial vehicles, a second order actuator for the aircraft's control surfaces has been added. The approximate plant model we have selected for our comparative analysis is defined as:

$$\ddot{x} = \frac{M_d \psi \sin(y)}{I} - \frac{M_t \sin(x)}{I} + \frac{M_q \dot{x}}{I} \tag{32}$$

Where M_d, M_t, M_q, I are the aircraft constants, y is the position of the control surface actuator, and ϕ is the scaling factor between the actuator and control surface positions. The unmodeled actuator dynamics are represented as:

$$\ddot{y} = \Lambda \omega_n^2 \delta - 2\beta \omega_r \dot{y} - \omega_n y \tag{33}$$

Where Λ is the actuator's DC gain, ω_n is the natural frequency, β is actuator damping, and the control signal is δ. Due to the neural network adaptive element's capacity to accurately approximate unmodeled dynamics, the plant is simple enough for the MRAC to forgo the need of a detailed approximate inversion model. Therefore, the control signal will be directly developed from the pseudo-control:

$$v = \delta_{cmd} \tag{34}$$

The flight simulation performed on the approximate plant model involves of a pair of steep pitching maneuvers. The simulation is initiated with the UAV trimmed at a pitch of 0 degrees and is presented with the following commanded inputs.

1. Pitch 45 degrees and trim for 5 s
2. Return to a 0 degree pitch and remain trimmed for 5 s
3. Pitch to 45 degrees and trim for 5 s
4. Return to a 0 degree pitch and trim for 5 s

An important factor in the selection of a neural network based adaptive element is the NN's sensitivity to variations in the complexity of the system dynamics. An ideal neural network will be exhibit limited variations in tracking error as the complexity of the system changes. Both the RBF and SHL neural networks have been analyzed using the above flight simulation as additional sinusoidal nonlinearities and exponential nonlinearities are added to the plant dynamics.

6 Analysis of Neural Network Adaptive Elements

6.1 Analysis Metric

The performance comparison between the SHL and RBF neural network adaptive elements are based on the following metrics;

1. model tracking error
2. processing time
3. tuning robustness

The model tracking error metric is an analysis of the neural networks' ability to precisely model the higher order dynamics throughout the simulation. It is found by taking the Euclidean norm of the error between the system model and the reference model signal generated by the MRAC in Section 2. The mathematical representation of the metric is:

$$Tracking\,Error = \left\lVert \sqrt{\sum |v_{ad} + \Delta_{cmd} - \ddot{x}|^2} \right\rVert \tag{35}$$

Where v_{ad} is the adaptive element control signal, and $\ddot{x}$ is the second order plant dynamic. As the tuning parameters in the RBF and SHL neural networks are modified, the tracking error of the simulation is impacted. Since there are 6 tuning parameters for the SHL neural network and 10 for the RBF neural network, it is important to identify how the parameters impact one another as well as the controller performance characteristics. This analysis was facilitated through multiple simulations using the aircraft plant discussed in Section 5. The tuning parameters for the neural network in each simulation were varied so that a sample of all possible controller configurations could be analyzed. In order to reduce the

Fig. 4 Tracking error topography for SHL adaptive element as number of neurons and control matrix training weight are modified

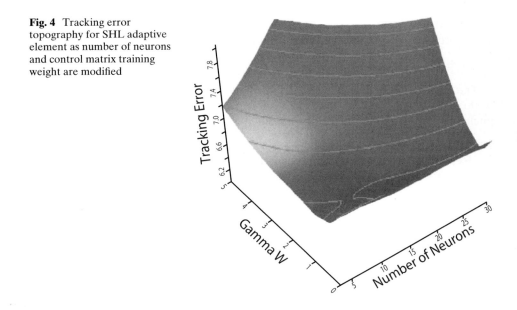

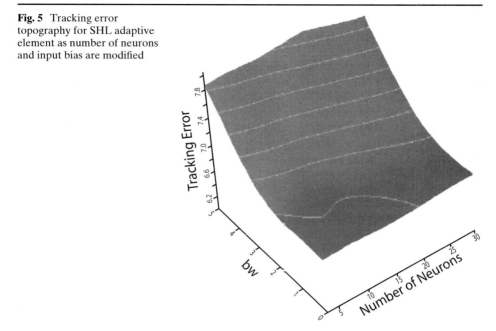

Fig. 5 Tracking error topography for SHL adaptive element as number of neurons and input bias are modified

computational complexity of the simulation, the commanded maneuver from Section 5 was simplified to implement only a 45 degree pitching maneuver lasting 5 s. The results were then analyzed using a statistical analysis tool known as JMP, which drew correlations between tuning parameters and the controller's performance characteristics. Simulation runs with tracking errors beyond 8 were removed because these values fall out of the region of optimal tracking. Runs where the number of neurons applied to the hidden layer exceeded 30 neurons were also removed. A large

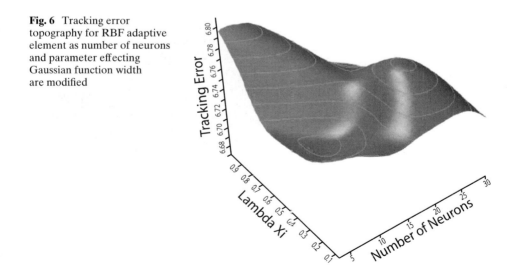

Fig. 6 Tracking error topography for RBF adaptive element as number of neurons and parameter effecting Gaussian function width are modified

Fig. 7 Tracking error
topography for RBF adaptive
element as number of neurons
and parameter effecting
Gaussian function center
location are modified

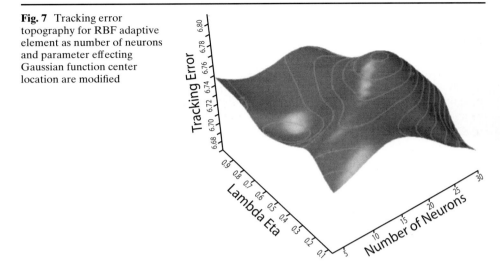

number of neurons in the hidden layer increase the processing time requirement
causing the controller to perform unfavorably in online applications.

The results of the correlation analysis between the tuning parameters and the
performance characteristics are displayed in Figs. 9 and 10. Positive values in the
correlation column indicate direct correlations, negative values indicate inverse correlations, while the magnitude indicates the strength of a correlation. Topographic
plots of the tracking error for the simulation runs were also generated. (Figures 4,
5, 6, 7, and 8) In each of the plots, the tracking error is the z-axis and the number
of neurons applied to the hidden layer is the x-axis. The plots only display simulated
runs where the tracking error remains below 8 and the y-axis ranges were selected to

Fig. 8 Tracking error
topography for RBF adaptive
element as number of neurons
and parameter effecting
control matrix training weight
are modified

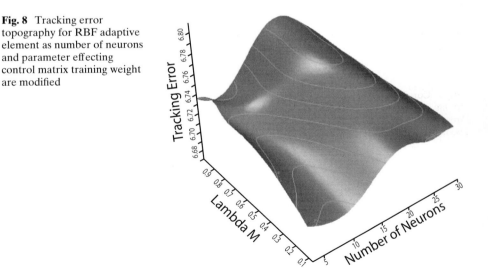

be as wide as possible without significantly impacting the tracking error. In certain instances where the tracking error was highly sensitive to a tuning parameter, the tuning parameter's range was reduced to include only values containing decent tracking error.

The processing time metric is an analysis of the update rate for the adaptive element at each time step throughout the simulation. This metric combines the time required to develop the adaptive element's control law with the time required to update the neurons and compute the adaptive element control signal, ad, and is a qualitative analysis where the values will change based on the processing power of the computer used to perform the simulation. The flight simulation using the RBF and SHL neural network were performed on computers with identical processor specifications to guarantee the simulation conditions were identical. In order to further guarantee the veracity of the results, each of the data points in Fig. 13 are the results of multiple simulation runs.

6.2 Evaluation of Simulation Results for RBF and SHL Adaptive Elements

The results of the flight simulation for the simplified plant containing a 2nd order actuator have yielded several significant findings. A striking difference between the two adaptive controllers is the shape of the tracking error topographic plots. Figures 4 and 5 illustrate the effects modifying the control matrix training weight and hidden layer bias have on the tracking error for controller configurations with a different number of neurons in the hidden layer. The simulated results show the tracking error for the SHL configuration contains a pronounced local minima region largely dependent on; the number of neurons selected, gamma W training weight, and the input bias value. This is in contrast with the tracking error topographic plot for the RBF controller, which contains several local minima. Figures 6, 7, and 8 show the variation in tracking error for the RBF controller when modifications are made to the tuning parameters for the Gaussian function width, center location, and control matrix M. As these three parameters are coupled with one another, the RBF neural network is more difficult to optimally tune than the SHL neural network. Since the RBF contains more local minima for the tracking error than the SHL NN, it is also more susceptible to overfitting errors and improper learning.

Fig. 9 Correlation between tuning parameters and tracking error for SHL neural network

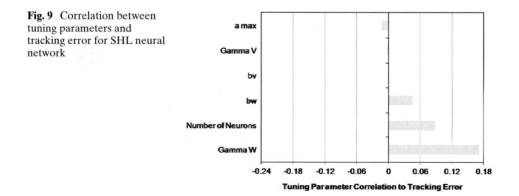

Fig. 10 Correlation between
tuning parameters and
tracking error for SHL neural
network

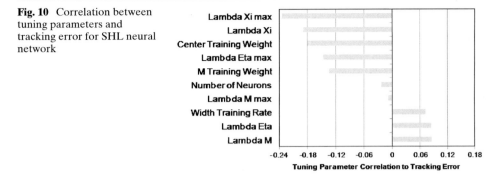

From tertiary inspection, the RBF neural network contains more tuning parameters than the SHL neural network, however this does not necessarily indicate
an increased tuning sensitivity. When the controller simulations are analyzed to
determine correlations between tuning variables and the tracking error, comparisons
between the tuning sensitivities of the RBF and SHL controllers can be drawn. An
ideal controller will contain only a few parameters with strong correlations while
a poorly-tunable controller contains multiple parameters with roughly equivalent
correlation values. Figures 9 and 10 show the impact each tuning parameter has
on the tracking error. From these results the RBF neural network is more difficult to
tune than the SHL neural network.

Another important result from the tracking error analysis is that for both neural
networks there exists an optimal value for the number of neurons used in the hidden
layer. Once this value is reached, provided optimal tuning parameters are used, there
is a negligible improvement to the tracking error as additional neurons are added.
Figures 11 and 12 are topographic plots for the tracking error of the RBF and
SHL neural networks when all tuning parameters aside from the number of neurons
and input control matrix are held constant. The y-axis for both plots was developed

Fig. 11 Norm tracking error
topography for the
approximate plant model with
second order actuator
dynamics using the SHL
adaptive element as the weight
training rate and number of
neurons are varied

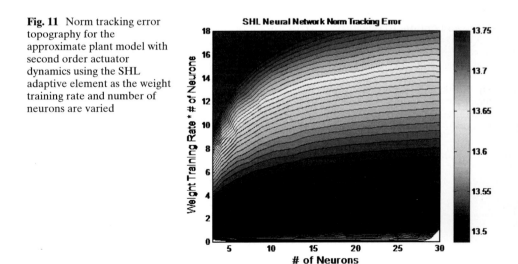

Fig. 12 Norm tracking error topography for the approximate plant model with second order actuator dynamics using the RBF adaptive element as the weight training rate and number of neurons are varied

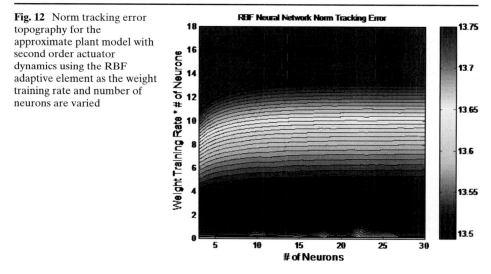

from the multiplication of the input matrix weight training weight and the number of neurons in the hidden layer. This was done to flatten the shape of the plots making them easier to interpret. The results of these plots show that the optimized tracking performance for both the RBF and SHL neural networks are nearly identical once the number of neurons used exceeded 10. Beyond this point, the addition of neurons to the hidden layer exhibited only a limited effect on reducing the tracking error.

The results of the processing time metric, Fig. 13, demonstrate the SHL neural network's improved update rate over the RBF neural network for the simplified plant containing second order actuator dynamics. As the quantity of neurons is increased for both neural networks, their processing time also increase in a linear manner. This relationship holds true as the complexity of the system is increased.

Fig. 13 Processing time required to compute v_{ad} for the SHL (*left*) and RBF (*right*) adaptive elements as additional neurons are applied

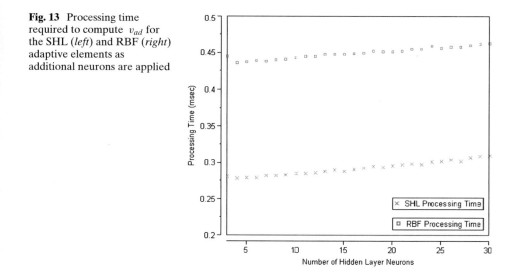

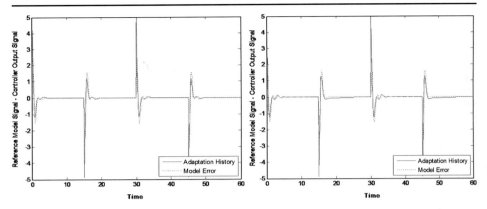

Fig. 14 Comparison of controller with optimized SHL adaptive element (*left*) and optimized RBF adaptive element (*right*) modeling the second order dynamics of the approximate longitudinal plant

When additional sinusoidal nonlinearities and exponential nonlinearities are added to the system dynamics, the processing time continues to increase in a linear manner with the number of neurons and the SHL controller maintains a faster update rate over the RBF controller.

A visual comparison of the ability for an optimally tuned RBF neural network and optimally tuned SHL neural network to model nonlinear dynamics reveals their shapes are nearly identical. Figure 14 displays side by side overlay plots of the higher order plant dynamics and the adaptive element's approximation control signal for both of the optimally tuned RBF and SHL neural networks. Figure 14 displays side by side plots of the aircraft model's pitch response for a commanded input when optimally tuned RBF and SHL neural networks are used. Overall, the optimized SHL NN was capable of modeling the tracking error of the plant model slightly better than the RBF NN (Fig. 15).

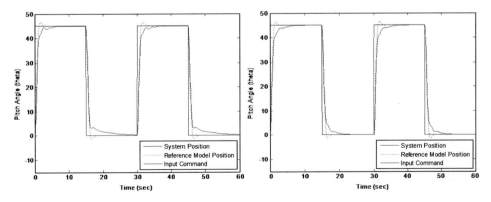

Fig. 15 Comparison of the UAV system response to a commanded input using a controller with optimized SHL adaptive element (*left*) and optimized RBF adaptive element (*right*)

7 Conclusion

The analysis of the performance characteristics for the RBF and SHL neural network yielded results that were both clear and important. Through the results of the tracking error ease of tuning, and processing time analyses, a quantifiable justification can be made for the selection of one neural network over the other. Since the performance characteristics were analyzed through the use of a simplified aircraft plant containing unmodelled 2nd order dynamics, the results developed in this paper will be similar to those found in an actual aerial vehicle.

The following is a brief restatement of the comparison results. From the analysis results for of the tracking error topography plots and parameter correlation table, the RBF controller is more difficult to tune than the SHL controller, and is additionally more susceptible to overfitting errors and improper learning. Furthermore, the control laws used to update the neuron weights of the RBF neural network result in a slower update rate than the SHL neural network as evidenced in Fig. 12. When the RBF and SHL neural networks are optimally tuned, their tracking errors are approximately identical. Based on these results the SHL neural network provides the controller with better performance characteristics and therefore makes it a better choice for the controller's adaptive element. As the performance demands for future UAVs increase, the results in this paper help to ensure engineers maximize the effectiveness of MRAC.

References

1. Johnson, E.N.: Limited authority adaptive flight control. Ph.D. Dissertation, School of Aerospace Engineering, Georgia Institute of Technology, Atlanta, GA (2000)
2. Calise, A.J., Lee, S., Sharma, M.: The development of a reconfigurable flight control law for tailless aircraft. In: AIAA Guidance Navigation and Control Conference, AIAA-2003-5741. Austin, August (2003)
3. Brinker, J.S., Wise, K.A.: Testing of reconfigurable control law on the X-36 tailless aircraft. AIAA J. Guid. Control Dyn. **24**(5), 896–902 (2001)
4. Johnson, E.N., Schrage, D.P.: The georgia tech unmanned aerial research vehicle: GTMax. In: AIAA Guidance Navigation and Control Conference, AIAA-2003-5741. Austin, August (2003)
5. Johnson, E.N., Kannan, S.K.: Adaptive trajectory control for autonomous helicopters. J. Guid. Control Dyn. **28**(3), 524–538 (2005)
6. Johnson, E.N., Proctor, A., Ha, J., Tannenbaum, A. Visual search automation for unmanned aerial vehicles. IEEE Trans. Aerosp. Electron. Syst. **41**(1), 219–232 (2005)
7. Lewis, F.L.: Nonlinear network structures for feedback control. Asian J. Control **1**(4), 205–228 (1999)
8. Nardi, F.: Neural network based adaptive algorithms for nonlinear control. Ph.D. Dissertation, School of Aerospace Engineering, Georgia Institute of Technology, Atlanta, GA (2000)
9. Johnson, E.N., Oh, S.M.: Adaptive control using combined online and background learning neural network. In: AIAA Guidance Navigation and Control Conference. Providence, August (2004)
10. Shin, Y.: Neural network based adaptive control for nonlinear dynamic regimes. Ph.D. Dissertation, School of Mechanical Engineering, Georgia Institute of Technology, Atlanta, GA (2005)
11. Hovakimyan, N., Nardi, F., Calise, A., Kim, N.: Adaptive output feedback control of uncertain nonlinear systems using single-hidden-layer neural networks. IEEE Trans. Neural Netw. **13**(6), 1420–1431 (2002)
12. Chowdhary, G., Johnson, E.: Adaptive neural network flight control using both current and recorded data. In: AIAA Guidance, Navigation, and Control Conference, AIAA-2007-6505. Hilton Head, August (2007)
13. Christophersen, H.B., Pickell, R.W., Neidhoefer, J.C., Koller, A.A., Kannan, S.K., Johnson, E.N.: A compact guidance, navigation, and control system for unmanned aerial vehicles. AIAA J. Aerosp. Comput. Inf. Commun. **3**(5), 187–213 (2006)

Small Helicopter Control Design Based on Model Reduction and Decoupling

Ivana Palunko · Stjepan Bogdan

Originally published in the Journal of Intelligent and Robotic Systems, Volume 54, Nos 1–3, 201–228.
© Springer Science + Business Media B.V. 2008

Abstract In this paper a complete nonlinear mathematical model of a small scale helicopter is derived. A coupling between input and output variables, revealed by the model, is investigated. The influences that particular inputs have on particular outputs are examined, and their dependence on flying conditions is shown. In order to demonstrate this dependence, the model is linearized in various operating points, and linear, direct and decoupling, controllers are determined. Simulation results, presented at the end of the paper, confirm that the proposed control structure could be successfully used for gain scheduling or switching control of a small scale helicopter in order to provide acrobatic flight by using simple linear controllers.

Keywords Small scale helicopter · Model reduction · Decoupling · Multivariable control

1 Introduction

Helicopters are known to be inherently unstable, nonlinear, coupled, multiple-input–multiple-output (MIMO) systems with unique characteristics. Moreover, small-scale helicopters have inherently faster and more responsive dynamics than full-scale helicopters which makes the stabilizing controller design a challenging problem. The stabilizing controller may be designed by the model-based mathematical approach, or by heuristic nonlinear control algorithms.

Due to the complexity of helicopter dynamics, linear control techniques are mostly based on model linearization. When dealing with non-aggressive flight scenarios,

I. Palunko · S. Bogdan (✉)
Laboratory for Robotics and Intelligent Control Systems,
Department of Control and Computer Engineering,
Faculty of Electrical Engineering and Computing,
University of Zagreb, Zagreb, Croatia
e-mail: stjepan.bogdan@fer.hr
URL: http://flrcg.rasip.fer.hr/

hovering, up–down, forward–backward flights, linearized models offer a technically sound alternative for controller design, testing and implementation. Different methods for model-based autonomous helicopter control have been presented in literature. In [1] control is maintained using classical PD feedback for altitude control, and as for lateral and longitudinal dynamics, design was based on successive loop closure. Other approaches are using simple nested PID loops [2] or the control logic based on low-order linear–quadratic regulator that enabled a small-scale unmanned helicopter to execute a completely automatic aerobatic maneuver [3]. In [4], a multivariable state–space control theory such as the pole placement method is used to design the linear state feedback for the stabilization of the helicopter in hover mode, because of its simple controller architecture. What connects [1–4] is that the control design is based on model linearization or model reduction. Furthermore, the coupling effect is either neglected or simply was not included in the model design. It was demonstrated that these simplifications were adequate for non-aggressive flight conditions.

Often there are comparisons between classical control theory and new control techniques. Due to that, the results of the most effective behavioural cloning approach are compared to PID modules designed for the same aircraft [5]. It has been found that behavioural cloning techniques can produce stabilizing control modules in less time than tuning PID controllers. However, performance and reliability deficits have been found to exist with the behavioural cloning, attributable largely to the time variant nature of the dynamics due to the operating environment, and the pilot actions being inadequate for teaching. In [6] a PID and fuzzy logic controller are compared. Proposed controllers were designed for the yaw, the pitch, the roll and the height variables for hover and slow flight. In both controllers an IAE criterion was used for parameter optimization, and both controllers handled flight trajectory with minimal error. System stability can not be determined from Bode plots due to neglected coupling, but it is possible that, with the initial optimized gains, the system is stable. The conclusion drawn here is that tuning PID modules remains superior to behavioural cloning for low-level helicopter automation [5] and as for the fuzzy controllers and the PID/PD controllers [6], both handled given flight trajectories with minimal error.

Nonlinear controllers are more general and cover wider ranges of flight envelopes but require accurate knowledge about the system and are sensitive to model disparities. Nonlinear H_∞ controller designed for hovering helicopter [7] can be exactly separated into two three degree-of-freedom controllers, one for translational motion and the other for rotational motion, even though longitudinal and lateral dynamics of helicopters in hovering are highly nonlinear and severely coupled. Substantial progress has been made towards designing a full envelope, decoupled helicopter flight control system and initial designs have been successfully 'flown' on the RAE Bedford flight simulator. Results had shown that the μ-synthesis controller design method [8] offers significant performance robustness to changes in the system operating point. In [9] a methodology in which neural network based controllers are evolved in a simulation, using a dynamic model qualitatively similar to the physical helicopter, is proposed. Successful application of reinforcement learning to autonomous helicopter flight is described in [10]. A stochastic, nonlinear model of the helicopter dynamics is fitted. Then the model is used to learn to hover in place, and to fly a number of maneuvers taken from an RC helicopter competition. Work presented in [11] is focused on designing practical tracking controller for a small scale helicopter following predefined trajectories. The basic linearized equations of motion, for model helicopter dynamics, are derived from the Newton–Euler equations for a rigid body that has six degrees of freedom to move in Cartesian space. The control system literature presents a variety of suitable techniques with well known strengths and weaknesses. These methods are known to produce effective controllers, but the insight of the control system and designers knowledge of the vehicle characteristics often play a major role in achieving successful results.

In this paper a complete nonlinear mathematical model of a small scale helicopter is derived. The resulting model is adjusted to a miniature model of remotely controlled helicopter. The nonlinear model is linearized and afterwards decoupled. For the controller design a linear control theory is used because of its consistent performance, well-defined theoretical background and effectiveness proven by many practitioners. Due to the tightly coupled characteristic of the model, a multivariable controller is designed ensuring decoupling and stability through all six degrees of freedom. The control algorithm is determined and its validity, as well as validity of derived model, is tested by simulation in Matlab *Simulink*.

2 Nonlinear Mathematical Model

In this section the derivation of a nonlinear mathematical model of the helicopter is presented. Six equations of motion constitute the nonlinear mathematical model of a helicopter. The aerodynamic lift of the main rotor $\overrightarrow{F}_{zR}$ is determined from aerodynamic lift of an airplane wing [12]:

$$\left\| \overrightarrow{F}_{zR} \right\| = \frac{1}{2} c_{zR} \rho_{\mathbf{air}} S_{zR} \left\| \overrightarrow{\omega}_R \right\|^2 \int_0^R \left\| \overrightarrow{r} \right\|^2 \mathrm{d}r = \frac{1}{6} c_{zR} \rho_{\mathbf{air}} S_{zR} \left\| \overrightarrow{\omega}_R \right\|^2 R^3 \tag{1}$$

$$\overrightarrow{F}_{zR} = \left\| \overrightarrow{F}_{zR} \right\| \frac{\overrightarrow{\omega}_R}{\left\| \overrightarrow{\omega}_R \right\|} \times \frac{\overrightarrow{r}}{\left\| \overrightarrow{r} \right\|} \tag{2}$$

Lift-induced drag $\overrightarrow{F}_{tR}$ is used for calculation of the main rotor revolving torque $\overrightarrow{M}_{ok}$. The expression for lift-induced drag is derived from relation given in [12]:

$$\left\| \overrightarrow{F}_{tR} \right\| = \frac{1}{2} c_{tR} \rho_{\mathbf{air}} S_{tR} \left\| \overrightarrow{\omega}_R \right\|^2 \int_0^R \left\| \overrightarrow{r} \right\|^2 \mathrm{d}r = \frac{1}{6} c_{tR} \rho_{\mathbf{air}} S_{tR} \left\| \overrightarrow{\omega}_R \right\|^2 R^3 \tag{3}$$

$$\overrightarrow{F}_{tR} = \left\| \overrightarrow{F}_{tR} \right\| \frac{\overrightarrow{\omega}_R}{\left\| \overrightarrow{\omega}_R \right\|} \times \frac{\overrightarrow{r}}{\left\| \overrightarrow{r} \right\|} \tag{4}$$

The main rotor model is designed according to Hiller system. The main rotor plane acting force is generated by blades deflected for an angle α. Dependence of lift $\overrightarrow{F}_{uz}$ on angle of attack, α, is described as:

$$\overrightarrow{F}_{uz} = c_{uz} \rho_{\mathbf{air}} \alpha \overrightarrow{\jmath}_R^2 \tag{5}$$

Torques $\vec{\tau}_{uz\alpha}$ and $\vec{\tau}_{uz\beta}$,

$$\left\|\vec{\tau}_{uz\alpha}\right\| = \sum_i \Delta \left\|\vec{\tau}_i\right\| = \left\|\vec{F}_{uz}\right\|_{\alpha} \int_0^{l_2} \left\|\vec{r}\right\| dr = \frac{1}{4} c_{uz} \rho_{\text{air}} \left\|\alpha \vec{\omega}_R\right\|^2 l_2^4 \qquad (6)$$

$$\vec{\tau}_{uz\alpha} = \left\|\vec{\tau}_{uz\alpha}\right\| \frac{\vec{\omega}_R}{\left\|\vec{\omega}_R\right\|} \times \frac{\vec{r}}{\left\|\vec{r}\right\|} \qquad (7)$$

$$\left\|\vec{\tau}_{uz\beta}\right\| = \sum_i \Delta \left\|\vec{\tau}_i\right\| = \left\|\vec{F}_{uz}\right\|_{\beta} \int_0^{l_2} \left\|\vec{r}\right\| dr = \frac{1}{4} c_{uz} \rho_{\text{air}} \beta \left\|\vec{\omega}_R\right\|^2 l_2^4 \qquad (8)$$

$$\vec{\tau}_{uz\beta} = \left\|\vec{\tau}_{uz\beta}\right\| \frac{\vec{\omega}_R}{\left\|\vec{\omega}_R\right\|} \times \frac{\vec{r}}{\left\|\vec{r}\right\|} \qquad (9)$$

produce main rotor forward and lateral tilt. These torques, are used later in this section in equations of motion.

By changing the rotational plane, the main rotor torque is changed, as well. To be able to equalize all torques, it is necessary to transform a rotated coordinate system to a non-rotated coordinate system. The effect of the main rotor in direction of change of the rotational axis is expressed through angular momentum. The general expression for angular momentum acting on a rigid body with fixed axis can be stated as:

$$\vec{L} = \sum \vec{r}_i \times \vec{p}_i \qquad (10)$$

which gives

$$\vec{L} = \mathbf{R}_{z,\psi} \mathbf{R}_{y,\theta} \mathbf{R}_{x,\varphi} I_R \vec{\omega}_R. \qquad (11)$$

The torque is defined as:

$$\vec{\tau} = \frac{d\vec{L}}{dt} = \vec{r} \times \frac{d\vec{p}}{dt} + \frac{d\vec{r}}{dt} \times \vec{p} = \vec{r}_i \times \vec{F}_i, \qquad (12)$$

from which the main rotor total torque, taking into account the change of rotational plane, is derived. As a result of the main rotor motion, two main rotations, around x and y axis, are considered (rotation around z axis is neglected). Torque generated by change of the rotational plane can be presented as:

$$\vec{\tau} = \frac{d}{dt} \left(\mathbf{R}_{y,\theta}\right) \mathbf{R}_{x,\varphi} I_R \vec{\omega}_R + \mathbf{R}_{\psi,\theta} \frac{d}{dt} \left(\mathbf{R}_{x,\varphi}\right) I_R \vec{\omega}_R + \mathbf{R}_{y,\theta} \mathbf{R}_{x,\varphi} I_R \dot{\vec{\omega}}_R \qquad (13)$$

where:

$$\mathbf{R}_{x,\varphi} = \begin{pmatrix} 1 & 0 & 0 \\ 0 & \cos\varphi & -\sin\varphi \\ 0 & \sin\varphi & \cos\varphi \end{pmatrix}, \mathbf{R}_{y,\theta} = \begin{pmatrix} \cos\theta & 0 & \sin\theta \\ 0 & 1 & 0 \\ -\sin\theta & 0 & \cos\theta \end{pmatrix}, \mathbf{R}_{z,\psi} = \begin{pmatrix} \cos\psi & -\sin\psi & 0 \\ \sin\psi & \cos\psi & 0 \\ 0 & 0 & 1 \end{pmatrix}.$$

$$\qquad (14)$$

Fig. 1 Lateral stability in y–z plane

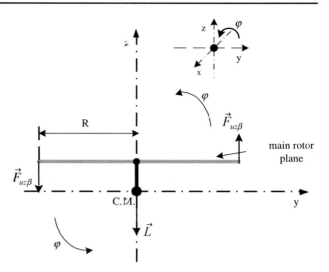

Finally, torques can be given in the following form:

$$\vec{\tau}_{Dx} = \vec{\tau} \begin{bmatrix} 1 \\ 0 \\ 0 \end{bmatrix} = I_R \, \dot{\vec{\omega}}_R \cos\varphi \sin\theta + I_R \vec{\omega}_R \left(-\dot{\varphi} \sin\varphi \sin\theta + \dot{\theta} \cos\varphi \cos\theta \right), \quad (15)$$

$$\vec{\tau}_{Dy} = \vec{\tau} \begin{bmatrix} 0 \\ 1 \\ 0 \end{bmatrix} = -I_R \dot{\vec{\omega}}_R \sin\theta - I_R \omega_R \dot{\varphi} \cos\varphi, \quad (16)$$

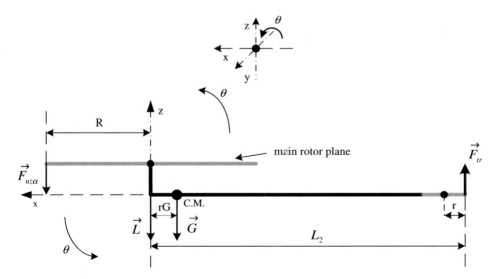

Fig. 2 Lateral stability in x–z plane

Fig. 3 Directional stability
x–y plane

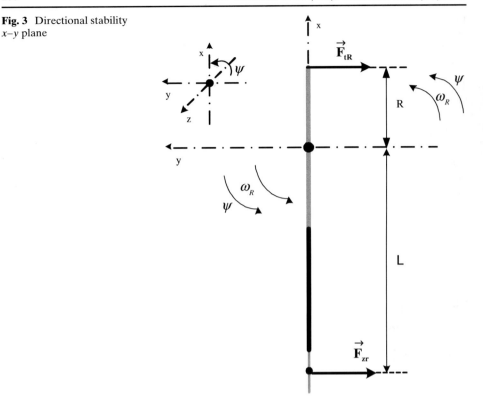

$$\vec{\tau}_{Dz} = \vec{\tau} \begin{bmatrix} 0 \\ 0 \\ 1 \end{bmatrix} = I_R \, \dot{\vec{\omega}}_R \cos\varphi \cos\theta - I_R \omega_R \left(\dot{\varphi} \sin\varphi \cos\theta + \dot{\theta} \cos\varphi \sin\theta \right). \quad (17)$$

All physical principles and approximations applied to the main rotor are applied
to the rear rotor, as well. Aerodynamic force of the rear rotor is derived from
expression:

$$\left\| \vec{F}_{zr} \right\| = \frac{1}{2} c_{zr} \rho_{\text{air}} S_{zr} \left\| \vec{\omega}_r \right\|^2 \int_0^r \left\| \vec{r} \right\|^2 \, dr = \frac{1}{6} c_{zr} \rho_{\text{air}} S_{zr} \left\| \vec{\omega}_r \right\|^2 r^3 \quad (18)$$

$$\vec{F}_{zr} = \left\| \vec{F}_{zr} \right\| \frac{\vec{\omega}_r}{\left\| \vec{\omega}_r \right\|} \times \frac{\vec{r}}{\left\| \vec{r} \right\|} \quad (19)$$

This force is used for stabilization of the main rotor revolving torque; therefore, it is
later applied in equations of motion. The rear rotor induced drag is determined as:

$$\left\| \vec{F}_{tr} \right\| = \frac{1}{2} c_{tr} \rho_{\text{air}} S_{tr} \left\| \vec{\omega}_r \right\|^2 \int_0^r \left\| \vec{r} \right\|^2 \, dr = \frac{1}{6} c_{tr} \rho_{\text{air}} S_{tr} \left\| \vec{\omega}_r \right\|^2 r^3 \quad (20)$$

$$\vec{F}_{tr} = \left\| \vec{F}_{tr} \right\| \frac{\vec{\omega}_r}{\left\| \vec{\omega}_r \right\|} \times \frac{\vec{r}}{\left\| \vec{r} \right\|} \quad (21)$$

Table 1 Experimentally determined coefficients	c_{zR}	c_{tR}	c_{uz}	c_{zr}
	0.0053	2.3×10^{-5}	0.00048	0.59

This force induces a rear rotor revolving torque which, as a consequence, generates lateral instability in helicopter motion.

2.1 Dynamical Equations of Helicopter Motion

As a consequence of the acting forces and torques described in previous analysis, a helicopter motion is induced. Rotational motion of the main rotor blades is described by equation:

$$I_R \, \dot{\vec{\omega}}_R = \vec{\tau}_{RR} - \vec{\tau}_{\mu R} \tag{22}$$

A DC motor torque is derived from [13]:

$$\vec{\tau}_{RR} = C_{MR} i_R = C_{MR} \frac{U_R - C_{GR} \vec{\omega}_R}{R_R}, \tag{23}$$

while the main rotor air friction torque can be defined as

$$\vec{\tau}_{\mu R} = C_{\mu R} \vec{\omega}_R \tag{24}$$

The same concept is used to describe the rear rotor motion:

$$I_r \, \dot{\vec{\omega}}_R = \vec{\tau}_{rr} - \vec{\tau}_{\mu r} \tag{25}$$

Developed torque is determined in the same way as in the case of the main rotor. Same principle is used for rear rotor air friction:

$$\vec{\tau}_{rr} = C_{Mr} \frac{U_r - C_{Cr} \vec{\omega}_r}{R_r} \tag{26}$$

$$\vec{\tau}_{\mu r} = C_{\mu r} \vec{\omega}_r \tag{27}$$

Fig. 4 RC helicopter model *Ecco piccolo*

2.1.1 Motion Around x-axis, y–z Plane

Figure 1 shows the motion around x-axis, in y–z plane. Rotation around x-axis is caused by torque $\vec{\tau}_{uz\beta}$, a consequence of stabilizing bar blade lift force. An additional influence on rotation is produced by main rotor torque $\vec{\tau}_{Dx}$ and is associated with angular momentum through expressions (11), (13) and (15).

A motion equation around x-axis is derived from Fig. 1:

$$I\ddot{\varphi} = \vec{\tau}_{uz\beta} - \vec{\tau}_{\mu x} + \vec{\tau}_{Dx} \tag{28}$$

$$\vec{\tau}_{\mu x} = c_{\mu x}\,\dot{\varphi} \tag{29}$$

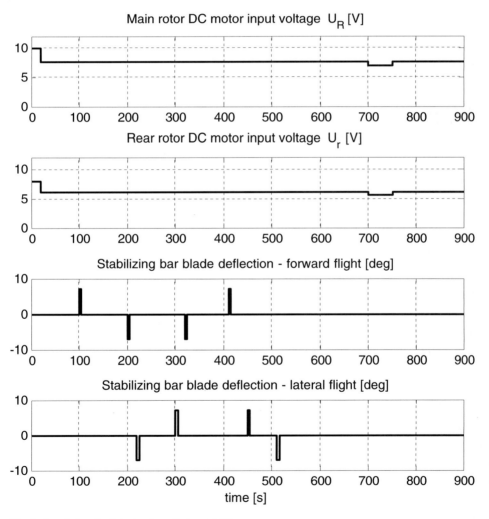

Fig. 5 Input signals of derived nonlinear model

2.1.2 Motion Around y-axis, x–z Plane

A motion in x–z plane is presented in Fig. 2. Stabilizer bar blade lift torque (caused by the change of the rotational plane of the main rotor) is balanced with the rear rotor rotational drag momentum. In standard helicopter construction the helicopter body is balanced so that the centre of gravity is placed on the main rotor axis. Displacement of the centre of gravity (for r_G from the main rotor axis) induces torque $\vec{\tau}_G$.

The lateral motion is described by the following equations:

$$I\ddot{\theta} = \vec{\tau}_{uz\alpha} - \vec{\tau}_{\mu y} + \vec{\tau}_{tr} + \vec{\tau}_{Dy} - \vec{\tau}_G \tag{30}$$

$$\vec{\tau}_{tr} = \vec{F}_{tr} \int_0^{L_2} \vec{r} \, dr = \frac{1}{2} L_2^2 \vec{F}_{tr} \tag{31}$$

$$\vec{\tau}_G = \vec{G} \int_0^{r_G} \vec{r} \, dr = \frac{1}{2} r_G^2 \vec{G} \tag{32}$$

$$\vec{\tau}_{\mu y} = c_{\mu y} \dot{\theta} \tag{33}$$

2.1.3 Motion Around z-axis, x–y Plane

A directional stability of the small helicopter is obtained by balancing lift torque of rear rotor and rotational drag momentum of the main rotor (Fig. 3).

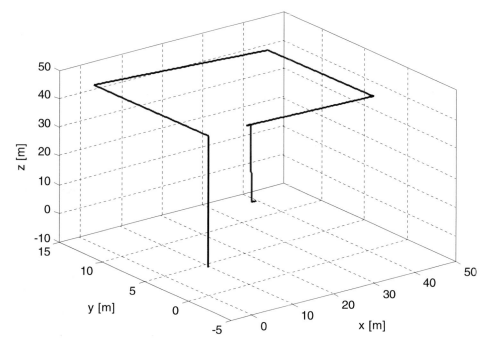

Fig. 6 Testing path sequence of the helicopter

The differential equation describing directional motion has the form:

$$I \ddot{\psi} = -\vec{\tau}_{tR} - \vec{\tau}_{\mu z} + \vec{\tau}_{zr} + \vec{\tau}_{Dz} \tag{34}$$

where:

$$\vec{\tau}_{zr} = \vec{F}_{zr} \int_0^{L_2} \vec{r} \, dr = \frac{1}{2} \vec{F}_{zr} L_2^2 \tag{35}$$

$$\vec{\tau}_{tR} = 2\vec{F}_{tRL} \int_0^R \vec{r} \, dr + 2\vec{F}_{tRl} \int_0^R \vec{r} \, dr = \vec{F}_{tRL} R^2 + \vec{F}_{tRl} R^2 \tag{36}$$

$$\vec{\tau}_{\mu z} = c_{\mu z} \dot{\psi} \tag{37}$$

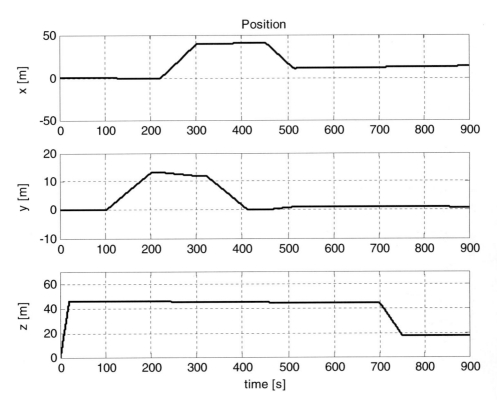

Fig. 7 Helicopter model position

2.2 Kinematical Equations of Motion

The total aerodynamic force is comprised of three components which represent acting forces of helicopter vertical, forward and lateral motion. As in the case of main rotor torques, a coordinate transformation is required.

$$\overrightarrow{F} = \mathbf{R}_{z,\psi}\mathbf{R}_{y,\theta}\mathbf{R}_{x,\varphi} \begin{bmatrix} 0 \\ 0 \\ F_z \end{bmatrix} = \begin{pmatrix} \cos\varphi\sin\theta\cos\psi + \sin\varphi\sin\psi \\ \cos\varphi\sin\theta\sin\psi - \sin\varphi\cos\psi \\ \cos\varphi\cos\theta \end{pmatrix} = \begin{bmatrix} F_{zx} \\ F_{zy} \\ F_{zz} \end{bmatrix} \tag{38}$$

where φ and θ are the main rotor tilt angles.

Kinematic equations of motion obtain the following form:

$$m\ddot{x} = \overrightarrow{F}_{zx} - \overrightarrow{F}_{\mu x} \tag{39}$$

$$m\ddot{y} = \overrightarrow{F}_{zy} - \overrightarrow{F}_{\mu y} \tag{40}$$

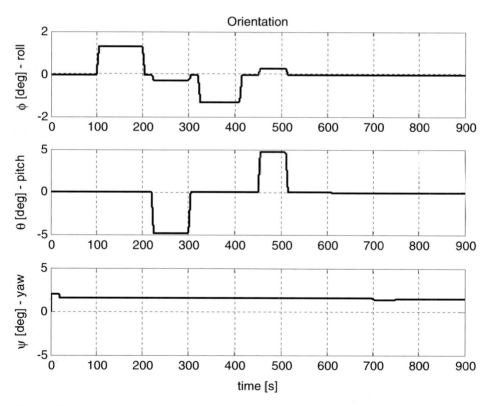

Fig. 8 Helicopter model orientation

Fig. 9 MIMO system

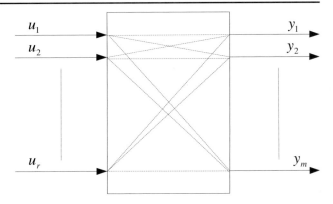

$$m\ddot{z} = \vec{F}_{zz} - \vec{F}_{\mu z} - \vec{G} \tag{41}$$

$$\vec{F}_\mu = c_\mu \vec{v} = c_\mu \begin{bmatrix} \dot{x} \\ \dot{y} \\ \dot{z} \end{bmatrix} = \begin{bmatrix} F_{\mu x} \\ F_{\mu y} \\ F_{\mu z} \end{bmatrix} \tag{42}$$

In order to calculate values of the aerodynamical lift, the main rotor induced drag, the main rotor stabilizer bar blade lift and rear rotor lift, coefficients c_{zR}, $c_{tR}c_{\mu z}$ and c_{zr} (Table 1) are determined experimentally from the RC helicopter model (Fig. 4).

The DC motor coefficients for the main and the rear rotor are determined from their input–output characteristics. Air friction coefficients are determined according to [14].

2.3 Model Testing by Simulation

Derived nonlinear model is tested by simulation in *MathWorks Matlab*. In order to show the model validity set of maneuvers has been conducted. Input signals, namely, the main rotor DC motor voltage, the rear rotor DC motor voltage, and sliding bar deflections are shown in Fig. 5.

Predefined input signals force the helicopter to execute the following set of maneuvers:

vertical up ⇒ backward ⇒ lateral left ⇒ forward ⇒ lateral right ⇒ vertical

down ⇒ hover.

Figure 6 shows the path accomplished by the simulation. From results shown it can be concluded that the model is physically correct, so that it can acceptably well execute motion in all directions, and has no problem with changes in course of motion.

Following two figures show the helicopter model position (Fig. 7) and orientation (Fig. 8).

From detailed view of position and orientation it can be seen that there is some slight drift in motion through x and y axes. These problems will be resolved by implementation of the control algorithm.

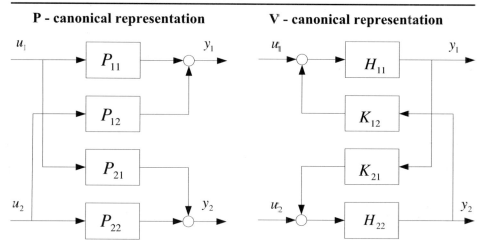

Fig. 10 Canonical representations of MIMO system

3 Control Algorithm Synthesis

The complete nonlinear model of the small scale helicopter, derived in previous section, has four input and six output variables, which makes it a multivariable system. In the text that follows decomposition of the model and design of decoupling controllers are presented.

3.1 Multiple Input Multiple Output Systems

In a multiple input multiple output (MIMO) system, it might happen that one manipulated variable affects more than one controlled variable (Fig. 9). Due to this

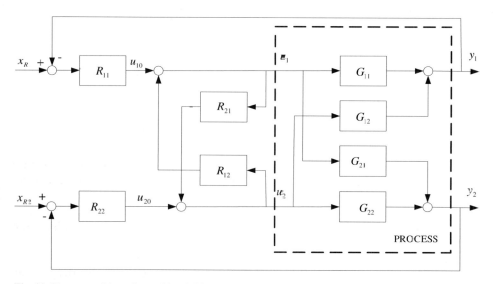

Fig. 11 Decomposition of a multivariable system represented in P-canonical form

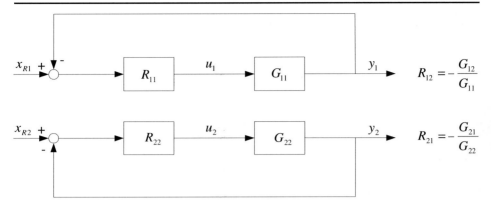

$$R_{12} = -\frac{G_{12}}{G_{11}}$$

$$R_{21} = -\frac{G_{21}}{G_{22}}$$

Fig. 12 Decoupled control loops

fact, MIMO systems are mostly coupled and nonlinear; therefore, they are difficult to control by conventional controllers.

Linearization of those processes is tied with essential practical limitations.Hence, universal theory for multivariable controller design is not entirely developed. Due to various structures of complex technical processes usage of standard forms of mathematical models is questionable. However, high-dimension systems can be decomposed to sets of 2×2 subsystems.

Two commonly used input–output models of multivariable systems are the P- and V-canonical representations [15–17].

According to Fig. 10, P-canonical representation, in general, can be written as:

$$Y(s) = P(s) U(s) \tag{43}$$

P-canonical representation has the following characteristics:

- Every input is acting on every output, summation points are on outputs.
- Change in some dynamic element is sensed only in corresponding output.
- Number of inputs and outputs can differ.

V-canonical representation, in general, can be expressed as:

$$Y(s) = G_H(s) \left[U(s) + G_k(s) Y(s) \right] \tag{44}$$

Since V-canonical representation is defined only for processes with same number of inputs and outputs, which is not true in our case; further discussion is related to P-canonical representation.

The multivariable controller design is based on process decomposition, i.e. the main goal is to compensate the influence that coupled process members have on controlled system. The method is presented on the example of a 2×2 system in P-canonical representation (Fig. 11).

Controllers R_{11} and R_{22} are called the main controllers, and R_{12}, R_{21} decoupling controllers. Controllers R_{12} and R_{21} are determined from decoupling conditions, i.e.

from the requests that y_1 depends only on u_{10}, and y_2 depends only on u_{20}. Hence, from Fig. 11 it follows that:

$$G_{12} + G_{11} R_{12} = 0 \Rightarrow R_{12} = -\frac{G_{12}}{G_{11}}, \tag{45}$$

$$G_{21} + G_{22} R_{21} = 0 \Rightarrow R_{21} = -\frac{G_{21}}{G_{22}}, \tag{46}$$

With decoupling controllers determined according to Eqs. 45 and 46, an ideal compensation of the process coupling effect is achieved. However, realization of ideal decoupling controllers could be in conflict with practical limitations. A coupled

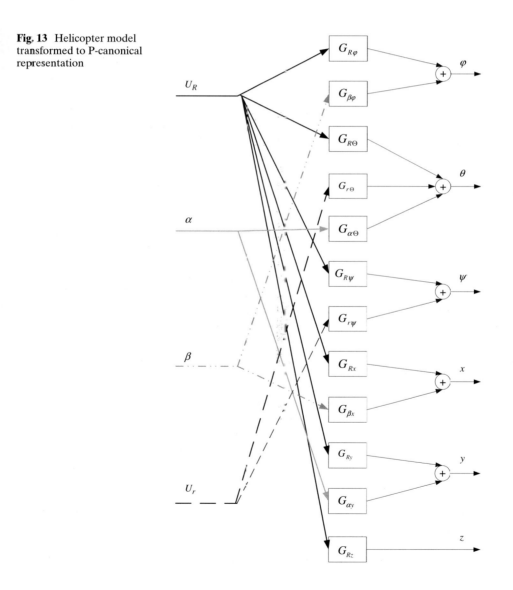

Fig. 13 Helicopter model transformed to P-canonical representation

system (Fig. 11) is decomposed in two equivalent independent control loops which can be considered separately (Fig. 12). Control synthesis of both control loops is carried out with single-loop control methods. The analogy holds in the case of $n \times n$ processes [17].

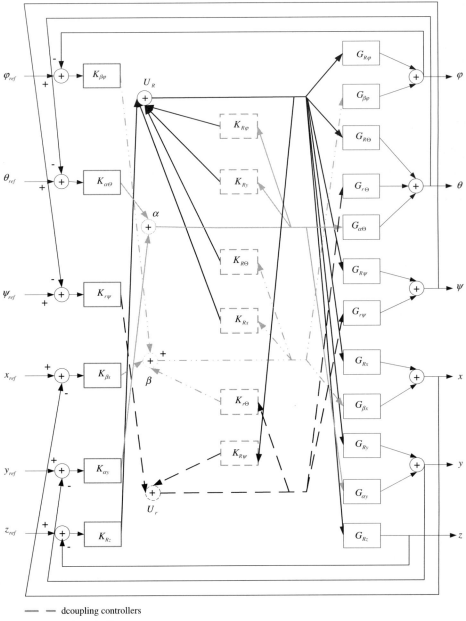

Fig. 14 Decomposition of the small helicopter model

In the following section, the influence that manipulated variables have on controlled variables in derived nonlinear model of the helicopter is investigated.

3.2 Mathematical Model Decomposition

From six equations of motion, representing six controlled variables, the influence of every manipulated variable on a certain state variable is determined.

Lateral stability equation:

$$I\ddot{\varphi} = \overrightarrow{\tau}_{uz\beta} - \overrightarrow{\tau}_{\mu x} + \overrightarrow{\tau}_{Dx}$$

$$\overrightarrow{\tau}_{uz\beta} \qquad \Rightarrow \bar{U}_R, \beta$$

$$\overrightarrow{\tau}_{\mu x} \qquad \Rightarrow \bar{G}_{R\varphi}, G_{\beta\varphi}$$

$$\overrightarrow{\tau}_{Dx}. \qquad \Rightarrow \bar{U}_R \qquad (47)$$

The effects of input variables on torques that participate in equation presenting lateral stability of the helicopter, given in expression (47), can be seen from Eqs. 8 and 15. After determining the input variables that influence output variable φ, transfer functions $G_{R\varphi}$ and $G_{\beta\varphi}$ are derived. The same principle could be used for longitudinal stability and directional stability equations.

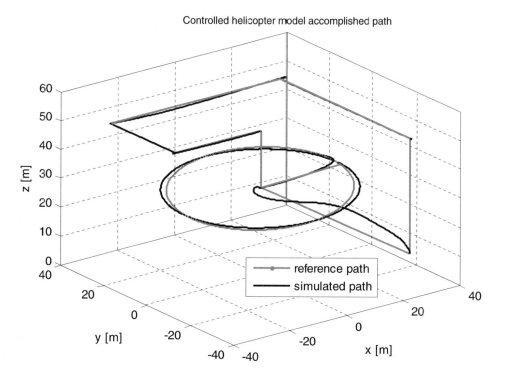

Fig. 15 A path accomplished by the controlled helicopter

Longitudinal stability equation:

$$I\ddot{\theta} = \vec{\tau}_{uz\alpha} - \vec{\tau}_{\mu y} + \vec{\tau}_{tr} + \vec{\tau}_{Dy} - \vec{\tau}_G$$

$$\vec{\tau}_{uz\alpha} \Rightarrow U_R, \alpha$$

$$\vec{\tau}_{\mu y}$$

$$\vec{\tau}_{tr} \Rightarrow U_r \qquad\qquad \Rightarrow G_{R\Theta}, G_{r\Theta}, G_{\alpha\Theta}$$

$$\vec{\tau}_{Dy} \Rightarrow U_R$$

$$\vec{\tau}_G \tag{48}$$

Directional stability equation:

$$I\ddot{\psi} = -\vec{\tau}_{tR} - \vec{\tau}_{\mu z} + \vec{\tau}_{zr} + \vec{\tau}_{Dz}$$

$$\vec{\tau}_{tR} \Rightarrow U_R$$

$$\vec{\tau}_{zr} \Rightarrow U_r \qquad\qquad \Rightarrow G_{R\psi}, G_{r\psi}$$

$$\vec{\tau}_{Dz} \Rightarrow U_R$$

$$\vec{\tau}_{\mu z} \tag{49}$$

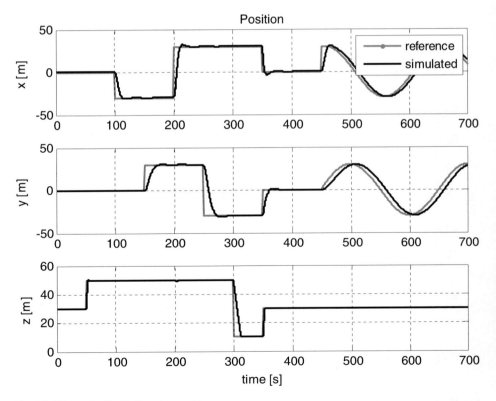

Fig. 16 The controlled helicopter position

Influence that inputs U_R, β and α have on outputs x, y and z can be determined from the following equations:

$$m\ddot{x} = \overrightarrow{F}_{zx} - \overrightarrow{F}_{\mu x}$$
$$\overrightarrow{F}_{zx} = (\cos\varphi\cos\theta\cos\psi + \sin\varphi\sin\psi)\left(\overrightarrow{F}_z\right) \Rightarrow U_R, \beta \Rightarrow G_{Rx}, G_{\beta x} \qquad (50)$$

$$m\ddot{y} = \overrightarrow{F}_{zy} - \overrightarrow{F}_{\mu y}$$
$$\overrightarrow{F}_{zy} = (\cos\varphi\sin\theta\sin\psi - \sin\varphi\cos\psi)\left(\overrightarrow{F}_z\right) \Rightarrow U_R, \alpha \Rightarrow G_{Ry}, G_{\alpha x} \qquad (51)$$

$$m\ddot{z} = \overrightarrow{F}_{zz} - \overrightarrow{F}_{\mu z}$$
$$\overrightarrow{F}_{zz} = \cos\varphi\cos\theta\,\overrightarrow{F}_z \Rightarrow U_R \Rightarrow G_{Rz} \qquad (52)$$

The final result of derived helicopter model decomposed into P-canonical representation is shown in Fig. 13.

The model decoupling is conducted according to equations derived for P-canonical structures 45 and 46. A complete scheme of multivariable closed-loop process is presented in Fig. 14. Having determined the complete decomposed model of a small helicopter it is easy to investigate the influence that a particular input variable has on output variables under various flying conditions.

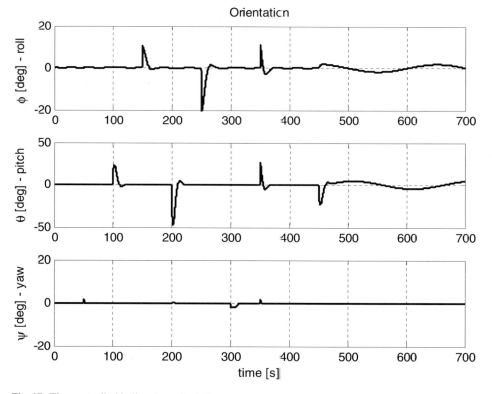

Fig. 17 The controlled helicopter orientation

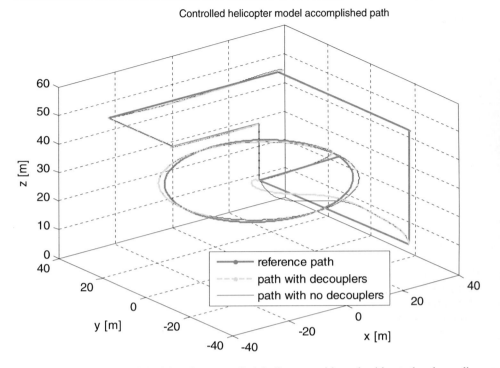

Fig. 18 A path accomplished by the controlled helicopter with and without the decoupling controllers

The closed loop system, presented in Fig. 14, is composed of six decoupling and four main controllers which ensure system decomposition and stability. In order to determine decoupling controllers around point (0,0,30,0,0,0) – hovering, the presented nonlinear model has been linearized. For this particular flying condition results are as follows:

φ controllers:

- Main controller: $K_{\beta\varphi}$
- Decoupling controller:

$$K_{R\varphi} = -\frac{G_{R\varphi}}{G_{\beta\varphi}} \approx 0 \tag{53}$$

With the decoupling controller, the influence of the main rotor DC motor input voltage on roll angle is annulled. Controller design is conducted according to Eq. 45. Because the gain of the derived controller is negligible, it is not taken into account, i.e. the decoupling controller transfer function is set to zero.

θ controllers:

- Main controller: $K_{\alpha\theta}$
- Decoupling controllers:

$$K_{R\theta} = -\frac{G_{R\theta}}{G_{\alpha\theta}} \approx 0 \tag{54}$$

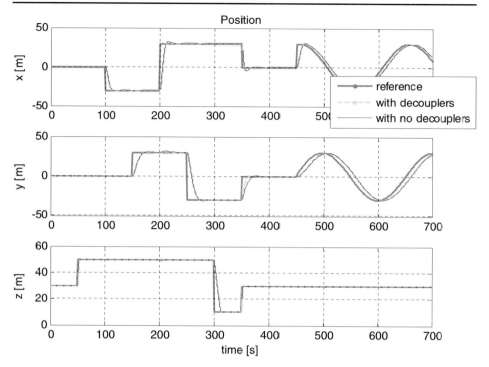

Fig. 19 The controlled helicopter position with and without decoupling controllers

$$K_{r\theta} = -\frac{G_{r\theta}}{G_{\alpha\theta}} = -3.6 \times 10^{-5} \times \frac{s + 150.2}{s + 22.4} \tag{55}$$

With controller $K_{r\theta}$, the influence of the rear rotor DC motor input voltage is decoupled from pitch angle θ. Controller 55 is derived according to Eq. 45.

ψ controllers:

- Main controller: $K_{r\psi}$
- Decoupling controller:

$$K_{R\psi} = -\frac{G_{R\psi}}{G_{r\psi}} = -0.789 \cdot \frac{(s + 22.4)(s - 60.82)}{(s + 18.52)(s + 100.4)} \tag{56}$$

Decoupling controller $K_{R\psi}$ decouples the main rotor DC motor input voltage and yaw angle ψ. Controller 56 is designed according to Eq. 45 and represents the ideal decoupling between those two variables.

x controllers:

- Main controller: $K_{\beta x}$
- Decoupling controller:

$$K_{Rx} = -\frac{G_{Rx}}{G_{\beta x}} \approx 0 \tag{57}$$

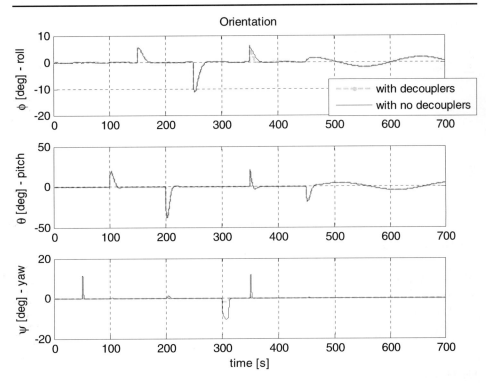

Fig. 20 The controlled helicopter orientation with and without decoupling controllers

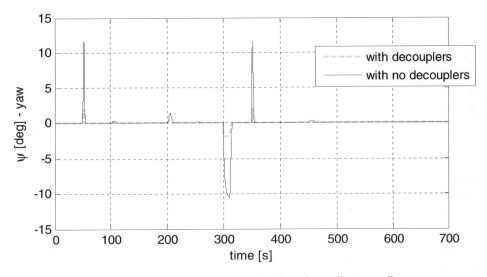

Fig. 21 The controlled helicopter yaw angle with and without decoupling controllers

y controllers:

- Main controller: $K_{\alpha y}$
- Decoupling controller:

$$K_{Ry} = -\frac{G_{Ry}}{G_{\alpha y}} \approx 0 \tag{58}$$

z controllers

- Main controller: K_{Rz}

It should be noticed that only two decoupling controllers (out of six) are required for the system decomposition in hovering mode of operation.

3.3 Closed-loop Control System Testing

The control algorithm derived in previous section has been tested by simulation in *Matlab*. The closed-loop system consists of outer position control loops (x, y, z) that generates set points for inner orientation control loops (φ, θ, ψ). The

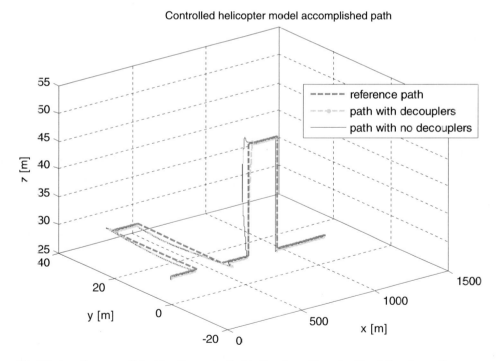

Fig. 22 A path accomplished by the controlled helicopter with and without the decoupling controllers at $(0,0,30,0,-50,0)$

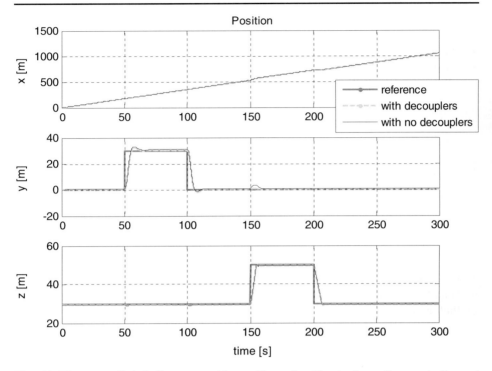

Fig. 23 The controlled helicopter position with and without decoupling controllers at (0,0,30,0,−50,0)

helicopter should complete a path in 3D space by executing a predefined sequence of maneuvers:

vertical up ⇒ backward ⇒ lateral left ⇒ forward ⇒ lateral right ⇒ vertical

down ⇒ ⇒ back in initial point ⇒ circulate.

The accomplished path is shown in Fig. 15.

The following figures (Figs. 16 and 17) present position and orientation of the helicopter. It can be seen that there is a slight overshoot in positioning. Except from that, the model executes the sequence with great precision and shows a high level of maneuverability.

As can be seen in Fig. 17 changes in angles are produced as a consequence of lateral and forward motion.

The same path in 3D space, executed without decoupling controllers, is presented in Fig. 18. Figure 19 presents helicopters position with and without decoupling controllers. The coupling effect can be clearly seen in yaw angle response, Fig. 20. Due to the change of altitude at $t = 300$ [s] and $t = 350$ [s], the tilt of yaw angle appears (enlarged in Fig. 21) in range from −10° to 12°.

Effects that other coupled loops have on the system dynamics are negligible.

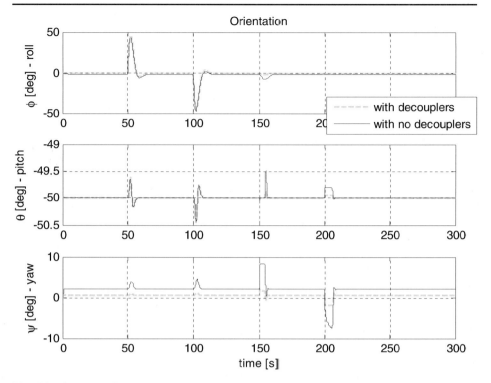

Fig. 24 The controlled helicopter orientation with and without decoupling controllers at (0,0,30,0,−50,0)

The next test has been performed at new working point (0,0,30,0,−50,0). By using the same principle as in the previous case the new control algorithm is derived. The derived decoupling controllers are given by the following equations:

φ decoupling controllers:

$$K_{R\varphi} = -\frac{G_{R\varphi}}{G_{\beta\varphi}} \approx 0 \qquad (59)$$

θ decoupling controllers:

$$K_{R\theta} = -\frac{G_{R\theta}}{G_{\alpha\theta}} \approx 0 \qquad (60)$$

$$K_{r\theta} = -\frac{G_{r\theta}}{G_{\alpha\theta}} = -3.6 \times 10^{-5} \times \frac{s + 150.2}{s + 22.4} \qquad (61)$$

ψ decoupling controllers:

$$K_{R\psi} = -\frac{G_{R\psi}}{G_{r\psi}} = -0.789 \times \frac{(s + 22.4)(s - 60.82)}{(s - 18.52)(s + 100.4)} \qquad (62)$$

x decoupling controllers:

$$K_{Rx} = -\frac{G_{Rx}}{G_{\beta x}} = -993.876 \times \frac{s + 20.18}{s + 18.52} \qquad (63)$$

y decoupling controller:

$$K_{Ry} = -\frac{G_{Ry}}{G_{\alpha y}} \approx 0 \qquad (64)$$

The control algorithm derived for a helicopter model around working point (0,0,30,0,−50,0) has been tested by simulation in *Matlab*. The helicopter is supposed to complete a path in 3D space by fsmoving forward (inclination of 50° in forward direction) and in the same time executing a predefined sequence of maneuvers:

lateral left ⇒ lateral right ⇒ vertical up ⇒ vertical down

To show the coupling effect at working point (0,0,30,0,−50,0), the simulation has been done with and without the decoupling controllers, as it was for working point (0,0,30,0,0,0). The accomplished paths are shown in Fig. 22.

From Figs. 22 and 23 it can be seen that there is no significant deviations in position between simulations preformed with and without decoupling controllers.

However, the coupling effect appeared for roll and yaw angles as it was expected (Fig. 24).

4 Conclusion

In this paper, a complete nonlinear mathematical model of small scale helicopter is derived. The coupling effect between input and output variables, revealed by the model, is investigated. Nonlinear model is linearized and afterwards decoupled. For controller design linear control theory is used because of its consistent performance and proved effectiveness. It is shown that influence of particular input on a particular output highly depends on flying conditions. Simulation results, presented at the end of the paper, confirm that the proposed control structure could be successfully used for gain scheduling or switching control of a small scale helicopter in order to provide acrobatic flight by using simple linear controllers.

5 List of Symbols

φ	roll angle, [°]
θ	pitch angle, [°]
ψ	yaw angle, [°]
x	position through x-axis, [m]
y	position through y-axis, [m]
z	position through z-axis, [m]
τ_{RR}	main rotor motor torque, [N m]
τ_{tR}	main rotor rotational drag momentum, [N m]
$\tau_{\mu R}$	main rotor drag torque, [N m]
$\tau_{uz\beta}$	stability blade torque, due to an attack angle change, [N m]
$\tau_{uz\alpha}$	secondary bar torque, due to change of an attack angle [N m]
τ_{Dx}	torque due to rotational plane change of the main rotor, x-axis, [N m]
τ_{Dy}	torque due to rotational plane change of the main rotor, y-axis, [N m]
τ_{Dz}	torque due to rotational plane change of the main rotor, z-axis, [N m]

τ_{rr}	rear rotor motor torque, [N m]
τ_{zr}	rear rotor torque, [N m]
τ_{tr}	rear rotor rotational drag momentum, [N m]
$\tau_{\mu r}$	rear rotor drag torque, [N m]
$\tau_{\mu x}$	air friction torque motion around x-axis, [N m]
$\tau_{\mu y}$	air friction torque motion around y-axis
$\tau_{\mu z}$	air friction torque motion around z-axis, [N m]
τ_G	torque due to the centre of gravity displacement, [N m]
F_{zR}	aerodynamical lift of the main rotor, [N]
F_{iR}	lift-induced drag of the main rotor, [N]
F_{tRL}	main rotor blade induced drag, [N]
F_{tRl}	main rotor stabilizing blade induced drag, [N]
$F_{\mu z}$	lift force because of change of a wing attack angle [N]
F	total aerodynamical force of the main rotor [N]
F_{zx}	towing force causing motion through x-axis, [N]
F_{zy}	towing force causing motion through y-axis, [N]
F_{zz}	towing force causing motion through z-axis, [N].
F_{tr}	rear rotor induced drag, [N]
F_{zr}	rear rotor aerodynamical force, [N]
c_{zR}	aerodynamical lift coefficient of the main rotor
$c_{\mu x}$	air friction coefficient of motion around x-axis
$c_{\mu y}$	air friction coefficient of motion around y-axis
$c_{\mu z}$	air friction coefficient of motion around z-axis
ρ_{air}	air density, [kg/m^3]
R	length of the main rotor blades, [m]
S_R	main rotor disk surface area, [m^2]
I_R	moment of inertia of the main rotor, [kg m^2]
I_r	moment of inertia of the main rotor, [kg m^2]
I	moment of inertia of helicopter body, [N m]
L_2	distance between rear rotor and centre of gravity, [m]
U_R	main rotor motor voltage, [V]
R_R	main rotor motor resistance, [Ω]
C_{GR}	main motor constant, [V s/rad]
C_{MR}	main motor constant, [N m/A]
$c_{\mu R}$	main motor drag constant, [N/rad]
U_r	rear rotor motor voltage, [V]
R_r	rear rotor motor resistance, [Ω]
C_{Gr}	rear rotor constant, [V s/rad]
C_{Mr}	rear rotor motor constant, [N m/A]
$c_{\mu r}$	rear rotor drag constant, [N/rad]

References

1. Woodley, B., Jones, H., Frew, E., LeMaster, E., Rock, S.: A contestant in the 1997 international aerial robotics competition. In: AUVSI Proceedings. Aerospace Robotics Laboratory, Stanford University, July (1997)
2. Buskey, G., Roberts, J., Corke, P., Wyeth, G.: Helicopter automation using a low-cost sensing system. In: Proceedings of the Australian Conference on Robotics and Automation (2003)

3. Gavrilets, V., Martinos, I., Mettler, B., Feron, E.: Control logic for automated aerobatic flight of a miniature helicopter. In: Proceedings of the AIAA Guidance, Navigation, and Control Conference, (Monterey, CA). Massachusetts Institute of Technology, Cambridge, MA, August (2002)
4. Ab Wahab, A., Mamat, R., Shamsudin, S.S.: Control system design for an autonomous helicopter model in hovering using pole placement method. In: Proceedings of the 1st Regional Conference on Vehicle Engineering and Technology, 3–5 July. Kuala Lumpur, Malaysia (2006)
5. Buskey, G., Roberts, J., Wyeth, G.: A helicopter named Dolly – behavioural cloning for autonomous helicopter control. In: Proceedings of the Australian Conference on Robotics and Automation (2003)
6. Castillo, C., Alvis, W., Castillo-Effen, M., Valavanis, K., Moreno, W.: Small scale helicopter analysis and controller design for non-aggressive flights. In: Proceedings of the IEEE SMC Conference (2005)
7. Yang, C.-D., Liu, W.-H.: Nonlinear decoupling hover control of helicopter H_{00} with parameter uncertainties. In: Proceedings of the American Control Conference (2003)
8. Maclay, D., Williams, S.J.: The use of μ-synthesis in full envelope, helicopter flight control system design. In: Proceedings of the IEEE International Conference on Control (1991)
9. De Nardi, R., Togelius, J., Holland, O.E., Lucas, S.M.: Evolution of neural networks for helicopter control: why modularity matters. In: Proceedings of the IEEE Congress on Evolutionary Computation, July (2006)
10. Ng, A.Y., Jin Kim, H., Jordan, M.I., Sastry, S.: Autonomous helicopter flight via reinforcement learning. In: Proceedings of the NIPS Conference, ICML, Pittsburgh, PA (2006)
11. Budiyono, A., Wibowo, S.S.: Optimal tracking controller design for a small scale helicopter. In: Proceedings of the ITB, March (2005)
12. Piercy, N.A.V.: Elementary Aerodynamics. English University Press, London (1944)
13. Kari, U.: Application of model predictive control to a helicopter model. Master thesis, Norwegian University of Science and Technology, Swiss Federal institute of Technology Zurich (2003)
14. NASA: Shape Effects on Drag. www.grc.nasa.gov
15. Seborg, D.E., Edgar, T.F., Mellichamp, D.A.: Process Dynamics and Control. Wiley, New York (1989)
16. Ogunnaike, B.A., Ray, W.H.: Process Dynamics, Modelling and Control. Oxford University Press, Oxford (1994)
17. Tham, M.T.: Multivariable Control: An Introduction to Decoupling Control. University of Newcastle upon Tyne, Newcastle upon Tyne (1999)

Fuzzy Logic Based Approach to Design of Flight Control and Navigation Tasks for Autonomous Unmanned Aerial Vehicles

Sefer Kurnaz · Omer Cetin · Okyay Kaynak

Originally published in the Journal of Intelligent and Robotic Systems, Volume 54, Nos 1–3, 229–244.
© Springer Science + Business Media B.V. 2008

Abstract This paper proposes a fuzzy logic based autonomous navigation controller for UAVs (unmanned aerial vehicles). Three fuzzy logic modules are developed under the main navigation system for the control of the altitude, the speed, and the heading, through which the global position (latitude–longitude) of the air vehicle is controlled. A SID (Standard Instrument Departure) and TACAN (Tactical Air Navigation) approach is used and the performance of the fuzzy based controllers is evaluated with time based diagrams under MATLAB's standard configuration and the Aerosim Aeronautical Simulation Block Set which provides a complete set of tools for rapid development of detailed six-degree-of-freedom nonlinear generic manned/unmanned aerial vehicle models. The Aerosonde UAV model is used in the simulations in order to demonstrate the performance and the potential of the controllers. Additionally, FlightGear Flight Simulator and GMS aircraft instruments are deployed in order to get visual outputs that aid the designer in the evaluation of the controllers. Despite the simple design procedure, the simulated test flights indicate the capability of the approach in achieving the desired performance.

Keywords Fuzzy logic based autonomous flight computer design · UAV's SID · TACAN visual simulation

S. Kurnaz (✉) · O. Cetin
Turkish Air Force Academy, ASTIN,
Yesilyurt, Istanbul 34807, Turkey
e-mail: skurnaz@hho.edu.tr

O. Cetin
e-mail: o.cetin@hho.edu.tr

O. Kaynak
Department of Electrical and Electronic Engineering,
Bogazici University, Bebek, 80815 Istanbul, Turkey
e-mail: o.kaynak@ieee.org

1 Introduction

In the literature, it can be seen that the research interests in control and navigation of UAVs has increased tremendously in recent years. This may be due to the fact that UAVs increasingly find their way into military and law enforcement applications (e.g., reconnaissance, remote delivery of urgent equipment/material, resource assessment, environmental monitoring, battlefield monitoring, ordnance delivery, etc.). This trend will continue in the future, as UAVs are poised to replace the human-in-the-loop during dangerous missions. Civilian applications of UAVs are also envisioned such as crop dusting, geological surveying, search and rescue operations, etc.

One of the important endeavors in UAV related research is the completion of a mission completely autonomously, i.e. to fly without human support from take off to land on. The ground station control operator plans the mission and the target destination for reconnaissance and surveillance. UAV then takes off, reaches the target destination, completes the surveillance mission, and turns back to the base and lands on autonomously. In literature, many different approaches can be seen related to the autonomous control of UAVs; some of the techniques proposed include fuzzy control [1], adaptive control [2, 3], neural networks [4, 5], genetic algorithms [7] and Lyapunov Theory [8]. In addition to the autonomous control of a single UAV, research on other UAV related areas such as formation flight [6] and flight path generation [9] are also popular.

The approach proposed in this paper is fuzzy logic based. Three fuzzy modules are designed for autonomous control, one module is used for adjusting the roll angle value to control UAV's flight heading, and the other two are used for adjusting elevator and throttle controls to obtain the desired altitude and speed values.

The performance of the proposed system is evaluated by simulating a number of test flights, using the standard configuration of MATLAB and the Aerosim Aeronautical Simulation Block Set [11], which provides a complete set of tools for rapid development of detailed six-degree-of-freedom nonlinear generic manned/unmanned aerial vehicle models. As a test air vehicle a model which is called Aerosonde UAV [10] is utilized (shown in Fig. 1). Basic characteristics of Aerosonde UAV are shown in Table 1. The great flexibility of the Aerosonde, combined with a sophisticated command and control system, enables deployment and command from virtually any

Fig. 1 Aerosonde UAV

Table 1 UAV specifications

UAV specifications	
Weight	27–30 lb
Wing span	10 ft
Engine	24 cc, 1.2 kw
Flight	Fully autonomous/base command
Speed	18–32 m/s
Range	> 1,800 mi
Altitude range	Up to 20,000 ft
Payload	Maximum 5 lb with full fuel

location. GMS aircraft instruments are deployed in order to get visual outputs that aid the designer in the evaluation of the controllers.

The paper is organized as follows. Section 2 starts with the basic flight pattern definition for a UAV and then explain a sample mission plan which includes SID (Standard Instrument Departure) and TACAN (Tactical Air Navigation) procedures. A basic introduction to fuzzy control is given and the design of the navigation system with fuzzy controllers used for the autonomous control of the UAV is explained in Section 3. The simulation studies performed are explained and some typical results are presented in Section 4, and finally the concluding remarks and some plans about future work are given in Section 5.

2 UAV Flight Pattern Definition

A reconnaissance flight to be accomplished by a UAV basically contains the following phases; ground roll, lift off, initial climb, low altitude flight, climb, cruise, loiter over target zone, descent, initial and final approach and finally landing, as shown in Fig. 2. The basic flight maneuvers of the UAV during these phases are climb, descent, level flight and turns.

In this study, UAV is considered to take off and land on manually. Autonomous navigation starts when UAV reaches 2 km away form ground control station and climbs 100 m (MSL). It reaches way points in order and finishes when UAV reaches the ninth waypoint. The ninth point is the midpoint of the airfield where the landing is going to be. The definition of each point includes speed, altitude and position (longitude and longitude coordinates) values. The dashed line in Fig. 2 represents

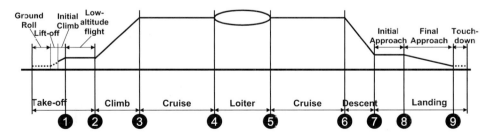

Fig. 2 Basic UAV's reconnaissance flight parts

manual control; the continuous line represents autonomous control. Character A shows the target zone location.

In this study, the test pattern includes an airport departure (Istanbul Ataturk Airport (LTBA) Turkey Runway 36L SID departure) and TACAN approach (to Yalova Airport (LTBP) runway 08) to show that UAV can fly autonomously a pattern which is designed for aircrafts if it has enough performance parameters. In classical SID and TACAN maps, a waypoint is defined with the radial angle and the distance between the VOR (VHF Omni-directional Radio Range) station and the waypoint. After transformation of the waypoints as GPS coordinates, UAV can apply SID departure and TACAN approach as a mission plan without a VOR receiver (Fig. 3).

It is presumed in this study that UAV takes off from Runway 36 and land on Runway 08. In the test pattern followed, there is a simulated target zone which is defined with three GPS coordinates. UAV visits these three points in order. While UAV is applying the flight plan, it records video when it is over the simulated target zone. After completing mission over the simulated target zone, it will try to reach IAF (Initial Approach Fix) to perform TACAN descending. UAV must obey the altitude commands for each waypoint in the plan because there are some minimal descending altitudes to avoid the ground obstacles.

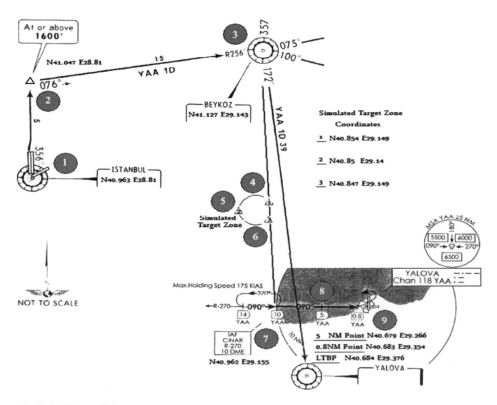

Fig. 3 UAV test flight pattern

For the test flight pattern shown in Fig. 3; the first point of the path is Istanbul Ataturk Airport which is the departure airport. After take off, the flight continues 5 NM at 356° heading while climbing to arrive at the second fix at a minimum of 1,600 ft altitude. After reaching the second fix, it turns 076° heading and fly 15 NM at YAA 1D way. When UAV reaches the BEYKOZ fix it heads south to reach the first fix of the simulated target zone. Then UAV makes a nearly half round through the three fixes of the target zone. After the completion of three target zone points, UAV turns to the seventh fix which is IAF (Initial Approach Fix) for TACAN approach for YALOVA Airport, 10 NM to the airport at MSA (Minimum Safe Altitude) of 4,000 feet. After reaching IAF, UAV turns 090° heading while descending. The eight fix is 5 NM to airport and MSA 2000'. The last fix (the ninth one) is 0.8 NM to airport; the last point for autopilot navigation. After this point if UAV operator sees the aircraft switches to the manual control, if not UAV makes a circle over airport.

3 Navigation Computer Design

As shown in Fig. 4, there are two computers in an autonomous UAV. One of them is the flight computer and the other is the mission (navigation) computer. UAV flight computer basically sets the control surfaces to the desired positions by managing the servo controllers in the defined flight envelope supplied by the UAV navigation computer as a command, reads sensors and communicates with the mission computer and also checks the other systems in the UAV (engine systems, cooling systems, etc.). In fact the navigation computer is a part of the mission computer. Because there are many duties beside navigation, like payload control, communication with GCS, etc., the navigation computer is used for flying over a pattern which is designed before the

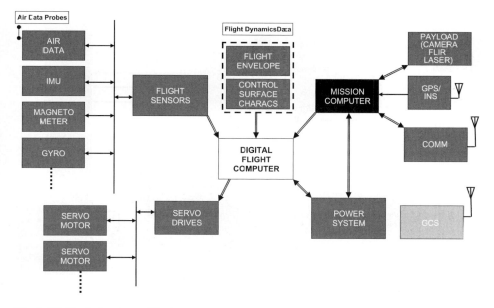

Fig. 4 UAVs electronics architecture

flight or while flying in LOS (Line Of Sight). The flight and the mission computers are always active but the navigation computer is used especially in autonomous flight. When GCS (Ground Control Station) gets control of UAV, the navigation computer goes into a passive state. During autonomous flight, the navigation computer gets the position values from sensors (GPS receiver, altimeter, INS (Internal navigation System), etc.) and then matches these values (the current position) with the desired position values (the waypoint values). The navigation computer then determines the required maneuvers of the UAV to reach the goal position and sends these data to the flight computer to apply them to control surfaces (Fig. 5).

The operation of the navigation computer proposed in this paper is fuzzy logic based. This is the main difference of the reported study with the other works seen in the literature.

Basically, a fuzzy logic system consists of three main parts: the fuzzifier, the fuzzy inference engine and the defuzzifier. The fuzzifier maps a crisp input into some fuzzy sets. The fuzzy inference engine uses fuzzy IF–THEN rules from a rule base to reason for the fuzzy output. The output in fuzzy terms is converted back to a crisp value by the defuzzifier.

In this paper, Mamdani-type fuzzy rules are used to synthesize the fuzzy logic controllers, which adopt the following fuzzy IF–THEN rules:

$$R^l:\text{If } \left(x_1 \text{ is } X_1^l\right) \text{ AND} \ldots \text{AND } \left(x_n \text{ is } X_n^l\right) \text{ THEN } y_1 \text{ is } Y_1^l, \ldots, y_k \text{ is } Y_k^l \quad (1)$$

where R^l is the lth rule $x = \left(x_{l_1}, \ldots, x_n\right)^T \in U$ and $y = (y_l, \ldots, y_k)^T \in V$ are the input and output state linguistic variables of the controller respectively, $U, V \subset \Re^n$ are the universe of discourse of the input and output variables respectively, $\left(X_{l_1}, \ldots, X_n\right)^T \subset U$ and $(Y_1, \ldots, Y_k)^T \subset V$ are the labels in linguistic terms of input and output fuzzy sets, and n and k are the numbers of input and output states respectively.

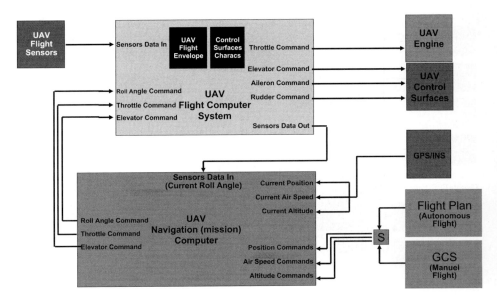

Fig. 5 Communication between UAVs Flight and Navigation Computers

We consider a multi-input and single-output (MISO) fuzzy logic controller ($k = 1$), which has singleton fuzzifier. Using triangular membership function, algebraic product for logical AND operation, product–sum inference and Centroid defuzzification method, the output of the fuzzy controller has the following form:

$$y_j = \frac{\sum_{l=1}^{M} \left(\prod_{i=1}^{N} \mu x_i^l (x_i) \right) y_j}{\sum_{l=1}^{M} \prod_{i=1}^{N} \mu x_*^* (x_i)} \tag{2}$$

where N and M represent the number of input variables and total number of rules respectively. μx_i^l denote the membership function of the lth input fuzzy set for the ith input variable.

Three fuzzy logic controllers are designed for the navigation computer in order to control the heading, the altitude and the air-speed. These three controllers acting in combination enable the navigation of the aerial vehicle (Fig. 6).

The navigation computer of UAV has four subsystems, namely heading, speed, altitude and routing subsystems. The routing subsystem calculates which way point is the next. The way point definition includes the position (the longitude and the latitude coordinates in GPS format), the speed and the altitude information of the point. When the UAV reaches the waypoint position $\pm 01°$, it means that UAV has checked that target waypoint and passes on to the next one. If the UAV cannot come within $\pm 01°$ of the position of the target waypoint, it will make a circle in pattern and retry to reach that point. The routing subsystem applies this check procedure and supplies the definitions of the next waypoint to other subsystems. The inputs to the heading subsystem are the current position of the UAV (longitude and latitude coordinates in GPS format), the current roll angle (this information is received from sensors by the flight computer) and the next waypoint position which is defined by the routing system. The duty of the heading subsystem is turning the UAV to the target waypoint. The speed subsystem maintains the air speed of the UAV at the desired value. It uses the current speed and the speed command (defined by the routing system) values as inputs and its output is the throttle command. In this study, while controlling speed, the pitching angle is not used to see the effectiveness

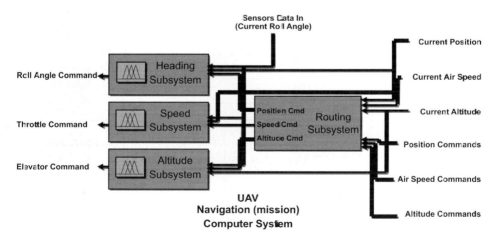

Fig. 6 Navigation computer design

of the throttle fuzzy control over UAV's air speed and to calculate the response time. The last subsystem is the altitude subsystem and it aims to maintain the altitude of the UAV at the desired value. It has the current altitude and the altitude command values as inputs and the elevator command as output.

If the fuzzy controller types in literature are reviewed, it can be seen that there are two main classes of fuzzy controllers: one is position-type fuzzy controller which generates control input (u) from error (e) and error rate (Δe), and the other is velocity-type fuzzy logic controller which generates incremental control input (Δu) from error and error rate. The former is called PD Fuzzy Logic Controller and the latter is called PI Fuzzy Logic Controller according to the characteristics of information that they process. Figure 7a and b show the general structure of these controllers.

PI Fuzzy Logic Controller system has two inputs, the error $e(t)$ and change of error $\Delta e(t)$, which are defined by

$$e(t) = y_{ref} - y \tag{3}$$

$$\Delta e(t) = e(t) - e(t-1) \tag{4}$$

Where y_{ref} and y denote the applied set point input and plant output, respectively. The output of the Fuzzy Logic Controller is the incremental change in the control signal $\Delta u(t)$. Then, the control signal is obtained by

$$u(t) = u(t-1) + \Delta u(t) \tag{5}$$

As stated earlier, there are three fuzzy logic controllers in the heading, the speed and the altitude subsystems. The heading subsystem has the roll angle fuzzy logic controller, the speed subsystem has the throttle fuzzy logic controller and the altitude subsystem has the elevator fuzzy logic controller.

The throttle fuzzy logic controller has two inputs: the speed error (i.e. the difference between the desired speed and the actual speed) and its rate of change. The

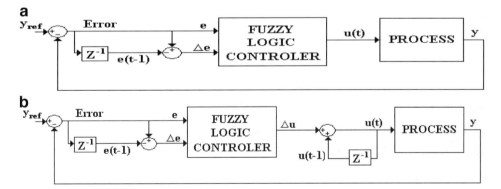

Fig. 7 **a** PD type fuzzy logic controller. **b** PI type fuzzy logic controller

Fig. 8 PID type fuzzy logic controller

latter indicates whether the UAV is approaching to the desired speed or diverging away. Like the throttle controller, the elevator control has two inputs, the altitude error and its derivative. The control output of the block is the elevator, responsible for the head going up or down. The elevator and the throttle fuzzy logic controllers are similar to the PI type fuzzy logic controller shown in Fig. 7. Because of the highly nonlinear nature of the UAV model and the inference between the controlled parameters, it is easier to aim for the required change in the control input rather than its exact value. This is the main reason for the choice of PI type fuzzy controllers.

The PI controller (also PI FLC) is known to give poor performance in transient response due to the internal integration operation. Wit a PD controller (also PD FLC) it is not possible to remove out the steady state error in many cases. Then, when the required goal cannot be reached by using only a PD or PI type of controller, it is better to combine them and construct a PID Fuzzy Controller. This is the case for roll angle control (which is used for the control of heading), and then PID type fuzzy controller was utilized for the control of heading in this work. Most commonly used PID Fuzzy Controller schematic can be seen in Fig. 8. More detailed information about PID Fuzzy Controllers can be found in [13].

While developing the fuzzy logic controllers, triangular membership functions are used for each input of the fuzzy logic controllers and simple rule tables are defined by taking into account the specialist knowledge and the experience. The control output surfaces shown in Fig. 9a, b and c are the typical ones.

As output membership functions, the throttle control output was represented with seven membership functions equally spaced in the range of [−0.02, 0.02] (frac). The membership functions used for the elevator control were rather special however, as shown in Fig. 10.

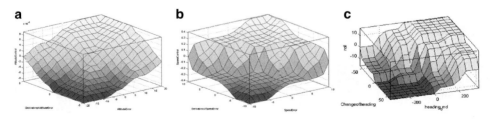

Fig. 9 a Rule surface of altitude. **b** Rule surface of air speed. **c** Rule surface of heading

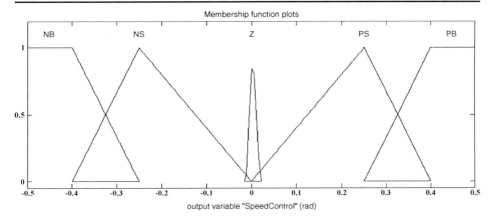

Fig. 10 Elevator control output

4 Simulation and Simulation Results

The performance and the potential of the control approach proposed are evaluated by using MATLAB's standard configuration and the Aerosim Aeronautical Simulation Block Set, the aircraft simulated being Aerosonde UAV. Additionally FlightGear Flight Simulator [12] is deployed in order to get visual outputs that aid the designer in the evaluation of the controllers (Fig. 14). Despite the simple design procedure, the simulated test flights indicate the capability of the approach in achieving the desired performance. The Simulink models that are used during the simulation studies is depicted in Fig. 11.

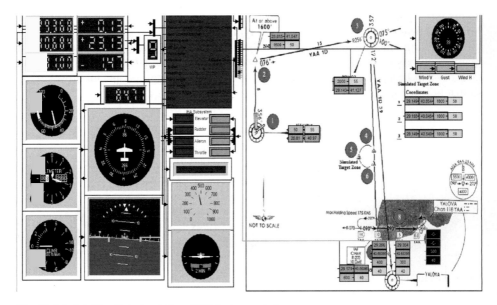

Fig. 11 Matlab Simulink simulation GMS instrument view

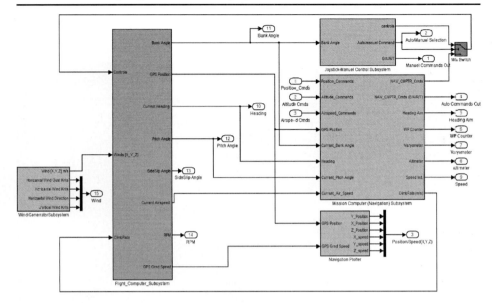

Fig. 12 Flight computer and navigation computer in simulation

In Fig. 11, the test flight pattern which includes SID and TACAN procedures is an input for the simulation. This is a kind of mission for Aerosonde UAV which includes; take of from LTBA runway 36 L and then perform SID to reach the related fixes, take video over the simulated target zone and then reach IAF to apply TACAN approach to LTBP and land on. Aerosonde UAV's current attributes can be traced over GMS aircraft instruments. These instruments are like typical aircraft instruments. A ground station control operator can manually fly the UAV by using these instruments.

As shown in Fig. 12, there are some extra modules used in simulation studies in addition to the flight computer and the mission computer modules. One of them is Joystick and Manuel control subsystem. This subsystem is used for controlling UAV by a joystick in manual selection. If the controller presses and holds the first button of the joystick, it means that the UAV is in the manual mode. When he leaves the

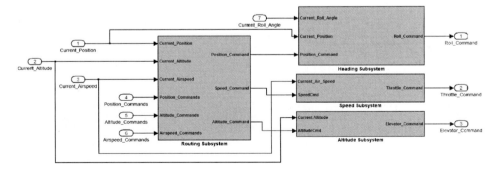

Fig. 13 Navigation computer in Simulink

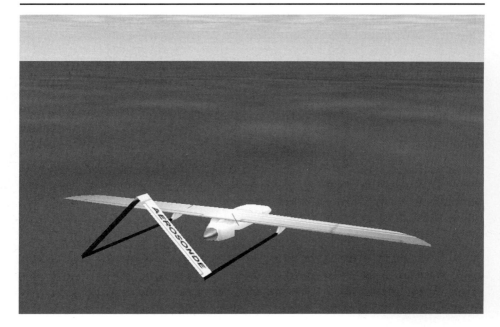

Fig. 14 Aerosonde UAV view in FlightGear while simulation is running

button, UAV assumes autonomous flight position. If the operator takes the UAV into an undesired state in manual control; after switching to autonomous control, the mission computer brings UAV to a stable position first of all and then aims to reach next waypoint. A stable position means to set the UAV's control surfaces to bring the aircraft within its limitations.

There are some limitations while controlling the UAV. These limitations are like the maximum climb rate (600 m/min), the maximum descent rate (800 m/min), the maximum speed (60 kn/h), the maximum angle of climb (25°), the maximum angle of descent (25°), the maximum roll angle (30°), etc..

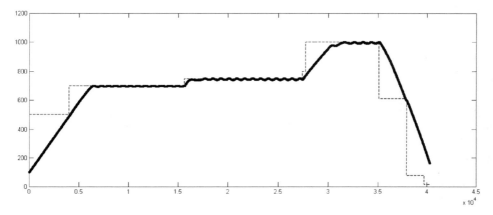

Fig. 15 Altitude–time diagram (meter/simulation time)

Another extra module used in simulation studies is the wind subsystem. The subsystem generates random wind effects as a canker to simulate the environment of UAV. The wind is represented by three dimensional vectors $[x,y,z]$. The limitations imposed in this subsystem are maximum wind speed 8 kn/h; maximum wind gust is ±2 kn/h for all altitudes. The use of the wind module enables us to see the effectiveness of the mission computer under simulated wind effects.

The last extra module in used is the navigation plotter subsystem. This module is used for generating diagrams synchronously while simulation is running to see the performance of the other modules especially that of the mission computer. These diagrams are the altitude–time diagram (Fig. 15), the heading–time diagram (Fig. 16), the speed–time diagram (Fig. 17) and the UAV position diagram (Fig. 18). By using these diagrams, one can evaluate the performance of the fuzzy controllers.

The Simulink diagram of the navigation computer is shown in Fig. 13. There are four subsystems under the navigation computer block namely routing, heading, speed and altitude subsystems. These subsystems are designed as described in Section 3 of this paper (Fig. 14).

Figure 15 depicts the change of altitude with time during test flight pattern, together with the desired altitude. It can be seen that despite the random wind effects, UAV can reach the desired altitude within range (±50 m). While doing this, the maximum angle of attack is 25° and the minimum one is 20° so that stall and possible over speeding conditions are avoided. It can therefore be stated that the fuzzy controller is successful in holding the altitude at desired levels using elevator control.

Figure 16 shows the heading response under random canker wind effects together with the desired heading command for the test flight pattern. The response time is limited because of the limitations imposed on the roll and the yaw angles. There again

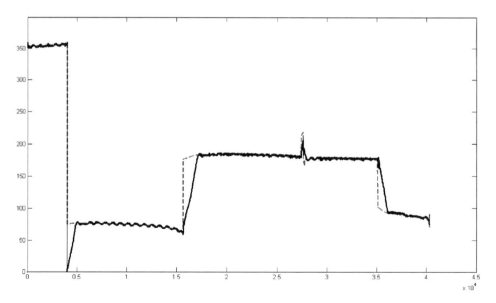

Fig. 16 Heading–time diagram (degree/simulation time)

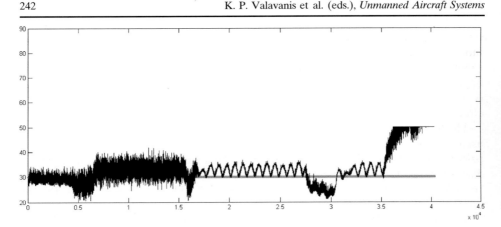

Fig. 17 Speed–time diagram (meter/second/simulation time)

the fuzzy logic controller is successful in holding the desired heading by controlling the roll angle.

The change in the speed of UAV over the test flight pattern under random canker wind effects is shown in Fig. 17, the desired speed being 30 m/s over the whole duration of the flight. The mission computer controls the speed only by throttle control not with the angle of attack. This is purposefully done to see how the throttle effects the speed and how a fuzzy logic controller controls throttle. So large errors can be seen while applying climb and descent maneuvers.

While the simulation is running a trajectory of the vehicle is generated and plotted on the navigation screen. By this way one can observe and monitor the aircraft position over a 100 km × 100 km map or a GPS coordinate based map (Fig. 18). It

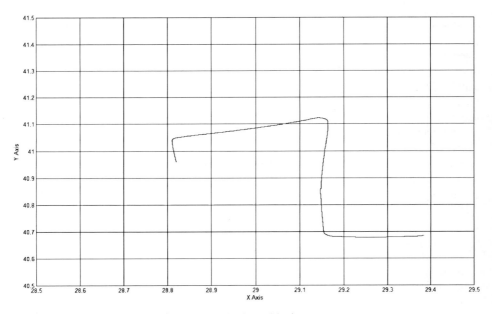

Fig. 18 UAV position diagram (latitude/longitude position)

can be seen from Fig. 18 that UAV gas completed the test flight pattern successfully under random canker wind effects. That is to say UAV reached every fix point which is defined in test flight pattern in $01°$ $(00'\ 00''\ 010$ as a GPS definition) error range.

5 Conclusion

The main purpose of the navigation computer is to enable the UAVs to accomplish their mission autonomously, without any (or with minimal) input from the operator. Mission computer design in this paper provides autonomy to the UAV in all phases of a typical UAV's mission except take off and land on. These provide the airplane with improved dynamic stability by regulating the flight parameters within limited ranges, at the same time tracking of UAV mission plan.

Although there are many control law architectures, the classic PID control approach augmented with online gain scheduling provides the ideal mix of robustness and performance for typical aircraft dynamics. The stability and control loops can be tuned to provide the desired performance and robustness specifications by adjusting a set of autopilot parameters or gains. But this is done through linear analysis—the nonlinear aircraft model is linearized for a representative set of flight conditions that cover the operating envelope of the aircraft. The linear dynamics of the closed-loop system (aircraft + autopilot) are analyzed in terms of stability and control responses (overshoot, settling time). By using fuzzy controllers, this difficult design process is avoided; nevertheless stable control and fast reaction time over conventional autonomous UAVs can be achieved as shown in this paper. The capability to do a dynamic planning of the desirable flight pattern is also important and this is done in this paper by using the current position of the moving UAV and the stationary target positions.

There are three key attributes for performing autonomous navigation; *perception, intelligence* and *action*. Perception means the ability of the UAV to acquire knowledge about the environment and itself. If UAV completes its mission without human support after takes off it means that UAV has intelligence attribute. And the last one is action which is the ability of vehicle to travel from point A to point B. How capable is a vehicle to perform these different functions is the metric to evaluate the degree of its autonomy. In this paper intelligence and action attributes applied by using fuzzy logic controllers successfully.

This paper also demonstrates that an UAV can apply a SID and TACAN approach if it has enough performance like an aircraft without human control. A series of SID for an airport can be planned just before the mission and the SID allowed by ATC (Air traffic controller) and meteorology can be applied autonomously. UAVs can apply a TACAN approach too if they are properly instrumented with VOR and TACAN receivers. Without a VOR receiver, a SID can be applied by using GPS. Based on these attributes, fuzzy logic has been identified as a useful tool for developing controllers for UAVs so that they will be able to perform autonomous navigation.

In this paper, a fuzzy logic based autonomous flight controller for UAVs is proposed. The simulation studies presented verify that the UAV can follow the predefined trajectories despite the simplicity of the controllers. However, as seen by the simulation results, there exist some oscillations and errors when wind effects are added to the simulation environment. In future studies, the goals are to develop algorithms for a better tuning of membership functions which utilize the well known ANFIS (Adaptive Neuro Fuzzy Inference Systems) approach or possibly type-2 fuzzy sets. Autonomous take off and land on will also be tried.

References

1. Doitsidis, L., Valavanis, K.P., Tsourveloudis, N.C., Kontitsis, M.: A framework for fuzzy logic based UAV navigation and control. In: Proceedings of the International Conference on Robotics Automation, vol. 4, pp. 4041–4046 (2004)
2. Schumacher, C.J., Kumar, R.: Adaptive control of UAVs in close-coupled formation flight. In: Proceedings of the American Control Conference, vol. 2, pp. 849–853 (2000)
3. Andrievsky, B., Fradkov, A.: Combined adaptive autopilot for an UAV flight control. In: Proceedings of the 2002 International Conference on Control Applications, vol. 1, pp. 290–291 (2002)
4. Dufrene, W.R., Jr.: Application of artificial intelligence techniques in uninhabited aerial vehicle flight. In: The 22nd Digital Avionics Systems Conference vol. 2, pp. 8.C.3–8.1-6 (2003)
5. Li, Y., Sundararajan, N., Sratchandran, P.: Neuro-controller design for nonlinear fighter aircraft maneuver using fully tuned RBF networks. Automatica 37, 1293–1301 (2001)
6. Borrelli, F., Keviczky, T., Balas, G.J.: Collision-free UAV formation flight using decentralized optimization and invariant sets. In: 43rd IEEE Conference on Decision and Control vol. 1, pp. 1099–1104 (2004)
7. Marin, J.A., Radtke, R., Innis, D., Barr, D.R., Schultz, A.C.: Using a genetic algorithm to develop rules to guide unmanned aerial vehicles. In: Proceedings of the IEEE International Conference on Systems, Man, and Cybernetics, vol. 1, pp. 1055–1060. (1999)
8. Ren, W., Beard, R.W.: CLF-based tracking control for UAV kinematic models with saturation constraints. In: Proceedings of the 42nd IEEE Conference on Decision and Control, vol. 4, pp. 3924–3929 (2003)
9. Dathbun, D., Kragelund, S., Pongpunwattana, A., Capozzi, B.: An evolution based path planning algorithm for autonomous motion of a UAV through uncertain environments. In: Proceedings of the 21st Digital Avionics Systems Conference vol. 2, pp. 8D2-1–8D2-12 (2002)
10. Global Robotic Observation System, Definition Of Aerosonde UAV Specifications: http://www.aerosonde.com/aircraft/
11. Unmanned Dynamics, Aerosim Aeronautical Simulation Block Set Version 1.2 User's Guide: http://www.u-dynamics.com/aerosim/default.htm
12. FlightGear Open-source Flight Simulator. www.flightgear.org
13. Qiau, W., Muzimoto, M.: PID type fuzzy controller and parameter adaptive method. Fuzzy Sets Syst. 78, 23–35 (1995)

From the Test Benches to the First Prototype of the muFly Micro Helicopter

Dario Schafroth · Samir Bouabdallah ·
Christian Bermes · Roland Siegwart

Originally published in the Journal of Intelligent and Robotic Systems, Volume 54, Nos 1–3, 245–260.
© Springer Science + Business Media B.V. 2008

Abstract The goal of the European project muFly is to build a fully autonomous micro helicopter, which is comparable to a small bird in size and mass. The rigorous size and mass constraints infer various problems related to energy efficiency, flight stability and overall system design. In this research, aerodynamics and flight dynamics are investigated experimentally to gather information for the design of the helicopter's propulsion group and steering system. Several test benches are designed and built for these investigations. A coaxial rotor test bench is used to measure the thrust and drag torque of different rotor blade designs. The effects of cyclic pitching of the swash plate and the passive stabilizer bar are studied on a test bench measuring rotor forces and moments with a 6–axis force sensor. The gathered knowledge is used to design a first prototype of the muFly helicopter. The prototype is described in terms of rotor configuration, structure, actuator and sensor selection according to the project demands, and a first version of the helicopter is shown. As a safety measure for the flight tests and to analyze the helicopter dynamics, a 6DoF vehicle test bench for tethered helicopter flight is used.

Keywords MAV · Test bench · Coaxial · Aerodynamics · Stability · Design · Flight test

D. Schafroth (✉) · S. Bouabdallah ·
C. Bermes · R. Siegwart
Autonomous System Lab,
Swiss Federal Institute of Technology Zurich (ETHZ),
Zurich, Switzerland,
e-mail: dario.schafroth@mavt.ethz.ch

S. Bouabdallah
e-mail: samir.bouabdallah@mavt.ethz.ch

C. Bermes
e-mail: christian.bermes@mavt.ethz.ch

R. Siegwart
e-mail: rsiegwart@ethz.ch

K. P. Valavanis et al. (eds.), *Unmanned Aircraft Systems.* DOI: 10.1007/978-1-4020-9136-0_14   245

1 Introduction

The state of the art in aerial robotics has moved rapidly from simple systems based on RC models, only able to do basic hover or cruise flights using inertial sensors, to robotic vehicles able to navigate and perform simple missions using GPS and/or vision sensors. In terms of fixed wing aircraft, this state of the art evolution concerns UAVs as large as 3 m like the Sky–sailor solar airplane [1], as well as vehicles as small as the Procerus MAV [2] or the Aerovironment's Black widow MAV [3]. Rotary wing systems also follow this state of the art evolution, with the largest available MAVs like the quadrotor developed at Cambridge university [4]. However, the research focus for palm size helicopters is still on vehicle design and flight stabilization. Here, examples are the muFR helicopter [5], the CoaX developed at EPF Lausanne [6] and the MICOR developed by the University of Maryland [7]. These MAV's are among the most advanced, but their capabilities are still limited to automatic hovering or human aided navigation. The European project muFly was launched in July 2006. Its project consortium consists of six partner institutions working on different fields as sensors, actuators and power supply. The goal of the project is the development of a fully autonomous micro helicopter, comparable in size and mass to a small bird. At this scale, a lot of challenges arise, such as the low efficiency of the rotor system [8] or the low thrust to weight ratio. The muFly project tackles these problems by investigating the different components of the system. Therefore, test benches are built to acquire experimental data. Together with simulation results, this data is used to design the actual helicopter. But even before these problems can be attacked, the general concept for the helicopter has to be chosen.

Today there exist many different configurations of rotary wing aircrafts, such as quadrotor, conventional single rotor, axial and coaxial helicopters (Fig. 1), and each one possesses advantages and drawbacks. Thus, a careful evaluation of each of them is necessary before the rotor configuration for muFly can be chosen. This includes important criteria like compactness, mass, power consumption and payload. Those criteria are listed, weighted and used for grading the different configurations in Table 1. Note that the two first criteria are muFly specific. *Compatibility 1* constrains the selection to the system specifications defined in the project proposal, while the *Compatibility 2* constrains the selection to the technology available from the project partners.

All these five different configurations are virtually designed at a scale comparable to that of muFly, and their respective masses and power consumptions are calculated. Every criterion is graded on a scale from 1 (worst) to 10 (best), only mass, power consumption and index indicator are calculated quantities. They are scaled to fit into the grading range. While the coaxial and axial concept mainly convince with their compactness and compatibility with the muFly specifications, the tandem and quadrotor achieve a high payload. The conventional helicopter configuration suffers from a lack of compactness and the fact that not all the power is used for propulsion (tail rotor). The evaluation shows that a coaxial setup is the best choice for the muFly application.

Next question is how to steer the helicopter, so different steering concepts are evaluated such as moving the center of gravity [6] or using flaps to change the orientation of the down wash [7]. For our project, a simplified swash plate mechanism

Fig. 1 Different helicopter configurations: a) Quadrotor, b) Axial, c) Conventional, d) Coaxial and e) Tandem

allowing only cyclic pitch is chosen. This is mainly due to its fast response to steering inputs.

The focus of this paper is how to build a micro helicopter from scratch by first designing different test benches for understanding the different problems affecting the design and then using this knowledge to design the helicopter itself. Designing an MAV is a very challenging task. There are many important aspects to look at, such as having an efficient propulsion group to achieve high thrust and high maneuverability

Table 1 Evaluation summary. Compatibility 1: Compatibility with the system specifications, Compatibility 2: Compatibility with the available technology

Criteria	Weight	Conventional	Axial	Coaxial	Tandem	Quadrotor
Compatibility 1	5	7	9	9	5	4
Compatibility 2	6	5	8	9	5	6
Compactness	8	5	10	10	4	3
Mass (/10)	8	5.96	6.26	6.45	6.45	7.99
Power consumption (/2)	8	5.54	5.75	6.21	6.21	7.95
Index indicator ($\times$10)	7	8.81	8.17	7	7	4.28
Realization simplicity	5	6	8	7	6	10
Control simplicity	5	6	7	8	7	9
Payload	4	6	4	6	8	8
Maneuverability	4	9	7	7	6	7
Reliability	6	5	4	7	6	7
Total		214.67	277.11	**295.76**	191.76	179.44

with a low power consumption. For a first estimation and layout of the different components, calculations and simulations are useful, but at the end the calculations have to be verified by experimental data. Therefore building test benches prior to designing the MAV itself is important at this state of the art, where a lot of research problems are still open.

The paper is organized as follows: in Section 2 the problem of low rotor efficiency is investigated using a coaxial rotor test bench to measure the resulting thrust and torque for different rotor blade design parameters. Section 3 is investigating the resulting forces obtained by the steering mechanism and furthermore the effects of the stabilizer bar used on the helicopter to obtain passive stability. After investigating the aerodynamical effects, the design of the first prototype is presented in Section 4, and a test bench for safe flight tests and for analyzing the helicopter dynamics is shown in Section 5. Finally the conclusion of the work and an outlook are given in Section 6.

2 Understanding the Propulsion

Aerodynamics is one of the major challenges faced in MAV design. In fact, the power required for the propulsion of a micro helicopter is more than 90% of the total power consumption and is the most limiting factor of the flight duration. It is important to understand the aerodynamic effects to have an efficient propulsion group and moreover to achieve appropriate control.

In the low Reynolds number regime the muFly rotors are operating in (about $Re \approx 60000$), viscous effects start to play an important role. Phenomena like laminar separation bubbles strongly affect the aerodynamic efficiency, which is much lower than for full scale helicopters. Their Figures of Merit (FM, ratio between induced power and total power) can reach up to 0.8, while for an MAV the FM is up to 0.5 [9]. Unfortunately, there is not a lot of literature and experimental data available for this range, one exception is [10]. The lack of trustful aerodynamical data makes it very important to have an own rotor measurement setup. Therefore, a coaxial rotor test bench has been designed and built for measuring thrust and torque of different parameter combinations on the rotor blades. Aside from the experimental investigation, simulation models are developed. Three common approaches are used for a better understanding of the problem:

1. Blade Element Momentum Theory (BEMT) using X–Foil software [11],
2. Free Vortex Wake Approach [12],
3. Computational Fluid Dynamics (CFD).

The complexity of the approach is increasing in the order of appearance. The BEMT simulation is used as a very fast first layout tool. The free vortex approach simulates the rotor wake which is not included in the BEMT and gives more information on the velocity field in the down wash, which strongly affects the lower rotor. The CFD simulation is then used for simulating all the 3D effects. Here, the commercial CFD code Ansys CFX is used.

The rotor test bench setup and its components are shown in Fig. 2, and its block diagram is shown in Fig. 3. The whole system is controlled by a PC through the Virtual Com Ports (VCP), which are connected to the motor controllers and to the data acquisition module. Thus, the rotor can be run at any desired angular velocity, and the respective thrust and torque are measured. The rotor heads on the test bench are designed such that the blades can be mounted rapidly and precisely at any pitch angle. For the experiments, it usually ranges from 10° to 20°. The motor controller module provides information about the motor current, which relates to the motor load torque. The rotor blades are designed in CAD and are directly manufactured on a rapid prototyping machine. This provides flexibility in testing the desired profiles and is only limited by the strength of the material. However, since the aerodynamic and centrifugal forces are relatively small, material strength

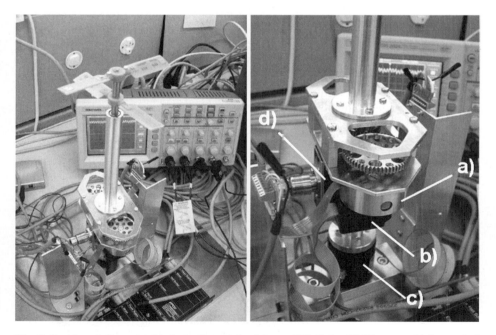

Fig. 2 Test bench for rotor blade testing: a) 2× Maxon EC 45 flat 30 W motor, b) 2× optical encoders, c) RTS 5/10 torque sensor, d) FGP FN 3148 force sensor

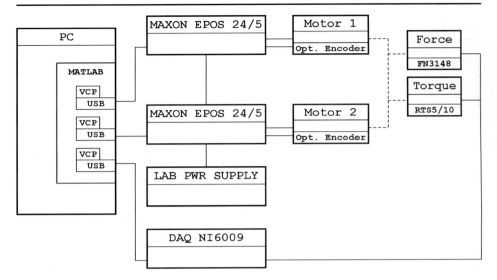

Fig. 3 Block diagram of the rotor test bench

is usually not a problem. Moreover, production on the rapid prototyping machine allows the testing of different aerodynamic enhancements like winglets or Gurney flaps. After designing the blades in CAD, the manufacturing takes only one hour of production time.

Some printed blades are displayed in Fig. 4.

Different parameters like radius, chord length, maximum camber, position of the maximum camber and twist are varied throughout testing. It is beyond the scope of this paper to discuss all the aerodynamic results, but for illustration Fig. 5 shows example results for different rotor radii. The data has been created using two *NACA0012* blades with a chord length of $c = 0.02$ m. The first two plots show the resulting thrust and torque at a pitch angle of $\Theta = 16°$ for different rotational speeds. As expected, a higher RPM leads to more thrust and also to higher torque,

Fig. 4 Different blades with different profiles, length and taper used on the test bench. The blades are manufactured using a rapid prototyping machine

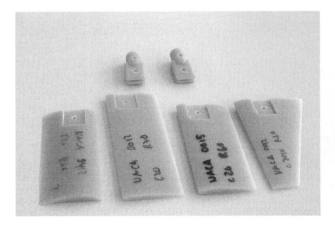

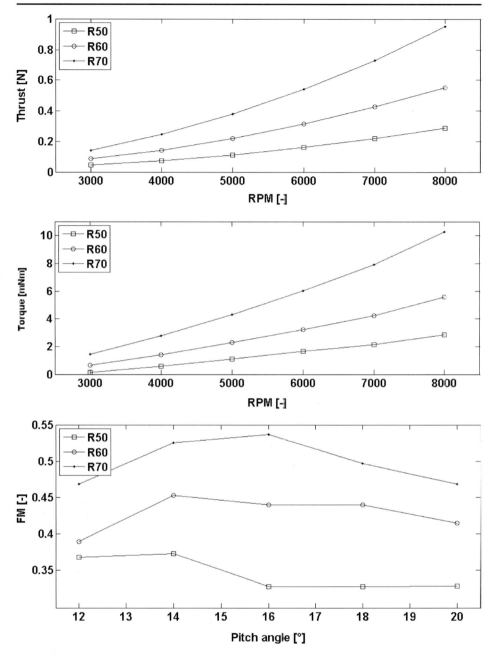

Fig. 5 Experimental results for rotors with different radii (*NACA0012, c* = 0.02 m)

since torque and thrust are related by the induced drag. Due to this dependency, the Figure of Merit *FM* [13] is used to evaluate the aerodynamic efficiency of the rotor. The third plot shows the *FM* of the different configurations, and it is obvious that a larger radius leads to a higher *FM*.

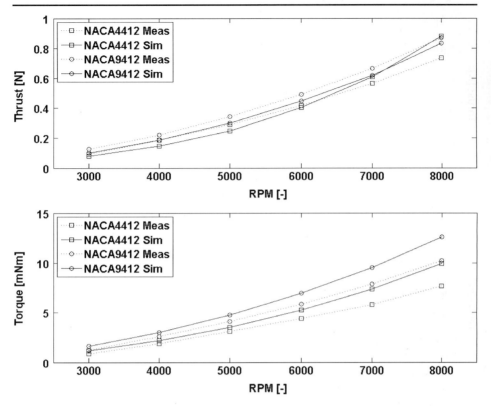

Fig. 6 Comparison between BEMT simulation and measurement ($R = 0.06$ m, $c = 0.02$ m)

On the simulation side, first results of the BEMT and CFD simulations show already good correlation, the in house development of a vortex approach code is still in development. In Fig. 6, some results for the BEMT simulation are shown. The results show the same behavior and order of magnitude as the measurement, but differ from the experimental data up to 20%. This is still in an acceptable range and is mainly a result of the inaccuracy of the BEMT model, unmodeled 3D effects (tip losses) and the aerodynamical coefficients obtained by X–Foil.

3 Understanding the Steering and the Passive Stability

Apart from the efficiency and the thrust of the rotors, there exists a strong interest in studying the behavior of the rotor during swash plate cyclic pitching. Since one goal of the project is to develop the whole propulsion system, it is necessary to quantify the forces and torques available for steering. In addition, the muFly helicopter uses a stabilizer bar for passive stabilization, thus it is important to know how to dimension this device. These two needs motivate the design of a swash plate test bench (STB) using a 6–axis sensor to measure the resulting torques and forces from swash plate inputs and helicopter motions, respectively (Fig. 7).

Fig. 7 Swash plate test bench.
a) Servo motor, b) Rotor,
c) Swash plate, d) 6–axis
sensor, e) Control board,
f) Electric motor

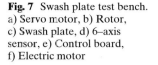
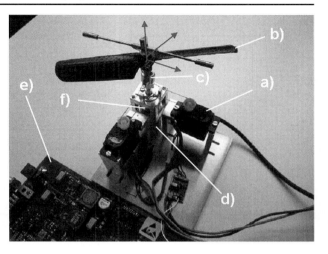

Two servo–motors actuate the swash plate, the resulting forces and torques are measured by the 6–axis sensor mounted beneath the motor. Since the side forces are very small (up to $\approx 0.15N$), and the sensor is very sensitive, obtaining meaningful results is not trivial. Nevertheless, after some changes in the mechanics and careful calibration, it is possible to measure the thrust vector direction and magnitude of a teetering rotor setup under cyclic pitch. This is still an ongoing work.

For the investigation of the passive stabilization, the STB is mounted on a pivoted platform driven by an electric motor (Fig. 8). The servo motors and swash plate are dismounted, and instead a rotor with stabilizer bar is mounted on the motor (see Fig. 7).

In the experiment, the STB is tilted with the platform, emulating a helicopter roll or pitch motion. As a result of the motion, the stabilizer bar exerts a cyclic

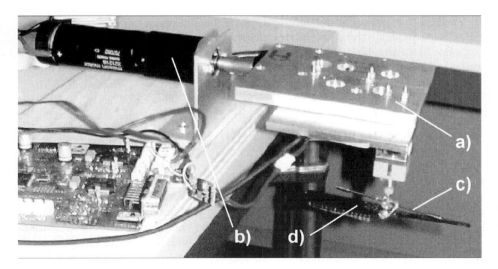

Fig. 8 STB with stabilizer bar mounted on pivoted platform: a) Platform, b) Electric motor for pivoting, c) Stabilizer bar, d) Rotor

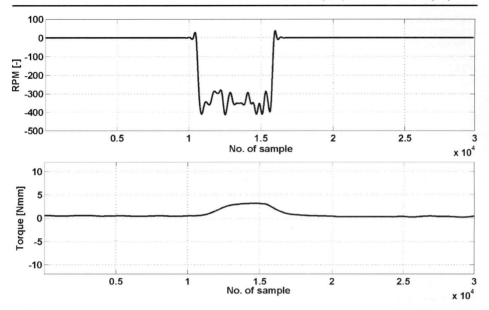

Fig. 9 Pivoted platform's angular velocity (top), and measured rotor moment along the pivoting axis (bottom)

input on the rotor, which in turn results in a change of the rotor's tip path plane orientation with respect to the body fixed coordinate system. The resulting forces and moments are measured by the 6–axis sensor. The investigation of different stabilizer bar inertias and constant phase angles with respect to the blade feathering axis is ongoing work. Figure 9 shows a first result of the experiments.

The two graphs show the pivoted platform's angular velocity and the measured moment along the pivoting axis, which corresponds to the stabilizer bar's response to a roll motion. With the start of the pivoting, the stabilizer bar reacts, however, it reaches mechanical saturation rather quickly. The result shows that the stabilizer bar response to an angular motion of the helicopter is measurable. Once the magnitude and direction of the response moment is known, this information can be used for the tuning of the stabilizer bar on the actual helicopter.

4 Designing the First Prototype

Among the different decisions related to MAV design, sensor selection is of high importance. It strongly influences the overall configuration and the performance of such vehicles. This is especially true because the designer is often constrained to few sensor variants to choose from. Table 2 shows possible combinations of different sensors for the five basic functionalities of muFly, namely: attitude and altitude control, take–off and landing, 3D navigation and obstacle avoidance. The evaluation of the different possibilities suggests the use of an omnidirectional camera with a laser for obstacle avoidance and navigation, an ultrasonic sensor for altitude control and an Inertial Measurement Unit (IMU) for attitude control.

Table 2 The concept selected for the first muFly prototype

	Attitude control	Altitude control	Take-off and landing	Navigation	Obstacle avoidance	Complexity
IMU	+					low
Laser Omnicam		+		+	+	high
Down looking camera		+	+	+		high
Down looking sonar		+	+			low
Side looking sonar				+	+	average
Forward looking ste. cam.		+		+	+	high
Down looking ste. cam.		+	+	+		high

The IMU and the omnidirectional camera are specifically designed for the purpose of the project. In fact, the IMU is an extremely lightweight piece of electronics combining state of the art 2D gyroscope and 3D accelerometer, for a total mass of 2 g. The novelty with the omnidirectional camera is the polar radial arrangement of pixels, which, in combination with a conic mirror and a 360° laser plane, provides an extremely lightweight (3.5 g) range finder based on triangulation.

On the structural side, a lightweight, robust and reliable frame is required for the muFly helicopter.

A first concept that incorporates all of these aspects is shown in Fig. 10.

The rotor system is surrounded by a cage–like carbon structure which offers several advantages: it ensures not only protection of and from the rotor system, but also provides the possibility to place the motors face to face on top and bottom of the cage. Thus, despite the coaxial rotor system no gear box is needed. Subsequently, losses due to gear box efficiency are reduced and the mechanical reliability is increased. Furthermore, with the cage the helicopter possesses a non–rotating surface on top, where the laser plane generator can be placed. This allows for

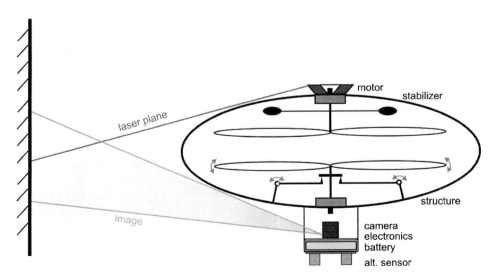

Fig. 10 The concept of the first prototype of the muFly helicopter

Fig. 11 Bimorph piezoelectric actuators: a) Swash plate, b) Piezo actuator, c) Rotor blade, d) BLDC motor

a sufficiently high distance between camera and laser plane to increase the resolution of the triangulation.

Concerning the actuators for propulsion, brushless DC (BLDC) outrunner motors are presently the best solution available in terms of power to mass ratio and thermal behavior. Adversely, the selection of appropriate steering actuators, where high bandwidth, stroke and force are needed, is much more difficult. Several actuation mechanisms have been looked at, and the decision to use piezoelectric elements has been made mainly because of their high bandwidth and precision. The four piezoelectric actuators used are operating at 150 V and their initial stroke is amplified by a lever arm. Placed in cross configuration, each opposite pair of actuators operates

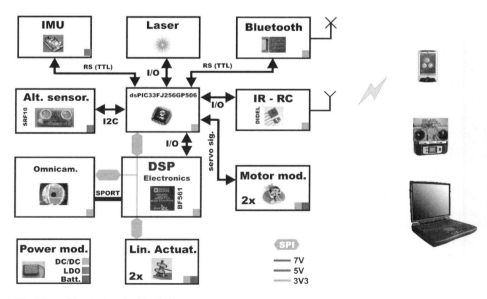

Fig. 12 muFly electronics block diagram

Fig. 13 The muFly helicopter
with a) carbon fiber sandwich
cage, b) stabilizer bar,
c) carbon fiber rotor blades
(in–house made), d) swash
plate, e) BLDC motors,
f) linear actuators, g) PCB
main board with double core
DSP, h) lithium–polymer
battery, i) ultra–sound sensor.
Not shown are the
omnidirectional camera, the
inertial measurement unit and
the laser plane generators
(not mounted)

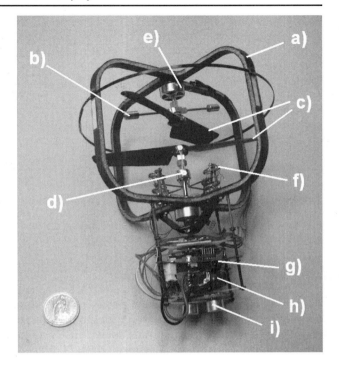

one of the swash plates tilting axes. While one actuator pulls, the other one pushes (push–pull mode) which results in a higher actuation force (Fig. 11).

On the electronic side, muFly uses a double core DSP and a micro controller for the sensor/actuator interface, an infrared receiver for manual control and a bluetooth module for communication with the ground station. The block diagram is shown in Fig. 12.

All the electronics are accommodated in a pod structure attached under the cage structure. This design meets the objectives in terms of compactness, mass and processing power, achieving the five capabilities listed before. Altogether, the first prototype of muFly is a coaxial helicopter with an overall mass of 78 g, 12 cm span and 15 cm height (Fig. 13).

5 Analyzing the Dynamics

The instability inherent to helicopters in general makes it very difficult to analyze the system behavior in flight, and thus to validate the dynamic simulation model. A training platform that eliminates the risk of a crash is a solution to this problem. However, it has to provide sufficient space for the helicopter to operate in, and, more importantly, it should only exert minimal external forces and moments on the helicopter in normal flight mode.

The vehicle test bench (VTB) created during the project is an original 6 DoF cable–based system on which one can mount a muFly helicopter on the central structure as shown in Fig. 14.

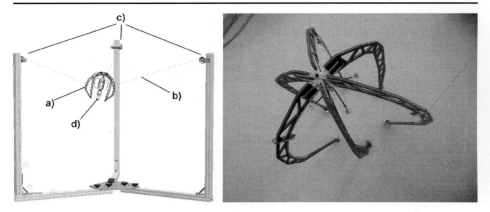

Fig. 14 CAD design (left) and manufactured carbon structure (right) of the vehicle test bench:
a) Carbon structure, b) Wires, c) Electric motors with pulleys, d) muFly helicopter

This structure is supported by three wires, and actively controlled by three electric
motors, which permits permanent gravity compensation. The control is done feed–
forward by using the geometric properties of the test bench and the motor controllers
(Fig. 15). Thus, for the three translations, muFly endures only the motors' and the
pulleys' inertia when translating in an arbitrary direction.

Concerning the three rotations, the central carbon–made structure (see Fig. 14
right) is equipped with three rolls allowing rotational motions thanks to its three sets
of arms. The three sets are mounted on a common axis and can rotate independently
of each other. Each set is composed of two arms, one for the gravity compensation

Fig. 15 Vehicle Test Bench
control block diagram

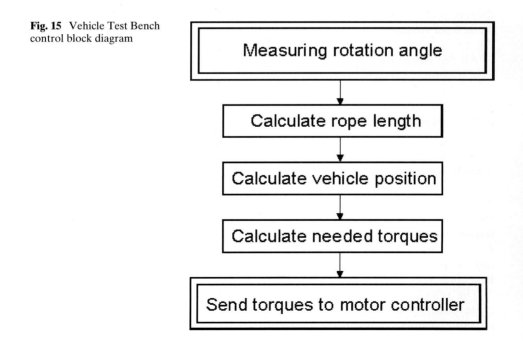

with a counter weight, and one for the wire connection to the motor pulleys. Thus, for the three rotations muFly endures only the low friction torques on the rolls. At the end, the VTB is mechanically almost transparent to the helicopter. This feature allows to fly the helicopter freely in the working space of the VTB, almost like in normal flight. At a later stage of the project, the VTB will be used for parameter identification experiments with the muFly helicopter.

6 Conclusion and Outlook

This paper presents the European project muFly and the approach advocated for the design of the micro coaxial helicopter, which consists first in the development of several test benches before developing the flying system itself.

The challenges of designing an efficient rotor in this scale is investigated by measuring the thrust and torque for different rotor configurations on a coaxial test bench. The test bench is explained in detail and first results are shown as an illustration. Furthermore the experimental research is supported by simulation and a first comparison is given. After looking at the efficiency of the rotor the forces and moments on the rotor during cyclic pitch are measured on a swash plate test bench (STB). In addition, the STB is used to investigate the effect of the stabilizer bar used on the helicopter.

The development of the first prototype is explained, especially the choice of the sensors, the actuators and the structural design. Finally, a vehicle test bench (VTB) system has been designed and built in order to test the passive stability of muFly and also to fly it safely in a confined work space. The VTB compensates all the gravity effects and emulates an almost free flight of the helicopter in safe conditions. Several investigations are still ongoing. On the aerodynamical side, more rotor configuration will be tested and compared with the further developed simulations with the goal of proposing an efficient propulsion system for a micro coaxial helicopter in the size of mufly. Additionally, the rotor system will be optimized to achieve high steering and control performance by extensive testing on the swash plate test bench. An intermediate design of the helicopter is already started for testing and optimization of the whole system.

Acknowledgements The authors would like to thank Mr. M. Buehler, Mr. D. Fenner and Mr. J. Nikolic for technical support. muFly is a STREP project under the Sixth Framework Programme of the European Commission, contract No. FP6-2005-IST-5-call 2.5.2 Micro/Nano Based Sub-Systems FP6-IST-034120. The authors gratefully acknowledge the contribution of our muFly project partners BeCAP at Berlin University of Technology, CEDRAT Technologies, CSEM, Department of Computer Science at University of Freiburg and XSENS Motion Technologies.

References

1. Noth, A., et al.: Autonomous Solar UAV for Sustainable Flights. Springer, ch. 12, pp. 377–405 (2007)
2. Taylor, C.N.: Techniques for overcoming inaccurate pose estimates in mavs. In: Proc. (IEEE) International Conference on Intelligent Robots (IROS'07), San Diego, USA (2007)
3. Grasmeyer, J., Keennon, M.: Development of the black widow micro air vehicle. In: Proc. 39th AIAA Aerospace Sciences Meeting and Exhibit, Reno, USA (2000)

4. Kemp, C.: Visual control of a miniature quad-rotor helicopter. Ph.D. dissertation, University of Cambridge (2006)
5. Wang, W., et al.: Autonomous control for micro-flying robot and small wireless helicopter x.r.b. In: Proc. (IEEE) International Conference on Intelligent Robots (IROS'06), Beijing, China (2006)
6. Bouabdallah, S., et al.: Design and control of an indoor coaxial helicopter. In: Proc. (IEEE) International Conference on Intelligent Robots (IROS'06), Beijing, China (2006)
7. Bohorquez, F.: Design, Analysis and Performance of a Rotary Wing Mav. Alfred Gessow Rotorcraft Center, USA (2001)
8. Pines, D., Bohorquez, F.: Challenges facing future micro air vehicle development. AIAA J. Aircraft, **43**(2), 290–305, (2006)
9. Bohorquez, F., et al.: Hover performance of rotor blades at low reynolds numbers for rotary wing micro air vehicles. An experimental and cfd study. AIAA Paper 2003-3930 (2003)
10. Coleman, C.P.: A Survey of Theoretical and Experimental Coaxial Rotor Aerodynamic Research. NASA Technical Paper 3675 (1997)
11. Drela, M.: X-foil subsonic airfoil development system. http://raphael.mit.edu/xfoil (2004)
12. Griffiths, D., Leishman, J.: A study of dual-rotor interference and ground effect using a free-vortex wake model. In: Proc. of the 58th Annual Forum of the American Helicopter Society, Montréal, Canada (2002)
13. Bramwell, G.D.A.R.S., Balmford, D.: Helicopter Dynamics. Butterworth-Heinemann (2001)

Modeling and Global Control of the Longitudinal Dynamics of a Coaxial Convertible Mini-UAV in Hover Mode

J. Escareño · A. Sanchez · O. Garcia · R. Lozano

Originally published in the Journal of Intelligent and Robotic Systems, Volume 54, Nos 1–3, 261–273.
© Springer Science + Business Media B.V. 2008

Abstract The aim of this paper is to present a configuration for a Convertible Unmanned Aerial Vehicle, which incorporates the advantages of the coaxial rotorcraft for hover flight and the efficiencies of a fixed-wing for forward flight. A detailed dynamical model, including the aerodynamics, is obtained via the Newton-Euler formulation. It is proposed a nonlinear control law that achieves global stability for the longitudinal vertical-mode motion. Indeed, we have performed a simulation study to test the proposed controller in presence of external perturbations, obtaining satisfactory results. We have developed an embedded autopilot to validate the proposed prototype and the control law in hover-mode flight.

Keywords Convertible UAV · Coaxial rotorcraft · Longitudinal control · Global stability · Embedded architecture

This work was partially supported by Mexico's National Council of Science and Technology (CONACYT).

J. Escareño (✉) · A. Sanchez · O. Garcia · R. Lozano
Universite de Technologie de Compiegne,
Centre de Recherches de Royallieu B.P. 20529,
60205 Compiegne, France
e-mail: juan-antonio.escareno@hds.utc.fr

A. Sanchez
e-mail: asanchez@hds.utc.fr

O. Garcia
e-mail: ogarcias@hds.utc.fr

R. Lozano
e-mail: rlozano@hds.utc.fr

1 Introduction

In recent years, the interest on Convertible Unmanned Aerial Vehicles (CUAVs) has increased since the operational scope is wider than that provided by conventional designs. These vehicles combine the maneuverability of the helicopter to deal with confined spaces (hover, vertical take-off and landing) and the forward-flight advantages of the airplane (speed and range). The flight profile of CUAVs improves the performance of missions like: the search and rescue of people in hazardous locations or circumstances (earthquakes, spills and fires); and surveillance or information collection on important installations in either highly sensitive locations (borders, ports and power-plants) or remote or uninhabitable places (polar zones, deserts and toxic areas). A logical drawback arising from single-propeller aircrafts is the vulnerability to reacting torque; to counteract this effect a complementary input of angular momentum is needed. Therefore, helicopters vehicles waste some power to feed a tail-rotor, while fixed-wing aircrafts, in hover flight, use differential control surfaces (elevon or ailerons). An alternative way to face the reacting torque, without using complementary controls, is the use of counter-rotating propellers. This configuration increases the thrust and reduces the size of the vehicle but it increases the mechanical complexity. In the CUAV's field, the tailsitter configuration stands out among other CUAVs configurations (tiltrotor, tiltwing, etc.) because it lacks of swashplate or tilting elements, instead it uses the propeller to wash the normal aircraft surfaces (elevator, aileron and rudder) so that the aircraft can generate the torques to handle vertical flight. However, in vertical-flight regime, this configuration is very vulnerable to wind gusts, which reveals the importance of robust hover flight for this kind of vehicles. We have fused the previous concepts to obtain a vehicle that offers the mechanical simplicity of tail-sitter and the operational flexibility of coaxial rotorcrafts. The topic of CUAVs is currently addressed by numerous research

Fig. 1 Experimental convertible coaxial UAV: Twister

groups. In [1], the authors provide a simulation study on trajectory tracking for the RC model of the Convair XFY-1[1]. This paper presents a quaternion-based control for autonomous hover flight and a feedback linearization control of the longitudinal dynamics in order to perform the transition. Another approach is presented in [2], where the authors present an autonomous hover flight and perform the transition to forward flight manually. Also, in [3] is presented the T-wing configuration, which is a twin-engine tail-sitter Unmanned Aerial Vehicle (UAV) that performs vertical flight, using linear control techniques to stabilize the linearized vertical dynamics.

This paper contributes with the proposition of an alternative tail-sitter configuration capable of performing hover flight (see Fig. 1). A detailed mathematical model, including the aerodynamics, is obtained via the Newton-Euler formulation. In terms of control, we propose a control algorithm that achieves global stability for the longitudinal underactuated dynamics during vertical flight.

The paper is organized as follows: Section 2 presents the overall equations of motion of the mini-UAV. The control design to stabilize the longitudinal dynamics, during hover mode, is described Section 3. A simulation study to observe the performance of the propose control law is provided in Section 4. The experimental setup, including the real time results of Twister's attitude stabilization in hover mode are presented in Section 5. Finally some concluding remarks and perspectives are given in Section 6.

2 Twister's Description

Our prototype is mechanically simpler than the classical coaxial design since we use two fixed rotors instead of a rotor hub, which is composed of linkages and swashplates for two rotor discs assembled around the rotor shaft. Since the structure of most tail-sitter UAVs has a predominant fixed-wing structure, consequently a weaker torque generation is obtained to achieve hover flight. To overcome this drawback, we have placed a rotor near the control surfaces to increase the air flow, improving the aerodynamic-based torque. Additionally, the fact that the propellers are mounted over the same axis of center of gravity (CG), represents a mechanical advantage because the inherent adverse torque gyroscopic and blade's drag torques) is considerably reduced. The vertical roll motion is controlled the rudder's deflection . The vertical pitch is controlled by the elevator deflection . Due to the counter-rotating capabilities the yaw motion is auto-compensated, however the blade geometry is not exactly the same, therefore we handle the yaw remanent by propellers differential angular speed (see Fig. 2).

2.1 Longitudinal Dynamic Model

In this section we obtain the longitudinal equations of motion of the Twister through the Newton-Euler formulation. Let $\mathcal{I}=\{i_x^{\mathcal{I}}, k_z^{\mathcal{I}}\}$ denote the inertial frame, $\mathcal{B}=\{i_x^{\mathcal{B}}, k_z^{\mathcal{B}}\}$

---

[1] hobby-lobby.com/pogo.html.

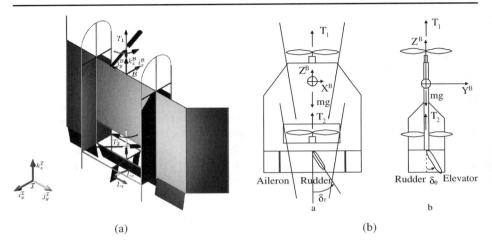

(a)   (b)

Fig. 2 Twister mini UAV. **a** Free-body diagram. **b** Actuators of the vehicle

denote the frame attached to the body's aircraft whose origin is located at the CG and $\mathcal{A}=\{i_x^A, k_z^A\}$ represent the aerodynamical frame (see Fig. 3). Let the vector $q = (\xi, \eta)^T$ denotes the generalized coordinates where $\xi = (x, z)^T \in \Re^2$ denotes the translation coordinates relative to the inertial frame, and $\eta = \theta$ describes the vehicle attitude. The general rigid-body equations of motion, based on the Newton-Euler approach [4, 5], are given by

$$\begin{cases} V^B = \dot{\xi}^B \\ m\dot{V}^B + \Omega \times mV^B = F^B \\ \Omega = q = \dot{\theta} \\ \mathbf{I}\dot{\Omega} + \Omega \times \mathbf{I}\Omega = \Gamma^B \end{cases} \tag{1}$$

Fig. 3 Forces of the vehicle

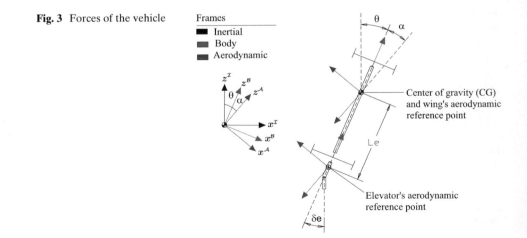

where $F^B \in \Re^2$ and $\Gamma^B \in \Re^2$ are respectively the total force and torque applied to the aircraft's CG, $m \in \Re$ denotes the vehicle's mass, $\Omega = q$ is the body frame angular velocity, $V^B = (u, w)^T$ is the translational velocity of the aircraft's center of mass, $\mathbf{I} \in \Re$ contains the moment of inertia about B.

The orientation of the mini-UAV is given by the orthonormal rotation matrix $\mathcal{R}^{B \to \mathcal{I}}$.

$$\mathcal{R}^{B \to \mathcal{I}} = \begin{pmatrix} c_\theta & s_\theta \\ -s_\theta & c_\theta \end{pmatrix} \tag{2}$$

where we have considered the notations $s = \sin$ and $c = \cos$.

2.1.1 Aircraft's Aerodynamics

In order to continue with the dynamic analysis, it is worth providing several aerodynamic notions.

The aerodynamic frame $\mathcal{A}$, in longitudinal motion, may be seen as the result of a rotation provided by the attack angle α, which is the aircraft-relative angle between the center line and the velocity of the aircraft relative to the air mass through which it flies (Relative wind or free-stream). The transformation matrix to transform a the coordinate of a vector in the aerodynamic frame to the body and inertial frame are shown respectively by the following (for details see [6] and [7])

$$R^{A \to B} = \begin{pmatrix} c_\alpha & s_\alpha \\ -s_\alpha & c_\alpha \end{pmatrix} \text{ and } R^{A \to \mathcal{I}} = \begin{pmatrix} c_{(\theta + \alpha)} & s_{(\theta + \alpha)} \\ -s_{(\theta + \alpha)} & c_{(\theta + \alpha)} \end{pmatrix} \tag{3}$$

The frame $\mathcal{A}$ is formed using as reference the direction of the wind velocity vector; it contains the principal aerodynamic forces: the lift L and the drag D

$$F_a^A = \begin{pmatrix} -L \\ -D \end{pmatrix} = \begin{pmatrix} \frac{1}{2} \rho S \left(C_{D_c} + k C_L^2 \right) V_r^2 \\ \frac{1}{2} \rho S \left(C_{L_\alpha} \alpha \right) V_r^2 \end{pmatrix} \tag{4}$$

where ρ represents the air density, S represents the wing surface, including the elevator, α is the attack angle, V_r is wind velocity vector and finally C_L, C_{D_0}, C_{L_α} and $C_{L_{\delta_e}}$ are typical aerodynamical non-dimensional coefficients of drag and lift.

For future control analysis it is convenient to provide several assumptions which establish the aerodynamic scenario where the UAV will operate in hover flight:

- **A1.** The aircraft's control surfaces are submerged by the propeller air slipstream
- **A2.** The magnitude of the drag force is smaller compared with the lift L and thrust T.
- **A3.** Since the air slipstream coincides with zero-lift line of a symmetric wing profile, the aerodynamic terms (L,D) vanish at the vertical equilibrium point.
- **A4.** The lift L_w is provided by the wing for translation motion and the lift L_e is provided by the elevator for rotational motion
- **A5.** The CG and the aerodynamic center of the wing are located at the same point

2.1.2 Translational Motion

Body frame: The translation motion of the UAV is described by the following vectorial equation

$$m\dot{V}^B + \Omega \times mV^B = R^{\mathcal{I} \to \mathcal{B}} mG^{\mathcal{I}} + T^B + R^{\mathcal{A} \to \mathcal{B}} F_a^{\mathcal{A}} \qquad (5a)$$

where $G^{\mathcal{I}} \in \mathfrak{R}^2$, $G^{\mathcal{I}} = (0, -g)$ is the gravity vector, $T^B \in \mathfrak{R}^2$, $T^B = (0, T_1 + T_2)$ is the thrust provided by motors and $F_a^{\mathcal{A}} \in \mathfrak{R}^2$, $F_a^{\mathcal{A}} = (-L, -D)^T$ is the aerodynamic force vector. The previous equations may be rewritten in scalar form as

$$\begin{cases} m\dot{u} = -mqw + mg\sin(\theta) - L\cos(\alpha) - D\sin(\alpha) \\ m\dot{w} = mqu - mg\cos(\theta) + L\sin(\alpha) - D\cos(\alpha) \end{cases} \qquad (6)$$

Inertial frame The dynamic behavior of the vehicle relative to the inertial frame is described by the following expression

$$V^{\mathcal{I}} = \dot{\xi}$$
$$m\dot{V}^{\mathcal{I}} = mG^{\mathcal{I}} + R^{\mathcal{B} \to \mathcal{I}} T^B + R^{\mathcal{A} \to \mathcal{I}} F_a^{\mathcal{A}} \qquad (7)$$

and the scalar form of the previous equation

$$\begin{cases} m\ddot{X} = T\sin(\theta) - L\cos(\theta + \alpha) - D\sin(\theta + \alpha) \\ m\ddot{Z} = T\cos(\theta) + L\sin(\theta + \alpha) - D\cos(\theta + \alpha) - mg \end{cases} \qquad (8)$$

2.1.3 Rotational Motion

The rotational motion of the longitudinal dynamics is described by the following equation

$$I_y\ddot{\theta} = L_e\cos(\alpha)l_e = U(\delta_e) \qquad (9)$$

Parasitic torques Usually the propeller introduces parasitic torques, such as the gyroscopic effect and the blade's drag. However, for this configuration, the counter-rotating propellers reduce considerably the gyroscopic torque and the fact that the propellers are located over the CG's line avoids the moment generation of the propeller's drag.

3 Longitudinal Control

3.1 Attitude Control

In order to design a controller that stabilizes the attitude dynamics (Eq. 9) of the mini-UAV we propose the following Lyapunov function

$$V = \frac{1}{2}I_y\dot{\theta}^2 + k_{s1}\ln(\cosh(\theta)) \qquad (10)$$

where $V \geq 0$ and $k_s > 0$. The corresponding time derivative is given as

$$\dot{V} = I_y\dot{\theta}\ddot{\theta} + k_{s1}\tanh(\theta)\dot{\theta} \qquad (11)$$

the previous can be rewritten as

$$\dot{V} = \dot{\theta} \left(I_y \ddot{\theta} + k_{s1} \tanh(\theta) \right) \tag{12}$$

now, in order to render $\dot{V}$ negative definite, we propose the following control input

$$U(\delta_e) = -k_{s1} \tanh(\theta) - k_{s2} \tanh(\dot{\theta}) \tag{13}$$

where $k_{s2} > 0$, then substituting the control input in Eq. 12 leads to the following expression

$$\dot{V} = -\dot{\theta} k_{s2} \tanh(\dot{\theta}) \tag{14}$$

finally, it follows that $\dot{V} < 0$, it proves that the proposed control input (Eq. 13) stabilize the attitude of the vehicle around the origin.

3.2 Position and Attitude Control

The hover flight represent a critical stage for fixed-wing vehicles that perform hover, since there exist aerodynamic forces that act as perturbations during the vertical flight mode. Therefore, is important to provide a robust enough algorithm to handle large amount of angular displacements.
From Eqs. 8 and 9 we can obtain the longitudinal dynamics

$$\ddot{X} = T \sin(\theta) - L \cos(\theta + \alpha) - D \sin(\theta + \alpha) \tag{15}$$

$$\ddot{Z} = T \cos(\theta) + L \sin(\theta + \alpha) - D \cos(\theta + \alpha) - 1 \tag{16}$$

$$I_y \ddot{\theta} = u_\theta(\delta_e) \tag{17}$$

Recalling **A1-A4** we can rewrite Eqs. 15, 16 and 17 as the following reduced model:

$$\ddot{z} = T \cos \theta - 1 \tag{18}$$

$$\ddot{x} = T \sin \theta \tag{19}$$

$$\ddot{\theta} = u_\theta(\delta_e) \tag{20}$$

However, we will show that our control law performs satisfactorily even if we include these aerodynamic terms (perturbation) in the simulation study.

3.2.1 Control Strategy

Our control strategy consists in bounding the pitch attitude (θ) via a tracking error involving at the same time the equation that will be used, after accomplishing certain design conditions, for the stabilization of the translational positions (z, x).
For design purposes let us introduce the following altitude control input whose stabilization will be reached when the tracking error vanishes.

$$T \triangleq (r_z + 1) \sqrt{(k_1 \dot{x} + k_2 x)^2 + 1} \tag{21}$$

where $z_s = -l_1\dot{z} - l_2(z - z_d)$, z_d is the desired setpoint and l_i for $i = 1, 2$ are positive constants.

$$\theta = \arctan(-k_1\dot{x} - k_2 x)$$
$$\ddot{e}_x = \ddot{\theta} + \gamma$$
$$\ddot{\theta} = \tau_\theta = -k_{e1}\dot{e}_x - k_{e2}e_x - \gamma$$

Let us introduce a tracking error

$$e_x \triangleq \theta - \arctan(-k_1\dot{x} - k_2 x) \tag{22}$$

where k_1 and k_2 are positive constants.

From the previous, we obtain the error dynamics

$$\ddot{e}_x = \tau_y + \frac{k_1\dot{T}\sin\theta + k_2 T\sin\theta + k_1 T\dot{\theta}\cos\theta}{(k_1\dot{x} + k_2 x)^2 + 1}$$
$$- \frac{2(k_1\dot{x} + k_2 x)(k_2\dot{x} + k_1 T\sin\theta)(\dot{e}_x - \dot{\theta})}{(k_1\dot{x} + k_2 x)^2 + 1} \tag{23}$$

where

$$\dot{e}_x = \dot{\theta} + \frac{k_2\dot{x} + k_1 T\sin\theta}{1 + (k_1\dot{x} + k_2 x)^2} \tag{24}$$

and

$$\dot{T} = \frac{(k_1\dot{x} + k_2 x)(k_2\dot{x} + k_1 T\sin\theta)(z_s + g)}{\sqrt{(k_1\dot{x} + k_2 x)^2 + 1}}$$
$$- (l_2\dot{z} + l_1(T\cos\theta - g))\sqrt{(k_1\dot{x} + k_2 x)^2 + 1}$$

In order to stabilize the error dynamics, the following controller is proposed

$$\tau_y = -k_3\dot{e}_x - k_4 e_x$$
$$- \frac{k_1\dot{T}\sin\theta + k_2 T\sin\theta + k_1 T\dot{\theta}\cos\theta}{(k_1\dot{x} + k_2 x)^2 + 1}$$
$$+ \frac{2(k_1\dot{x} + k_2 x)(k_2\dot{x} + k_1 T\sin\theta)(\dot{\theta} - \dot{e}_x)}{(k_1\dot{x} + k_2 x)^2 + 1}$$

where k_3 and k_4 are positive constants such that the polynomial $s^2 + k_3 s + k_4$ is stable.

Table 1 Twister simulation parameters

Parameter	Value
m	$1kg$
g	$1\frac{m}{s^2}$
length	$0.8m$
width	$0.25m$
I_y	$0.00069kg \cdot m^2$

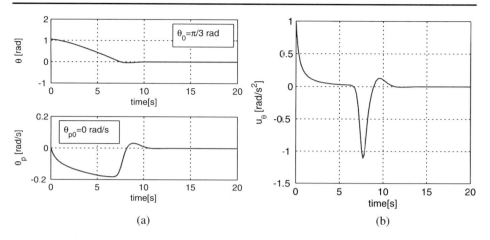

Fig. 4 Pitch. **a** Angular position and rate behavior. **b** Control input

Consequently, Eq. 23 becomes

$$\ddot{e}_x = -k_3 \dot{e}_x - k_4 e_x$$

and therefore $\dot{e}_x \to 0$ and $e_x \to 0$ asymptotically.

Fig. 5 Simulation results

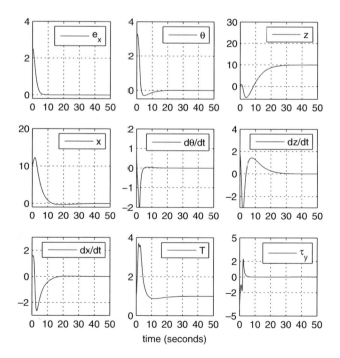

For a time $t_1 > 0$ large enough $\dot{e}_x$ and e_x are arbitrarily small, hence we get

$$\cos\theta = \frac{1}{\sqrt{(k_1\dot{x} + k_2 x)^2 + 1}} \tag{25}$$

$$\sin\theta = \frac{-k_1\dot{x} - k_2 x}{\sqrt{(k_1\dot{x} + k_2 x)^2 + 1}} \tag{26}$$

From (18), (21) and (25) we obtain

$$\ddot{z} = -l_1\dot{z} - l_2(z - z_d)$$

Fig. 6 Process of the embedded system

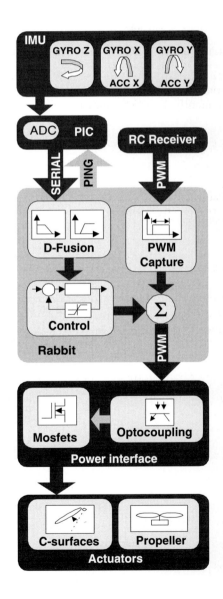

(a)   (b)

Fig. 7 Embedded system. **a** Embedded IMU. **b** Embedded microntroller

Choosing l_1 and l_2 such that the polynomial $s^2 + l_1 s + l_2$ is stable, $\dot{z} \to 0$ and $z - z_d \to 0$, then for a time large enough $t_2 > t_1$ the Eq. 21 reduces to

$$T = g\sqrt{(k_1\dot{x} + k_2 x)^2 + 1} \tag{27}$$

From Eqs. 19, 26 and 27 the dynamics of the horizontal position is described by

$$\ddot{x} = -gk_1\dot{x} - gk_2 x$$

Proposing k_1 and k_2 such that the polynomial $s^2 + gk_1 s + gk_2$ is stable, then $\dot{x} \to 0$ and $x \to 0$. Now, from Eqs. 22 and 24 leads to $\dot{\theta} \to 0$ and $\theta \to 0$. Finally all the states converges to the desired values as $t \to \infty$.

4 Simulation Study

In order to validate the control strategy described in the previous section, we have run simulations to observe the performance of the vehicle. In Table 1 are depicted the parameters employed in the simulations.

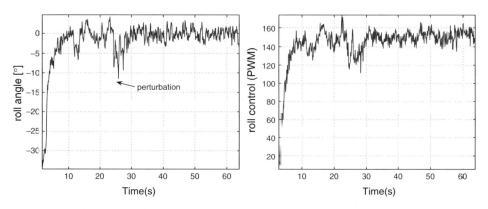

Fig. 8 Roll performance

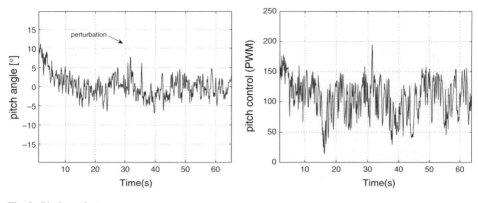

Fig. 9 Pitch performance

4.1 Attitude

The vehicle starts with initial angle of $\theta = \frac{pi}{3}$ and an angular velocity of $\dot{\theta} = 0$. With unitary saturation gains, i.e. $k_{s1} = 1$, $k_{s2} = 1$ (Fig. 4).

4.2 Position and Attitude

The vehicle starts, first from the initial conditions $(x, z, \theta) = (12, 8, \pi)$, the control objective is to regulate the attitude and the $x-$ position, and to track the altitude to a desired value ($z^d = 10$).

Figure 5 shows the error tracking behavior e_x and its influence over the whole dynamics. Also, observe the relation between θ and x, as well the good performance of the control law to stabilize a distant initial conditions.

5 Experimental Setup

In this section we describe the low-cost embedded [see Fig. 6] system to carry out an autonomous hover flight. This embedded platform includes three basic modules: The inertial measurement unit (IMU) and the onboard computer.

1. *IMU:* We have built an IMU that includes a dual-axis accelerometer, which senses the angular position of the vehicle (ϕ, θ), and three gyros, which sense the angular rate of the vehicle ($\dot{\phi}, \dot{\theta}, \dot{\psi}$).
2. *Embedded Control:* The IMU (analog signals) feeds the PIC microcontroller which sends this information to the Rabbit microcontroller throughout the serial port. The inertial information is filtered to get rid of the electrical and mechanical noise (mainly due to rotor's gearbox). Finally, the control signal is sent to the motors (propellers) and servomotors (ailerons-elevator) via the PWM ports (see Figs. 6 and 7).

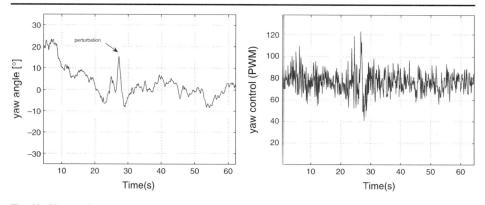

Fig. 10 Yaw performance

5.1 Experimental Performance

In the experimental stage of this undergoing pro ect, the experiments were focused to stabilize the 3DOF attitude dynamics (roll, pitch, yaw). The overall performance is depicted in Figs. 8, 9 and 10.

6 Concluding Remarks

We have detailed the longitudinal dynamics of the vehicle based on the Newton-Euler formulation. We have proposed a CUAV design that blends the capabilities of fixed-wing vehicles (range, endurance) with the hover flight advantages of a coaxial helicopter (reactive torque cancelation). A global control law is proposed to stabilize the longitudinal dynamics of the Twister in the vertical flight. An embedded system (IMU and microcontroller) was developed to implement the proposed control law. We have performed an autonomous hover flight in hover mode.

References

1. Knoebel, N., Osborne, S., Snyder, D., Mclain, T., Beard, R., Eldredge, A.: Preliminary modeling, control, and trajectory design for miniature autonomo s tailsitter. In: Proceedings of the AIAA Guidance, Navigation and Control Conference and Exhibit, Keystone, August 2006
2. Green, W.E., Oh, P.Y.: A MAV that flies like an a rplane and hovers like a helicopter. In: Proceedings of the 2005 IEEE/ASME International Conference on Advanced Intelligent Mechatronics Monterey, 24–28 July 2005
3. Stone, H.: Aerodynamic modelling of a wing-in-slipstream tail-sitter UAV. In: 2002 Biennial International Powered Lift Conference and Exhibit, Williamsburg, 5–7 November 2002
4. Bedford, A., Fowler, W.: Dynamics. Addison-Wesley, Reading (1989)
5. Goldstein, H., Poole, C.P., Safko, J.: Classical Mechanics. Adison-Wesley, Reading (1980)
6. Etkin, B., Reid, L.: Dynamics of Flight. Wiley, New York (1991)
7. Stevens, B.L., Lewis, F.L.: Aircraft Control and Simulation 2ed. Wiley, New York (2003)

Subsonic Tests of a Flush Air Data Sensing System Applied to a Fixed-Wing Micro Air Vehicle

Ihab Samy · Ian Postlethwaite · Dawei Gu

Originally published in the Journal of Intelligent and Robotic Systems, Volume 54, Nos 1–3, 275–295.
© Springer Science + Business Media B.V. 2008

Abstract Flush air data sensing (FADS) systems have been successfully tested on the nose tip of large manned/unmanned air vehicles. In this paper we investigate the application of a FADS system on the wing leading edge of a micro (unmanned) air vehicle (MAV) flown at speed as low as Mach 0.07. The motivation behind this project is driven by the need to find alternative solutions to air data booms which are physically impractical for MAVs. Overall an 80% and 97% decrease in instrumentation weight and cost respectively were achieved. Air data modelling is implemented via a radial basis function (RBF) neural network (NN) trained with the extended minimum resource allocating network (EMRAN) algorithm. Wind tunnel data were used to train and test the NN, where estimation accuracies of 0.51°, 0.44 lb/ft^2 and 0.62 m/s were achieved for angle of attack, static pressure and wind speed respectively. Sensor faults were investigated and it was found that the use of an autoassociative NN to reproduce input data improved the NN robustness to single and multiple sensor faults. Additionally a simple NN domain of validity test demonstrated how the careful selection of the NN training data set is crucial for accurate estimations.

Keywords Micro (unmanned) air vehicle · Flush air data sensing systems · Extended minimum resource allocating neural networks · Fault accommodation

1 Introduction

Air data measurements are a crucial part of any flight data acquisition system. The basic air data required to compute most other air data quantities of interest (e.g. airspeed, rate of climb etc.) are freestream pressure (this consists of total and static pressure) and aerodynamic orientation (angle of attack and angle of sideslip). They

I. Samy (✉) · I. Postlethwaite · D. Gu
Department of Engineering, University of Leicester, Leicester, UK
e-mail: isarl@le.ac.uk

are essential for flight performance evaluation and are also required in flight control systems onboard the aircraft. Conventionally, critical air data are measured using *air data booms* protruding from the aircraft local flow field. Freestream pressure is measured using a pitot-static tube while angle of attack (α) and angle of sideslip (β) are measured using small vanes mounted on the boom. Different designs and applications may exist, however the basic air data boom remains the most popular method for air data measurements.

Despite their popularity, air data booms are known to have measurement disadvantages in addition to possible malfunctions: accuracy may be adversely affected by boom bending and vibration, probe size and geometry, and by the flow interference due to the probe itself. Furthermore, in military-related applications, external instrumentation is undesirable in stealth vehicles. As a result, in recent years more research has been carried out to find alternative solutions to air data booms. An example is the use of optical air data systems, which measure the atmosphere outside of an air vehicle and provide information regarding the environment ahead of the flight vehicle [1]. These systems are very accurate and more importantly are not affected by weather external to aircraft such as icing or plugging. However with the primary goal of most air vehicle manufacturers being the reduction of costs, researchers found the concept of air data measurements using a matrix of pressure orifices/ports to be a cheaper alternative to optical systems and booms.

The measurement of flush surface pressures to estimate air data parameters has been known for some time and is referred to as a flush airdata sensing (FADS) system. The first FADS system was developed and tested on the NASA X-15 hypersonic aircraft [2, 3]. It considered a hemispherical nose, mounted with 4 pressure ports, which was steered into the wind vector to measure the air data. Results were promising, however the concept of the steered nose was considered too complex. Consequently, over the years the FADS system experienced many modifications and successful applications. It would be difficult to discuss all of them here. However, most aeronautical applications of the FADS system originate from the initial tests carried out by NASA in the early 1980s. Examples include [4–6]. Recently the FADS system was implemented on the NASA Dryden F-18 Systems Research Aircraft [7]. This system uses 11 pressure ports in the radome of the aircraft and was tested at speed up to Mach 1.6, α up to 80° and $\beta \pm 20°$. Other applications of the FADS system include [8–10].

From the literature we realize that few examples have been extended to a micro (unmanned) air vehicle (MAV) and this motivated the need to investigate such an application. MAVs are found within the spectrum of unmanned air vehicles (UAVs) and are characterised by their small size. Several military missions have taken advantage of this feature to use them in applications which can only be attained at great risks. It is perhaps important to note that the FADS system is an invaluable alternative to air data booms specifically for MAV applications. This is because current air data booms can be too heavy and expensive for use on a MAV. Additionally, due to the dangerous and secretive environments that they can be exposed to, external instrumentation is best avoided.

One way to derive air data states from the surface pressure measurements is to define and solve a suitable aerodynamic model. It involves defining a nonlinear aerodynamic model with known structure and unknown coefficients, using complex flow physics theory. The unknown coefficients can then be derived via nonlinear regression algorithms [7, 11]. Simpler approaches consider a basic lookup table, where calibration data is used to create the lookup table. This is the least efficient of methods as it can result in high memory usage and slow execution times. The model-based method is considerably faster. However not only does it require detailed knowledge of the system, but the iterative algorithms used to solve the model can be computationally slow.

As an alternative, we investigate the feasibility of using neural network models. *Neural networks* (NNs) provide a means of modelling linear or nonlinear systems without the need of detailed knowledge of the system. They primarily rely on sufficient training data from which they can develop their structure. This makes them an attractive modelling solution to applications where the theory is poor but the relevant data is plentiful [8]. An advantage of NNs is that it is possible to implement them in just a few lines of code which is suitable in MAV applications where computational power may be limited. A disadvantage however of NNs is that they suffer from the ability to incorporate known system properties into the model structure. Its structure is purely based on available system input/output data. Therefore the training algorithm and its implementation is a crucial step in NN design. The two most popular NN architectures are the *multilayer perceptron* (MLP) and the *radial basis function* (RBF) NN. MLP NNs trained with error backpropagation algorithms have been successfully implemented in a FADS system by several authors [8, 9, 12, 13]. Instead, we consider a RBF NN trained with the *extended minimum resource allocating network* (EMRAN) algorithm [14]. Unlike the conventional MLP NN and the RBF NN, the EMRAN RBF NN does not generally suffer from dimensionality problems. This is because the number of hidden neurons is not fixed *a priori* but is instead continuously updated based on set criteria. For example a hidden neuron will be completely removed during training if its contribution to the NN estimates is insignificant. This property and other global approximation capabilities made the EMRAN RBF NN a strong candidate for our application.

Most applications tend to mount the FADS system near the aircraft nosetip mainly for two reasons. Firstly the aerodynamic model relating the surface pressure and air data states is derived around a blunt body, and so is most valid at the vehicle stagnation point. Secondly the nosetip has been used traditionally as the air data measurement location. Unfortunately many MAVs (as it is in our case) are driven by a nose-propeller which obstructs the mounting of the FADs system. Therefore we consider the wing leading edge as an alternative location.

In conclusion, our work is distinct from mentioned studies in that: (1) A FADS system is designed and implemented on the wing of a MAV which flies at speeds as low as Mach 0.07 and (2) an EMRAN RBF NN is configured to model the aerodynamic relationships in the FADS system. The report is organized as follows. Section 2 introduces the general structure of the aerodynamic model relating pressure and air data. The EMRAN RBF NN used to model the system is then discussed in detail. Section 3 presents the experimental apparatus. We also briefly discuss the *computational fluid dynamic* (CFD) analysis done in order to decide on suitable locations of the FADS system. The exact location of the pressure orifices is then concluded. Section 4 is the methodology section of our work. Section 5 presents the results. Detailed analysis of the results is discussed. This includes analyzing and improving the robustness of the FADS system to sensor faults. We also discuss and prove the importance of carefully selecting the NN training data. Conclusions to the work done here can be found in Section 6. Last but not least, Section 7 outlines the future work which will be carried out to verify the wind tunnel results.

2 System Modelling Using Neural Networks

2.1 Air Data System Model

The minimum air data required to calculate the relevant flow field properties include; temperature (T), freestream static pressure (P_∞), total (stagnation) pressure (P_0), angle of attack (α) and sideslip angle (β). Using these basic air data parameters ($T, P_\infty, P_0, \alpha, \beta$) most other air data of interest can be directly calculated [8]. Conventionally an air data boom consisting of a pitot-static tube and wind directional vanes is used to accurately measure $P_\infty, P_0, \alpha, \beta$. However for reasons discussed earlier we aim to estimate these four parameters using a FADS system. The FADS system installed on an air vehicle essentially converts a vector of surface pressure measurements to the air data states $P_\infty, P_0, \alpha, \beta$.

The pressure distribution on the surface of a fixed vehicle geometry is not uniform and is generally a function (denoted by f) of wind speed (V) and vehicle aerodynamic orientation [9]:

$$\mathbf{p} = f(V, \alpha, \beta) \tag{1}$$

For incompressible flow (applicable at low-speed flight):

$$P_0 - P_\infty = \frac{1}{2}\rho V^2 \tag{2}$$

where ρ is the air density and (along with temperature) is assumed invariant [15]. Therefore we can also re-write Eq. 1 as a function of the following variables:

$$\mathbf{p} = f(P_\infty, P_0, \alpha, \beta) \tag{3}$$

The inverse of Eq. 3 is:

$$(P_\infty, P_0, \alpha, \beta) = f'(\mathbf{p}) \tag{4}$$

In a FADS system, the air data states $P_\infty, P_0, \alpha, \beta$ are estimated by solving Eq. 4 using the available surface pressure measurements $\mathbf{p}$.

There are several ways to implement the function f', and one popular technique used by NASA on their research aircrafts is via nonlinear regression analysis, where f' is derived from advanced flow physics theory [7]. Another approach is via lookup tables. However both these techniques can either suffer from the requirement of detailed knowledge of the complex function f' or potentially slow execution times. As an alternative we demonstrate the possibility of using NNs to model f'.

2.2 EMRAN RBF NN

Neural networks have been used in a wide range of engineering applications. Examples of such applications and a good general introduction to NNs can be found here [16, 17]. With NNs, the designer requires minimal knowledge of the system. Instead, the model is developed via neural network training/learning.

The two most popular NN architectures are the MLP NN and the RBF NN. Over the years there has been a particular interest in RBF NNs due to their good generalisation performance and simple network structure [14]. A good review of RBF NNs can be found here [18]. A typical two-layered RBF NN consisting of input units, a hidden layer and an output layer is shown in Fig. 6. The input units are not counted as a layer because they perform no computation. This is why weights are not found between the input units and hidden layer. The input units simply transfer the system input vector to the hidden units which form a localized response to the input pattern. The hidden layer includes the nonlinear radial basis functions. Theoretical analysis and practical results have shown that the choice of nonlinearity is generally not crucial to the RBF NN performance [19]. Gaussian functions are typically used. The output layer then performs a simple linear combination of the hidden layer outputs.

A typical single output RBF NN with N hidden units can be expressed as follows (at sample instant i):

$$ys_i = \lambda_{i0} + \sum_{n=1}^{N} \lambda_{in} \exp\left(\frac{-\|\mathbf{x}_i - \mu_n\|^2}{\sigma_n^2}\right) \tag{5}$$

where ys_i is the current NN output, $\mathbf{x}_i$ is the current input vector, λ are the individual weights found between the hidden and output layer, and μ and σ are the centres and widths (fixed) of the Gaussian functions respectively. $\|\cdot\|$ is the Euclidean norm. In the conventional implementation of the RBF network, N is assumed to be known a priori and a training algorithm is used to update the weights of the RBF NN. In other words the centres and widths of the Gaussian functions are fixed. This approach was appealing as conventional linear optimisation methods could be used to update the weights.

From Eq. 5 we realise that this basic approach requires an exponentially increasing number of hidden units N versus the size of the input space. The NN is said to suffer from the curse of dimensionality. To overcome this problem and the need to heuristically identify a suitable number of hidden units, a more advanced network structure was developed, known as the EMRAN algorithm [14]. Let us now briefly present the mathematical foundations of the EMRAN RBF NN.

A single output RAN RBF NN starts with zero hidden units. It then *only* adds hidden units if *all three* of the following criteria are met:

$$e_i = y_i - ys_i \qquad > \text{E1} \tag{6}$$

$$e_{iRMS} = \sqrt{\sum_{j=i-(M-1)}^{i} \frac{e_j^2}{M}} \quad > \text{E2} \tag{7}$$

$$d_i = \|\mathbf{x}_i - \mu_{ir}\| \qquad > \text{E3} \tag{8}$$

where y_i is the current target (ideal) output, ys_i is the current NN estimation and μ_{ir} is the centre of the hidden unit *closest* to the current input vector $\mathbf{x}_i$. E1, E2 and E3 are fixed, pre-defined thresholds. E1 ensures that the estimation error is below a set threshold. E2 checks if the root mean square (RMS) of the past M errors is also below a set threshold. E3 checks if the minimum distance between the current input vector and the centres of the hidden units is significantly small. Only if *all* criteria

(6–8) are met is a new hidden unit added. If *one or two* of the criteria (6–8) are not met, then a training algorithm updates all existing free parameters; centres, widths and weights. The resulting NN is referred to as a RAN RBF NN and was originally developed by Platt [20] to overcome the dimensionality problems associated with the conventional RBF NN.

This can be extended to a MRAN RBF NN by pruning the hidden units which contributed the least to the network output. This is done by recording the output of each hidden unit and normalizing them to the highest hidden unit output at sample instant *i*:

$$r_{in} = \frac{o_{in}}{\max \{o_{i1}, o_{i2}, \ldots, o_{iN}\}} \tag{9}$$

where for N hidden units, o_{in} is the nth hidden unit output at sample instant i. If r_{in} remains below a set threshold for several consecutive data inputs, the corresponding hidden unit can be removed.

The MRAN RBF NN attempts to use a minimum number of hidden units, which speeds up the NN processing time. The weakness of the MRAN RBF NN however, is that all the free parameters (centres, widths and weights) are updated in every step which is impractical especially if there are many hidden units, i.e. N is large. This was found to be dangerous especially in online applications where the NN processing time must be less than the input sampling time to avoid any time delays. Consequently an EMRAN RBF NN was developed which updates the parameters of only one 'winner' neuron in an attempt to speed up the training process while maintaining the same approximation characteristics of the MRAN RBF NN. The neuron with the centre closest to the data input vector $\mathbf{x}_i$ is chosen as the 'winner' neuron as it will probably contribute the most to the network output.

2.3 NN Training

Many NN training algorithms exist today, and the designer generally tends to choose a specific algorithm based on its speed, accuracy and compact structure. Probably the most popular NN training algorithm is the *error back-propagation* (BP) algorithm. This method (which is also referred to as the method of *gradient descent*) updates the free parameters of the NN in an iterative fashion along the NN error surface with the aim of moving them progressively towards the optimum solution. Other training algorithms which are typically used in the RBF NN are the *least mean square* (LMS) algorithm and the *extended kalman filter* (EKF). In the conventional RBF NN 5, where the hidden layer is fixed, the weights connecting the hidden and output layers are updated via LMS algorithms. This technique benefits from linear optimisation but as we have mentioned earlier, can sometimes require a large RBF NN structure. The non-linear EKF can be used to train the EMRAN RBF NN and has shown to produce good results [21].

In our case we will use a slightly modified gradient descent algorithm to train the EMRAN RBF NN. The training algorithm proceeds as follows [22]:

$$\mathbf{\theta_{i+1}} = \mathbf{\theta_i} - \eta \frac{\partial ys_i}{\partial \mathbf{\theta_i}} \bigg|_i e_i - \eta \sigma \mathbf{\theta_i} \tag{10}$$

where θ is the vector of free parameters (made up of centres, widths and weights), η is the learning rate, the differential is the gradient of the RBF output with respect to the free parameters, e is as in Eq. 6, and σ is a positive constant known as the stabilizing factor chosen by the designer. The second term on the right hand side of Eq. 10 is simply the delta rule associated with the gradient descent algorithm. The third term is designed to counteract casual parameter drifting by slowly driving the parameters towards zero.

Training the NN structure can be done either offline or online (i.e. onboard the vehicle). The basic principle of NN training is that the error between the NN estimate and the real measurement is back-propagated to update the NN structure appropriately. Therefore during online learning the NN requires a reference measurement of its estimates onboard the vehicle, which is generally obtained via sensors. In our case this means additional instrumentation (e.g. an air data boom) which directly measure the air data states. This is impractical as we aim to minimize costs and overall vehicle weight and so most FADS systems tend to train/calibrate the NN offline using wind tunnel or flight data. The final NN structure is then frozen and used onboard the air vehicle.

A difficult decision during NN training is deciding when to stop. In general there are no well-defined criteria for stopping [17]. Rather, there are some reasonable criteria, each with its own practical merit, which may be used to terminate NN training. In our case we will implement two popular criteria simultaneously. The first criterion checks for model convergence and is based on the *rate of change in the root mean squared error per epoch* (ΔRMS), where 1 epoch represents 1 pass through the whole training data. The second criterion is based on the generalisation performance of the NN. A common problem during NN training is over-fitting the structure. In this scenario, the NN has been trained to best fit the *training set* and can consequently produce poor estimations when queried with new data which is within the bounds of the original training set, i.e. generalisation is poor. To tackle this problem, the NN must be simultaneously queried with an independent data set (which we will refer to as the *testing set*) during the offline training process. Training is then stopped if the estimation errors of the testing set start to increase.

The correct choice of the NN learning rate is crucial as a high learning rate guarantees good estimations but it also degrades the global approximation capability of the network. Currently there is no formal guideline to defining the optimum learning rate and other tuning parameters in a NN a priori nor is there one optimum NN structure and training algorithm which suits all applications [8]. The designer must apply a heuristic-based approach when designing the NN. Satisfactory performance is then generally concluded based on the estimation characteristics and execution speed of the NN.

2.4 Summary

An EMRAN RBF NN trained with a modified gradient descent algorithm 10 will be implemented to construct the inverse function f' (Eq. 4). NN inputs will include the vector of pressure measurements from the FADS system, and the NN outputs will include the air data state estimates (all data obtained from wind tunnel tests). In this report we consider only three air data state estimates; $\hat{\alpha}, \hat{P}_\infty, \hat{V}$. Note that P_0 can then be calculated from P_∞ and V using Eq. 2.

To avoid over-fitting the NN structure, two independent data sets will be considered. The first data set (*training set*) will be used to construct the NN structure while the second data set (*testing set*) will be simultaneously used to query the NN with learning switched off. Training will then be safely stopped once the testing set RMS estimation error increases for 100 consecutive epochs. In parallel to this, to check for NN convergence, training can also be stopped if the ΔRMS of the training set is less than 0.1% for 100 consecutive epochs.

3 Equipment

3.1 The MAV

The MAV was obtained from the Blue Bear Systems Research (BBSR) Ltd. The aim was to design and install a FADS system on the wing of the MAV to estimate three airdata states $\hat{\alpha}$, $\hat{P}_\infty$, $\hat{V}$. The MAV uses a MH64 wing section. Some properties are shown in Table 1.

3.2 Wind Tunnel Set-Up

Wind tunnel experiments were carried out at Leicester University. The wind tunnel used is an open-ended subsonic wind tunnel capable of reaching speeds of 40 m/s. It has a sufficiently large working section of 0.46 × 0.46 m. The MAV wing section was sting mounted in the wind tunnel (Fig. 1). The *external balance* supporting the sting bar included a scaled-turning knob (calibrated in deg) which allowed the variation and measurement of wing α. To allow for sideslip variations the standard sting bar was slightly modified (Fig. 3). The part allowing the sideslip comprises two rectangular pieces of metal with one end rounded to a semicircle. The straight edge of one is connected to the sting bar and the straight edge of the other is connected to a bar which is fixed to the wing (near the wing *centre of gravity*). The two plates can rotate about a hole through which passes a bolt to lock them together at the chosen angle. The rounded ends carry a scale to indicate the set wing β. A pitot-static tube was mounted in the wind tunnel ahead of the wing to measure P_0 and P_∞. Using these two measurements, wind speed V can then be calculated using Eq. 2. Therefore overall, the four air data states P_∞, P_0, α, β were measurable.

3.3 FADS System-Matrix of Pressure Orifices (MPO)

The FADS system consists of a matrix of pressure orifices (MPO) and a suitable data acquisition (DAQ) system. The location and design of the MPO is vehicle-dependant, i.e. there is no standard place of installing them. The general rule is

Table 1 MAV characteristics

Characteristics	Value
Speed range	8–20 m/s
Mass	450 g
Wing Span	488 mm
Wing root chord	250 mm
Wing tip chord	200 mm
Wing thickness	8.61%

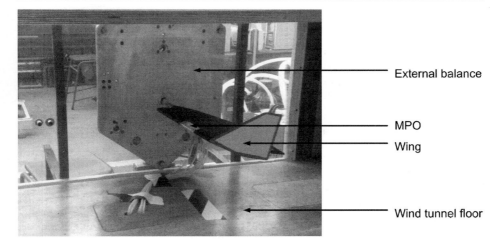

External balance

MPO
Wing

Wind tunnel floor

Fig. 1 Wing *sting* mounted in wind tunnel (shown at positive α and zero β)

that the MPO must experience large pressure gradients when air data quantities are changed. From a typical *pressure coefficient* (Cp) plot we already know that this generally happens near the wing leading edge. However, the exact MPO locations must be derived from CFD analysis of the airfoil. The MAV wing built in Gambit is shown in Fig. 2. It is important to note that we are only interested in identifying pressure gradients and therefore accurate estimates of surface pressure magnitudes are not necessary, i.e. strict CFD convergence criteria are not applied. CFD analysis was implemented in Fluent, and the final pressure ports arrangement is shown in Fig. 3. The number of pressure ports in the MPO was chosen as a compromise between the need to accurately estimate the air data states and the cost of instrumentation. As there are three air data states $\hat{\alpha}$, $\hat{P}_\infty$, $\hat{V}$ to be estimated it was reasonable to assume that a minimum of three pressure ports will be required. Two extra ports were added in order to improve the redundancy options and also decrease the overall noise

Fig. 2 MAV 3D-Wing section built in Gambit for CFD analysis

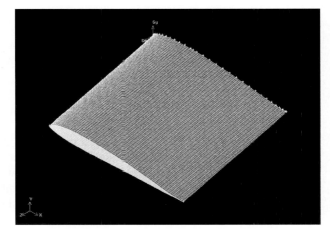

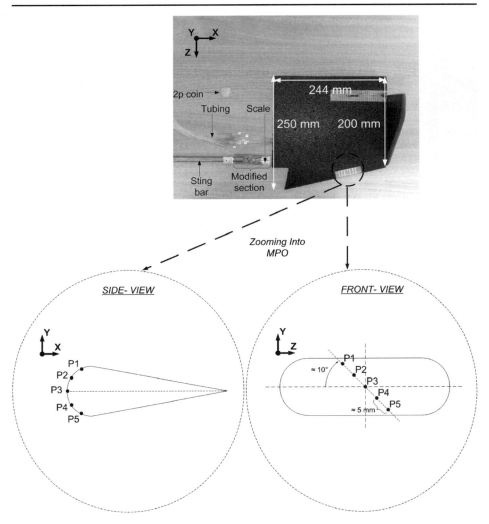

Fig. 3 Top view of wing and MPO (with pressure ports P1, P2, P3, P4 and P5). MPO not shown to scale

sensitivity of the FADS system. Overall, the MPO consisted of five pressure ports (P1, P2, P3, P4, P5), 0.5 mm orifice diameter each.

3.4 FADS System-Data Acquisition (DAQ)

Each orifice was connected to a calibrated differential (with respect to room pressure) pressure sensor via pressure tubing. It is important to keep the tubing distance to a minimum to avoid any time lags. The sensor board was therefore placed as close as possible to the wind tunnel giving an approximate tubing length of 0.38 m. The pressure sensors have a pressure range of ± 5 in. H_2O and were read in by a PC-based

16 bit DAQ card at a sampling frequency of 50 Hz. The DAQ software used to log the pressure data for each test was Labview. This concludes the DAQ section of the FADS system.

The remaining wind tunnel measurements consisted of the four air data states P_∞, P_0, α, β which are used as a reference for calibrating the FADS system. The pitot-static tube is connected to the DAQ card and P_0, P_∞ are recorded in Labview (wind speed V is then calculated in Labview). A potentiometer was connected to the turning knob of the external balance and calibrated to measure applied α, and β was simply recorded from the scale on the sting bar. Digital acquisition of β settings was not necessary as dynamic tests were only done for changes in α but fixed β. To compensate for any noise corruption, wind tunnel data was initially filtered. A second order Butterworth low pass filter was implemented in Labview. To comply with the Nyquist sampling theorem, the cutt-off frequency chosen must be smaller than half the sampling frequency. With a sampling frequency of 50 Hz, the cutt-off frequency was heuristically chosen to be 1.5 Hz.

4 Tests

The tests are divided into two groups. Group 1 considered static tests where angle of attack rate ($\dot{\alpha}$) is zero. It consisted of two data sets, one for training the NN (Tr$_1$) and the other to query it (Te$_1$) with learning switched off. Group 2 considered dynamic tests where the concluded NN (from group 1) was implemented on a new test set Te$_2$ (where $\dot{\alpha}$ is nonzero and NN learning is switched off).

4.1 Static Tests

Static tests were run for an α range of $-9°$ to $13°$, in $2°$ increments. This was done at three different speeds V; 12, 15 and 20 m/s, and seven different β settings; $-9°$, $-6°$, $-3°$, $0°$, $3°$, $6°$, $9°$. Overall there were 252 different static tests. As only static tests are considered here, pressure data was logged only when steady state measurements were achieved.

Following DAQ the recorded data was used to train or test the NN. The 252 static tests were divided into a NN training set and a NN testing set by taking α slices, giving Tr$_1$ at $-9°$, $-5°$, $-1°$, $3°$, $7°$, $11°$ and Te$_1$ at $-7°$, $-3°$, $1°$, $5°$, $9°$, $13°$. In conclusion, 126 static tests were used to train the NN, and the remaining 126 were used to query it.

4.2 Dynamic Tests

The concluded NN structure from the static tests was implemented on Te_2. Dynamic tests were run at a fixed speed V of 15 m/s and β of $0°$. The wing angle of attack $\alpha(t)$ was randomly varied at different $\dot{\alpha}(t)$ and with different waveforms; square-wave, sine-wave and ramp-type. Pressure data was simultaneously logged from each pressure port at a sampling frequency of 50 Hz. Note that certain time evolutions of $\alpha(t)$ were not feasible in real flight, but were necessary to analyze the estimation characteristics of the trained NN.

5 Results and Discussion

5.1 Static Tests

5.1.1 Wind Tunnel Data

Wind tunnel data was recorded prior to any NN implementation. Overall there were 252 different static tests. Figure 4 shows the MPO pressure distribution for example

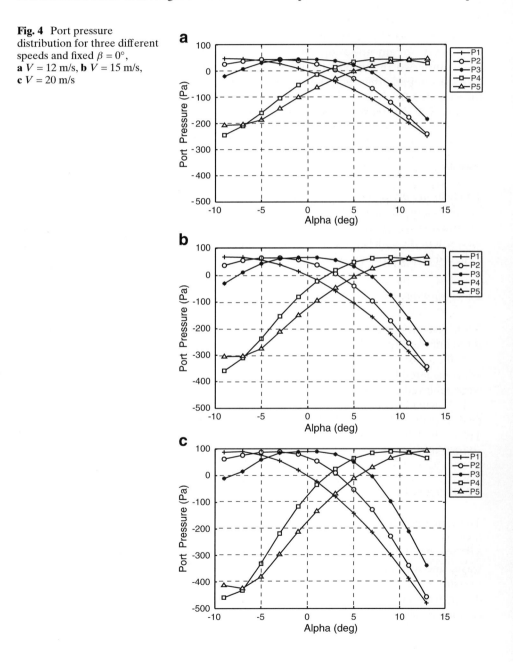

Fig. 4 Port pressure distribution for three different speeds and fixed $\beta = 0°$, **a** $V = 12$ m/s, **b** $V = 15$ m/s, **c** $V = 20$ m/s

scenarios. From Fig. 4 we notice: (1) Ports have different pressure distributions and (2) large pressure gradients occur for different wind tunnel settings. These properties facilitate NN mapping of pressure data to the air data states. We also notice how the variation in pressure with wing angle of attack is relatively linear which makes the air data estimation problem simpler.

5.1.2 NN Training and Testing Stage

It is perhaps important to restate that there is no universally accepted NN structure that is guaranteed to produce optimum estimates for all applications. Furthermore, the network developer must use a trial and error approach to define suitable NN tuning parameters. This is because parameters contained in the NN do not represent meaningful quantities in terms of the system being modelled. The main NN tuning parameters used in the EMRAN RBF NN are shown in Table 2. All the NN inputs were normalized between 0 and 1 before implementation in the NN.

NN training was stopped if the ΔRMS was less than 0.1% for more than 100 consecutive epochs and/or the RMS estimation error for Te_1 increased for more than 100 consecutive epochs. These criteria checked for NN structural convergence and overfitting respectively. Figure 5 shows the NN training stage. Note that the estimation error units are not shown as it is made up of mixed NN outputs. Training was eventually stopped after 582 epochs. The NN structure was then frozen. The resultant NN is a fully connected 5–3–3 NN (Fig. 6).

The NN testing stage simply involves recording the NN estimation errors after training is stopped. The 5–3–3 NN was queried with Te_1 and the results recorded (Fig. 11, Appendix). Overall, the NN RMS estimation errors were 0.51°, 0.44 lb/ft^2 and 0.62 m/s for $\hat{\alpha}$, $\hat{P}_\infty$, $\hat{V}$ respectively. From Fig. 11 we notice how estimation errors tend to be smaller for tested speeds of 15 m/s. This is because testing data for speeds of 12 and 20 m/s are more likely to be outside the limits of the training data set Tr_1, and therefore NN extrapolation was necessary which can lead to larger errors. This observation is related to the NN domain of validity which will be discussed in detail later.

5.1.3 Fault Accommodation

Faults in a FADS system can be caused by pressure orifices blockage, sensor errors and electrical wiring failures. It is important that the system maintains satisfactory performance under these conditions. Fault detection has been the subject of many engineering applications over the years. The most straightforward method being the use of physical redundancy where faults are detected based on a voting scheme. Following fault detection, faulty measurements must then be accommodated in the

Table 2 EMRAN RBF NN main tuning parameters (refer to Eqs. 6, 7, 8 and 10)

Tuning parameter	Value
E1	0.2
E2	0.1
E3	0.3
η	0.007
σ	1e-6

Fig. 5 NN training/testing
RMS estimation errors

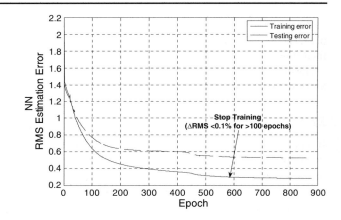

system to avoid any performance degradation. In our study we will not be discussing fault detection due to the extensive literature available covering this topic. Instead we will be investigating the effects on the NN estimation performance once a fault is detected and ways to reduce these effects, i.e. we will be discussing fault accommodation.

To investigate this, we artificially introduce faults in the NN testing data set Te_1. The type of fault we will consider is *total sensor failure*. This is a catastrophic failure where the sensor stops working and outputs a constant zero reading. As we are so far only considering static tests, the time evolution of the fault is not considered.

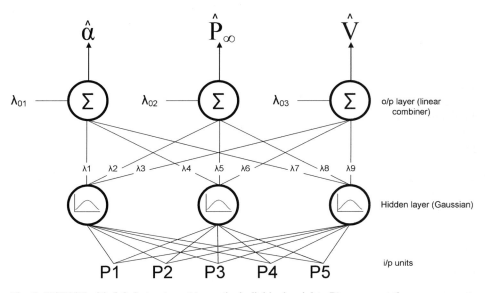

Fig. 6 RBF NN with 5–3–3 structure. λ's are the individual weights. P's represent the pressure ports in Fig. 3

The trained 5–3–3 NN structure (Fig. 6) is queried with Te₁ but this time one, two or three of the NN inputs are set to zero. Three fault accommodation scenarios are then studied:

1. *No correction*: Faulty inputs are not accommodated.
2. *Next port*: In this case we simply accommodate the fault by replacing its measurement with the neighbouring non-faulty port. So for example in Fig. 3 if P1 measurements are faulty, then pressure from P2 replaces them. The NN will therefore have pressures ports; P2, P2, P3, P4 and P5 as inputs. If P1, P3 and P5 are faulty then the NN inputs would be P2, P2, P2, P4 and P4.
3. *AA–NN*: The faulty port reading is replaced with the corresponding output from an *autoassociative* NN (AA–NN). An AA–NN is one which simply reproduces its inputs at its outputs [17]. In our case, a fully connected 5:10:5 EMRAN RBF NN was developed to do so.

Figure 7 shows the results for the fault accommodation tests. In general we notice that if the fault is not accommodated, the NN RMS estimation errors increase significantly especially in the presence of multiple faults (with $\hat{\alpha}$ RMS error reaching 14° for three faults). However when fault accommodation is implemented by replacing the faulty port measurement with its neighbouring port, estimation errors are reduced significantly. In our case using the 'Next port' option resulted in an average 50.39% decrease in the NN RMS estimation errors. On the other hand using the 'AA–NN' option resulted in a larger reduction of 69.6%. The use of the autoassociative NN greatly improved the robustness of the FADS system to faults.

A drawback of using the AA–NN is the further memory usage required onboard the air vehicle. The 'Next Port' option does not suffer from this which is why in many applications of the FADS system, the designer tends to include redundant pressure orifices. In our design we can see that the pairs P1, P2 and P4, P5 could be considered as redundant port locations as they have similar pressure distributions (Fig. 4).

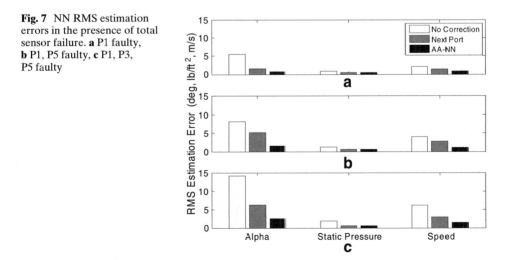

Fig. 7 NN RMS estimation errors in the presence of total sensor failure. **a** P1 faulty, **b** P1, P5 faulty, **c** P1, P3, P5 faulty

In conclusion, including redundant pressure ports in the FADS system improves the robustness of the overall system to faults. Furthermore it can mitigate the effects of system noise. However in applications where space onboard the vehicle is limited or instrumentation costs need to be low, an AA–NN can be used to reproduce pressure measurements which are faulty or corrupted by noise.

5.2 Dynamic Tests

The concluded NN structure from the static tests (Fig. 6) was implemented on Te_2 where $\dot{\alpha}(t) \neq 0$. At a fixed β of $0°$ and V of 15 m/s, the wing angle of attack was varied continuously and the NN estimate $\hat{\alpha}(t)$ recorded. An overall NN RMS estimation error of $0.58°$ was achieved for Te_2.

Figures 8 and 9 show the NN estimation characteristics for two examples of the dynamic tests carried out. Let us refer to them as *dynamic test 1* (DT1) and *dynamic test 2* (DT2) respectively. Let us now investigate the estimation characteristics of DT1 in detail (Fig. 8). It can be seen that at some time frames the NN estimations are significantly poor. For example at around 120 seconds, where $\alpha > 15°$, the NN underestimates α by a large amount. At first it may seem that this is simply due to random estimation error patterns. However if we observe the estimations from DT2 (Fig. 9) we notice that at around 75 s, where again $\alpha > 15°$, the NN underestimates α. These observations show that NN performance is poor for specific α settings.

The cause of this performance degradation is related to the domain of validity of the NN. It is suggested that the accuracy of the response of any multidimensional interpolator to new data is dependent on the location of that point relative to the domain of validity of the model being used. The domain of validity is in turn related to the boundaries of the training set [23]. A general rule is that when the NN is queried with data which lies outside the boundaries of the training set, NN estimations are poor. In simpler terms, the extrapolation properties are poorer than the interpolation properties. In our case, the method of dividing the training and testing data set was chosen for ease of presentation. However, ideally a more robust approach is needed to define Tr_1, so that the domain of validity encompasses all possible input data patterns.

Fig. 8 RBF NN estimation ($\hat{\alpha}(t)$) for DT1

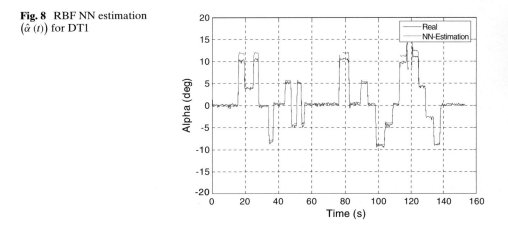

Fig. 9 RBF NN estimation $(\hat{\alpha}\,(t))$ for DT2

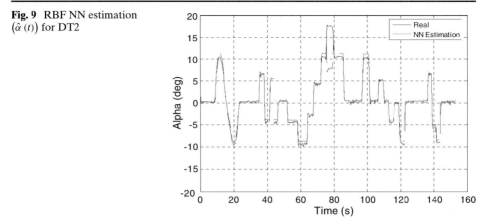

There is plenty of literature describing different ways of robustly defining the domain of validity. A good survey paper can be found in [24]. The simplest technique is by defining the maximum and minimum of each input parameter (in our case that would be each pressure port). Testing data which lies outside these limits is said to be outside the domain of validity. Despite its simplicity, unfortunately this method overestimates the domain of validity. More complex methods involve the definition of the convex hull polytope (CHP) which tightly encompasses the set of multidimensional data. Once the CHP is defined, the exteriority of each input pattern is then calculated. Patterns which lie outside the CHP have an exteriority > 0, and ones which are in the CHP have an exteriority $= 0$.

For our task, to define the domain of validity we would require all possible pressure input patterns. Patterns which lie *on* the convex hull (i.e. the outermost points) must then be included in Tr_1. This greatly reduces the need for NN extrapolation.

To demonstrate the importance of carefully defining the NN training set Tr_1, we will reconsider the NN estimations for DT1 in Fig. 8. For simplicity let us define the domain of validity using the maximum/minimum approach described earlier. We will assign pressure from Tr_1 and DT1 the subscripts '$\_Tr_1$' and '$\_DT1$' respectively. The following logic statement (shown for P1 only) has been exploited to indicate (on a graph) when a pressure input is outside the domain of validity:

$$\{P1\_DT1 < \min(P1\_Tr_1)\} \vee \{P1\_DT1 > \max(P1\_Tr_1)\} \Leftrightarrow \{plot = 25\} \quad (11)$$

Equation 11 basically states that if pressure from P1 in DT1 exceeds the bounds defined by P1_ Tr_1 then plot a value of 25. The plot value is not meaningful, it is simply chosen for ease of presentation. Similarly Eq. 11 is implemented for the remaining pressure ports P2, P3, P4 and P5 but with plot values of 20, 15, 10 and 5 respectively. Figure 10 displays the results from this task. We have also included the residual from DT1 which was simply calculated as the difference between the NN estimates $\hat{\alpha}\,(t)$, and the measured value $\alpha(t)$. Note also that the plots, in Fig. 10, for P4 and P5 are not included as they did not exceed any limits.

From Fig. 10 we notice how the residual significantly increases only when the pressure input patterns lie outside the NN domain of validity. So for example at 120 s, the residual is at its maximum when pressure from ports P1, P2 and P3 exceed the limits.

Fig. 10 Domain of validity test for P1, P2 and P3. If P1, P2, P3 lie outside the domain of validity, a value of 25, 20, 15 is plotted respectively, otherwise a value of zero is plotted. NN residual also shown (*solid blue*)

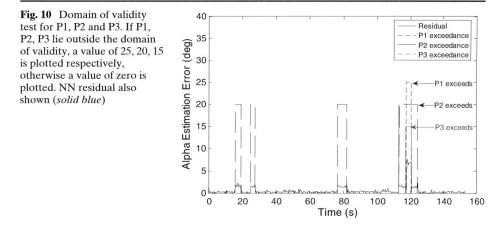

The simple domain of validity test presented here has shown the general importance of appropriately selecting the NN training set before any application. Different approaches exist for doing so and it is up to the designer to choose a suitable method. It is perhaps also important to note that training data is best obtained from real flight tests as the FADS system will eventually be used during flight. The NN domain of validity must then be updated from flight test runs, i.e. it must be updated with every new pressure pattern which is exterior to the existing domain of validity.

6 Conclusions

The findings and experience obtained in this study are summarized as follows:

- A FADS system was designed and tested in a wind tunnel to perform estimates of angle of attack, static pressure and wind speed $\left(\hat{\alpha}, \hat{P}_{\infty}, \hat{V}\right)$.
- The FADS system was designed for a MAV with 5 pressure orifices placed on the wing leading edge.
- A NN approach is used to model the aerodynamic relationship between vehicle surface pressure and air data to avoid the slow executions times associated with conventional lookup table approaches and the complexity associated with nonlinear regression methods.
- An EMRAN RBF NN was used due to its compact structure
- NN training was terminated according to two criteria. Criterion 1 considered the *rate of change of the RMS training estimation error* (ΔRMS). If the ΔRMS remained below 0.1% for more than 100 consecutive epochs, the NN is said to have converged and training is stopped. Criterion 2 considered the increase in the RMS testing estimation error. If this error increased for more than 100 consecutive epochs, training is stopped in order to avoid over-fitting the network structure.
- NN training was eventually stopped after 582 epochs resulting in a 5–3–3 network structure.
- Air data RMS estimation accuracies of 0.51°, 0.44 lb/ft^2 and 0.62 m/s for $\hat{\alpha}$, $\hat{P}_{\infty}$, $\hat{V}$ respectively were achieved for static tests where $\dot{\alpha}(t) = \dot{\beta}(t) = \dot{V}(t) = 0$.

- A fault accommodation test on the 5–3–3 RBF NN revealed that a 50% reduction in NN estimation errors is possible if redundant pressure ports were considered. Similar tests showed that a 70% reduction is possible if an autoassociative NN was used instead to reproduce the faulty measurements.
- Dynamic tests where $\dot{\alpha}(t) \neq 0$ were carried out in the wind tunnel and the 5–3–3 NN structure was used to output $\hat{\alpha}(t)$. An RMS estimation accuracy of 0.58° was achieved.
- A simple logic test showed that the NN estimation residual is large when NN input patterns lie outside the NN domain of validity defined by the maximum and minimum limits of the training data set. This observation proved the importance of carefully selecting the NN training data set so that NN extrapolation is minimized.
- An important design criterion for real time implementation of NNs is to ensure that the processing time is less than the input data sampling time. This is generally checked by comparing the time it takes for the NN to process one sample of data to the sampling time T. For our tests, the NN was run on a 1.6 GHz Pentium processor which gave a NN processing time of 0.32 ms. This is considerably lower than the sampling time T of 20 ms.

UAVs are currently ineligible for a standard airworthiness certificate, and are only assigned a special airworthiness certificate in the experimental category for research and development purposes [25]. However it is highly feasible that these restrictions will eventually be removed and UAVs will be integrated into the National Airspace System (NAS). One of the policies that the FAA adopts for regulators to issue an airworthiness certificate is based on the air-vehicle's 'potential to do damage'. This categorises the air-vehicles in terms of weight, size, speed etc. Ultimately weight has relevance for airworthiness risks, and the FADS system suggested here takes this into consideration. Our FADS system weighed approximately 35 g while the mini airdata boom typically used by BBSR for their UAVs, weighs 170 g. In this case, a reduction in weight of 135 g may not seem significant, but, relatively speaking, an 80% reduction in weight can be crucial in large unmanned air vehicles for both flight and airworthiness purposes. In addition, the FADS system's overall cost is about £75 in comparison to the airdata boom which costs £2,200. This large reduction in cost is beneficial for several reasons, but is mainly cost effective for military applications where UAVs are more likely to be destroyed during the mission.

7 Future Work

The current work was done to investigate the performance of a FADS system mounted on the wing leading edge of a MAV. It must be noted that further developments need to be considered such as implementing a more robust approach to training the NN which may result in a different NN structure than the one outlined here. However preliminary results presented in this report were promising to consider implementation in real flight where the NN structure can be developed further using real flight data. This part of the project will be carried out at the Blue Bear Systems Research (BBSR). A mini vane capable of measuring α has already been purchased [26]. It will be suitably mounted on the MAV and the FADS system can then be compared to the conventional sensors which include the mini vane and a pitot static tube.

Acknowledgements   The authors would like to sincerely thank the BlueBear Systems Research (BBSR) Ltd for providing the MAV and installing the FADS system. Special thanks are also given to Paul William (Leicester University) for help with the wind tunnel tests. The first author is grateful for the financial support from the Overseas Research Students (ORS) award.

Appendix

Figure 11

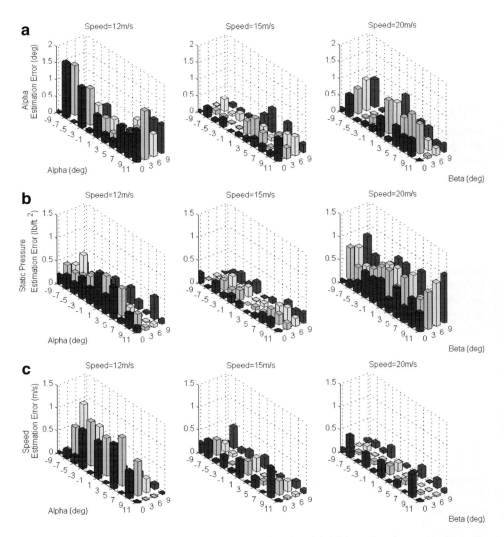

Fig. 11 NN estimation errors at different wind tunnel setting: (**a**) Alpha estimation errors, (**b**) static pressure estimation errors, (**c**) speed estimation errors. Note only positive β range shown here

References

1. Andersen, D., Haley, D.: NASA tests new laser air data system on SR-71 Blackbird. NASA. http://www.nasa-usa.de/home/hqnews/1993/93-163.txt (1993). Accessed 17 September 1993
2. Cary, J.P., Keener, E.R.: Flight evaluation of the X-15 Ball-Nose Flow-Direction sensor as a airdata System. NASA TN D-2923 (1965)
3. Wolowicz, C.H., Gosett, T.D.: Operational and performance characteristics of the X-15 spherical hypersonic flow direction sensor. NASA TN D-3076 (1965)
4. Larson, T.J., Siemers III, P.M.: Subsonic tests of an All-Flush-Pressure-Orifice air data system. NASA TP 1871 (1981)
5. Larson, T.J., Whitmore, S.A., Ehernberger, L.J., Johnson, J.B., Siemers III, P.M.: Qualitative evaluation of a flush air data system at transonic speeds and high angles of attack. NASA TP-2716 (1987)
6. Larson, T.J., Moes, T.R., Siemers III, P.M.: Wind tunnel investigation of a flush airdata system at Mach numbers from 0.7 to 1.4. NASA TM-101697 (1990)
7. Whitmore, S.A., Davis R.J., Fife J.M.: In flight demonstration of a real time flush airdata sensing system. NASA TM 104314 (1995)
8. Rohloff, T.: Development and evaluation of neural network flush air data sensing systems. Ph.D. thesis, Dept. of Mechanical Engineering, Univ. of California (1998)
9. Crowther, W.J., Lamont, P.J.: A neural network approach to the calibration of a flush air data system. Aeronaut. J. **105**, 85–95 (2001)
10. Brown, E.N., Friehe, C.A., Lenschow, D.H.: The use of pressure fluctuations on the nose of an aircraft for measuring air motion. J. Clim. Appl. Meteorol. **22**, 171–180 (1983)
11. Wenger, C., Devenport, W.: Seven-hole pressure probe calibration method utilizing look-up error tables. AIAA J. **37**, 675–679 (1999)
12. Rediniotis, O., Vijayagopal, R.: Miniature multihole pressure probes and their neural network based calibration. AIAA J. **37**, 666–674 (1999)
13. Rediniotis, O., Chrysanthakopoulos, G.: Application of neural networks and fuzzy logic to the calibration of the seven-hole probe. J. Fluid. Eng.—Trans. ASME **120**, 95–101 (1998)
14. Lu, Y., Sundararajan, N., Saratchandran, P.: Analysis of minimal radial basis function network algorithm for real-time identification of nonlinear dynamic systems. IEE Proc. Control Theory Appl. **4**, 476–484 (2000)
15. Houghton, E.L., Carpenter, P.W.: Aerodynamics for Engineering Students. Butterworth Heinemann, Burlington, MA (2003)
16. Gallinari, F.: Industrial Applications of Neural Networks. World Scientific Publishing Co. Pte. Ltd., Great Britain (1998)
17. Haykin, S.: Neural Networks: A Comprehensive Foundation. Macmillan College Publishing Company, New York (1994)
18. Powell, M.J.D.: Radial basis function for multivariable interpolation: a review. In: Mason, J.C., Cox, M.G. (eds.) Algorithms for Approximation, pp. 143–167. Clarendon Press, Oxford (1987)
19. Chen, S., Cowan, F.N., Grant, P.M.: Orthogonal least squares learning algorithm for radial basis function networks. IEEE Trans. Neural Netw. **2**, 302–309 (1991)
20. Platt, J.C.: A resource allocating network for function interpolation. Neural Comput. **3**, 213–225 (1991)
21. Kadirkamanathan, V., Niranjan, M.: A function estimation approach to sequential learning with neural networks. Neural Comput. **5**, 954–975 (1993)
22. Fravolini, M.L., Campa, G., Napolitano, M., Song, Y.: Minimal resource allocating networks for aircraft SFDIA. Adv. Intell. Mechatron. **2**, 1251–1256 (2001)
23. Courrieu, P.: Three algorithms for estimating the domain of validity of feedforward neural networks. Neural Netw. **7**, 169–174 (1994)
24. Helliwell, I.S., Torega M.A., Cottis, R.A.: Accountability of neural networks trained with 'Real World' data. In: Artificial Neural Networks, Conference Publication No. 409 IEE, pp. 218–222. 26–28 June (1995)
25. FAA: Unmanned Aircraft Systems (UAS) Certifications and Authorizations. US Dept. of Transportation. http://www.faa.gov/aircraft/air_cert/design_approvals/uas/cert/ (2007). Accessed 5 November 2007
26. mini vane. SpaceAge Control. http://www.spaceagecontrol.com/pm/uploads/Main.Adpmain/100386.pdf (2008). Accessed 13 February 2008

Testing Unmanned Aerial Vehicle Missions in a Scaled Environment

Keith Sevcik · Paul Oh

Originally published in the Journal of Intelligent and Robotic Systems, Volume 54, Nos 1–3, 297–305.
© Springer Science + Business Media B.V. 2008

Abstract UAV research generally follows a path from computer simulation and lab tests of individual components to full integrated testing in the field. Since realistic environments are difficult to simulate, its hard to predict how control algorithms will react to real world conditions such as varied lighting, weather, and obstacles like trees and wires. This paper introduces a methodic approach to developing UAV missions. A scaled down urban environment provides a facility to perform testing and evaluation (T&E) on control algorithms before flight. A UAV platform and test site allow the tuned control algorithms to be verified and validated (V&V) in real world flights. The resulting design methodology reduces risk in the development of UAV missions.

Keywords Unmanned aerial vehicles · Testing and evaluation · Verification and validation

1 Introduction

The robotics community is faced with an ever increasing demand for robots that operate in cluttered outdoor environments. To perform tasks such as search and rescue and surveillance, the robots must operate in unstructured, dynamic environments. The nature of these environments often drives development to focus on sensing and control algorithms.

This work is funded in part by the National Science Foundation CAREER award IIS 0347430.

K. Sevcik (✉) · P. Oh
Drexel Autonomous Systems Laboratory,
Department of Mechanical Engineering,
Drexel University, Philadelphia, PA 19104, USA
e-mail: kws23@drexel.edu

P. Oh
e-mail: paul@coe.drexel.edu

The current design paradigm begins with laboratory development and testing. Sensors are characterized in sterile, structured environments with the claim that the results are extensible to real life objects and conditions. While it is true that rigorous testing such as that presented in [1] and [2] helps one to understand the limitations of hardware, it is difficult to determine how the sensor and sensing algorithms will perform in unpredictable field conditions.

Similarly, computer simulations aid in the design of control algorithms. As the control code is refined, greater and greater detail can be incorporated into the model to approximate real world conditions. Orr et al. [3] investigated methods for simulating environmental conditions such as wind gusts. Sensors have also been incorporated into computer simulation, as shown in [4]. However, present day computer models are unable to incorporate unstructured environments. In particular, objects such as trees and bushes are exceedingly difficult to accurately integrate into simulation.

Following lab development, the sensing hardware and control software are transferred to the robotic platform in order to perform real world tests. Many times the first test of the integrated sensing hardware and control software occurs in the field during these flights.

This design methodology invites costly, time consuming failures. Errors in programming, unforeseen design challenges, and unpredictable real world conditions lead to catastrophic crashes. To mitigate these risks, we propose a step in between lab development and real world flights where sensing and control can be tested and evaluated without having to fly the robotic platform.

The authors' previous work in this area [5] involved a full scale mock urban environment inside a dof gantry. Sensor suites attached to the end effector of the gantry could be virtually flown through the environment. The motions of the gantry were governed by a high fidelity math model of the robotic platform. This allowed hardware-in-the-loop testing of the robot's sensing and control algorithms.

This approach proved useful for evaluating the robot's reactions to varying environmental conditions. However, physical limitations confined the testing area to a small slice of the urban environment. This limited testing to low-speed, low altitude maneuvers. While this technology could be implemented on a larger scale, it was ultimately unfeasible to evaluate entire missions, which could occur over several thousand square meters.

To solve these issues, inspiration was drawn from the early development of flight simulators. As described in [6], some of the first flight simulators utilized scaled models of terrain to provide visual feedback for pilots. These systems provided high fidelity, realistic visual cues for pilots. However, simulations were limited to the area of the models. This approach was abandoned in favor of computer based simulators which provided endless terrain maps, albeit at the sacrifice of realism.

The problem faced by simulation of UAV missions is quite the opposite. Missions are often confined to a defined region such as a town or group of buildings. Computer simulations attempt to model real world effects, but fail to capture the caveats of operating in real world environments. Scaled models such as that shown in Fig. 1 provide a means to test sensors and control algorithms against realistic environments.

This paper presents the design of a testing facility for UAV missions and its use to guide the development of a robotic helicopter. Section 2 describes the facility and its integration into the design process. Section 3 describes the robotic platform. Section 4

Fig. 1 A scaled model
environment for testing UAV
missions. Things such as trees
and unstructured lighting that
are difficult to capture
in computer simulation are
easily incorporated here

describes the mission and algorithms being tested in the facility. Section 5 describes experimental results to date. Finally, conclusions and future work are presented in Section 6.

2 Testing Facility

The goal of this research is to introduce a more sound design methodology to the field of UAV research. Through testing and evaluation (T&E), sensors and control algorithms can be tuned before flight. The refined hardware and software can then go through verification and validation (V&V) on board the actual robotic system. This affords a more robust end product and better management of risk during development. To guide the design of these T&E and V&V setups, the mission profiles must first be defined.

The types of missions under investigation are those typically executed by UAVs on-station after being deployed from a remote location. Such missions include reconnaissance, perch-and-stare and payload delivery. Many of these missions require investigation of more fundamental capabilities such as autonomous mapping and landing zone identification. Areas of interest are typically urban environments containing obstacles such as buildings, poles, trees and thin wires.

Such missions have been investigated in both [7] and [8]. In these experiments, the operational area was as large as 220×220 m flown at altitudes in the 10's of meters. The craft in these missions traverse the environment at speeds ranging from $4 - 10$ m/s. Furthermore, the ascent velocity in [7] is limited to 3 m/s while the descent velocity is limited to 1 m/s. These requirements are compiled in Table 1.

Table 1 Constraint velocities

Axis	Gantry (m/s)	Scaled (m/s)	Mission required (m/s)
X	0.012–0.61	1.04–53.0	4–10
Y	0.019–0.61	1.65–53.0	4–10
$+Z$	0.030–0.61	2.61–53.0	0–3
$-Z$	0.101–0.61	8.79–53.0	0–1

From these criteria, the V&V environment selected by the authors was the helicopter development range at Piasecki aircraft. As can be seen in the satellite photo in Fig. 2, the area encompasses several hundred meters. Buildings and wooded regions provide a variety of terrain to test UAVs.

The focus of this design methodology is to create a continuous path from laboratory research to real world flights. The transition from T&E to V&V should therefore be as seamless as possible. As such, the T&E environment was created to closely approximate the Piasecki facility, as shown in Fig. 3. The facility was recreated at 1/87th scale, which is a common modeling scale. This permits access to a wide range of obstacles and terrain features which can be added in the future.

Assessment of UAV control algorithms required a testing facility capable of repeatable and controllable simulation of UAV dynamics and flight paths. The Systems Integrated Sensor Test Rig (SISTR), shown in Fig. 4, is a National Science Foundation funded UAV testing facility that provides this capability. SISTR measures $19 \times 18 \times 20$ ft enclosing the scaled T&E environment.

As described in [5], the facility is surrounded by a dof computer controlled gantry. Using the math model of the UAV and model adaptive control, the gantry can be programmed to mimic the flight of an aerial vehicle. UAV sensor suites can be attached to the end effector of the gantry to provide real-time sensor feedback for testing sensor and control algorithms.

In mimicking the flight of a UAV, one of the most important design factors is that the velocities of the UAV can be accurately matched in the scaled down model. To accomplish this, the translational motions of the gantry must scale appropriately to fall within the operational velocity ranges of the UAV. Table 1 displays the maximum and minimum velocities achievable by the gantry, the scaled values of those velocities, and the corresponding required mission velocities. As can be seen, the velocity ranges required for the X-axis and Y-axis are easily achieved by the gantry. However, the Z-axis velocities of the gantry are faster than those

Fig. 2 Satellite image of the helicopter development range at Piasecki Aircraft. The complex contains typical terrain for UAV missions, such as urban and wooded environments. The area spans several hundred meters, allowing ample room for UAV flight tests

Fig. 3 The V&V environment compared to the 1/87th scale T&E environment. The T&E facility was created to closely approximate the helicopter development range in order to draw a continuous path between laboratory testing and real-world flights

Fig. 4 Systems integrated sensor test rig (SISTR). SISTR provides a stage for testing and evaluating sensor and control algorithms in a scaled environment. The dof gantry that comprises SISTR can be programmed through model adaptive control to mimic the flight of UAVs

required by the mission. This issue exists under the current software solution for controlling the gantry. The authors believe the gantry hardware is capable of achieving slower motions. This issue is currently being addressed by the authors.

Finally, the position in all translational axes of the gantry can be controlled to within ±1 cm. This scales up to a resolution of ±0.87 m. This position accuracy is well within the ±2 m accuracy of the typical GPS system. This provides a complete facility which can accommodate T&E of sensor and control algorithms for many different UAV platforms. To show the validity of this approach, the authors use a specific robotic system to show the complete design process incorporating T&E and V&V.

3 Robotic Platform

To perform V&V, a Rotomotion SR100 electric UAV helicopter was used, shown in Fig. 5. The SR100 is sold as a fully robotic helicopter capable of performing autonomous take off, landing, and GPS waypoint navigation when controlled from a laptop base station. Control from the base station to the helicopter is routed through an 802.11 wireless network adapter.

The SR100 has a rotor diameter of 2 m allowing it to carry a payload of up to 8 kg. For these experiments, we outfitted the helicopter with custom landing gear, a custom camera pan/tilt unit, the SICK LMS200, a serial to Ethernet converter, and two 12 V batteries for payload power. In total we added approximately 7 kg of payload. This greatly reduces the flight time, which is up to 45 min without a payload.

The biggest attraction of this platform, however, is the fact that it is already outfitted with all of the necessary sensors to calculate its pose. Gyros, an inertial measurement unit, and a magnetometer provide the craft's attitude and heading. This information is fused with a Novatel GPS system to provide position data. The position is reported as Cartesian coordinates relative to a global frame, who's origin is at the location where the helicopter was activated.

In selecting hardware to perform initial tests, the authors looked to previous experience designing UAV sensor suites. As a Future Combat Systems (FCS) One

Fig. 5 The SR100 helicopter from Rotomotion, Inc. The SR100 is sold as a fully robotic package capable of automated take off, landing, and GPS waypoint following

team member, the authors have gained extensive experience designing sensor suites for robots flying in near-Earth environments. The FCS Class II program focused on building a UAV to fly missions in areas such as urban terrain and forests. This project identified a few fundamental requirements for these sensor suites.

The sensor must detect a wide range of obstacles. In urban terrain, object size and composition can vary drastically, from buildings to telephone poles to thin wires and clothes lines. In particular, sparse objects such as trees and bushes are troublesome to detect.

The sensor must also be able to detect obstacles from far away and at oblique angles. The speed that a UAV can travel at is directly related to how far away it can detect obstacles. The greater the detection distance, the more time the UAV has to react and plan a new flight path.

Our experience has shown that computer vision is a good candidate when judged against these criteria. Additionally, computer vision is a well understood sensing method and one of the most common sensing methods utilized on UAVs. Commercial UAVs are equipped with cameras for surveillance. Research UAVs use cameras to accomplish tasks such as autonomous landing, target tracking and obstacle avoidance. This makes computer vision a very attractive problem to investigate.

To illustrate the feasibility of this design methodology, a sensing algorithm must be exhibited on the scaled model, and the results must be replicated in the real world. A sensing algorithm must be utilized that tests the capabilities of the scaled T&E environment.

4 Missions and Algorithms

One of the fundamentals of most computer vision algorithms is feature detection. Techniques such as optic flow, object recognition, and target tracking all rely on detecting features in the image. Feature tracking is therefore a good representative technology for testing the feasibility of scaling UAV missions.

One of the most common feature tracking methods is the Kanade–Lucas–Tomasi (KLT) feature tracker. A KLT tracker works by first performing feature detection across the entire image to determine the strongest features to track. The area of the image neighboring these features is recorded. This feature window is then tracked through successive images.

By looking at the direction of optic flow in the feature window, the KLT tracker estimates the direction that the feature has moved. Several iterations are then performed in which the feature window is shifted and the successive frames are compared. This process continues until the feature is located in the next frame. Depending on implementation, the tracker can then keep the original feature, discard it in favor of the feature found in the current image, or search for an entirely new feature if the current feature has become weak.

By searching only in the feature window as opposed to the entire image, and by using optic flow to guide the direction of search, the KLT tracker allows for features to be detected and tracked very quickly. An efficient implementation can often track features in real time.

The notional mission we wished to evaluate was a helicopter performing surveillance near a group of buildings. As the helicopter flies past the buildings, the KLT

Fig. 6 Kanade–Lucas feature
detection implemented
on video of the scaled T&E
environment. The algorithm
successfully detected features
such as the corner of buildings
and windows. These features
are consistent between the
model and the real world,
allowing for a more strict
comparison of the scaled
and full sized environments

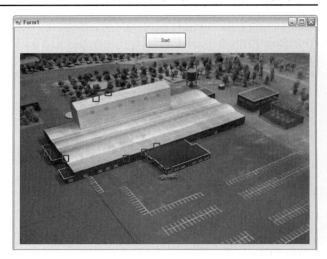

feature tracker will process video from an on-board camera. The KLT tracker will find the strongest features in the visible field and track them as the helicopter moves. This scenario would be executed in the scaled T&E environment and in the full sized V&V environment. The features detected and tracked in the scaled flight should match those detected in full sized flight.

5 Experimental Results

Kanade–Lucas feature detection was implemented on video taken of the scaled model. Preliminary results are shown in Fig. 6. Boxes drawn around an area represent a feature window containing a detected feature. It can be seen that features such as the corners of buildings and windows were detected.

The features detected on the model were made to closely approximate those of the real world environment. This allows for a consistent comparison between the model and the real world.

6 Conclusions and Future Work

Preliminary results indicate that it will be possible to directly compare the scaled model against the real world setting. In order to make this comparison, the SR100 must be flown through the testing facility at Piasecki Aircraft. This flight must then be duplicated in the scaled environment.

The SR100 is equipped with the correct sensors to localize its position and measure its pose. This allows for the path of the helicopter to be recorded. To use SISTR to trace this path through the scaled environment, the dynamics of the helicopter must be appropriately scaled.

Another issue that must be addressed is distortion of the image because of the camera lens curvature. There are also issues associated with correcting the image for perspective. To make an accurate comparison, these distortions must be removed from the image.

The results from these experiments will provide a measure for how closely a scaled environment approximates the real world. The result will be a continuous path from laboratory development to real world implementation.

Acknowledgements The authors wish to thank Jesse Greenberg of Simulab Studios for the construction of the scaled testing environment. Thanks also go to Piasecki Aircraft for their continued support and use of their testing facilities.

References

1. Alwan, M., Wagner, M.B., Wasson, G., Sheth, P.: Characterization of infrared range-finder PBS-03JN for 2-D mapping. In: International Conference of Robotics and Automation (ICRA) (2005)
2. Ye, C., Borenstein, J.: Characterization of a 2-D laser scanner for mobile robot obstacle negotiation. In: International Conference of Robotics and Automation (ICRA) (2002)
3. Orr, M.W., Rasmussen, S.J., Karni, E.D., Blake, W.B.: Framework for developing and evaluating MAV control algorithms in a realistic urban setting. In: American Control Conference (ACC), pp. 4096–4101 (2005)
4. Netter, T., Franceschini, N.: A robotic aircraft that follows terrain using a neuromorphic eye. In: International Conference on Intelligent Robots and Systems (IROS), pp. 129–134 (2002)
5. Narli, V., Oh, P.: A hardware-in-the-loop test rig for designing near-earth aerial robotics. In: International Conference on Robotics and Automation (ICRA), pp. 2509–2514 (2006)
6. Allerton, D.J.: Flight simulation: past, present and future. The Aeronaut. J. **104**(1042), 651–663, December (2000)
7. Scherer, S., Singh, S., Chamberlain, L., Saripalli, S. Flying fast and low among obstacles. In: International Conference on Robotics and Automation (ICRA), pp. 2023–2029 (2007)
8. Hsieh, M.A., Cowley, A., Keller, J.F., Chaimowicz, L., Grocholsky, B., Kumar, V., Taylor, C.J., Endo, Y., Arkin, R.C., Jung, B., Wolf, D.F., Sukhatme, G.S., MacKenzie, D.C.: Adaptive teams of autonomous aerial and ground robots for situational awareness. J Field Robot **24**(11–12), 991–1014, November (2007)

A Framework for Simulation and Testing of UAVs in Cooperative Scenarios

**A. Mancini · A. Cesetti · A. Iualè · E. Frontoni ·
P. Zingaretti · S. Longhi**

Originally published in the Journal of Intelligent and Robotic Systems, Volume 54, Nos 1–3, 307–329.
© Springer Science + Business Media B.V. 2008

Abstract Today, Unmanned Aerial Vehicles (UAVs) have deeply modified the concepts of surveillance, Search&Rescue, aerial photogrammetry, mapping, etc. The kinds of missions grow continuously; missions are in most cases performed by a fleet of cooperating autonomous and heterogeneous vehicles. These systems are really complex and it becomes fundamental to simulate any mission stage to exploit benefits of simulations like repeatability, modularity and low cost. In this paper a framework for simulation and testing of UAVs in cooperative scenarios is presented. The framework, based on modularity and stratification in different specialized layers, allows an easy switching from simulated to real environments, thus reducing testing and debugging times, especially in a training context. Results obtained using the proposed framework on some test cases are also reported.

Keywords UAVs · Simulation · Cooperative scenarios

1 Introduction

During last years, in addition to ground vehicles, mobile robotics is broadening to innovative branches as *Unmanned Surface/Underwater Vehicles* and *Unmanned Aerial Vehicles* (UAVs). Missions of various kinds are in most cases performed by a fleet of cooperating autonomous and heterogeneous vehicles. Interaction, cooperation and supervision are the core problem of these complex systems. The complexity correlated to today challenges in terms of missions and tasks sets up the necessity of simulating, debugging and testing. Simulation activities are fundamental because different methodological approaches can be easily implemented and evaluated to reduce developing times. This is particularly true in an educational context.

A. Mancini (✉) · A. Cesetti · A. Iualè · E. Frontoni · P. Zingaretti · S. Longhi
Dipartimento di Ingegneria Informatica, Gestionale e dell'Automazione,
Università Politecnica delle Marche, Via Brecce Bianche, 60131 Ancona, Italy
e-mail: mancini@diiga.univpm.it

K. P. Valavanis et al. (eds.), *Unmanned Aircraft Systems*. DOI: 10.1007/978-1-4020-9136-0_18   307

In the case of ground robots a lot of simulation and test frameworks have been developed. Probably, Player / Stage / Gazebo [1] is actually the most complete framework owing to advanced features like the emulation of 2D–3D environments, sensor simulation (Laser Range Finder (LRF), sonar,...) and integration with commercial robotic platforms (i.e., MobileRobots [2], irobot [3]). The framework proposed by Frontoni et al. [4] is particularly suited in an educational context allowing an easy switching from simulated to real environments. Other simulation environments are taking the attention of the scientific community for the full integration with a lot of commercial platforms, for example Carmen [5], Microsoft Robotics Studio [6] and USARsim (for RoboCup) [7].

For the UAV branch of robotics the state of the art is a bit different. First, it is more fragmented because the set of aerial vehicles is more heterogeneous. In the case of simulated ground vehicles, common robots are differential wheeled or car like. On the contrary, among aerial vehicles there are blimps, gliders, kites, planes, helicopters, etc. Each vehicle has a particularity that makes the difference in a mathematical description of physical phenomena. Mathematical models of aerial vehicles are really complex because an aerodynamic description is necessary for a realistic modelling.

In this paper a framework for simulation and testing oriented to rotary-wings aerial vehicles is presented. The framework allows UAV simulation (as stand-alone agents or exchanging data for cooperation) owing to a Ground Control Station (GCS) that supervises the tasks of each agent involved in the mission. The paper is organized as follows. Next session introduces our framework; a Unified Modelling Language (UML) representation is first introduced to synthetically describe concepts that inspired our framework. The use of a UAV CAD modelling for parameter extraction and simulation aids is proposed in Section 3; the modelling activity is contextualized to the Bergen Twin Observer Helicopter. In Section 4, a test case involving take off, landing and navigation is presented; a cooperative scenario that involves two helicopters in an exploration mission is included. In Section 5 conclusions and future works are outlined.

2 Framework

In setting up a framework for the simulation of complex and multi-agent scenarios we identified the following major aspects:

- high fidelity mathematical model of vehicles;
- multi-agent management;
- extended set of simulated sensors;
- modularity;
- reduced time for updating/adding new modules;
- Virtual Reality rendering;
- easy switching from simulated to real world and vice versa;
- educationally oriented.

Till today, game engines and flight simulators are the only available frameworks to simulate UAVs. Game engines (like FlightSimulator [8] or Flight Management System (FMS) [9]) are optimal for visualization, while flight simulators (like JSBSim,

YASim and UUIU [10]) are characterized by a high-fidelity mathematical model, but are lacking in high quality rendering. Most of them are developed for planes. A good, but expensive exception, is the RotorLib developed and commercialized by RTDynamics [11]; in the helicopter context, frameworks with the requirements listed above are almost absent [12]. The framework here proposed aims at overtaking this lack.

In Fig. 1 a graphical abstraction with the main modules of the developed framework is shown.

The stratification of the framework permits to identify five layers: Supervision, Communication, Dynamics, Agent, User Interaction. All the tasks that involve one or more vehicles are managed and supervised by the GCS, and data are sent to agents using the communication layer. A socket based interface allows the data exchange between GCS and agents in the case of simulated agents, while the communication makes use of a dedicated long-range radio modem if a real vehicle (e.g., helicopter) is used [13].

Detailed descriptions of more relevant modules of the proposed framework are presented in the following subsections. First relations among classes are easily represented using an UML diagram and, then, agent structure, simulated dynamics, basic control laws, GCS and virtual reality and world representation are analyzed. Without loss of generality, each aspect is contextualized to a particular class of UAVs, i.e., helicopters.

All the modules are implemented in Matlab/Simulink; the main motivation of this choice is the reduced complexity for code development. In particular, the end-user of the framework can easily integrate his code for developing and testing an algorithm,

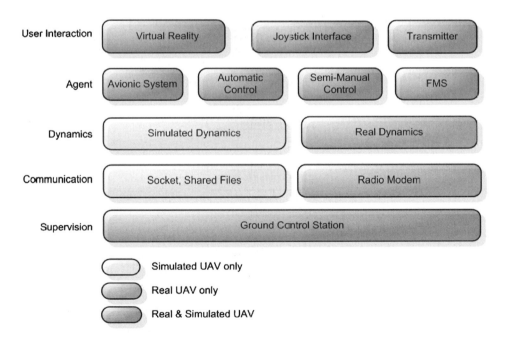

Fig. 1 Framework for UAV simulation; a new simulator engine instance is generated for each agent to be simulated

e.g., for obstacle avoidance, without the necessity of re-compiling other activities. An additional motivation for the adoption of Matlab is the capability to interface the AeroSim toolbox released by Unmanned Dynamics [14]. The AeroSim Blockset is a Matlab/Simulink block library which provides components for rapid development of nonlinear 6-DOF aircraft dynamic models. In addition to aircraft dynamics the blockset also includes environment models such as standard atmosphere, background wind, turbulence and earth models (geoid reference, gravity and magnetic field). These blocks can be added to the basic framework to increase the realism of simulation.

2.1 Framework Description by UML

Before introducing the description of each layer, the proposed framework is presented making use of an UML diagram, according to the Object Management Group's specification; UML allows to model not only application structures, behaviors and architectures, but also business processes and data structures [15]. The most significant diagram to model the proposed framework is the *class* diagram, which is useful to represent hierarchy, networks, multiplicity, relations and more. Due to complexity of relations, first, main-classes (Fig. 2) are presented and, then, diagrams of sub-classes and objects (instances of classes) follow.

The *Communication* class allows the connection among agents and GCS. Specialized classes, as shown in Fig. 3, implement different communication methods by socket and radio modem.

Agents can be also monitored and remotely controlled owing to the *User Interaction* Class, as shown in Fig. 4. *Active and Passive* stand for the user implication; an active interaction implies the remote control of an agent; a passive one regards cockpit virtualization and virtual reality.

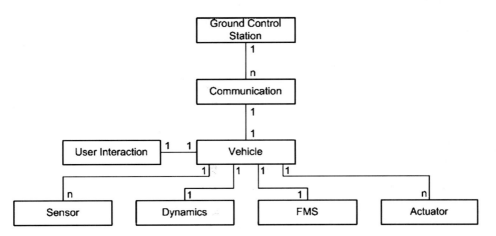

Fig. 2 The class diagram. Only parent classes are shown; the GCS can supervise and monitor a variable number of agents. Each agent is then autonomous in terms of control and mission execution

Fig. 3 Communication Class. Simulated agents use the socket paradigm; real communication makes use of a long range radio modem

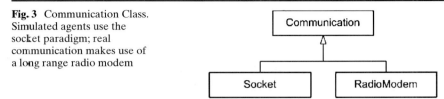

The *Vehicle* class is modelled in Fig. 5 taking into account the heterogeneous agents. An interesting analysis of aerial vehicles can be found in [16]. Vehicles have a series of dependencies with other classes as actuators, sensors, dynamics and the Flight Management System (FMS) as shown in Fig. 2.

Figures 6 and 7 show the complete class diagram including the parent and derived classes. Sensors are fundamental accomplishing successfully and secure autonomous missions; environment must be sensed to execute autonomous tasks like take-off, navigation and landing. Specialized classes are necessary to interface sensors in providing data. If a limited set of sensors is available in the simulated scenario, a new class (*Generic*) for the management of new sensors can be implemented. Sensors and actuators form the *Avionic System*. In general, each agent can have a set of i sensors and j actuators with i possibly different from j; for example in the case of a helicopter, five classes are derived from the *Sensor* class and three classes (or four if the *Servo Motor* class is splitted into *Analog* and *Digital*) from the *Actuator* class (see Figs. 6 and 7).

The diagram of the *Dynamics* class is proposed in Fig. 8. This class makes sense only in a simulation scenario. Considering the membership class of an agent, the mathematical model used to describe the dynamics varies significantly. In the case of a helicopter, models as *Disk Actuator* and *Blade Element* are widely adopted [17].

The FMS, formed by a set a classes, is described in more details in the next sub-section.

2.2 Agent Structure

In a simulated or real case, the structure of an agent in the context of UAVs is based on a complex interaction of different specialized modules. In the real case, the FMS is implemented as real-time code running on high performance architectures as PC104+, FPGA, DSP; in the simulation environment, FMS is a complex set of

Fig. 4 User Interaction Class. User(s) can interact with agents owing to active interfaces as joystick (simulated scenario) or transmitter (real case)

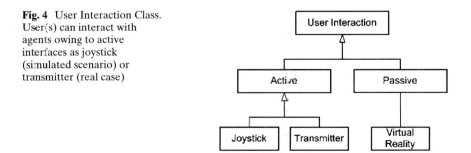

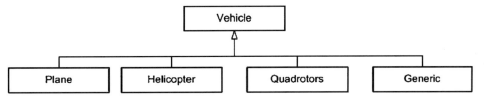

Fig. 5 Vehicle class. The nature of UAVs can be strongly different

S-functions to reduce the simulation complexity. However, in both cases FMS has a series of basic packages as (see Fig. 9):

–   Communication Module
–   Queue of Tasks
–   Guidance Module
–   Fast Path Re-planner
–   Attitude and Pose Estimator
–   Auto and/or Semi-manual Pilot
–   Obstacle Avoidance
–   Fault Diagnosis Identification and Isolation

FMS exchanges data continuously with GCS for telemetry and task assignments/supervision. Its *Communication module* makes use of sockets or functions to interface the radio modem according to the *Communication* class used.

References about position are generated by the *Guidance Module*, which decides step by step what references should be passed to the controllers (*Auto-Pilot*). This module takes into account the actual position of the agent with respect to the local inertial frame and the goal to reach. Tasks like take-off, landing, navigation point to point or waypoints are currently available in the developed framework.

The *Fast Path Replanner* (FPR) provides a real-time re-calculation of path according to information provided by the *Obstacle Avoidance* package. FPR provides also for correcting the path if external disturbances (e.g., wind) generate a high error in position.

The *Attitude and Position Estimator*, using the inertial data obtained by an *Attitude Heading Reference System* (AHRS) and an Inertial Navigation System (accelerations, speed of rotation,...) calculates the position and attitude of the vehicle; inertial

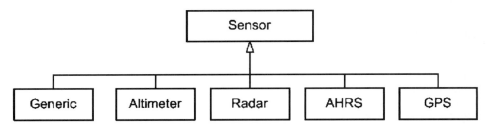

Fig. 6 Sensor class. The main difference between real and simulated scenario is the possible set of sensors that can be tested and/or evaluated; the cost of simulation in only related to the code writing time, while the cost of real device can be significantly different. The main drawback of simulated sensors is the difficulty to integrate video sensors

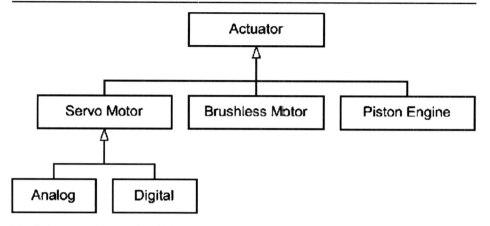

Fig. 7 Actuator class. Each vehicle has a series of actuators to accomplish different tasks like lift generation, heading variation and power generation. The main advantage of simulated actuators is the possibility to change on the fly the type of actuator used for evaluating the effects of different choices

strapdown equations are currently implemented and solved [18]. To enhance the localization process in the case of a real helicopter, a bayesian or other approaches on video data are planned to be used [19].

The *Auto-Pilot and/or Semi-manual control* are the core of vehicle's control. The adopted philosophy tries to emulate the training process of a novel-pilot, who usually controls directly only a limited set of vehicle's axes, while the teacher supervises the activities. Other axes are generally controlled by an electronic flight stabilizer, providing for the vehicle stabilization on a desired point in terms of attitude and/or position or speed. The user can control a set of axes by a joystick/transmitter interface. This feature is especially suitable in the field of photogrammetry, where the user concentrates only on forward and/or lateral movements, while the control of altitude and heading (heading lock) is performed by inline controllers.

Both in simulated and real cases the control of each axis can be manual or automatic; this feature expands the set of possible missions and allows decoupling the control of a vehicle so that the user can concentrate only on a restricted set of tasks. Obviously, an embedded hardware needs to be developed to interface the automatic control with the manual control; then the pilot can reserve his attention to a limited set of axes, while the remaining ones are delegated to inline controllers.

Controllers can be easily updated or modified by changing their code; no additional activity is required. Controllers can be simple PIDs or PIDs with gain scheduling and fuzzy logic. Feedback linearization is available in the framework, with

Fig. 8 Dynamics class exists only in the simulation context; different models can be tested and implemented as Matlab m-file(s) or S-function(s)

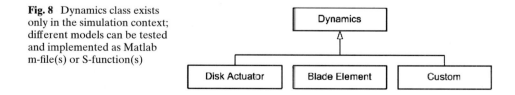

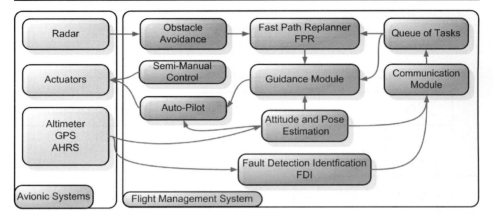

Fig. 9 The FMS is the core of an agent; a set of modules are responsible of navigation, communication, task execution and control

some tricks to increase its robustness: too time consuming is a major drawback of this technique. Other control techniques, e.g., based on H_∞ can be included.

The *Obstacle Avoidance* module tries to avoid obstacles owing to information obtained by the Avionic System, e.g., by radar and LRF sensors; actually, a set of modules based on fuzzy logic are available in the framework to improve the safety of vehicles during navigation [20, 21].

The *Avionic System*, in the case of simulated or real vehicles, is formed by actuators and sensors. Actuators are usually analog or digital servos in reduced-scale helicopters; a second order model to represent the servo dynamics is adopted in simulated environments. Sensors provide information for a large set of aspects as navigation, obstacle avoidance, mapping and other. The framework is equipped with some useful navigation sensors that provide information for the FMS. Using a radar sensor new tasks become feasible, as flight/operate at a given altitude or avoid an unexpected obstacle. In fact, the *Radar* altimeter provides the altitude above the ground level, calculated as the difference between height above the sea level and ground elevation (Digital Elevation Model DEM) maps; geometric corrections, due to pitch and roll angles, are then applied. Noise is added to make the simulation more realistic; failure occurrences are also simulated.

An approach similar to *Radar* is used to emulate the *Global Position System* (GPS). The geographic coordinates of the aircraft are computed from the knowledge of data about a starting point; noise is then added to match the performance of a common GPS receiver able to apply EGNOS corrections.

Simulated sensors as IMU and AHRS are also available in the framework; in this case an error model of each sensor is implemented (misalignment, temperature drift, non-linearity and bias).

In Table 1, an analysis of main differences between real and simulated case is presented; table summarizes some features extracted from the previous descriptions. The switch from virtual to real world and vice versa is relatively easy; mainly FMS is the module that requires different set-up especially for the automatic control (control laws of each servo installed on the real helicopter). The supervision in the case of simulated or real scenario is a bit similar; the GCS is responsible to supervise each

Table 1 Many elements (features) are shared between real and simulated scenario. The main difference concerns the communication and avionic module

Aspect	Simulated Scenario	Real Scenario
Supervision	Similar	Similar
Communication	Socket	Radio modem
Dynamics	Blade Element, Actuator Disk	Real phenomena
FMS	Similar (different control laws)	Similar
Avionic System	Simulated sensors & actuators	Real HW
User Interaction	Similar	Real streaming video

agent involved in a cooperative scenario where two or more agents cooperate to accomplish a specific task. The communication, for simulation scenarios, makes use of a socket paradigm between two or more PCs connected on a LAN; real helicopters exchange data using a dedicated long range radio-modem. The FMS, in both cases (simulated and real scenario) is similar; the main differences are the code written in different programming languages (real helicopter needs hard-real time software) and the control laws (different set of parameters). The *Avionic System* is different in terms of sensors and actuators, but the software interface is the same. User interaction, provided by GCS and a manual control interface (by using a joystick or an RC transmitter), is similar; in the case of a real helicopter, manual control is performed using an RC transmitter, while in the simulated scenarios, pilots are more familiar with joystick.

Keeping in mind the similarities between simulated and real cases, the switch from simulated to real vehicles is easy owing to the adoption of an approach focused on the stratification in specialized layers.

2.3 Helicopter Dynamics Simulation

In this sub-section attention is aimed at the modelling aspects of agents.

The framework is actually provided with a helicopter mathematical model. It is derived from Newton-Euler equations and applicable to helicopters with the common configuration of one main rotor and one tail rotor.

Employing the principles of modularity and standardization, the complete model is broken down into smaller parts that share information and interact among themselves. In particular, we identified four subsystems describing actuator dynamics, rotary wing dynamics, force and moment generation processes and rigid body dynamics, respectively [22]. The connections between subsystems, state and control variables are defined in Fig. 10.

The classical control inputs for a helicopter include lateral, longitudinal, collective and pedal control, denoted u_{lat}, u_{long}, u_{col}, and u_{ped}, respectively.

The actuator dynamics is much faster than that of the helicopter, so it is possible to ignore it when modelling the system. This makes the control signals affect the physical quantities directly: swash-plate tilt in lateral (A) and longitudinal (B) direction, main rotor blade pitch (θ_{MR}) and tail rotor blade pitch (θ_{TR}), respectively.

In the second block thrust magnitudes of main (T_{MR}) and tail (T_{TR}) rotor are determined as function of pitch inputs, while T_{MR} direction, defined by lateral (β_{1s}) and longitudinal (β_{1c}) flapping angles, is function of the swash-plate tilting. Equations

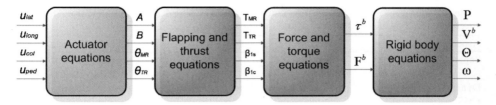

Fig. 10 Helicopter dynamics: the diagram of the components in the helicopter model is formed by four main blocks. It is common practice to assume that the blade pitch angles can be controlled directly, ignoring actuators dynamics (*green block*)

are also influenced by factors such as translatory movement and attitude of the helicopter body.

The thrust force magnitudes and directions are then split into sub components of forces and moments affecting the helicopter in the following block, which outputs three dimensional force (F^b) and the moment (τ^b) vectors. In the last block the final equations that describe the translational and rotational movement of the helicopter are derived. For this purpose the helicopter is regarded as a 6-DOF rigid body. The equations of motion are written with respect to the body coordinate frame (BF), which is attached to the Centre Of Gravity (COG). The complete model presents four output vectors describing the position P, the translatory velocity V^b, the attitude Θ, and the angular velocity ω of the helicopter. Θ, is formed by the so-called Euler angles (ϕ, θ, ψ), i.e., the angles after a roll, pitch and yaw movement, respectively. According to the top-down design, the modelling progression is now described following the block diagrams from right to left.

2.3.1 Rigid Body Equations

The equations of motion for a rigid body subjected to an external wrench applied at the centre of mass and specified with respect to the BF is given by Newton-Euler equations [37]:

$$F^b = m \cdot \dot{V}^b + \omega \times \left(m \cdot V^b \right)$$

$$\tau^b = I \cdot \frac{d\omega}{dt} + \omega \times (I \cdot \omega)$$

where $\mathbf{F}^b$ and τ^b are the vectors of forces and external torques, V^b and ω^b are the translatory and angular velocity, m is the mass and $\mathbf{I}$ is the inertia matrix.

The previous equations yield the matrix describing the motion of the rigid body:

$$\begin{bmatrix} \dot{V}^b \\ \dot{\Theta} \\ \dot{\omega} \end{bmatrix} = \begin{bmatrix} \frac{1}{m} \cdot F^b - \omega \times V^b \\ P_{sb}\left(\Theta\right) \cdot \omega \\ I^{-1}\left(\tau^b - \omega \times (I \cdot \omega)\right) \end{bmatrix}$$

where $\dot{\Theta}$ is the time derivative of the Euler angles and the P_{sb} is the frame transformation matrix.

The main problem is the determination of the inertia matrix, which is difficult to calculate due to the complexity of the vehicle. A new part added to original model must take in count that fuel, hence helicopter mass, is time-varying.

2.3.2 Forces and Torque Equations

The translatory forces acting on the helicopter and included in the modelling consist of:

- $\mathbf{F}^b_{MR}$: forces caused by the main rotor thrust;
- $\mathbf{F}^b_{TR}$: forces caused by the tail rotor thrust;
- $\mathbf{F}^b_g$: forces caused by the gravitational acceleration.

The main rotor and tail rotor thrusts act in the centre of the main rotor disc and tail rotor disc respectively, while the gravitational force acts in the COG. The resulting force $\mathbf{F}^b$, stated in the BF, is:

$$
\mathbf{F}^b = \mathbf{F}^b_{MR} + \mathbf{F}^b_{TR} + \mathbf{F}^b_g = \begin{bmatrix} f^b_{x,MR} \\ f^b_{y,MR} \\ f^b_{z,MR} \end{bmatrix} - \begin{bmatrix} f^b_{x,TR} \\ f^b_{y,TR} \\ f^b_{z,TR} \end{bmatrix} + \begin{bmatrix} f^b_{x,g} \\ f^b_{y,g} \\ f^b_{z,g} \end{bmatrix}
$$

$$
= \begin{bmatrix} -T_{MR} \cdot \sin(\beta_{1c}) - \sin(\theta) \cdot m \cdot g \\ T_{MR} \cdot \sin(\beta_{1s}) + T_{TR} + \sin(\phi) \cdot \cos(\theta) \cdot m \cdot g \\ -T_{MR} \cdot \cos(\beta_{1s}) \cdot \cos(\beta_{1c}) + \cos(\phi) \cdot \cos(\theta) \cdot m \cdot g \end{bmatrix}
$$

The torques are primarily caused by three components:

- τ^b_{MR}: torques caused by main rotor;
- τ^b_{TR}: torques caused by tail rotor;
- τ^b_D: counter-torque caused by drag on the main rotor.

An accurate drag is complex to model and therefore a simple model is used, while the torque generated by the tail rotor drag is disregarded due to the relative small influence it has on the model. Torques are defined positive in the clockwise direction. The resulting torque τ^b, stated in the BF, is:

$$
\tau^b = \tau^b_{MR} + \tau^b_{TR} + \tau^b_D = \begin{bmatrix} f^b_{y,MR} \cdot h_m - f^b_{z,MR} \cdot y_m + f^b_{y,TR} \cdot h_t + Q_{MR} \cdot \sin(\beta_{1c}) \\ -f^b_{x,MR} \cdot h_m - f^b_{z,MR} \cdot l_m - Q_{MR} \cdot \sin(\beta_{1s}) \\ f^b_{x,MR} \cdot y_m - f^b_{y,MR} \cdot l_m - f^b_{y,TR} \cdot l_t + Q_{MR} \cdot \cos(\beta_{1c}) \cdot \cos(\beta_{1s}) \end{bmatrix}
$$

where l_m, y_m and h_m are the distances between the rotor hub and the COG along the x^b, y^b and z^b axes, respectively, and l_t and h_t the distances from the centre of the tail rotor to the COG along the x^b and z^b axes. Q_{MR} is a coefficient expressing the relationship between the main rotor thrust and the drag.

The main problem in this context concerns the determination of the COG. Its position changes due to loads that can be added to the helicopter (e.g., a sensor) and fuel consumption. In Section 3 an approach to solve this kind of problem is proposed.

2.3.3 Flapping and Thrust Equations

To determine thrust magnitude two complimentary methods are used: momentum and blade element theory. The momentum theory yields an expression for thrust based on the induced velocity through the rotor disk. Because this is a single equation with two unknowns a second expression is needed to make a solvable set of equations. This second equation is generated using blade element theory, which is based on the development of thrust by each blade element. The result of this is a set of thrust equations that is solved by a recursive algorithm.

The thrust generated by the main rotor can be described by the following equation:

$$T_{MR,TR} = (w_b - v_i) \cdot \frac{\rho \cdot \Omega \cdot R^2 \cdot a \cdot B \cdot c}{4}$$

where ρ is the density of the air, Ω is the rotor angular rate, R is the rotor radius, a the constant lift curve slope, B is the number of blades, c is the blade chord, w_b is the velocity of the main rotor blade relative to the air, and v_i is the induced wind velocity through the plane spanned by the rotor. The main rotor thrust equations are recursively defined, with T_{MR} depending on v_i and vice versa, so the main rotor thrust T_{MR} is calculated by the use of a numerical method. For more details see [17, 23].

2.3.4 Flapping

A part of the lateral and longitudinal inputs from the swash plate, A and B, is fed directly to the main rotor, while the remaining part is fed to the main rotor through the control rotor. The result is a lateral and longitudinal blade flapping on the main rotor, denoted β_{1c} and β_{1s}.

The flapping equations used to calculate the flapping angles consider as inputs the moments affecting each blade element about the effective flapping hinge, i.e.:

- Gyroscopic Moments;
- Aerodynamic Moment;
- Centrifugal Moments;
- Spring Moments;
- Inertial Moments;
- Gravitational Moments.

β_{1c} and β_{1s} can be found by computing the equilibrium point between all moments affecting the blade; for more details see [24].

2.4 Basic Control Laws

The implemented controllers here presented are based on a classical PID nested loop structure [25, 26]. In Fig. 11 an UML diagram of controllers is shown; the diagram evidences that each controller belongs to *Auto-Pilot* module. The *Auto-Pilot* module uses a feedback from *Attitude and Pose Estimator* module and attempts to keep the helicopter at the position, heading and altitude commanded by the *Guidance Module*.

Considering the light coupling of the system the correlation among the state variables is neglected and the control system is divided into four decoupled loops: lateral, longitudinal, collective and yaw, one for each control input.

Both lateral and longitudinal controllers present the same configuration. A schematic block diagram of the nested loop structure is shown in Fig. 12. Because

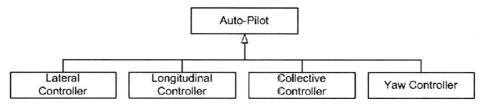

Fig. 11 The Auto-Pilot Module is constituted by four decoupled subsystems that compute control signal for every flight control input. Each controller follows the reference signals generated by the Guidance Module

of the under actuation of the system, in order to involve all the information coming from the sensors, the structure is composed from three control loops:

– the inner performs an adjustment of angle;
– the intermediate performs an adjustment of velocity;
– the outer performs an adjustment of position.

Peculiarity of the structure is that the error of a control loop is used to generate the reference for the following.

The yaw controller presents a simpler two-levels nested structure in which the outer loop performs an adjustment of the heading, while the inner loop performs an adjustment of the yaw rate, improving the stabilization of the system.

The collective controller structure is formed by a single loop in which only the information about altitude is taken into account.

2.4.1 Performances and Partial Results

The structure previously shown improves the system robustness to disturbances. Any outer control loop gives to the immediately following loop an important forward control action, guarantying a faster system adjustement.

In Figs. 13 and 14, a response to external unknown disturbances is reported. The simulation is focused on a take off phase and hovering with the presence of torque disturbances.

Such simulations evidence the utility of a framework. It allows simulating the helicopter in various operative conditions; in this case it is useful to test the robustness of control laws to external disturbances as wind. Another possible application is simulating a mass variation, e.g., caused by fuel consumption.

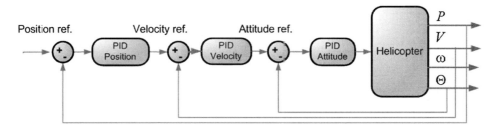

Fig. 12 Lateral and longitudinal action. These inputs constitute together the cyclic control. Swashplate angles are regulated by servos and leverages

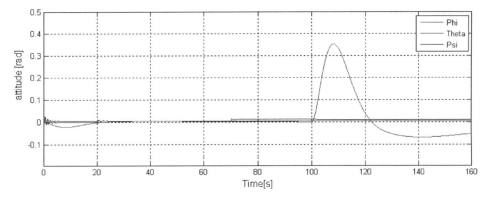

Fig. 13 Take off and hovering with torque disturbances—Attitude graph. $Z_{ref} = 1$ m; $T_{sim} = 160$ s. Disturbances: $C_x = 1$ Nm at T=60 s; $C_y = 1$ Nm at T=20 s; $C_z = .5$ Nm at T=100 s

2.5 Ground Control Station

The GCS has a lot of capabilities among which telemetry data acquisition and data logger for post flight analysis; in the cooperative context GCS is responsible for mission and task allocation/supervision. Data are collected and sent using the communication layer.

A GUI was developed to obtain a visual feedback for:

- single agent;
- all agents;
- mission status;
- telemetry.

A screenshot of the developed GCS GUI is shown in Fig. 15.

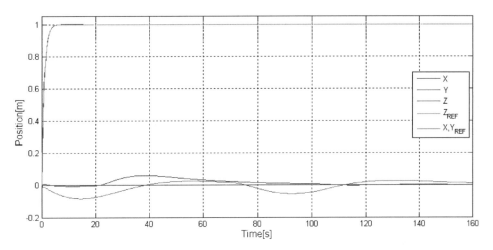

Fig. 14 Take off and hovering with torque disturbances—Trajectory graph

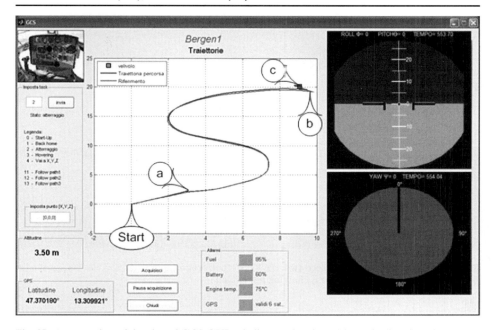

Fig. 15 A screenshot of developed GCS GUI; a helicopter is selected from the list of on-line agents that are joining the mission; telemetry data, attitude and position are shown

User can control the mission of each agent choosing the vehicles; the main panels allow to monitor in real-time the agent's status owing to Attitude Direction Indicator (ADI); information as global position (GPS coordinate), status of embedded electronics-batteries, fuel consumption are currently shown in the GUI.

An interesting feature of GUI is the capability to control directly a set of vehicle's axes using a joystick interface; in this case the interaction between human and machines (remote controlled vehicles) allows to control the vehicles taking into account the information provided by ADI indicators. In a simulation context, joystick is interfaced using the Matlab Virtual Reality Toolbox.

2.6 Virtual Reality and World Representation

A basic module allows a synthetic rendering of world and agent(s); as mentioned in the introduction section, market offers a series of different complex systems to virtualize world and agents. The choice adopted in the proposed framework was to integrate the Matlab Virtual Reality Toolbox. A VRML model of world (environment) and agents (aerial vehicles) can be easily controlled in a Simulink diagram. Students are often familiar with the Matworks software.

The mission area is represented as a digital grid map or DEM. The knowledge of a map is necessary only to simulate sensors and other modules; a map can be used by navigation modules to calculate admissible paths.

A set of different world scenarios are available in the developed framework. Scenarios are generated considering the DEM of mission area. DEM maps represent

Fig. 16 A sample of scenario where two UAVs are performing a mission. On the background a Bergen Twin Observer Helicopter is climbing a small hill

the mission area (real or simulated) as a grid map; a critical parameter is the cell resolution. The resolution of available maps is usually 10 m in the case of real scenarios. This value is too high in critical mission where an accurate localization is required. The GUI allows to edit/create a new DEM to overtake this limitation; data are obtained by exploration of mission's area.

Virtual Reality Toolbox is used to present in soft real-time the state of each vehicle involved in the mission. A VRML world can be customized in terms of textures, position of camera(s) (attached to vehicle or fixed), light(s). The above mentioned toolbox is also used to interface a joystick; this kind of device allows a manual control of the helicopter (user can select the set of axes that wants to control). This feature is really useful for novel pilot(s) during the training phases.

A 3D model of Bergen Twin Observer Helicopter was developed; a more detailed introduction to 3D CAD modelling is presented in Section 3. In Fig. 16 a basic virtual scenario is presented.

Currently, work is focused on the adoption of other virtual reality environments inspired to flight simulator games as FlightGear [27] and Microsoft Flight Simulator [8].

3 CAD Modelling

All subsystems composing helicopter mathematical model require some physical parameters like inertia matrix, mass, distances between COG and force attacking points, rotors geometry and leverage gains.

To perform a really useful simulation for a specific helicopter it is essential to insert in the model its real parameters. In this way an effective shift of results from simulated to real applications becomes possible.

Data as mass, inertia matrix and COG position are time-variant (due to fuel consumption), difficult to measure and have to be re-calculated every time the helicopter works in a different set up (for example after the installation of a new sensor).

Because of these considerations, a 3D CAD model implementation of a possible agent was carried out. A reverse engineering activity on existing reduced-scale helicopter, due to the lack of a technical data sheet, was performed. Actually, a full-detailed model of Bergen Twin Observer Helicopter, designed in Solid Edge environment, is available in the framework; the result of design is shown in Fig. 17.

Solid Edge represents a standard in 3D mechanical modelling. It is a powerful feature-based CAD software, quite simple to use and available in a cheap academic license. It allows an accurate and rapid design of the system and its geometric and inertial characterization.

The model can be exported to specific software, saved in VRML format or merely used for a rendering process. The obtained digital model can be used to Fig. 18:

- evaluate the effect of customization changes (e.g. addition of payloads, sensors.);
- simply extract geometrical and inertial parameters after any structural or set up variation;

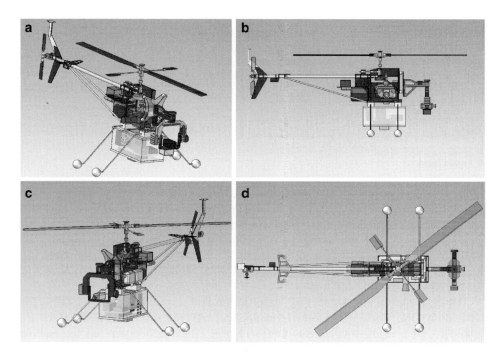

Fig. 17 A series of views of the CAD model of Bergen Twin Observer; a modified landing gear was designed to guarantee a better stability during the take-off and landing manoeuvres. The transparencies allow to see hidden parts, e.g., avionic box and fuel. **a** A front view of Bergen Twin Observer. **b** A lateral view. **c** Pan tilt system shown in foreground; the camera is mounted. **d** A top view

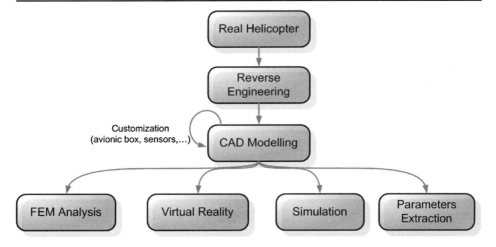

Fig. 18 Reverse engineering is useful in analysis and synthesis; starting from the real model it is possible to extract a digital model that will be then used to evaluate the effects of customitazion

– perform Finite Elements Analysis to highlight possible sources of cracks / breakages;
– visualize the agent in a Virtual Reality environment;
– perform simulations with external multi-body software.

The extracted CAD model is also useful for a Computational Fluid Dynamics analysis using software like Fluent [28].

4 Test Cases

In this section, some simulations using the presented framework are reported. Simulations are performed in scenarios with one or more (cooperating) helicopters to accomplish an exploration mission. Before introducing the results of simulation runs, an UML diagram is shown in Fig. 19 to highlight the relation among the instances of classes (objects). The UML diagram is useful to understand how the simulations are set up. A Bergen Twin Observer model is adopted, equipped with different sets of simulated sensors and actuators.

Sensors adopted are AHRS (for stabilization and movement control), GPS (for navigation) and Radar (for obstacle avoidance). All the simulated helicopters are linked to GCS using sockets for data exchange. Each helicopter has five servos (digital and analog) and one main engine (piston engine). Instances of bergen-TwinObserver class run on two different laptops connected with an Ethernet link.

The simulation time of reported simulations is strongly close to the real time (simulation is a bit slowly in particular conditions and the lag depends on the controllers complexity).

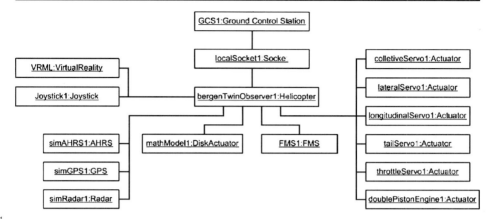

Fig. 19 UML object diagram for one helicopter in the proposed simulations; the structure of the other helicopter is the same

4.1 One Helicopter

The scenario is described by a DEM (0.5 m cell resolution). The GCS assigns the following tasks:

– follow path 1 until point *a*;
– reach waypoint *b*;
– reach waypoint *c*;
– follow path 2 until point *d*;
– back-home *e*.

In spite of a long and not very smooth path, both *Auto-Pilot* and *Guidance Module* actions allow carrying out with success a path following mission; maximum error from nominal track is 0.5 m. Figure 20 shows as the position error, in absence of external disturbances, is very low.

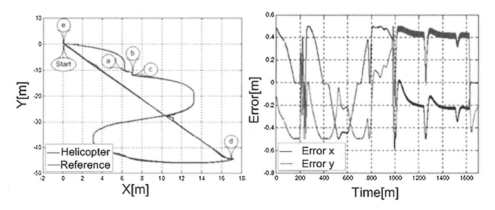

Fig. 20 Labels *a*, *b*, *c*, *d* and *e* represent the tasks performed by agent during mission; the position error, in absence of external disturbances, is very low

Fig. 21 Leader-follower
mission. Proportional
controllers tend to minimize
the trajectory tracking error;
radius of circle is lower than ϵ

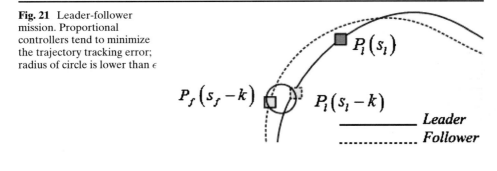

4.2 Two Helicopters: A Leader-follower Mission

This simulation presents two helicopters involved in a "leader-follower mission" [29, 30]. Leader-follower mission belongs to the problem of *coalition formation* inspired by the motion of bevy. The main objective of coalition formation is the clustering of a series of agents to reach a target or to cooperate, extending the capabilities of each agent ("union is strength") [31, 32].

Two identical simulated helicopters are instances of bergenTwinObserver Class. They run on different notebooks with an Ethernet link; socket is the communication method to exchange data.

Leader starts the mission tracking a simple circular path and follower tends to maintain a fixed distance minimizing the error in terms of the following expressions:

$$P(s) = \left[x(s)\ y(s)\ z(s) \right]^{T} \ \left\| P_f(s_f) - P_l(s_l - k) \right\|^{2} < \varepsilon\,k, \varepsilon \in \mathbb{R}$$

where subscript l stands for leader while f for follower, P stands for the helicopter position, k for the distance between helicopters evaluated along trajectory. In Fig. 21 a graphical representation of leader-follower mission is shown.

Follower estimates leader trajectory using leader position obtained by radar and GCS telemetry. Then, on the base of estimated trajectory, follower tends to track

Fig. 22 During the flight, the
helicopter-follower maintains
a fixed distance to leader.
The problem of leader is a
trajectory tracking

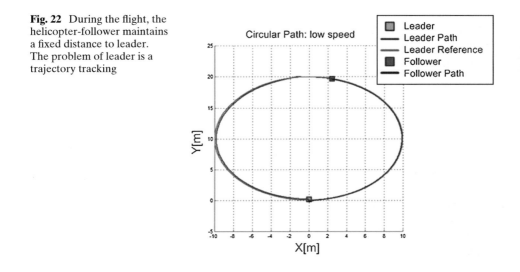

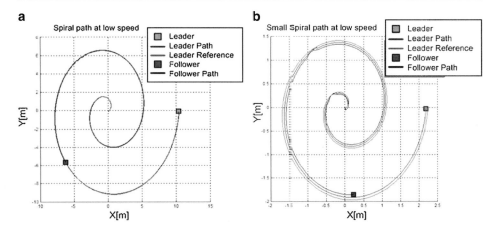

Fig. 23 Two helicopters are moving around the center of the spiral; this is a typical exploring mission. **a** Leader&Follower error tracking is always limited; main drawback is reduced translational speed. **b** In this case the tracking error due to limited manoeuvre are is larger, but limited (lower than 20cm)

leader trajectory minimizing the error. An interesting robust trajectory tracking approach for reduced-scale helicopters can be found in [33]. In Fig. 22, a graphical representation of simulation is shown.

4.3 Two helicopters: A Leader-follower Mission with a Complex Trajectory

In this sub-section the results of tracking two different spiral paths are reported (see Fig. 23). The choice of this kind of trajectory is not casual: it is a typical trajectory adopted in exploration or Search&Rescue missions. The difference between the two simulations is the amplitude of the spiral.

The small spiral has a peculiarity: the helicopter tracks the reference trajectory at very low speed (altitude is fixed); however the error is limited (< 20 cm). The simulation evidences that the framework enables the cooperation of two or more helicopters. In this case, cooperation stands for data exchange to maintain a fixed distance between two vehicles. More complex scenarios that involve more than two helicopters are easy to test.

5 Conclusions and Future Works

Today, UAVs have deeply modified the concepts of surveillance, Search&Rescue, aerial photogrammetry, mapping, etc. The kinds of missions grow continuously; missions are in most cases performed by a fleet of cooperating autonomous and heterogeneous vehicles.

In this paper a framework for UAV simulation in cooperative scenarios and testing was presented. The modularity of its architecture permits to update or rewrite a block in a short time; new controllers can be easily tested. This activity does not require re-compiling or deep rearrangement of the code.

Adding or changing the mathematical model, different aerial vehicles can be simulated; actually the research unit is working on simulating a quad-rotor helicopter. This kind of vehicle is versatile and useful for short range missions; due to these characteristics, the quad-rotor is widely used by researchers in the UAV context.

Moreover, the proposed approach allows an easy switching from simulated to real environments; this is possible owing to stratification of functions in specialized layers.

User interaction, e.g., training of novel-pilots, is supported by GCS, joystick or RC-transmitter interfaces.

Future works will be steered to improve the quality of the Virtual Reality module for an easy interaction with vehicles without a video streaming feedback. Integration of new kind of aerial vehicles will be the main activity. The adoption/integration of FlightGear or Microsoft Flight Simulator graphical engines will be then investigated. New robust non-linear control techniques to enhance the performance (in terms of energy consumption and Time To Task) of agents will be tested. Fault Diagnosis Identification and Isolation techniques, outlined in [34–36], will be then implemented to extend the capabilities of each agent, allowing navigation or landing also in presence of non-disruptive faults.

Acknowledgements   This work was partially supported by master students A. Pistoli and A. Saltari.

References

1. Collett, T.H.J., MacDonald, B.A., Gerkey, B.P.: Player 2.0: Toward a practical robot programming framework. In: Australasian Conf. on Robotics and Automation, ACRA'05, Sydney, 5–7 December 2005
2. Mobilerobots inc. home page. http://www.mobilerobots.com (2008)
3. irobot corporation home page. http://www.irobot.com (2008)
4. Frontoni, E., Mancini, A., Caponetti, F., Zingaretti, P.: A framework for simulations and tests of mobile robotics tasks. In: Proceedings of 14th Mediterranean Conference on Control and Automation, MED'06, Ancona, 28–30 June 2006
5. Carmen robot navigation tool kit. http://carmen.sourceforge.net/ (2008)
6. Microsoft: Microsoft robotics studio developer center. http://msdn.microsoft.com/robotics/ (2008)
7. Unified system for automation and robot simulation. http://sourceforge.net/projects/usarsim/ (2008)
8. Microsoft flight simulator. http://www.microsoft.com/games/flightsimulatorX/ (2008)
9. Fms project. http://www.flying-model-simulator.com/ (2008)
10. Jsbsim: Open source flight dynamics model in c++. http://jsbsim.sourceforge.net/ (2008)
11. Rotorlib home page. http://www.rtdynamics.com/v1.0/ (2008)
12. Taamallah, S., de Reus, A.J.C., Boer, J.F.: Development of a rotorcraft mini-uav system demonstrator. The 24th Digital Avionics Systems Conference, DASC'05, Washington, D.C., 30 October–3 November 2005
13. Frontoni, E., Mancini, A., Caponetti, F., Zingaretti, P., Longhi, S.: Prototype uav helicopter working in cooperative environments. In: Proceedings of IEEE/ASME International Conference on Advanced Intelligent Mechatronics, AIM'07, Zurich, 4–7 September 2007
14. Unmanned dynamics - aerosim toolbox. http://www.u-dynamics.com/aerosim/ (2008)
15. Uml resource page. http://www.uml.org (2008)
16. Bouabdallah, S., Murrieri, P., Siegwart, R.: Design and control of an indoor micro quadrotor. In: Proceedings of IEEE International Conference on Robotics and Automation, ICRA '04, New Orleans, 26 April–1 May 2004
17. Heffley, R.K., Mnich, M.A.: Minimum-complexity helicopter simulation math model. Technical report, Contractor report NASA 177476, Aeroflightdynamics Directorate, U.S. Army Research and Technology Activity (AVSCOM) (1988)

18. Jetto, L., Longhi, S., Venturini, G.: Development and experimental validation of an adaptive extended kalman filter for the localization of mobile robots. IEEE Trans. Robot. Autom. **15**, 219–229 (1999)
19. Zingaretti, P., Frontoni, E.: Appearance-based robotics - robot localization in partially explored environments. IEEE Robot. Autom. Mag. **13**, 59-68 (2006)
20. Kwag, Y.K., Chung, C.H.: Uav based collision avoidance radar sensor. In: IEEE International Symposium on Geoscience and Remote Sensing, IGARSS'07, Barcelona, 23–27 July 2007
21. Dong, T., Liao, X.H., Zhang, R., Sun, Z., Song, Y.D : Path tracking and obstacles avoidance of uavs - fuzzy logic approach. In: The 14th IEEE International Conference on Fuzzy Systems, FUZZ'05, Reno, 22–25 May 2005
22. Koo, T.J., Sastry, S.: Output tracking control design of a helicopter model based on approximate linearization. In: Proceedings of the 37th IEEE Conference on Decision and Control, Tampa, December 1998
23. Bramwell, A.R.S., Done, G., Balmford D.: Bramwell's Helicopter Dynamics, 2nd edn. Butterworth Heinemann, Boston (2001)
24. Mettler, B., Tischler, M.B., Kanade T.: System identification of small-size unmanned helicopter dynamics. In: Presented at the American Helicopter Society 55th Forum, Montreal, May 1999
25. Sanders, C.P., DeBitetto, P.A., Feron, E., Vuong, H.F., Leveson, N.: Hierarchical control of small autonomous helicopters. In: Proceedings of the 37th IEEE Conference on Decision and Control, Tampa, December 1998
26. Buskey, G., Roberts, J., Corke, P., Wyeth, G.: Helicopter automation a using a low-cost sensing system. Comput. Control Eng. J. **15**, 8–9 (2004)
27. Flightgear project http://www.flightgear.org (2008)
28. Fluent cfd flow modeling software http://www.fluent.com (2008)
29. Wang X., Yadav, V., Balakrishnan, S.N.: Cooperative uav formation flying with obstacle/collision avoidance. IEEE Trans. Control Syst. Technol. **15**, 672–679 (2007)
30. Lechevin, N., Rabbath, C.A., Sicard, P.: Trajectory tracking of leader-follower formations characterized by constant line-of-sight angles. Automatica **42**(12) (2006)
31. Merino, L., Caballero, F., Martinez de Dios, J.R., Ollero, A.: Cooperative fire detection using unmanned aerial vehicles. In: Proceedings of IEEE International Conference on Robotics and Automation, ICRA 05, (2005)
32. Beard, R.W., et al.: Decentralized cooperative aerial surveillance using fixed-wing miniature uav. Proc. IEEE **94**(7), 1306–1324 (2006)
33. Mahony, R., Hamel, T.: Robust trajectory tracking for a scale model autonomous helicopter. Int. J. Robust Nonlinear Control **14**(12) (2004)
34. Monteriù, A., Asthana, P., Valavanis, K., Longhi, S : Experimental validation of a real-time model-based sensor fault detection and isolation system for unmanned ground vehicles. In: Proceedings of 14th Mediterranean Conference on Control and Automation MED'06, Ancona, 28–30 June 2006
35. Monteriù, A., Asthana, P., Valavanis, K., Longhi, S. Model-based sensor fault detection and isolation system for unmanned ground vehicles: theoretical aspects (part i). In: Proceedings of the IEEE International Conference on Robotics and Automation, ICRA'07, Roma, 10–14 April 2007
36. Monteriù, A., Asthana, P., Valavanis, K., Longhi, S. Model-based sensor fault detection and isolation system for unmanned ground vehicles: Theoretical aspects (part ii). In: Proceedings of the IEEE International Conference on Robotics and Automation, ICRA'07, Roma, 10–14 April 2007
37. Dzul, A.E., Castillo, P., Lozano, R.: Modelling and Control of Mini-Flying Machines. AIC Advances in Industrial Control. Springer (2005)

Distributed Simulation and Middleware
for Networked UAS

Ali Haydar Göktoğan · Salah Sukkarieh

Originally published in the Journal of Intelligent and Robotic Systems, Volume 54, Nos 1–3, 331–357.
© Springer Science + Business Media B.V. 2008

Abstract As a result of the advances in solid state, electronics, sensor, wireless communication technologies, as well as evolutions in the material science and manufacturing technologies, unmanned aerial vehicles (UAVs) and unmanned aerial systems (UAS) have become more accessible by civilian entities, industry, as well as academia. There is a high level of market driven standardisation in the electronics and mechanical components used on UAVs. However, the implemented software of a UAS does not exhibit the same level of standardisation and compartmentalisation. This is a major bottleneck limiting software maintenance, and software reuse across multiple UAS programs. This paper addresses the software development processes adapted by the Australian Centre for Field Robotics (ACFR) in major UAS projects. The presented process model promotes software reuse without sacrificing the reliability and safety of the networked UAS with particular emphasis on the role of distributed simulation and middleware in the development and maintenance processes.

Keywords Unmanned aerial system (UAS) · Unmanned aerial vehicle (UAV) ·
Software frameworks · Middleware · Simulation · RMUS · HWIL

A. H. Göktoğan (✉) · S. Sukkarieh
ARC Centre of Excellence for Autonomous Systems Australian Centre for Field
Robotics School of Aerospace, Mechanical, and Mechatronic Engineering,
The University of Sydney, 2006, NSW Australia
e-mail: aligoktogan@acfr.usyd.edu.au

S. Sukkarieh
e-mail: salah@acfr.usyd.edu.au

K. P. Valavanis et al. (eds.), *Unmanned Aircraft Systems*. DOI: 10.1007/978-1-4020-9136-0_19   331

1 Introduction

The goal of this paper is to describe the middleware oriented software development methodology used for the networked unmanned aerial system (UAS) at the Australian Centre for Field Robotics (ACFR).

The software development methodology provides the capability of transitioning, testing and validating, complex research algorithms from simulation to real-time demonstration. Particular focus is on the need to minimise re-coding of the research algorithms between the various testing phases; and for the development of the framework to provide distributed, decentralised and scalable capabilities.

Establishment of an operational infrastructure of an UAS, incorporating a dedicated fleet of UAVs in a R&D environment involves complex and demanding system development processes. Success of this endeavour highly depends on understanding the system life cycle management (SLCM) in R&D environments, and the unique characteristics of the UAS and flight operations. This paper looks at the software development side of the SLCM.

The next section discusses the market conditions of the commercial-off-the-shelf (COTS) hardware and software components for UAS. Section 3 briefly presents the software development processes that we used in developing our networked UAS. Section 4 presents the high level architecture of a multi-UAV system. Section 5 introduces a framework for distributed autonomous systems (AFDAS). The Comm-LibX/ServiceX middleware and its novel concept of virtual channel are covered in Section 6. Section 7 examines the real-time multi UAV simulator (RMUS). Finally, conclusion and future works are covered in Section 8.

2 COTS Components for UAS

Advances in solid state and electronics technologies fit more processing power to unit volume while reducing the cost and the electrical power consumption. Consequently, the new developments in sensor technologies offer a wider range of more reliable, faster, multi-modal sensors (MMS) in smaller packages with lower price tags. Likewise, the wireless communication industry is providing wider bandwidth over longer distances through smaller devices while radiating lower Radio Frequency (RF) energy. Furthermore, recent electrochemical fuel cells and batteries with high power density offer new possibilities both for the payload's electrical power needs, and for electrically driven propulsion systems particularly used on small UAVs.

Thanks to the evolution in material science that, stronger, lighter, fatigue resistant, and easily machineable complex composite structural materials are now widely available. Similarly, development in the manufacturing technologies, such as wide spread of computerized numerical control (CNC) machinery and rapid prototyping systems, made it possible of fabricate more complex structures in shorter time, with higher precision.

Along with the other unmanned systems, the UAVs benefit the most from these positive trends. The radio controlled (R/C) hobby gadgetry market, and the light aircraft industry offer wide range of mechanical components that can be used in construction new UAVs. Acquisition of commercial-off-the-shelf (COTS) small but powerful computers, sensors, and wireless communication devices are not beyond the reach of many academic institutions. Therefore, increasing number of academic institutions are able to utilise UAVs in their research, teaching and learning programs.

There is a high level of standardisation in the electronics and mechanical components. Highly standardised and compartmentalised, modular hardware sub-systems and components offer flexibility to UAS developers. The UAS architects and engineers can easily configure their hardware and upgrade it when needed. Because of the availability of multiple vendors of the same or similar hardware components, many UAS developers feel free in selecting their suppliers.

However, the software side of UAS does not exhibit the same level of standardisation and compartmentalisation. This is a major bottleneck limiting the software maintenance, and software reuse across multiple UAS programs. It is not easy to find many successful examples neither in academia nor in industry in which the UAS developers can acquire COTS software components, in an equally flexible manner as if they were acquiring the hardware components.

The limited availability of the COTS software components for the UAS development might be natural result of demand-and-supply relationship in the software market. That is, the demand for the COTS software solutions generated by the UAS developments may be not strong enough to create its own market, yet. Even so, there are exceptions to this generalisation where a significant standardisation works in progress in the military domain. The "Standard interfaces of UAV control system (UCS) for NATO UAV interoperability", (STANAG 4586) [1] is one of them. This particular standard has created its own market in which a number of companies provide STANAG 4586 compatible software solutions. This example shows the obvious; the UAS software market could thrive if enough demand is generated. However, STANAG requirements are too complex for many research groups in academia for their relatively smaller scale research projects.

By definition, research projects address new concepts and therefore one may argue that majority of the new UAS development may need their own implementation of software modules. Although this argument is partially valid, there is still a place for the COTS software solutions if enough demand is generated.

One of the reasons why the civilian UAS domain could not generate enough demand is because; in this domain, there are many different types of UAVs. They are ranging from micro aerial vehicles (MAVs) which can carry few grams of payload to the larger UAVs with hundreds of kilograms of Maximum Take-Off Weight (MTOW) capabilities. These large varieties of UAVs are equipped with very different types of computational units, such as tiny microcontrollers for MAVs, or a network of high performance computers for the top scale UAVs.

The need for reusable and maintainable software modules, components, frameworks, architectures, and services are not new and not specific to the UAS domain. These fundamental computer science and software engineering issues are widely addressed, and numerous techniques have already been developed and successfully used in other application domains by industry and academia. However, the same techniques are being slowly accepted by the UAS community particularly in academia. This may be due to the fact that many UAS projects are initiated by highly qualified experts in aeronautical and/or control engineering fields. Inadvertently, due to their highly focused views, they often underestimate the importance and complexities associated with maintainable software development methodologies.

Increasing involvement of computer scientist and software engineers in the UAS development life cycle is expected to change this bottleneck.

There is a variety of literature, addressing software development methodologies for aerial vehicles. Ernst et al. in [2] presents a method of designing and building UAV flight controllers using C code generated from MatLab. They also address the validation of the resultant system on a COTS simulator X-Plane. Although, this method is highly efficient way of developing flight controllers for small UAVs in which the flight controller consists of a single microcontroller, it does not address widely distributed systems. Similarly, as they acknowledge in their paper, X-Plane based simulation may provide acceptable simulation for single UAV, but, in its current form, it can not address the needs of complex mission simulations incorporating multiple UAVs and other entities such as mission sensors, and ground vehicles. Shixianjun et al. in [3] also presents a MatLab-Simulink based code generation for a UAV and its hardware-in-the-loop simulation (HWIL) tests. Their system is also based on a single UAV, and does not address distributed systems.

There are also a number of examples in the literature covering issues of multi-UAV system software. In [4] Doherty et al. explains the WITAS UAV project in which their helicopter UAV is controlled by a flight control software built around a DOS based real-time kernel "RTkernel". The research methodology used in WITAS project acknowledged the importance of the simulation. Their flight experiments were supplemented with a great deal of simulated experiments. The simulator architecture was based on a real-time common object request broker architecture (CORBA) as the software communication infrastructure. Their intelligent vehicle control architecture (IVCA) also used the same real-time CORBA based infrastructure which allows transition of the software implementations from the simulation environment to the HWIL simulator and eventually to the actual platforms.

CORBA, from the object management group (OMG) is an extreme case of an open standard specification for general purpose, distributed middleware architecture which can be implemented by different vendors. There are a number of CORBA implementations from different vendors. However, it is widely accepted that, many implementations do not match all of CORBA's published specification [5, 6]. For many software professionals, this may not necessarily be a serious disadvantage.

Once, (and if) a satisfying CORBA implementation is acquired, the CORBA based distributed system development (at least in theory) should be straightforward. Even so, it is acknowledged by many [5, 7–9] that using CORBA for any nontrivial application is surprisingly difficult. CORBA has a very large and complex set of application programming interface (API). Employment of a dedicated team of CORBA experts is almost a necessity for the development of successful, large scale, and maintainable CORBA based systems. For some mid-sized research organisations with limited resources, like the ACFR, this is less than a desired option. Having said that, there are also a large number of gratified CORBA users from a variety of application domains, including robotics and autonomous systems [10–13].

Flexibility and performance are considered rivals in software systems [14]. Despite its wide acceptance, particularly by moderate size information technology (IT) companies, CORBA's alleged flexibility leads to its rather unpleasant reputation of being "too big and complex". Surely, the technological advances in both memory and processing power of computer systems are helping CORBA to change its reputation in a positive direction. The OMG's recently published draft specifications for "common object request broker architecture—for embedded" (CORBA/e) [15] is another positive step toward CORBA's wider acceptance in the future. The CORBA/e draft specification addresses the bottlenecks of the old/existing, enterprise-CORBA and acknowledges CORBA's implementation issues. It particularly references CORBA's resource-hungry nature and its limitations to be used in real-time applications on resource limited embedded computing platforms. This draft is yet another valuable, lengthy document which one should study to learn from CORBA's past experiences. "CORBA/e sacrifices some of the general purpose nature of CORBA in order to support the development of real-time systems." [15], this statement is a clear indication that OMG now has a very different vision for the future of CORBA.

Following section briefly presents the software development processes that we used in developing our networked UAS. It also highlights the importance of modelling and simulation.

3 Software Development Processes for UAS

A simplified version of the application software development processes that we have followed in developing software applications for the UAS is shown in Fig. 1. It emphasises that the Modeling and Simulation (M&S) is an integral part of each process.

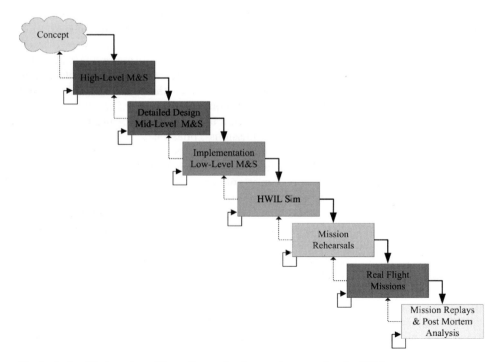

Fig. 1 A simplified version of the software development process flow for UAS

Due to its cascading form, the development process model in Fig. 1 is known as the waterfall model [16]. The waterfall model keeps iterations between neighbouring phases. In order to proceed from one phase to the next, often a review takes place to determine if the phase is completed successfully. The large downward solid arrows indicates the desired flow direction; from upper level phase to the lower level phase. However, sometime in practise, it become inevitable to go back to previous phases. This less desired upward flows are depicted by the dashed arrows. The small loop-back arrows on the lower left corner of the processes indicates that it is encouraged to iterate within the life-cycle phase before proceeding to the lower level phase.

The first stage in our software development process is the "high-level M&S" of the proposed system in Matlab/Simulink or similar high level simulation environments. After the *validation*[1] of the new concepts, techniques and algorithms, the class hierarchies, hardware and software components are defined in the "detailed design, mid-level M&S" phase. At this phase, unified modeling language (UML) is particularly useful for software design and documentation as well as for defining the test cases.

The "Implementation, Low-Level M&S" phase consists of coding the actual running software (often in C/C++) and testing individual modules. In these tests low-level M&S, such as simulation of software modules, interaction between software components, and sensor and control outputs are tested. Test results are used in the *verification*[2] of the developed algorithms.

All software applications, their interaction with each other, and the actual hardware that will be used in the real flight missions are tested in the "hardware-in-the-loop simulation (HWIL Sim)" phase. The HWIL Sim test results are used for the verification and *accreditation*[3] of the developed software, ie. results obtained at the high-level simulation phase are compared with the HWIL Sim test results.

Following successful completion of the Validation, Verification and Accreditation (VV&A) procedures in the HWIL Sim are assumed to be ready for the "mission rehearsals" phase. At this stage, the operation plans are prepared. These plans state the mission objective, define personnel work distribution and flight plans. The operational personnel are prepared for the mission through repeated mission rehearsals performed in the HWIL Sim environment.

After the operational personnel are reached to the required proficiency level for the proposed missions, the "real flight missions" phase can be performed in the actual flight area. During real flight missions, a number of data logging processes are performed both onboard the UAVs and on the ground stations. The logged data are used in "mission replays and post mortem analysis" phase. Soon after the flight operations, the logged data are replayed in the RMUS environment and the mission team assess if mission objectives are reached and decide if the flight mission needs to be repeated or not. The logged data is later used in the RMUS environment many times for post-mortem analysis of the missions.

---

[1] In the context of modelling and simulation (M&S), *validation* is the process of determining whether the selected simulation model accurately represents the system.

[2] *Verification*, answers the question of whether the selected simulation model is implemented correctly.

[3] *Accreditation* is "the official certification that a model or simulation is acceptable for use for a specific purpose" [17].

4 Multi-UAV System Architecture

This section presents the architecture of the multi-UAV system which is developed and successfully used by the ACFR. It also introduces the terminology used in this paper.

Traditional UAV applications involve relatively large UAVs with the capability of carrying multiple payload sensors in a single platform. Ever increasing availability, and accessibility of smaller UAVs to the research community encourage the development of multi-UAV applications, where the mission sensors would be distributed across a number of UAVs [18].

This is a major undertaking; it shifts the focus from the platform-centric to network-centric capabilities, and it requires the adoption of a network-centric operational philosophy; "The network-centric operational philosophy relies on the capability of shared situation awareness, through fusion of information disseminated from a wide range of networked sensors, and decentralized decision making to control the platforms and the sensors to maximize the information gain toward achieving the mission goals" [19].

Compared to the rather confined problem domain of the platform-centric capability, the network-centric capability has a much broader domain. In network-centric systems, the information dissemination often takes place across heterogenous, mobile information sources and information consumers. In such systems, it is not uncommon to have range and the bandwidth limited communication links with sporadic dropouts. However, in platform-centric systems, where all nodes are carried on the same platform, the information dissemination occurs on a much faster and more reliable local network.

The Brumby Mk III UAV (shown below) is capable of demonstrating the platform-centric system capabilities when a single-UAV is used in the flight missions [20]. More importantly, the network-centric system capabilities are also successfully demonstrated in flight missions in which, multiple Brumby Mk III UAVs are used [18, 21–25].

Brumby Mk III UAV The Brumby Mk III (Fig. 2) UAVs are the main flight platform used by the ACFR in multi-UAV experiments. They are delta wing UAVs with 2.9 m wing span, pusher type four blade propeller, and conventional tricycle undercarriage suitable for takeoff from and landing on short grass and asphalt surfaces. The Brumby Mk III UAVs are designed and manufactured at The University of Sydney as the third generation UAVs in the Brumby series. Their maximum take-off weight (MTOW) is approximately 45 kg. The replaceable nose cone of the Brumby Mk III UAVs is ideal for reconfiguration of the UAVs to carry different mission payloads in a plug-and-play manner.

Due to their delta wing design and the 16 Hp engine, the Brumby Mk III UAVs are very agile platforms. They can travel around 100 Knots (183.5 km/h) and have roll rates of approximately 180°/s. The agility of a UAV relates to the highly dynamic nature of the platform. Compared to slow, hovering vehicles like helicopters and blimps, highly dynamic, fixed wing UAVs pose additional challenges, especially in control and in communication. Although the software framework presented in this paper can be applied to a wide variety of systems, this paper is focused particularly on the multi-UAV systems of the agile, fixed wing UAVs.

Fig. 2 The Brumby Mk III UAV as it takes off. It has delta wing with 2.9 m wing span, pusher type propulsion, conventional tricycle undercarriage, and replaceable sensor nose cone

Avionics The Brumby Mk III UAVs are equipped with relatively complex avionics (Fig. 3). In a typical configuration, a single Brumby Mk III UAV carries six PC104+ computers, all networked through an onboard local area network (LAN), a flight and mission sensors, a microcontroller based fight mode switch (FMS), an optional digital signal processing (DSP) board, a spread spectrum radio modem, a 802.11b wireless Ethernet, a data acquisition module and power subsystem with two separate buses to power flight critical and non-flight critical payloads. The FMS is one of the most critical units in the Brumby Mk III UAV avionics architecture. It consists of a spread spectrum radio modem and an embedded microcontroller board. Depending on the flight mode activation strategies defined by the system engineering and the HWIL simulations, the embedded application software running on the microcontroller drives the servo actuators either based on the remote control commands received from the ground or commands received from the flight control computer (FCC). Thus, the FMS can switch the UAV operation mode between remotely piloted vehicle (RPV) and autonomous flight modes.

The FCC is a PC104+ computer and runs the QNX Real-Time Operating System (RTOS). The Brumby Flight Control System (BFCS) is the main high level application running on the FCC. The multi-threaded BFCS application calculates the real-time navigation solution and also generates low level control signals. Through the CommLibX/ServiceX middleware, the BFCS provides real-time navigation data to other onboard computers.

The Mission Computer (MC) also runs the QNX RTOS. As illustrated in Fig. 3, the FCC, and MC is linked via the onboard Ethernet hub. The MC is equipped with an 802.11b wireless Ethernet card. It provides air to air and air to/from ground communication links. Compared to the FMS's radio modem, the 802.11b provides a wider bandwidth. However, the 802.11b link has much shorter communication range and it is subject to frequent dropouts. Hence, flight critical data, such as RPV control commands are not transmitted via wireless Ethernet but via a radio modem.

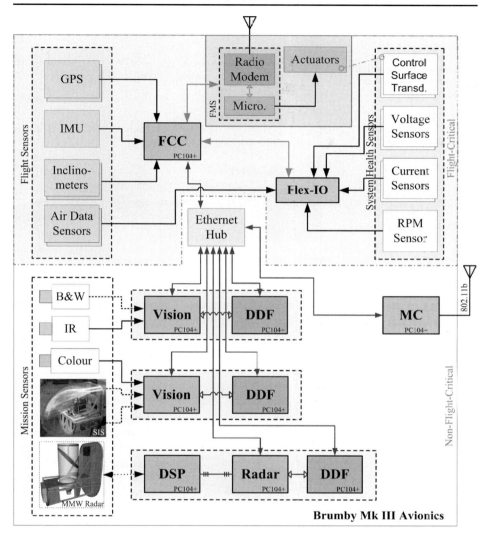

Fig. 3 A high level representation of the Brumby Mk III UAV avionics with the optional mission sensors

The mission sensor control computers run embedded Linux in soft-real-time while the decentralised data fusion (DDF) computers run QNX in hard real-time. All interact with each other via CommLibX/ServiceX middleware.

Unmanned Aerial System (UAS) With a number of interacting onboard subsystems UAVs are considered complex systems. However, a UAV is only one component of a larger whole; known as an *unmanned aerial system* (UAS), Fig. 4. This larger whole is comprised of the ground stations, communication, telemetry, control and navigation equipments, sensor payloads and all the other entities necessary to perform missions involving UAVs.

Fig. 4 Illustration of a
multi-UAV system consists
of heterogeneous UAVs
carrying mission sensors
and communicating with the
ground control station as
well as with each other

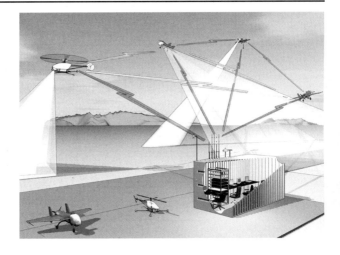

As illustrated in Fig. 3 and in Fig. 5, an operational multi-UAV system has
a number of computing and communication nodes. Each node concurrently runs
multiple processes with various levels of real-time requirements. In this context,
the term *real-time*, refers to satisfaction of timeliness requirements of individual
processes [26].

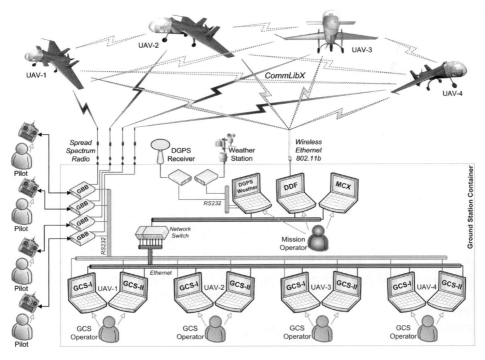

Fig. 5 An operational multi-UAV system consists of a number of computing and communication
nodes onboard the UAVs and the ground station

Multi-UAV systems have a number of components both on the ground and in the air. The communication links between these *distributed* components, particularly air to air and air to/from ground links are often subject to intermittent dropouts. Therefore, in practical terms, the overall system can not be controlled efficiently by a centralized computer. The nature of multi-UAV systems dictates the adoption of *decentralised* system architectures in which no single node is central for the operation of the entire system.

At the ACFR, the multi-UAV systems are often used for the demonstration of DDF [27], Simultaneous localisation and mapping (SLAM) [28] and various cooperative control [29] techniques. The decentralised system architecture is also an algorithmic requirement of these research areas [18].

Ground Station The UAV flights are monitored and controlled from the ground station which is within an impact-proof cargo container located next to the runway as shown in Fig. 6. All the ground based computing and the communication hardware is enclosed in the ground station. The system hardware architecture of the ground stations for four UAVs is shown in Fig. 5.

As shown in Fig. 5, the main components in the ground station are the ground control stations (GCS), mission control console (MCX), weather station, uplink controllers (also known as ground black box [GBB]), differential global positioning system receiver, and a decentralized data fusion node.

Often, during the taxi, take off and landing, human pilots control the UAVs with modified hand held radio control (R/C) units. Unlike the standard R/C transmitters, these modified R/C units, send the servo control signals to the uplink controllers instead of transmitting them directly to the UAVs as modulated radio frequency (RF) signals.

An uplink controller (GBB) is a unit built around an embedded microcontroller and a spread spectrum radio modem. Its main functions are to digitise the servo pulse streams coming from the R/C unit and pack them for transmission to the UAV and to the GCS. The GBB unit also acts as a full-duplex packet router and as such, it routes the telemetry data received from the UAV to its corresponding GCS and also sends the commands issued by the GCS to the UAV (Fig. 5).

Fig. 6 Photograph of two Brumby Mk III UAVs, next to the ground station container, are being prepared for a multi-UAV mission

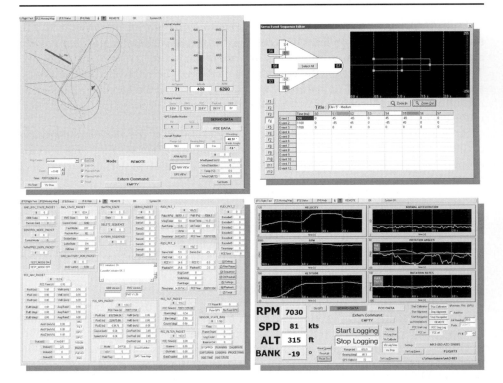

Fig. 7 Screen shots from the ground control station—I (GCS-I) application

The status and onboard instruments of each UAV is monitored in real-time by two GCS computers. GCS-I (Fig. 7) displays and logs the real-time telemetry data and GCS-II displays and logs the data related to mission and cooperative control. The telemetry data consists of more than a hundred types of data packages transmitted from the UAV to the ground at various data rates. These include position, air speed, altitude, engine RPM, and battery voltages which are also relayed by the GCS operator to the human pilot during the remote control of the UAVs. The logged telemetry data is used for post-mortem analysis of the flights. GCS-I also receives

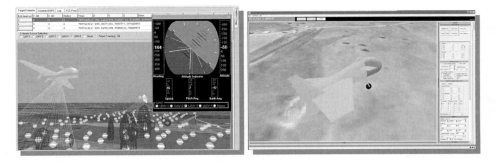

Fig. 8 Screen shots from the ground control station—II (GCS-II) application

the differential global positioning system (DGPS) messages and weather data via the local network in the ground station. It relays the DGPS messages to the UAV via the GBB. The local weather data is also displayed on GCS-I to aid in Situation Awareness (SA) [30] of the ground station operator and the pilot. As shown in Fig. 8, the GCS-II can optionally display, in real-time, the monitored UAV in a simulated 3D environment. This helps GCS operators to increase their SA.

Both the GCS-I and the GCS-II are multi-threaded applications and written for the MS-Windows operating system and they use CommLibX/ServiceX middleware.

5 A Framework for Distributed Autonomous Systems

The above mentioned multi-UAV system architecture exhibits the characteristics of highly decentralised, distributed, and networked systems. This networked system has heterogenous nodes. The nodes have a number of different hardware. They run a variety of software applications on a number of different operating systems with different levels of real-time requirements.

Typical multi-UAV missions, such as DDF, SLAM [18] and cooperative control demonstration missions [31], performed by the ACFR consists of recurring development phases with similar or identical tasks like platform, and sensor modeling, communication, data logging and replay, visualisation and time synchronisation. In order to minimise the repeated solutions, a systematic approach for design, development, deployment, and maintenance of the UAS system software is needed.

Component based software frameworks provide a suitable domain to apply systematic approach to the software development phases. The "Lego character" of components offers well defined, simple, yet powerful integration mechanisms. This leads to easily re/configurable and maintainable systems. A Framework for Distributed Autonomous Systems (AFDAS) is designed and developed with this vision in mind.

AFDAS aims to address both the low-level and high-level software requirements of a networked UAS. These include distributed, real-time simulator architectures, the ground station, and the flight critical applications of multi-UAV systems. AFDAS is leveraged by a large collection of software libraries, modules, and components that have been developed over the years and still continue to expand with refinement of the existing code as well as the new software being developed for the new UAS projects.

AFDAS uses a top-down decomposition approach to the domain of networked UAS and introduces a layered system software view. The operational and safety requirements of the networked UAS play an important role in these approaches. The domain partitioning, and framework layering increase the efficiency of managing the complex system requirements.

Control layers have different levels of task abstractions; higher control layers have more abstract tasks than the lower layers. Abstraction is also reflected to the real-time requirements. The real-time requirements ease at higher level of abstractions.

The networked UAS software system is designed in a hierarchical form. As it is illustrated in Fig. 9 and tabulated in Table 1 the system software hierarchy can be specified in four layers; actuator, subsystem, system, and system of systems (SoS) layers.

Fig. 9 Layers of the networked UAS software hierarchy

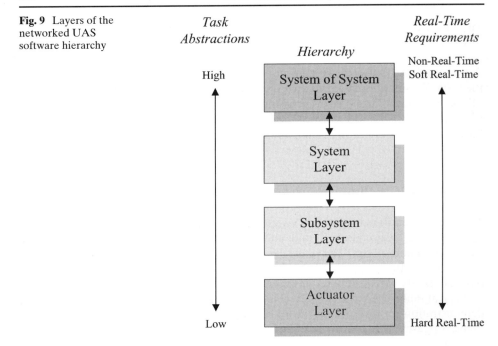

Actuator Layer   The actuator control layer provides access to the actuators via driver electronics and low-level device driver software modules. It is common to control the actuators by microcontroller based electronics boards. The small microcontroller boards often programmed with proprietary C compilers, and assemblers.

Table 1 Networked UAS Software Hierarchy

Control layers	HW&SW modules	SW Tools	RT requirements
SoS layer	WAN & LAN, IPC, comms.	Matlab, Simulink, Java, C/C++	Non real-time, soft real-time, event based tasks
System layer	Computers, WAN & LAN, IPC, comms.	Matlab, Simulink, Java, C/C++	Hard real-time, soft real-time, periodic tasks, event based tasks
Subsystem layer	Embedded computers, microcontroller boards, servo drivers, sensors, IPC, comms.	C/C++	Hard real-time, periodic tasks
Actuator layer	Microcontroller boards, PAL/PLD/FPGA peripheral devices, driver electronics, device drivers	Proprietary C, CUPL, HDL, VHDL, assembler	Hard real-time

The device driver software modules for these boards are carefully written, optimised codes with hard real time requirements tailored for particular hardware.

Subsystem Layer The subsystem layer provides low-level control and fusion behaviours. Low-level sensor data fusion, such as INS-GPS fusion for navigation solutions, and low-level servo controls such as PID servo loops are typical tasks at this level. This layer is built around multiple embedded computers, or microcontroller boards, connected to servo drivers and sensors.

The tasks at the subsystem layer are often programmed in C/C++ and run at a constant rate in hard real-time. The concurrent tasks running on a single embedded computer or on multiple networked embedded computers use Inter-Process Communication (IPC) mechanisms for data exchange.

System Layer The system layer provides high level control and decision making functionalities. Some of the tasks at this layer are the system level diagnostics and error handling, path planning, way-point navigation and guidance. These tasks could be periodic tasks or event based tasks with soft real-time or hard real-time requirements. Although most of the software code for the system layer tasks are hand written in C/C++, other higher level languages and graphical tools like MatLab and Simulink can also be used to auto generate the C/C++ codes [29, 31, 32].

The system layer consists of heterogenous set of computers, networked together via wide area network (WAN) and/or local area network (LAN). The networked computers in this layer also use various IPC and high level communication mechanisms for data exchange between multiple processes.

System of Systems Layer The System of Systems (SoS) layer is the highest level of control in the control hierarchy. Compared to the lower layers, the real-time requirements of this level are eased. The tasks at the SoS layer are concentrated on the interaction and interoperability of multiple systems. Humans may also be in the control loop in the SoS layer.

Many UASs have inherently distributed system architectures. In order to achieve interoperability at the SoS layer, the software associated with each individual system should meet a common set of requirements for their interfaces with each other. Often, multiple heterogeneous vehicles, which are complex systems themselves, and other systems, such as stationary sensors networks, operate together to fulfill the mission objectives in a dynamic environment. Communication plays a major role in their interactions since cooperation (often [33, 34]) requires communication between the vehicles [35, 36].

Depending on the communication requirements, layers can be linked together by using a number of different inter-process communication (IPC) methods, such as shared memory, message passing, pipes, sockets and remote procedure calls (RPC) etc. or higher level middlewares. Middleware is a suite of software that provides the connectivity between multiple otherwise isolated applications.

The difficulties of using different IPC mechanisms and middlewares should not be underestimated. Different IPC techniques have strengths, weaknesses, and traps that the software developers should manage in order to facilitate reliable and effective communication between tasks.

Mastering the intricacies of some of the traditional IPC techniques is a challenging task that requires time and effort, and above all experience. Developers who have the expertise in control of autonomous systems may not necessarily have the sufficient level of expertise in IPC. The CommLibX/ServiceX middleware [37, 38] which is detailed in the next section, addresses this problem and offers a simple, yet powerful way of IPC for the UAS software developers.

6 CommLibX/ServiceX Middleware

Increasing complexity of autonomous systems and the adoption of network-centric operational philosophy are motivating the development of large scale distributed systems. The non-distributed/stand-alone system approach is becoming a thing of the past.

Distributed system software components are glued together with middleware. Middleware is the systems software connecting the distributed software components, applications, and simulation modules together. As illustrated in Fig. 10 the middleware system often resides between the operating system and the high level distributed applications.

Middleware hides the complexities and the heterogeneities of the underlying operating systems, network protocols, and hardware of the target computers. It enhances the efficiency and quality of the distributed system development work and simplifies the maintainability.

There are many different middlewares available in the software market today. They can be grouped according to their application domains, language support, and real-time characteristics. Some of them are designed to have target specific requirements of a particular application domain, such as distributed simulation [39], and the others are general purpose [40].

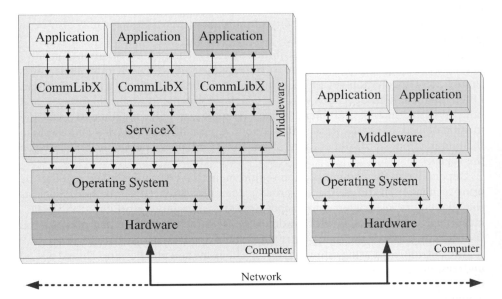

Fig. 10 CommLibX/ServiceX is the middleware, designed and developed to support the seamless network UAS research works at ACFR

The CommLibX/ServiceX is a novel, layered software framework designed as a middleware to provide hardware, operating system, and programming language abstraction for the communication, and other system services required for a networked UAS. The CommLibX/ServiceX middleware supports the seamless network UAS research works at the ACFR.

Originally, the CommLibX/ServiceX middleware is designed in UAS applications in mind. However, it is used on a number of other autonomous systems, and distributed simulation environments. The CommLibX/ServiceX middleware has very flexible and modular architecture. Its base code is small in size, and through the plug-in modules, its functionality can be tailored for the specific application domain. The CommLibX/ServiceX middleware supported on QNX, Linux, and Windows operating systems. A cut-down version of the CommLibX/ServiceX middleware is also demonstrated on low end small Atmel AVR microcontrollers with no operating system.

One of the most important objectives of the CommLibX/ServiceX middleware architecture is to enable the application developers to develop high-level distributed applications, much the same way, as if they were developing stand-alone, non-distributed applications. This assists the developers in focussing on their applications rather than the elaborately complex details of the underlaying operating systems and hardware.

The CommLibX/ServiceX middleware consists of two major parts; the CommLibX library and the ServiceX module (Fig. 10). The CommLibX is the interface between the high level distributed application and ServiceX. As its name implies ServiceX provides the middleware services including communication, task scheduling, and some limited direct hardware control. Each application initiates an instance of the CommLibX. The high level applications invoke operations on ServiceX via the CommLibX instance. And similarly, the ServiceX invokes operations on the high-level application through the same instance of the CommLibX.

The separation of the middleware into two parts as CommLibX and ServiceX has another benefit; CommLibX "isolates" ServiceX from the high-level applications. Execution of ServiceX does not depend on the state of the user application. What this means in practice is that an ill-behaved user application does not affect the performance of ServiceX. Hence, a faulty program does not jeopardize the operation of the rest of the distributed system.

In a CommLibX/ServiceX based distributed system, high-level applications may run on either one or multiple computers. Often, high-level applications are not aware if the other applications are running on the same computer or not. Regardless if they are running on the same computer or on separate computers, high-level applications do not directly interact with each other, but the interaction occurs through *virtual channels*. The virtual channel is a novel concept introduced by the CommLibX/ServiceX middleware. The virtual channels are the logical links between communicating nodes. Often they are used for the logical separation of different kinds of data packages. The virtual channels can be allocated either a single communication medium or they can be spread over all the communication mediums available to the distributed system.

Figure 11 illustrates a simplified version of a high-level information dissemination map of a networked UAS. Figure 4 and Fig. 5 would also help for the interpretation of the information dissemination process. The information flow between

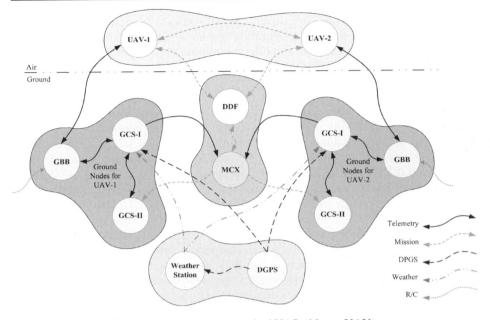

Fig. 11 Information dissemination map of a networked UAS with two UAVs

communicating entities are represented by the directed arrows. The line patterns of the arrow indicate the context of the information/data that will be communicated between nodes via virtual channels.

Intentionally, Fig. 11 does not show the physical properties of communication medium or the absolute physical locations of the applications constituting the communicating nodes. The information dissemination map is what the application developers need to know about the high level interactions between subsystems. Depending on the final physical configuration of the system, the virtual channels can be allocated and reallocated on any physical medium (cable or wireless Ethernet, Controller Area Network (CAN), shared memory etc.) available to the system. And the high-level application developers do not need to have any assumption about the physical nature of the communication channel. Furthermore, hardware configuration of the system can be changed, even at runtime, without changing the source code.

Virtual channels have event driven interfaces which ensures that upon receiving a data package/message from a virtual channel, the CommLibX invokes a user call-back function for data transfer.

As illustrated in Fig. 12, ServiceX provides middleware services and message transfer between CommLibX instances. ServiceX links the physical communication layers to the CommLibX library. ServiceX supports various networking devices including the CAN, standard serial ports (RS232/RS422/RS485), cable and wireless Ethernet, shared memory etc. It also supports multiple protocols including UDP, TCP/IP, and additional communication hardware and protocols can be interfaced to ServiceX through plug-in-modules.

For the applications running on the same computer, ServiceX maps virtual channels to the shared memory for maximum throughput. Virtual channels can be

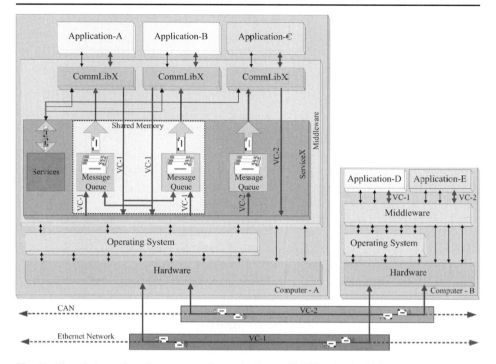

Fig. 12 Virtual channels and message exchange in CommLibX/ServiceX middleware

allocated on the same or different communication devices for applications running on separate networked computers.

The mapping of virtual channels to different physical mediums can be better explained by an example. Figure 12, shows two computers: "Computer-A", and "Computer-B". The first computer runs three high-level applications (A, B, C) and the other computer runs two high-level applications (D, E). Applications A,B and D communicate with each other over the virtual channel "VC-1". Similarly, applications C and E communicate via virtual channel "VC-2".

Computer-A, and Computer-B share two physical communication mediums; a CAN bus, and an Ethernet network. VC-1 is mapped to the Ethernet network while VC-2 is mapped to the CAN bus. Since application A, and B are running on the same computer (Computer-A), and using the same virtual channel (VC-1), then, regardless of the external communication mediums, the ServiceX links the application A and B over the shared memory. However, as shown in Fig. 12, VC-1 is also mapped to the Ethernet so that application D can also communicate with applications A and B.

Figure 13 shows ServiceX's multi-layered, hierarchical architecture. The layered architecture of ServiceX simplifies its port to different operating systems and different hardware. It consists of following three layers;

OS and HW Adaptation Layer This is the lowest level and provides interfaces to the operating system and to the hardware. A significant portion of this layer should be re-written when ServiceX needs to be ported to a new operating system or to a new computing hardware.

Fig. 13 ServiceX has
multi-layered architecture,
and allows its features to be
extended through plug-in
services

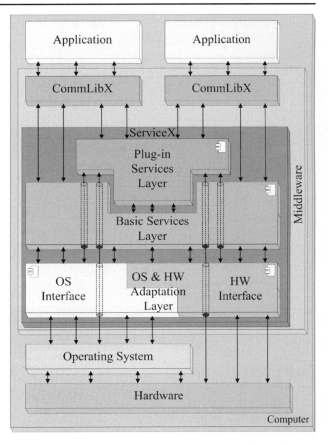

Basic Services Layer The basic services layer resides on top of the adaptation layer. As its name implies, the basic services layer provides bare minimum services for CommLibX/ServiceX middleware. These include, creation and utilisation of virtual channels, data logging, message packing/unpacking, basic communication statistics, support for basic communication hardware including Ethernet, CAN, standard serial ports and shared memory.

Plug-in Services Layer The plug-in services layer resides on top of the basic services layer. The features of ServiceX, hence the features of the overall Comm-LibX/ServiceX middleware, can be extended by incorporating plug-in services.

All the distributed software applications running on board the UAVs and on the ground station use CommLibX/ServiceX middleware. We also used Comm-LibX/ServiceX in our real-time mission simulator, and HWIL simulator. The next section gives an overview of our distributed real-time simulator architecture.

7 Real-Time Multi-UAV Simulator

In UAV experiments there are many avenues of failure: in software, hardware, and conceptual understanding and failures in algorithm development are just a few of many probable failure points. These pose significant burdens on the progress of the project as each flight test poses a risk to the surv vability. Hence, any new hardware or software module has to be thoroughly tested before used on real missions. A *real time multi-UAV simulator* (RMUS) system has been developed, at the earlier phase of ACFR's UAV research programs to address the problems associated with real life experiment [37].

The RMUS has been implemented as a testing and validation mechanism for our networked UAS. These mechanisms include off-line simulation of complex scenarios, HWIL tests, mission rehearsals, on- ine mission control for real UAS demonstrations, and validation of real test results. Prior to real flight operations, all algorithms, software implementations, their interaction with each other, and the actual hardware that will be used in the flight missions should be extensively tested in simulation [41]. The software and hardware, once validated, are then ported directly to the UAV platforms ready for real flight tests.

The RMUS has been extensively used in the various phases of the development of all single-UAV and multi-UAV flight operations conducted by the ACFR [18, 22, 24, 27, 31, 32, 38, 42–45]. These real flight operations include the demonstrations of various DDF, SLAM, and cooperative control techniques. The RMUS has been also used in a large number of other simulation-only UAV experiments [35, 46–48].

The RMUS architecture encompasses distinct architectural styles; it is distributed, it is multi-layered and it is event-based with message-passing foundations. The RMUS's architecture promotes component-based, modular, distributed simulation development where the simulation modules interact with each other with message passing mechanisms provided by the CommLibX/ServiceX middleware.

The key terminology and concepts comprising the RMUS architecture are as follows;

Simulator Cluster   A simulator cluster is a group of distributed simulation modules, and other hardware or software resources connected together that act as a single system to provide a wide span of resources and high processing power for complex simulations. The size and content of the cluster can be changed as the simulation requirements may vary.

Figure 14 illustrates a typical architecture of a RMUS cluster. More than one simulator cluster can be linked with each other via CommLibX/ServiceX middleware. Figure 15 illustrates two linked RMUS clusters. If a RMUS cluster should be interfaced with another simulation environment with a different protocol standard, it may require a "bridge" application on each ends of the clusters to convert from one protocol standard to the other.

Simulation Modules   Simulation modules are composed of either software applications or physical devices comprising the simulation cluster. A simulation module may be an implementation of an algorithm such as the generation of Voronoi diagrams, the DDF algorithm, SLAM code, or a flight control code, or a physical device such as GPS, R/C unit, radar electronics etc. interfaced to the simulation cluster. In Fig. 14 and in Fig. 15, major simulation modules are represented as square shaped blocks. The functionality of simulation modules can be enhanced through *Plug-in-Modules*. The rectangle blocks in Fig. 14 and in Fig. 15 show the plug-in-modules. The plug-in-modules are often created as dynamic libraries.

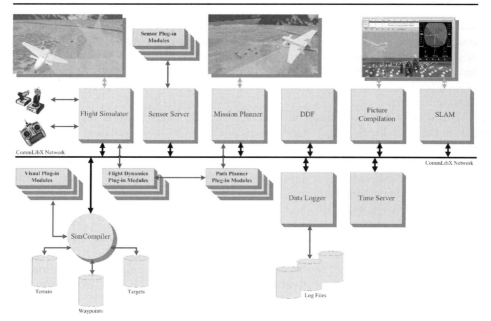

Fig. 14 An architectural view of a typical real time multi-UAV simulator (RMUS) cluster

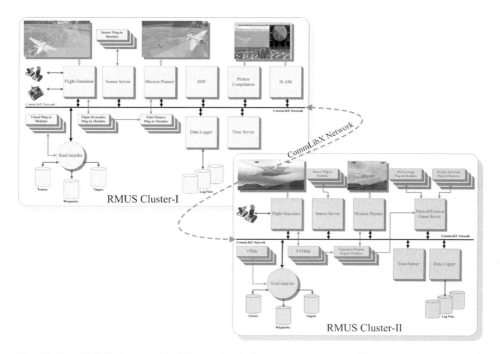

Fig. 15 Two RMUS clusters with different simulation modules, running different simulation sessions are linked with each other via CommLibX network

Simulation Objects Simulation objects are instances of simulation entities. The simulation objects interact with each other as well as interact with the physical devices, via message passing mechanism. Simulation objects may have 3D graphic appearances, or they may be non-visual. For example, in a RMUS simulation session, a simulated radar (a non-visual simulation-object) carried on a simulated UAV (visual simulation-object) may detect a simulated land vehicle (another visual simulation-object) on a terrain (yet another visual simulation-object) as a Feature of Interest (FoI). The 3D appearance of the simulation objects are defined by Visual-Plug-in-Modules (VPMs).

Simulation objects may be associated with pre-defined behaviors or their behavior can be (re)defined either through plug-in-modules or through the interaction with the other simulation objects. Consider a simple scenario; a UAV of a particular type may have a predefined flight dynamics model, and its flight can be simulated by using this model. The same UAV simulation object may receive a "kill the engine", and a "deploy parachute" pseudo commands from the GCS object. In such circumstances, the UAV object may load separate plug-in modules defining the flight dynamics of the deployed parachute. During the rest of the simulation, the flight path of the UAV can be calculated based on the flight dynamics model of the parachute rather than the flight dynamics model of the UAV.

Simulation Session Simulation session is the participation of simulation modules and execution of the simulation to achieve the aimed simulation tasks. Depending on the complexity and processing power requirements of the simulation, simulation sessions can be initiated on a single computer or on multiple networked computers.

Simulation session may be of the type HWIL, "human-in-the-loop" (HIL), or mission rehearsals etc. The physical devices are directly utilized in HWIL simulation sessions. Direct human interaction with the simulation is also possible. A typical example of HWIL and HIL simulation session would involve a human pilot using a real R/C unit to fly a simulated UAV object over a simulated terrain to capture aerial images of the FoIs.

The HWIL simulation environment used at the ACFR is constructed around the RMUS. Hence, both the single-UAV, and multi-UAV missions can be simulated in real-time.

Figure 16 shows the HWIL Sim system configuration for two Brumby Mk III UAVs. In order to reflect the real operational system configuration at a maximum level, the distributed HWIL Sim system is configured in such a way that it has almost identical architecture with the system hardware used for the real flight missions. By comparing Figure 16 with Figure 5 one can see the similarities and differences between the HWIL Sim architecture and the architecture of the ground station used in real flight trials. (The former figure shows a HWIL Sim for two UAVs and the latter figure shows the system architecture for four UAVs, hence comparison should be made by per-UAV basis.)

During the HWIL simulations, the real UAVs remain stationary. Therefore, the flight sensors can not produce flight data. The "HWIL-I" and "HWIL-II" computers concurrently run flight vehicle dynamics models to simulate the UAV motions. The sensor data (GPS, IMU, air data, RPM etc.) are generated in real-time based on these models. Synchronous operation capability is often required for the HWIL computers. The system wide synchronousity is achieved through the Time Management Services (TMS) of the CommLibX/ServiceX middleware.

Fig. 16 The hardware-in-the-loop simulation (HWIL Sim) system configuration for two Brumby Mk III UAVs

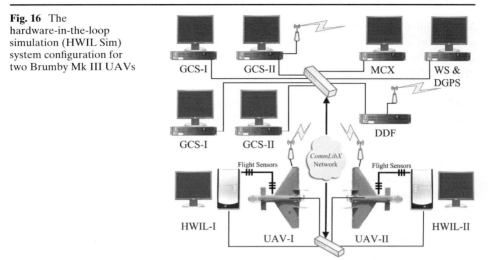

The software applications for the HWIL simulation environment has been developed based on AFDAS design drivers. Therefore, the HWIL Sim environment accommodates software applications running on multiple operating systems (QNX, Linux, and Windows) developed in multiple programming languages (MatLab, C/C++, and Delphi) and provide support for a number of communication hardware and software mediums (RS232, RS422, CAN, cable and wireless Ethernet, spread spectrum radio modem).

Figure 17 is the photograph of the HWIL Sim set-up for two Brumby Mk III UAVs. Note that the wings, the propeller, and the nose cone covers are removed from the UAVs and they are elevated on the custom made stands for the easy access to internals. The cables connected to the UAVs are providing the electrical power to the payload electronics, network connection, and flight sensor connections.

Fig. 17 The HWIL simulation set-up for two Brumby Mk III UAVs

8 Conclusion and Future Works

Due to the advancements in electronics, sensor, and communication technologies and evolutions in material science, and manufacturing technologies, UAVs are becoming more accessible by academic R&D institutions. The UAS developers can easily acquire highly standardised COTS modular hardware components from various suppliers. However, it is difficult to find well established standardised software components particularly for widely distributed, networked UASs.

This paper presented a software development methodology for a networked UAS. The methodology is strongly based on the use of real-time multi-UAV simulator (RMUS), distributed HWIL simulator, and CommLibX/ServiceX middleware. The HWIL simulator and the RMUS have been demonstrated on a large number of pre-mission planning tests as well as on the post mortem analysis of the flight missions by replaying the logged mission data, and comparing it with the simulation results for the re-evaluation of the system models.

The CommLibX/ServiceX is a novel middleware which provides hardware, operating system, and programming language abstraction for the communication, and other system services required for a networked UAS. The CommLibX/ServiceX middleware enables the application developers to develop high-level distributed applications, much the same way, as if they were developing stand-alone, non-distributed applications. Hence, the developers can focus on their applications rather than the low-level details of the underlaying operating systems and hardware.

All these software are based on A framework for distributed autonomous systems (AFDAS). AFDAS uses domain decomposition approach to the domain of networked UAS and introduce a layered system software view. It examines the UAS software applications in four layers; actuator, subsystem, system, and System of Systems (SoS) layers.

The presented software development methodology successfully has been used for many years on widely distributed, network UAS, consisting of multiple Brumby Mk III UAVs, a comprehensive ground station, containing network of computers. The Brumby Mk III UAV avionics is also comprised of a network of computers, sensors, and communication subsystems.

Both the RMUS and the CommLibX/ServiceX middleware are open to expansion. Development of additional new simulation modules for the RMUS is an ongoing process as more researchers use the RMUS in their research. Current activities include the building hovering UAVs (HUAVs) to our existing multi-UAV system to build widely distributed, and decentralised, networked UAS with heterogenous UAVs.

References

1. NATO, STANAG 4586 (Standard Interfaces of UAV Control System (UCS) for NATO UAV Interoperability), NATO Standardization Agency (NSA) (2004)
2. Ernst, D., Valavanis, K., Garcia, R., Craighead, J.: Unmanned vehicle controller design, evaluation and implementation: from MATLAB to printed circuit board. J. Intell. Robot. Syst. **49**, 85–108 (2007)
3. Song, S.J., Liu, H.: Hardware-in-the-loop simulation framework design for a UAV embedded control system. In: Control Conference CCC 2006, pp. 1890–1894 (2006)

4. Gösta, P.D., Kuchcinski, G.K., Sandewall, E., Nordberg, K., Skarman, E., Wiklund, J.: The WITAS unmanned aerial vehicle project. In: ECAI 2000. Proceedings of the 14th European Conference on Artificial Intelligence, pp. 747–755. Berlin (2000)
5. Henning, M.: The rise and fall of CORBA. ACM Queue **4**, 28–34 (2006)
6. Maffeis, S., Schmidt, D.C.: Constructing reliable distributed communication systems with CORBA. Commun. Mag., IEEE **35**, 56–60 (1997)
7. Henning, M., Vinoski, S.: Advanced CORBA programming with C++: Addison-Wesley (1999)
8. Brooks, A., Kaupp, T., Makarenko, A., Williams, S., Orebaeck, A.: Orca: a component model and repository. In: Brugali, D. (ed.) Software Engineering for Experimental Robotics, vol. 30 of STAR, pp. 231–251. Springer (2007)
9. Corke, P., Sikka, P., Roberts, J., Duff, E.: DDX: a distributed software architecture for robotic systems. In: Australasian Conference on Robotics and Automation. Canberra Australia (2004)
10. Paunicka, J.L., Corman, D.E., Mendel, B.R.: A CORBA-based middleware solution for UAVs. In: Fourth IEEE International Symposium on Object-oriented Real-time Distributed Computing, ISORC-2001, pp. 261–267. Magdeburg Germany (2001)
11. Jangy, J.S., Tomlinz, C.J.: Design and implementation of a low cost, hierarchical and modular avionics architecture for the DragonFly UAVs. In: Proceedings of the AIAA Guidance, Navigation, and Control Conference. Monterey (2002)
12. Wills, L., Kannan, S., Heck, B., Vachtsevanos, G., Restrepo, C., Sander, S., Schrage, D., Prasad, J.V.R.: An open software infrastructure for reconfigurable control systems. In: Proceedings of the American Control Conference, pp. 2799–2803 (2000)
13. Kuo, Y.-h., MacDonald, B.A.: Designing a distributed real-time software framework for robotics. In: Australasian Conference on Robotics and Automation (ACRA). Canberra (2004)
14. Croak, T.J.: Factors to consider when selecting CORBA implementations, CrossTalk. J. Def. Softw. Eng. **14**, 17–21 (2001)
15. Group, O.M.: Common object request broker architecture—for embedded, Draft Adopted Specification. http://www.omg.org/docs/ptc/06-05-01.pdf (2006)
16. Boehm, B.W.: Seven basic principles of software engineering. J. Syst. Software **3**, 3–24 (1983)
17. Department of Defense, D.5000.61, DoD modeling and simulation verification, validation, and accreditation. http://www.cotf.navy.mil/files/ms/DoDI500061{_}29Apr96.pdf (2003)
18. Sukkarieh, S., Nettleton, E., Kim, J.-H., Ridley, M., Göktoğan, A.H., Durrant-Whyte, H.: The ANSER Project: data fusion across multiple uninhabited air vehicles. Internat. J. Robot. Research **22**, 505–539 (2003)
19. Göktoğan, A.H.: A software framework for seamless R&D of a networked UAS. In: PhD Thesis, Australian Centre for Field Robotics, School of Aerospace, Mechanical and Mechatronic Engineering Sydney. The University of Sydney (2007)
20. Kim, J., Sukkarieh, S.: Autonomous airborne navigation in unknown terrain environments. IEEE Trans. Aeros. Electron. Syst. **40**, 1031–1045 (2004)
21. Nettleton, E.: ANSER past and present—multi UAV experimental research. In: The IEE Forum on Autonomous Systems, (Ref. No. 2005/11271), p. 9 (2005)
22. Nettleton, E., Ridley, M., Sukkarieh, S., Göktoğan, A.H., Durrant-Whyte, H.: Implementation of a decentralised sensing network aboard multiple UAVs. Telecommun. Syst. **26**(2–4), 253–284 (2004)
23. Ridley, M., Nettleton, E., Göktoğan, A.H., Brooker, G., Sukkarieh, S., Durrant-Whyte, H.F.: Decentralised ground target tracking with heterogeneous sensing nodes on multiple UAVs. In: The 2nd International Workshop on Information Processing in Sensor Networks (IPSN'03), pp. 545–565. Palo Alto, California, USA (2003)
24. Sukkarieh, S., Göktoğan, A.H., Kim, J.-H., Nettleton, E., Randle, J., Ridley, M., Wishart, S., Durrant-Whyte, H.: Cooperative data fusion and control amongst multiple uninhabited air vehicles. In: ISER'02, 8th International Symposium on Experimental Robotics. Sant'Angelo d'Ischia, Italy (2002)
25. Bryson, M., Sukkarieh, S.: Toward the real-time implementation of inertial SLAM using bearing-only observations. In: Journal of Field Robotics (2006)
26. Buttazzo, G.C.: Hard real-time computing systems: predictable scheduling algorithms and applications. In: 2nd ed. p. 425. New York, Springer (2005)
27. Nettleton, E.: Decentralised architectures for tracking and navigation with multiple flight vehicles. In: PhD Thesis, Department of Aerospace, Mechanical and Mechatronic Engineering Sydney. The University of Sydney (2003)
28. Kim, J.-H.: Autonomous navigation for airborne applications. PhD Thesis, Australian Centre for Field Robotics, School of Aerospace, Mechanical and Mechatronic Engineering Sydney. The University of Sydney, p. 237 (2004)

29. Cole, D.T., Göktoğan, A.H., Sukkarieh, S.: The implementation of a cooperative control architecture for UAV teams. In: 10th International Symposium on Experimental Robotics (ISER'06). Rio de Janeiro, Brazil (2006)
30. Endsley, M.R., Garland, D.J.: Theoretical underpinning of situation awareness: a critical review. In: NJ, M., Erlbaum, L. (eds.) Situation Awareness Analysis and Measurement: Analysis and Measurement, pp. 3–33 (2000)
31. Cole, D.T., Sukkarieh, S., Göktoğan, A.H.: System development and demonstration of a UAV control architecture for information gathering missions. J. Field Robot. **23**, 417–440 (2006)
32. Cole, D.T., Sukkarieh, S., Göktoğan, A.H., Hardwick-Jones, R.: The development of a real-time modular architecture for the control of UAV teams. In: Field and Service Robotics, FSR'05, pp. 321–332. Port Douglas, Australia (2005)
33. Michael, R.G., Matthew, L.G., Jeffrey, S.R.: Cooperation without communication. In: Distributed Artificial Intelligence, Morgan Kaufmann Publishers Inc., pp. 220–226 (1988)
34. Otanez, P.G., Campbell, M.E.: Hybrid cooperative reconnaissance without communication. In: 44th IEEE Conference on Decision and Control'05 and European Control Conference'05. CDC-ECC '05, pp. 3541–3546 (2005)
35. Chung, C.F., Göktoğan, A.H., Cheang, K., Furukawa, T.: Distributed simulation of forward reachable set-based control for multiple pursuer UAVs. In: SimTecT 2006 Conference Proceedings, pp. 171–177. Melbourne, Australia (2006)
36. Speranzon, A., Johansson, K.H.: On some communication schemes for distributed pursuit-evasion games. In: Proceedings of 42nd IEEE Conference on Decision and Control, pp. 1023–1028 (2003)
37. Göktoğan, A.H., Nettleton, E., Ridley, M., Sukkarieh S.: Real time Multi-UAV simulator. In: IEEE International Conference on Robotics and Automation (ICRA'03), pp. 2720–2726. Taipei, Taiwan (2003)
38. Göktoğan, A.H., Sukkarieh, S., Işıkyıldız, G., Nettleton, E., Ridley, M., Kim, J.-H., Randle, J., Wishart, S.: The real-time development and deployment of a cooperative multi-UAV system. In: ISCIS03 XVIII - Eighteenth International Symposium on Computer and Information Sciences, Antalya—Türkiye, pp. 576–583 (2003)
39. DMSO: High level architecture for simulation interface specification, defence modelling and simulation office (DMSO). https://www.dmso.mil/public/ (1996)
40. O.M.G. (OMG): CORBA: Common Object Request Broker Architecture. http://www.corba.org
41. Göktoğan, A.H., Sukkarieh, S.: Simulation of multi-UAV missions in a real-time distributed hardware-in-the-loop simulator. In: Proceeding of the 4th International Symposium on Mechatronics and its Applications (ISMA07). Sharjah, UAE (2007)
42. Göktoğan, A.H., Brooker, G., Sukkarieh, S.: A compact millimeter wave radar sensor for unmanned air vehicles. In: Preprints of the 4th International Conference on Field and Service Robotics, pp. 101–106. Lake Yamanaka, Yamanashi, Japan (2003)
43. Nettleton, E.W., Durrant-Whyte, H.F., Gibbens, P.W., Göktoğan, A.H.: Multiple platform localisation and map building. In: Sensor Fusion and Decentralised Control in Robotic Stystems III, pp. 337–347. Boston, USA (2000)
44. Sukkarieh, S., Yelland, B., Durrant-Whyte, H., Belton, B., Dawkins, R., Riseborough, P., Stuart, O., Sutcliffe, J., Vethecan, J., Wishart, S., Gibbens, P., Göktoğan, A.H., Grocholsky, B., Koch, R., Nettleton, E., Randle, J., Willis, K., Wong, E.: Decentralised data fusion using multiple UAVs - The ANSER Project. In: FSR 2001, 3rd International Conference on Field and Service Robotics, Finland, pp. 193–200 (2001)
45. Göktoğan, A.H., Sukkarieh, S.: Role of modelling and simulation in complex UAV R&D projects. In: 1st National Defense Applications, Modeling and Simulation Conference, (USMOS'05), pp. 157–166. Ankara-Türkiye (2005)
46. Göktoğan, A.H., Sukkarieh, S., Cole, D.T., Thompson, P.: Airborne vision sensor detection performance simulation. In: The Interservice/Industry Training, Simulation and Education Conference (I/ITSEC'05), pp. 1682–1687. Orlando, FL, USA (2005)
47. Bourgault, F., Göktoğann, A.H., Furukawa, T., Durrant-Whyte, H.: Coordinated search for a lost target in a Bayesian world. In: Advanced Robotics, Special Issue on Selected Papers from IROS 2003, vol. 18, pp. 979–1000 (2004)
48. Furukawa, T., Bourgault, F., Durrant-Whyte, H.F., Dissanayake, G.: Dynamic allocation and control of coordinated UAVs to engage multiple targets in a time-optimal manner. In: IEEE International Conference Proceedings on Robotics and Automation, pp. 2353–2358 (2004)

Design and Hardware-in-the-Loop Integration of a UAV Microavionics System in a Manned–Unmanned Joint Airspace Flight Network Simulator

Serdar Ates · Ismail Bayezit · Gokhan Inalhan

Originally published in the Journal of Intelligent and Robotic Systems, Volume 54, Nos 1–3, 359–386.
© Springer Science + Business Media B.V. 2008

Abstract One of the challenges for manned-unmanned air vehicles flying in joint airspace is the need to develop customized but scalable algorithms and hardware that will allow safe and efficient operations. In this work, we present the design of a bus-backboned UAV microavionics system and the hardware-in-the-loop integration of this unit within a joint flight network simulator. The microavionics system is structured around the Controller Area Network and Ethernet bus data backbone. The system is designed to be cross-compatible across our experimental mini-helicopters, aircrafts and ground vehicles, and it is tailored to allow autonomous navigation and control for a variety of different research test cases. The expandable architecture allows not only scalability, but also flexibility to test manned-unmanned fleet cooperative algorithm designs at both hardware and software layer deployed on bus integrated flight management computers. The flight simulator is used for joint simulation of virtual manned and unmanned vehicles within a common airspace. This allows extensive hardware-in-the-loop testing capability of customized devices and algorithms in realistic test cases that require manned and unmanned vehicle coordinated flight trajectory planning.

Keywords Microavionics · Hardware-in-the-loop testing · Flight network simulator

This work is funded partially by DPT HAGU program administered by ITU ROTAM.

S. Ates · I. Bayezit · G. Inalhan (✉)
Controls and Avionics Lab, Faculty of Aeronautics and Astronautics,
Istanbul Technical University, Istanbul, Turkey
e-mail: inalhan@itu.edu.tr

S. Ates
e-mail: serdar.ates@itu.edu.tr

I. Bayezit
e-mail: ismail.bayezit@itu.edu.tr

1 Introduction

The increasing use of unmanned vehicles in civilian (metropolitan traffic monitoring, rapid assessment of disaster areas) and military (reconnaissance, target identification, tracking and engagement) domains has driven critical requirements for joint operations of manned-unmanned systems in common airspace. Although the extent of joint operations can change from pure collision-avoidance (sense-avoid) to achieving common mission goals that require significant coordinated actions (mark-image) [11], it is essential to develop new algorithms and standardized hardware that allow manned-unmanned system interoperability [9]. In addition, the increasing complexities in missions and the safety-critical requirements on the avionics systems demand that all the joint airspace control and coordination algorithms, and standardized avionics hardware be tested on realistic testbeds [4, 14, 18] before using them in the real mission scenarios [10, 22].

In this paper, we provide the design and development of a cross-platform compatible and multiple bus backboned microavionics system that allows to test and develop such standardized hardware and software solutions in laboratory scale micro-air vehicles. The system is designed to be cross-compatible across our experimental mini rotary-wing and fixed-wing UAVs, and ground vehicles. It is also tailored to allow autonomous navigation and control for a variety of different research test cases. For joint manned-unmanned airspace operations, the hardware-in-the-loop (HIL) testing capability of our joint manned-unmanned flight network simulator environment [2] is utilized. This in-house developed simulator provides the unique capability of joint piloted flight simulation, virtual simulation of unmanned vehicles (and simulated manned vehicles) and real-time integration of flying unmanned vehicles within user defined flight scenarios. This simulator provides a realistic testing environment that minimizes expensive and failure problematic early-stage manned-unmanned vehicle joint in-flight testing.

Our microavionics system design is based on a research driven fundamental requirement to have a cross-compatible (i.e. the avionics can be used with minor modifications on different types of ground and air vehicles as depicted in Fig. 9) architecture that can support coordinated autonomy experiments across a heterogenous fleet of vehicles. Another major driver for this platform was that it should allow a collaborative hardware and software development environment, in which researchers can work on different topics (such as flight controls [21], image-driven navigation [14] or multiple vehicle coordination algorithms [15]) and could later integrate their work for flight and ground experiments [14] with minimal hardware and software redesign concerns. In comparison with elegant but monolithic microavionics architecture structured around different form factors such as single-board computers [8] or PC-104 stacks [12], we consider a scalable multi-processor architecture based on data bus backbones [6, 13]: Controller Area Network (CAN) as the control/mission bus and Ethernet as the payload bus.

The standardized microavionics hardware as seen in Fig. 1 includes sensors, a flight control computer (which is also denoted as autopilot), a flight management computer (which is also denoted as mission coordination computer) and a communication unit. The expandable architecture deploys a hybrid selection of COTS Motorola (MPC555), Arm processor boards (LPC2294), PC-104 processor stack and Gumstix modules, each with different operating systems and coding techniques (such as

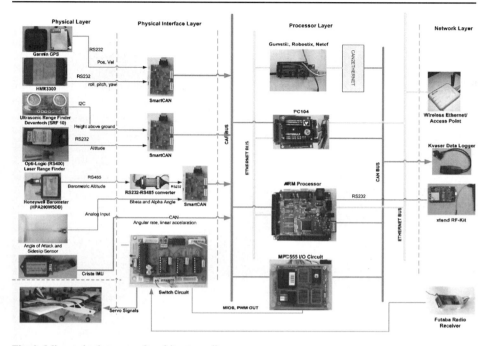

Fig. 1 Microavionics general architecture diagram

rapid algorithmic prototyping using automatic code generation via Matlab/Real Time Workshop Embedded Target). Specifically, MPC555 is used with Matlab/Real Time Workshop to rapidly design and prototype basic flight control algorithms using Simulink and the MPC555 embedded target automatic code generation feature. This fast-prototyping approach provides not only flexibility at coding level, but also provides seamless cross-platform compatibility. This is a major distinguishing factor of our architecture in comparison with general CAN Bus based architectures [6]. The microavionics system employs a complete sensor suite including Garmin GPS receiver, 3 axis accelerometer/gyro Crista Inertial Measurement Unit (IMU), Honeywell Altimeter and Digital Compass each used as the primary sensors. These units provide the real-time position, orientation and associated time-rate information. In addition, experiment specific units such as laser rangefinder, ultrasonic sensors and wireless IP cameras are integrated as plug-and-play add-on modules through custom interface boards. The microavionics system includes a X-Tend wireless transceiver for communication with the ground control station. The cross-platform compatible microavionics design provides an enabling technology to be used for a range of activities including autonomous take-off and landing flight research. This technology has been translated to micro-helicopter [14] and ground vehicle operations, and currently it is being translated to tailsitter VTOL (Vertical Take-off and Landing) aircraft [1] operations within civilian environments that involves agile maneuvering in urban environments [16].

The joint airspace network simulator, as seen in Fig. 2 allows rapid prototyping, software-in-the-loop and HIL testing of the microavionics system and the coor-

Fig. 2 Component views of the flight network simulator

dination algorithms across manned and unmanned fleets, and the mission control center. Open-source flight simulation software, FlightGear [19], is modified for networked operations and it is used as the 3D visualization element for the pilot and the mission controls. The manned vehicle dynamics, UAV dynamics and low-level control algorithms are embedded within the xPC computers using Matlab/Simulink rapid prototyping technique for real-time execution of the mathematical models and control algorithms. Equipped with touch-screen C2 (command and control) interface at the pilot station, the platform also allows us to rapidly prototype and test pilot-unmanned fleet supervisory control designs [5]. Real in-flight UAVs or in-the-lab complete hardware system can connect to the mission simulation using the communication link to the simulation server. The flight network simulator has the capability to provide not only the visualization, but also the communication layer for joint manned-unmanned vehicle operations and missions in a common airspace. For operations in common airspace [3], this platform has been serving as the main testing facility to test multiple vehicle coordination [10], communication protocol design [2] and concept-of-operation driven formation flight algorithms [15]. An extended description of the in-house developed flight network system, the hardware systems and the software architecture can be found in [2].

The organization of the paper is as follows: In Section 2, the design and the development of the cross-platform compatible microavionics hardware is given. Section 3 provides insight on the experimental mini flight and ground platforms, and the microavionics software implementations including autonomous control experiments and the HIL testing. In Section 4, following the treatment as given in [2], we provide an overview description of the joint airspace flight network simulator and its major components. In addition, HIL integration of the microavionics platform to the flight network simulator is illustrated.

2 Design of the Microavionics System

2.1 General Architecture

The architecture of the multiple bus-backboned microavionics system, as shown in Fig. 3, is composed of four different layers. The first layer, which is the physical layer, is composed of sensing devices, actuators and power circuits. This layer includes a IMU (Crista), a GPS (Garmin) unit, an altimeter (Honeywell), a magnetometer (Honeywell), a laser range finder (Opti-Logic), an ultrasonic range finder (Devantech), and angle of attack and sideslip sensors. This complete set of sensor provides us with considerable flexibility in design and testing of autonomous and cooperative control experiments across mini UAV platforms. The second layer is the physical interface layer which is used for the converting the sensing information to CAN Bus messages with unique IDs using the in-house designed and developed SmartCAN nodes. These nodes enable sensing information coming in serial or analog input data type to be converted to the CAN data type. In addition, switching operation between the autopilot and the remote RC pilot control is achieved with the help of the Switch Circuit as shown in Fig. 7. The switch circuit which is a flight critical circuit has been

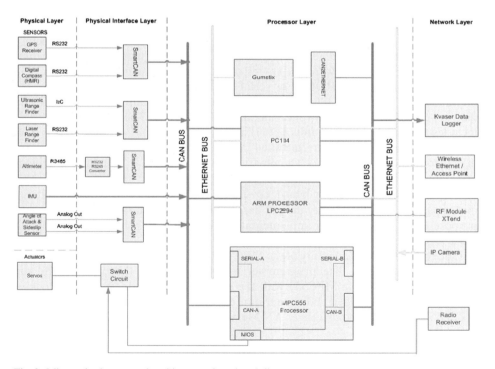

Fig. 3 Microavionics general architecture functional diagram

tested extensively on our ground and flight platforms as shown in Fig. 9 as a part of on going experiments.

The third layer is the processor layer structured around the Ethernet and CAN bus lines. Motorola MPC555 processor directs the control implementations of the vehicle. Higher level algorithms like task assignment, mission planning and collaborative control are implemented on the PC-104 processor stack. The ARM processor is used for communications with the Ground Station and for the coordination pattern computation. In addition, the expendable Gumstix single board platform is currently being tested as the standalone Link16 like unit acting as the mission-level communication and information distribution module.

The fourth layer is the Network Layer and it consists of Kvaser data logger which logs information streaming over the CAN bus, appending a time stamp to each of the messages. Second part of this layer includes the XTend RF module, an IP camera and the wireless Ethernet module. The ARM processor board (LPC2294) gathers all the flight critical sensor and servo signal information from the CAN bus. This board connected to the RF Module provides the critical link to the ground station during the flight and the ground tests. In addition, the IP Camera and wireless Ethernet module are used for real-time video and payload related data transmission to the ground station. In the next subsections, we will review the specific elements of the microavionics system in detail following a standard functional usage classification.

2.2 Processors

Motorola MPC555 processor is a 40 MHz processor having a PowerPC core with a floating point unit which allows acceleration and development advanced control algorithms. Motorola MPC555 processor with our printed MPC555 Carrier Board (as shown in Figs. 3 and 7), is the core processing unit in the microavionics architecture and it enables the programmer to design and prototype low-level control algorithms rapidly. In the basic hardware, 26 kbytes of fast RAM is included. There is also a 448 kbytes of Flash EEPROM. For serial communications, there are two queued multi-channel modules (QSMCM). MPC555 has two on-chip RS232 drivers and dual on-chip CAN 2.0B controller modules (TouCANs). The MPC555 core also includes two time processor units (TPU) and a modular input–output (I/O) system (MIOS1). 32 analog inputs are included in the system with dual queued analog-to-digital converters (QADC64).[1] The phyCORE555 single board computer module is integrated on our in-house developed MPC Carrier Board as in Figs. 3 and 7. As a consequence, the new carrier board, covers 1/3 the area and consumes less energy in comparison with the factory default carrier board while all the necessary general purpose IOs are available for implementations including in-system programming of MPC flash memory over the serial link.

One of the most critical utility of the MPC board is the support of Matlab/Simulink Real-Time Workshop. The programmer is able to use all the capacity and the devices of the Motorola MPC555 with the support of Embedded Target Blockset and Automatic Embedded code generation feature. Embedded Target Blockset for MPC555 includes CAN Drivers, CAN Message Blocks and all the MPC555 Driver

[1] www.freescale.com

Fig. 4 Processors - The ARM processor (LPC2294), MPC555 with in-house designed carrier board, Tiny886ULP Ultra Low Power PC104+ Computer, Gumstix with Robostix and netCF

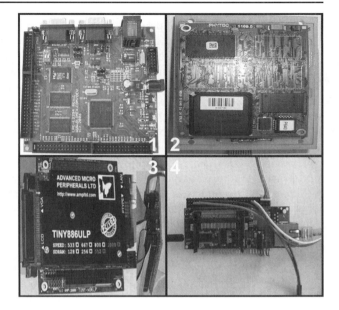

libraries. In addition, all configurations of MPC555 resources are controlled directly from the MPC555 Resource Configuration Block within Simulink MPC555 specific library. As a result, all the low-level control algorithm development is strictly isolated from the low-level device drivers of the processor.

In our control implementations, one of the most commonly used feature of the MPC555 hardware is the MIOS unit. MIOS unit includes digital input-output channels. These channels cover eight dedicated pins for pulse width modulation output for RC servos and ten pins for pulse width or pulse period measurement used in recording RC pilot inputs during system identification tests. Note that the Switch Circuit forwards 6 different PWM signals to the PWM Read channels of the MPC555. After capturing the PWM signals, MPC555 packs and streams this data as CAN messages with specific CAN IDs for data recording purposes. Another library that is commonly used is the CAN Message Blocks which enables transmission or receival of CAN data over CAN bus. These blocks include not only CAN Message Packing-Unpacking, but also feature splitting of the messages. The availability of simple and rapid shaping capability of any message in Matlab/Simulink environment allows us to parse the sensing information with basic mathematical operations. Complex algorithms and mathematical actions are also accelerated with the floating point unit of the MPC555. Finally TouCAN transmit and TouCAN receive blocks directly use two on-chip CAN communication units and configure the CAN message format (Fig. 4).

The Arm Processor (LPC2294)[2] *microcontroller* is based on a 32-bit CPU in ARM7 family with 256kbyte of embedded flash memory. 128-bit wide interface enables high speed 60 MHz operation. Arm processor board is mainly used in the communication link of the ground station with the avionics box and for the

---

[2]www.nxp.com

computation of coordination patterns. The ARM board includes four interconnected CAN interfaces with advanced filtering capability. The serial interfaces include two UARTs (16C550), fast I2C-bus (400 kbit/s) and two SPIs. In addition, there are eight channel 10-bit ADC units, two 32-bit timers (with four capture and four compare channels), six PWM outputs, Real-Time Clock, and a watchdog timer. TCP/IP communication is also supported by the Ethernet controller chip located at the bottom side of the ARM board. Various in-house developed driver codes are written for this board for CAN and serial communications, CAN to serial conversion, digital Kalman filtering and sensor message parsing.

Tiny886ULP - Ultra Low Power PC104+ Computer[3] is a % 100 PC-AT compatible single board computer having PC104 bus connectivity. The % 100 PC-AT compatibility makes it possible to develop code directly on Tiny886ULP and support for Windows, Linux and QNX operating systems makes it very easy to reuse and port existing codes which are developed on regular desktop PCs. We use this computer as the flight management computer to embed multiple vehicle coordination algorithms [15]. Tiny886ULP provides 1Ghz Crusoe processor (with FPU), 512Mb of SDRAM, IDE disk interface, 10/100 ethernet, 2xRS232, VGA output, 1xUSB, PC104+ bus and PS2 keyboard/mouse interface. With average power consumption of less than 5W, the processor provides easy integration and low-power dissipation (fanless) for our microavionic applications. We have run Windows, Linux and QNX operating systems on Tiny886ULP. The add-on PC104 boards that we use includes CAN104, PC104, CAN controller board and Firespeed2000 board. The CAN104 board provides 2 CAN 2.0b interfaces which gives the ability of the PC-104 stack to communicate with our present avionics bus. The Firespeed2000 board provides three IEEE-1394 (firewire) serial ports and high-speed data transfers from sources such as video and other avionic bus components.

Gumstix connex (http://www.gumstix.com/) *single board computer* measures just $80 \times 20 \times 6.3$ mm but according to its small size it has considerable process power which makes it ideal for microavionic systems. We are using this module to implement Link16 like communication and information distribution algorithms for mission operations. Linux 2.6 operating system runs on Gumstix and a free cross-compilation toolchain "buildroot" is provided by the manufacturer for software development. We port the desktop developed communication code to this unit. Gumstix connex board provides a 400 MHz Marvell XScale PXA255 processor (ARM V5 compatible), 64 MB of SDRAM and 16 MB on-board flash memory. Connectivity is achieved by 60-pin Hirose I/O header and a 92-pin bus header to additional expansion boards all powered from single 5V supply. We use "netCF" expansion board attached to our Gumstix board. This expansion board provides compact flash card and 10/100 ethernet interface for network connectivity. We also utilize the "Robostix" expansion card which provides 6xPWM output, 2xUART, 1xI2C and 8xADC and has been very useful for ground robotic applications in which it is operated in a stand-alone controller unit mode.

---

[3] http://amdltd.com

2.3 Sensor Suite

Sensor suite in the microavionics system design consists of an IMU, a GPS receiver unit, a magnetometer (digital compass), a barometric altimeter, a laser range finder, and an ultrasonic range finder.

As the GPS receiver unit, we use the readily available Garmin 15H. This device can track up to 12 satellites to determine the position and the velocity of the vehicle, and also outputs a PPS timing signal. This PPS signal is connected to the IMU for time stamping of IMU data. The NMEA sentences that will be transmitted by the GPS receiver unit are user selectable through its custom interface. The GPS data is received approximately every one second. This unit is able to output the acquired data via a RS232 based serial interface. As seen in Fig. 5, this unit provides precise global position and velocity information for vehicle navigation and control algorithms.

For inertial acceleration and angular velocity measurements, we use Crista IMU provided by Cloud Cap Technology. This unit is able to output the acquired data both via a RS232 based serial interface and a CAN interface. It also provides internal oversampling and averaging of the oversampled data automatically before transmission. The data transmission and oversampling rates are user selectable. The IMU is also able to time stamp the acquired signals via the PPS signal acquired from the GPS receiver unit. A data update rate of 50 Hz is used in this work with an oversampling rate of 20 Hz. Especially angular velocity measurements (p, q, r) are required by the autopilot control and navigation loops. Without these measurements, precise control of the vehicle is not possible.

The barometric altimeter unit Honeywell HPA200W5DB is able to provide temperature, pressure and barometric altitude measurements. We use the factory setting date update rate of 10 Hz for barometric altitude information.

The magnetometer (digital compass) used for inertial orientation aiding is the Honeywell HMR 3300. This unit provides three axis measurement of the platform orientation relative to the earth magnetic field. The orientation information of the heading angle is of 360 degrees and fully covers the range whereas the pitch and the roll are of 60 degrees. The data update for this device is 8 Hz. This unit is able to output the acquired data via a RS232 based serial interface. These Euler angles

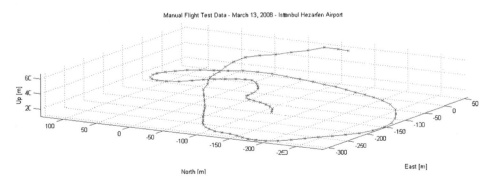
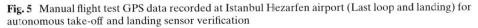

Fig. 5 Manual flight test GPS data recorded at Istanbul Hezarfen airport (Last loop and landing) for autonomous take-off and landing sensor verification

measurements are also vital and required by the autopilot control and navigation loops.

In order to develop autonomous landing and take-off algorithms, the air-vehicle's actual altitude has to be measured very precisely and accurately to prevent failures. To achieve this, a laser range finder and an ultrasonic range finder are used. The laser range finder unit used is the Opti-Logic laser range finder (RS400) which provides actual altitude of the vehicle relative to the terrain. The data update for this device is 10 Hz. This unit is able to output the acquired data via a RS232 based serial interface. The altitude information range is up to 400 Yards as seen in Fig. 6.

Devantech SRF10 ultrasonic range finder unit (URF) is used for measuring the actual altitude over terrain during take-offs and landings. This unit is able to output the acquired data via a I2C based interface. The data update rate for this device is 10 Hz. The altitude information ranges up to 255 inches with one inch resolution as seen in Fig. 6.

If the microavionics system is connected to the fixed wing UAV, angle of attack and sideslip sensors are fitted to the platform to measure the angle of attack and the sideslip angle. These sensors resemble vertical stabilizers of air-vehicles and are coupled with infinite rotary potentiometers to give analog outputs related to the angular position.

We use the Kvaser Memorator CAN data logger as the flight data recording unit. The flight data recording unit is used for acquiring data from the CAN bus and recording it to SD and MMC compatible cards. The memory can be up to 2 GBs which allow recording for hours for this work. The recorded data can easily be taken to a PC for analysis. This CAN based recording unit is also used as a CAN-USB converter, which can then be used for better debugging or system monitoring purposes.

We performed an execution profiling analysis after connecting all sensor devices to the MPC555 processor to test the capability of the processor. With 0.019994 s average sampling time, the maximum task turnaround time is 0.000559 sec in the worst case and the average turnaround time is 0.000542 s. With the execution profiling analysis, all the sensor data are read and parsed to calculate the elapsed time for control and

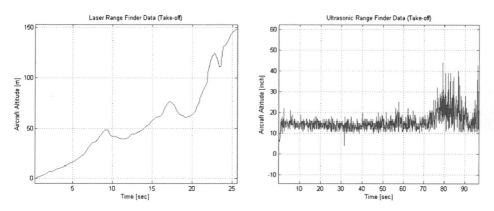

Fig. 6 Laser range finder and ultrasonic range finder data during take-off (Manual flight test data recorded at Istanbul Hezarfen Airport)

filtering. About 3% of processing capacity of the processor is used for parsing and with the remaining source of processor is used for control and filtering purposes.

2.4 Customized Boards: SmartCAN and Switch

SmartCAN: The distributed multi-processor design as shown in Fig. 3 is structured around CAN Bus line and the Ethernet Bus providing the ability to interoperate independent processors with different functionalities, native operating systems, and coding standards. This also streamlines the microavionics system allowing the re-configurability for different sensor sets through sensor customization via SmartCAN modules (as shown in Fig. 7).

Most of the sensors used in the microavionics system design are not compatible with CAN Bus and have different output such as RS232, UART, I2C, analog and digital. Our in-house developed SmartCAN modules have all sensors structured around CAN bus. The SmartCAN modules read, parse, and send all sensor information with predefined CAN IDs to the CAN bus. These CAN IDs are selected in a priority order to give priority to critical sensors. Every SmartCAN node design uses PIC18F458 to direct sensing information flow to and from CAN bus line. In case of sensors that have RS232 or UART output, MAX232 IC is used to convert TTL signals. After reading sensing information, it is processed by the PIC microcontroller and transmitted to the CAN bus via using PCA82C250 which is used for transition from the CAN protocol controller to the physical bus CAN High and CAN Low outputs. Four different sensors can be connected to a SmartCAN node. They can be switched and monitored with on-board physical switches and system alive/data streaming led indicators. The SmartCAN customized boards can be programmed with in-circuit debugger programming ports which eliminate the handicap of unmounting the PIC

Fig. 7 Autopilot hardware - SmartCAN Node, MPC555 with in-house designed carrier board, switch board, autopilot deck

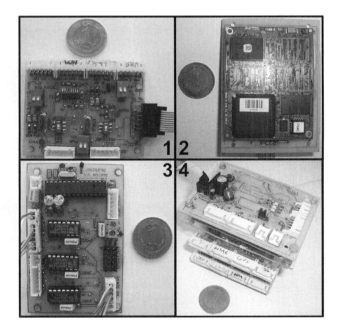

ICs for programming. PIC microcontrollers are programmed with the help of timer interrupt routines to reach the sampling rate of each sensor. In addition, watchdog timer software protection is activated to avoid sensor reading failures.

Switch: Eventhough the microavionic system is designed for autonomous flight tasks, it is very probable that unpredictable situations take form during flight tests, therefore a mechanism that will let the human factor intervene the platform during such conditions is a very critical functionality. For this purpose, a switch board as shown in Fig. 7 is designed. An extended feature of the switch board is the capability of switching the PWM outputs between the manual flight (by RC) and autonomous flight (by MPC555).

The Switch Board takes 8 PWM inputs from the RC radio transceiver and 6 inputs from the MPC555 (or any other board) and directs the control inputs to the servo actuators. 2 of the 8 PWM inputs are for channel selection and RC alive information. The remaining 6 PWM inputs are for human control. The switching operation is established by a PIC18F873 and the 74HCT125 three-state octal buffer is used as the interface of channel switches. When one of the control sources (human pilot or MPC555) is not available, the switch board assigns the other one as the actual controller automatically. In addition, for system identification and health-monitoring purposes, the signals going to the servo actuators are provided as another output which are read by the MPC555 in our architecture. These signals can be used by MPC555 to estimate the servo position at any desired time. This board has an individual battery supply which ensures that the switch board keeps working properly even if the microavionics system is out-of-order.

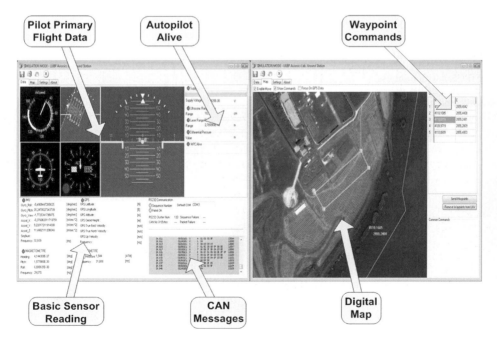

Fig. 8 Ground station graphical user interface

2.5 Ground Station

The ground station interface, which is shown in Fig. 8, is used not only to visualize the air-vehicle sensor data, but also to send waypoint data to the air-vehicle's mission planning processor. A XTend 900 MHz RF transceiver kit is used for the communication link between the air-vehicle and the ground station. The Ground Station graphical user interface is designed with Delphi 7.0 and has three tabs. At the first tab, pilot primary flight data, basic sensor reading with the sampling rates, CAN messages, MPC Alive state can be seen. At the second tab, the digital map is loaded and waypoint commands can be sent to the air-vehicle. At the third tab, initial parameters related to serial communication protocol and digital map file can be changed. This GUI interface can record all CAN messages just like the Kvaser Memorator data logger does, and can playback these log files simultaneously. It can also send a heartbeat signal to the air-vehicle to inform whether or not the communication link is still alive.

3 Microavionics Control Implementations

A collection of in-house modified real UAVs/UGVs (including Vario Benzine and Acrobatic Helicopters, Trainer 60 fixed-wing platform and a 1/6 scale Humvee [14]) are integrated with our cross-platform compatible microavionics systems to support autonomous flight research. The experimental micro-helicopter, fixed-wing and ground vehicles equipped with sensors and on-board computing devices, provide the necessary infrastructure (seen in Fig. 9) to support advanced research on autonomous flight including vertical take-off and landing [14], vision based control and distributed autonomy [15].

"Aricopter's" base structure is a COTS benzine helicopter manufactured by German model helicopter manufacturer Vario. The Benzine trainer model of Vario was selected as the flying platform, because of Vario's advantages on price, structural strength, and lifting capacity, over the other options like Century Heli's "Predator Gasser" and Bergen R/C's "Industrial Twin". "Microbee" is an advanced flight platform that is based on Vario's Acrobatic helicopter. This helicopter has an increased flight envelope and enables acrobatic maneuvers to be carried out. "Trainer 60" is a beginner level fixed-wing platform with a wingspan of 160*cm* and serves as the main aircraft platform for fixed-wing tests. "Humvee" is a ground vehicle which is used for testing the basic functionality of the avionics system. It is a 1/6 scale model of Hummer and it is manufactured by Nikko. These experimental flight and ground platforms provide a unique opportunity to prototype and demonstrate novel concepts in a wide range of topics covering agile maneuvering [8], advanced flight controls [13, 20], active vision sensing and fleet autonomy [7].

Algorithmic design and development of the multi-operational autonomous system are split into five different sections: flight/ground testing, system identification and model verification process, mathematical modeling of the flight/ground platform, controller design and simulation as depicted in Fig. 10. First of all, generic mathematical models of our UAV/UGV platforms [14] are realized around parametric models and these models are used for simulating the system and designing the appropriate controllers. Real flight and ground tests are used to develop the verified

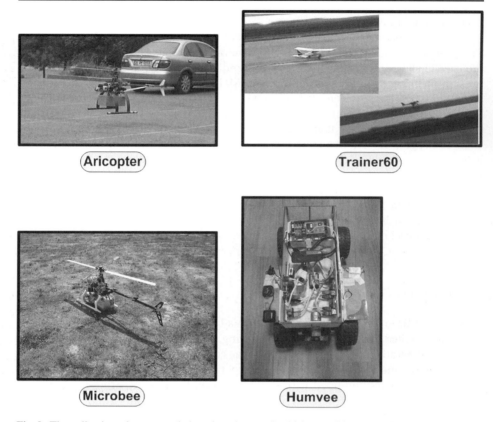

Fig. 9 The collection of unmanned aircraft and ground vehicles used for research

dynamic models of the flight or the ground platforms through system and parameter identification. In addition, the verified dynamic models are tested with hardware and software simulation setups to improve controllers and autonomous algorithms. Then the conceptual design loop returns back to the outside flight and ground experiments to test the modified controllers and algorithms to improve the autonomy of the system. In the next subsections, the specifics of the design approach are illustrated in detail for the ground and the air platforms.

Matlab/Simulink is the main software development environment for designing the basic autonomous control algorithms (including landing and take-off) and waypoint navigation. These designs can be rapidly embedded into the MPC555 microcontroller. All the inner and the outer control loops, and the navigation loops typical to mini UAVs are also realized on the MPC555 microcontroller. The microavionics system design also allows us to implement autonomous control algorithms for agile maneuvers [16]. With minor software modifications. the same system can also be used in formation flight experiments [2]. The same microavionics design is currently being transformed for usage in a VTOL tailsitter aircraft [1] which has distinct and switching flight regimes such as hover flight, level flight, and hover-level flight transitions.

Fig. 10 Design and development concept of autonomous flights

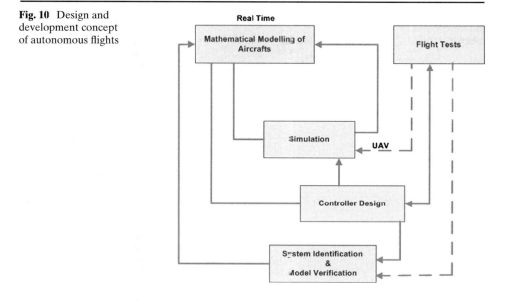

In the following sections, we will review two basic algorithmic development processes as examples. First, the autonomous control and a way-point navigation experiment on "Humvee" is illustrated. Through this, we will provide insight on the extensive usage of Matlab/Simulink environment for simulation and actual code generation for the outside tests. Second example details the HIL testing of the "Trainer 60" autopilot design before the actual flight tests. Both applications demonstrate the flexibility and the rapid-prototyping capability of the microavionics system.

3.1 Autonomous Control and Way-point Navigation Experiment on Humvee

The process that we followed to develop the ground vehicle closed loop control and navigation algorithms is illustrated in Fig. 11.

As the first step, we created a Matlab/Simulink simulation which emulates all the sensing data and control architecture. This simulation uses a non-holonomic car-like ground vehicle (as shown in Fig. 12) model as the dynamic model. Linear velocity of the ground vehicle is taken as constant in order to simplify the controller structure. A classical PD-controller is designed (as shown in Fig. 12) to control the heading of the vehicle. Later, this controller is tested on the dynamic model in the simulation. Before adding our waypoint navigation algorithm to the simulation, the PD-controller is tuned through outside tests.[4] The waypoint navigation algorithm, which is based on the primitive heading controller, tries to reach the given waypoints

---

[4]Initial values of the controller coefficients are selected based on the primitive three wheel car model. We tuned the controller performance (overshoot percent, settling time and rise time) using outside ground test data. Our simple waypoint navigation algorithm is added to the simulation after obtaining an acceptable performance. In addition, the tuned parameters are cross-validated by Matlab/Simulink simulations for our specific application as shown in Fig. 11.

Fig. 11 Microavionics control
implementation process flow
diagram

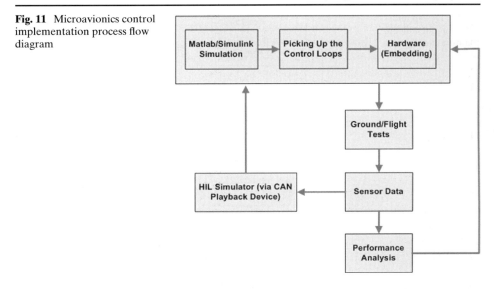

by changing the reference input angle between the vehicle heading and the target
waypoint. The waypoint navigation algorithm runs between two points, comparing
the distance between the given waypoint and the ground vehicle. If the distance is
equal to or less than a predefined limit, the algorithm accepts the following point
as the new target waypoint. All target waypoints are embedded in two vectors
(latitude and longitude vectors in waypoint navigation algorithm are implemented
as embedded Matlab Function Blocks) and the waypoint navigation algorithm is
fed by these latitude-longitude couples in a predefined order. Our microavionics
system design is also capable of accepting target waypoints via Ground Station as
depicted in Section 2.5. In real outside tests, a GPS receiver is used to update

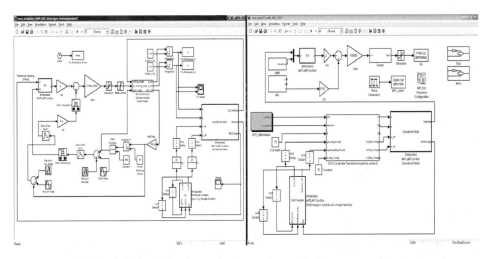

Fig. 12 Matlab/Simulink simulation for navigation and control of humvee, and implementation

the position measurements and a digital compass is used for updating the heading measurements. After the navigation algorithm is verified in simulation (as seen in Fig. 13), the design is adapted to be embedded on the MPC555 microcontroller. This new Matlab/Simulink code structure which is compatible with the Real-Time Workshop is illustrated in Fig. 12.

The adaptation process includes the replacement of some of the blocks of the initial simulation by the Matlab Real-Time Workshop compatible ones. For instance, ground vehicle's state blocks are replaced with TouCAN blocks (standard blocks of MPC555 Simulink library) which channel position information from the GPS receiver and heading information from the digital compass. Moreover PWM Out blocks are added to drive servo actuators on the ground vehicle, and the mathematical model of the ground vehicle is removed because it is unnecessary to use it in real outside tests. In addition, Embedded Matlab Function blocks in both standard C or m-file (Matlab Script Language) notation are created for algorithmically complex Simulink blocks. After this adaptation procedure, the C code required by the MPC555 microcontroller is automatically generated and embedded into the MPC555 microcontroller. The outside test results as illustrated in Fig. 13 are used to further tune the control and the navigation algorithm parameters. Note that the sensor data can also be played back for visualization as shown in Fig. 11. In addition, closed-loop system identification and model verification can be carried out, and the mathematical model of the unmanned vehicle can be improved by using the sensor data obtained from the outside tests.

Note that using our microavionics design automatic code generation feature, we observed that the control implementation process takes considerably less time than the conventional embedded control programming methods do [17]. This capability enables the researchers to focus on the design of control and navigation algorithms at a higher level and prevents the unnecessary effort of low-level driver programming

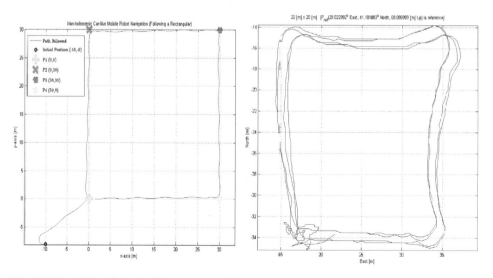

Fig. 13 Matlab/simulink simulation and outside test results for closed-loop navigation and control of humvee (Waypoints [(0;0), (0;20), (20;20), (20;0)] meters in ENU frame for outside testing)

during algorithm design. As an example, HIL testing of the autopilot system for "Trainer 60" fixed-wing UAV will be discussed in detail in Section 3.2 (Fig. 14).

3.2 HIL Testing of the Autopilot System for Trainer 60

The HIL testing of the autopilot system for Trainer 60 is one of the important steps in designing the autopilot algorithm as described in generic conceptual design approach shown in Fig. 10.

Autonomous waypoint navigation algorithm for Trainer-60 is designed with three outer loops designated as Altitude Hold, Airspeed Hold and Heading Hold (also defined as Coordinated Turn) modes [4]. This is illustrated in Fig. 15. Above these loops (as seen in Fig. 16), there is a navigation layer which generates the altitude, velocity and heading of the aircraft according to the waypoint map as described in detail at the previous section. Figure 16 illustrates the piloted take-off and the test of the autopilot algorithm.

HIL Setup in Fig. 14 illustrates the general distribution structure of all the hardware around CAN Bus and Ethernet Bus backbone. HIL platform can be split into four main layers. At the first layer, the sensing information is transmitted to MPC555 either via CAN Playback Device or via the state information from the xPC Target Box (xPC Target Box simulates the dynamic model). CAN Playback Device is capable of transmitting sensing information (GPS, Altimeter, HMR, IMU) over CAN Bus via Softing PC To CAN Interface Card. It is important to note that this sensory information can belong to real flight data recorded in the data logger

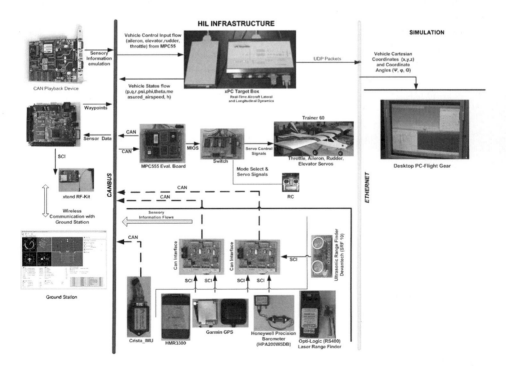

Fig. 14 HIL simulator design structure

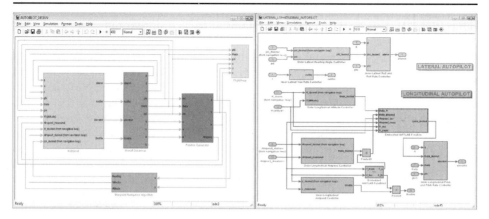

Fig. 15 Matlab/simulink simulation of Trainer-60 autopilot

(Kvaser Memorator). In addition, the wireless communication of the HIL setup with Ground Station is available in this layer. At the second the processor layer, we have an MPC555 processor to different types of control algorithms, and ARM processor to compute coordination patterns, transmit or receive information via XTend RF kit, and PC104 for high level control implementations, mission planning, and task assignment. Third layer is named as dynamical model layer and includes xPC Target Box. Identified and verified aircraft or ground vehicle dynamic models are embedded in xPC Target Box, and this box receives vehicle control inputs and transmit vehicle states by means of CAN channel after nonlinear state propagation. Furthermore, position and attitude information are transmitted via UDP packets from xPC Target to the visualization layer. Specifically, this data is converted to longitude, latitude and Euler data of the vehicle and sent to the visualization computer running the open-source FlightGear Flight Simulator. In addition, real vehicle HIL test is achieved as the control inputs can be sent to the aircraft servos as PWM signals. In this particular

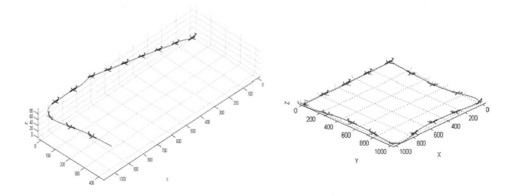

Fig. 16 Trainer 60 autopilot performance (Take-off by human pilot and waypoint navigation by autopilot)

setup, the HIL test of the autopilot system is realized in two different ways. The first method is named as "Option-A:Playback Mode" and it is illustrated in Fig. 17. CAN Playback Device streams all the sensing information in designated time slots with correct sensor IDs to the MPC555 over the CAN Bus. MPC555 uses this data as the sensing information as if it is in real flight, and generates PWM signals to the servos for the control signals of Trainer 60 model aircraft. Through this we can test in-flight anomalies and hardware problems. In addition, the real flight test of the Trainer 60 is animated using the xPC Target Box which converts the sensing information to vehicle coordinates and coordinate angles to visualize the recorded flight via FlightGear. Second HIL test method, "Option-B:Simulated Mode", is used for testing the autopilot algorithm and visualizing its performance via FlightGear Flight Simulator as depicted in Fig. 18. In this second mode, vehicle states stream over the CAN Bus to the MPC555 as the sensing information and MPC555 generates control inputs. During the test of our autopilot algorithm, the verified dynamic model is embedded in the xPC Target and the control loops are embedded in MPC555 processor. Control signals are transmitted to the xPC Target via CAN messages. xPC Target propagates the nonlinear dynamics of the aircraft to obtain the aircraft states. xPC Target transmits these states for the next control cycle to the MPC555 controller. In addition, xPC Target converts the state information into the Flight Gear native data packet form and it streams these packets via UDP to visualization computer for real-time visualization of the flight control algorithms.

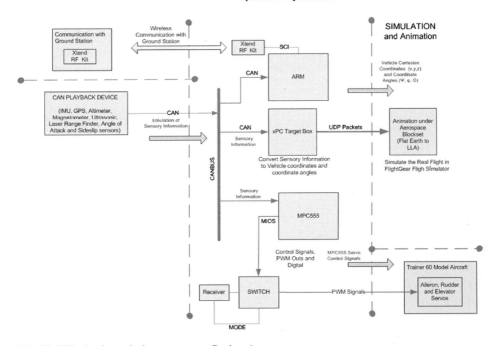

Fig. 17 HIL simulator design structure - Option A

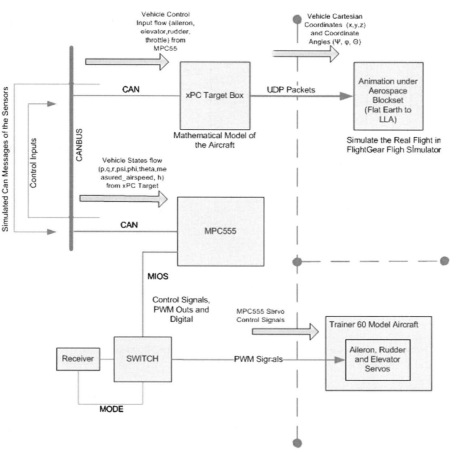

Fig. 18 HIL simulator design structure - Option B

In the next section, we illustrate how this microavionics structure is integrated to the flight network simulator for multiple vehicle coordination test scenarios that involve vehicles including manned flying vehicles in joint airspace.

4 HIL Integration

The Flight Network Simulator as in Fig. 19 has the capability of testing different kind of flight missions and cooperative flight algorithms for manned-unmanned platforms in joint airspace. Outside flight tests, as a part of day-to-day testing activities, need extensive logistics and may result in unnecessary risks (especially at the first trial tests) when there are complex collaborative applications. As a result, HIL integration and testing of the microavionics system in a joint airspace network simulator has critical importance. Specifically, the simulator serves as the first order

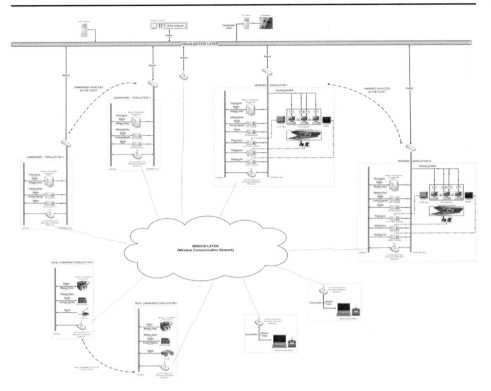

Fig. 19 Network diagram for the joint airspace manned-unmanned flight simulator

critical hardware verification and software validation step before the actual flight tests.

In addition, one unique feature of the in-house developed network simulator is that it allows the real in-flight UAVs to connect to a simulated mission scenario. This capability allows manned-unmanned formation flight test experiments to be carried while the manned platform is simulated and the actual UAV being in flight. As such, for UAV fleet experiments, the members of a large UAV fleet can be simulated while only a subset of them being in real flight.

Before going into the details of the HIL integration of the microavionics system (and the actual mini UAV platforms) to the network simulator, we will first review the general design and the architecture of the flight simulator covering the main components and data flow layers.

4.1 Flight Network Simulator

The general design of the simulator[5] is structured around two layers: visualization and mission layer. These two layers represent two different data bus structures

---

[5]The full and extensive treatment of the in-house developed flight network system, the hardware systems and the software architecture can be found in [2].

and data flows. As seen in Fig. 19, simulation elements such as piloted vehicle simulator, unmanned vehicles, real unmanned vehicles (ground vehicles and micro-UAVs), ground stations and the simulation control computers carry distinct wired and wireless connections to these two data layers.

Visualization Layer entails the passage of visualization and simulation related data packets (i.e. packets which result in a coherent visual picture of the whole scenario to the operators and the simulator pilot) across the wired ethernet network using UDP packets. The visualization layer uses the open-source FlightGear flight simulator packet structure to allow direct integration to the flight simulator visualization elements. These visualization elements include the three panel environment display for the pilot of the manned simulator (as shown in Fig. 2) and the pilot/operator panel for tactical/simulation displays. The *Mission Layer* is accomplished via wireless communications (serial and Ethernet) across each unique entity existing within the simulation using predefined data packet numbers and structures. Mission critical data such as target assignments, payload readings, commands and request are delivered through this wireless mission layer link. Because of mission-driven information flow requirements, a stand-alone and multi-functional module is needed for communication and information distribution with other entities in the overall network. Driven by this need, we have developed a standardized Command and Information Distribution Module (CID) as a part of the network simulator hardware. The module has several communication interfaces including Ethernet, CAN and serial link allowing rapid integration of a variety of additional piloted simulation vehicles, virtual or real unmanned vehicles to the existing simulation both at the visualization and the mission layer.

The reason for the layered approach for data flows stems from the need to differentiate the real operational wireless communication links in mission scenarios from the wired communication links to obtain visualization and simulation information for manned vehicle simulator and the operators. This break-up of layers not only represents a realistic mission implementation, but also allows HIL testing and analysis of wireless network structures (point-to-point, point-to-multi point, ad-hoc) that mimic the real operational communication links as it would be implemented and experienced in a real mission. The modular design of the network simulator elements is tailored according to the need for joint real-time simulation across groups of manned and unmanned fleets, and the mission control centers. As such, the hardware structure within the network simulator is tailored to mimic the distributed nature of each of the vehicle's processors and communication modules. Note that the layered bus approach to the visualization and the mission layer allows direct integration of new manned and unmanned simulation elements within the scenario once linked to the corresponding wired and wireless buses.

4.1.1 Simulation Control Center

Simulation Control Center consists of three elements : simulation operator computer, world computer and the FlightGear server. *Operator Panel* is used for configuring the network setting of the simulator. All information about the simulation (numbers of manned-unmanned computers, IP addresses of the machines) are input from this GUI. *World Computer* is used for controlling of the packet traffic of the visualization layer. For visualization and networked operations we use the Flight-Gear native data format. FlightGear native data format is identified by three distinct packet structures : Dynamic States Packet (D), FlightGear Multiplayer Packet (M) and Flight Dynamics Packet (FDM). D packets represent vehicle dynamic states such as position, orientation and velocity information supplied by the computers running the mathematical model of each of the manned and unmanned vehicles. M packets include information about position, orientation, velocity and the model of aerial vehicle. FDM packets include generic type and configuration information about the aerial or ground vehicles. Basically, an in-house developed program called State Sender runs on the World Computer and makes packet conversion from Dynamic (D) packet to the Multiplayer (M) or Flight Dynamic Model (FDM) packets. M and FDM packets are needed for visualization of the simulation. State Sender, being an add-on module to FlightGear, basically is a piece of C++ program that listens dynamic states information of a vehicle or any other simulation object and when it receives such information it makes the packet conversion. All the unmanned-manned computers which run the mathematical models of the vehicles (xPC computers) in the simulator send their states as Dynamic (D) packet to these programs at a specific frequency and State Sender programs convert the dynamic (D) packet to the multiplayer (P) packet and send them to the both FlightGear server and the 3D Multi-monitor visualization systems of manned simulator. In addition to that, for the manned simulator, the State Sender program makes a dynamic (D) to FDM packet conversion and sends the FDM packets to the Multi-monitor visualization system for cockpit-view visualization of the manned simulator. *FlightGear server (fg server)* in the network simulator runs in a separate computer, listening for incoming connections to simulator system. The protocol for exchanging packets in Fg server is UDP. Through the Fg server not only we create a complete list and picture of all the

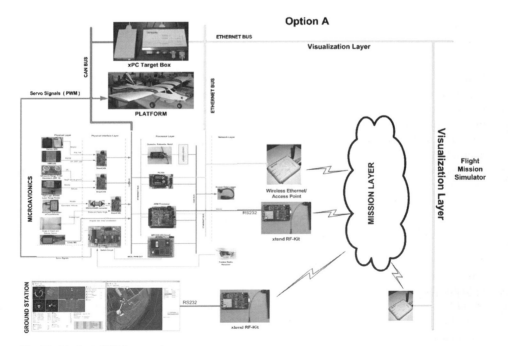

Fig. 20 Mode A HIL integration

simulation elements within the mission, but also generate tactical display information to the pilot/operator and the simulation controllers by linking to the server (Fig. 20).

4.1.2 Manned Vehicle Simulation

Manned vehicle simulation system as shown in Fig. 19 consists of a generic pilot station and a dedicated computer rack. The generic pilot station is equipped with both standard helicopter (cyclic, pedal, collective, throttle) and aircraft (thrust, flap) pilot input devices. In addition, the pilot could employ a four degrees of freedom joystick as a side stick. The pilot console includes a selection of switches (such as engine, autopilot, sas on-off) and warning lights (such as altitude, engine failure) that can be configured for different operational needs and conditions. The pilot station is equipped with a three-monitor visualization system that provides scenery information of the flight scenario from FlightGear. The computer rack system includes a set of dual bus (CAN and ethernet) connected computers for pilot input A/D conversion, real-time execution of the vehicle mathematical model, mission and flight controls, and C2 pilot/operator interface.

In addition to a huge selection of native FlightGear air vehicle models,[6] Matlab/Simulink dynamic models of several types of air vehicles such as helicopters (EC-120, Vario) and fighter aircrafts (F-16, F-4) have been developed in-house. In order to simulate the vehicle dynamics in real-time, Matlab/Simulink models xPC target technology is used. Matlab/Simulink simulation models are compiled to native code. The native code runs within xPC operating system on a target machine. After configuring the settings of xPC Target, a diskette including configured operating system is prepared. In our implementations, we use ordinary computers as an xPC Target computers by booting them with this diskette. Ethernet configuration enables connecting and uploading the Simulink model to the xPC Target computer. After uploading the model, simulation can be started using the remote control via Ethernet (Fig. 21).

4.1.3 Unmanned Vehicle Simulation

Unmanned vehicle simulation integration within the network simulator can be achieved through virtual vehicles and real unmanned vehicles. The virtual unmanned simulation is run on a rack of standard PCs which has four components: mission coordination computer, xPC computer/controller, xPC computer/dynamics, and CID module. As there are two types of information flow in the network, namely mission information and visualization information, there are two different communication buses: CAN and Ethernet bus. CAN bus is used for mission related information flow, while Ethernet is used for data flow to the visualization system. The components of the virtual unmanned vehicle simulation are shown in the Fig. 19. The computer system consists of four computers including the CID module. Each individual vehicle has to use CID module in order to join the mission network. This CID module implements a simple communication protocol that is used among components of the network for communication and information distribution. xPC Computer/Dynamics is used for executing the dynamics algorithm of the unmanned vehicle in real-time,

---

[6]Note that these models can also be used in the manned flight simulation.

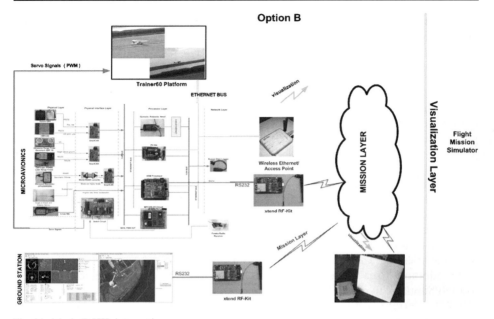

Fig. 21 Mode B HIL integration

while the control computer provides the closed-loop control and command functionality. Mission Coordination Computer is used for coordination between other manned-unmanned vehicles in the fleet when performing a mission collaboratively.

The real unmanned vehicles connect to the mission layer through their own CID module and vehicle information data is transported to the visualization layer through the mission simulator's own dedicated CID module which collects data from microair or ground vehicles naturally flying or driving outside the lab environment.

4.2 Hardware in-the-Loop Integration of the Microavionics System and the UAV Platforms

The hardware in-the-loop integration for the microavionics system is achieved in two modes: Mode A and Mode B. In mode A, the system is configured in the stand-alone HIL structure as illustrated in Section 3. In this setup, the xPC TargetBox computer, named as dynamical model layer, embeds our real time air vehicle dynamic model, and receives vehicle control inputs, transmit vehicle states by means of CAN channel with the microavionics system. For the network simulator scenario and visualization integration, xPC TargetBox is linked to the *World Computer* to which it transmits Dynamic Packet (D) position and attitude information via UDP packets over the wired ethernet line. At the *World Computer*, through StateSender, this is converted into (P) multiplayer packets for visualization and scenario integration.

In this particular setup, the microavionics module located at the lab, runs as if in real flight configuration implementing real-time controllers and algorithms in simulated joint airspace. The mission related communication is achieved through the original flight CID module with the wireless ethernet across all the other simulated manned and unmanned systems.

In mode B configuration, the microavionics system is installed on the UAV for outside flight tests. During the flight test, the UAV connect to the network simulator sending the Dynamic Packet (D) addressed to the *World Computer* through the CID module to the simulator's own CID module with external antennas over wireless ethernet. In addition, it receives 1 Hz updated scenario picture information from the *World Computer*. At the mission layer, the manned simulation or virtual unmanned simulation link to the real flight module through the simulator's CID module. Simulator's own CID module basically acts as a router with the flight test unit and the simulated units.

5 Conclusions

In this work, we have presented the design and the autonomous operation implementation details of a CAN and Ethernet bus back-boned microavionics system. In addition to mimicking the avionics structure of larger scale units used at high speed/high altitude UAVs, this system provides us with unique features such as rapid-prototyping/automatic code generation capability and cross-platform-compatibility to ground, fixed-wing and rotary mini unmanned vehicles. The bus-backbone also allows integration and testing of customized flight management computers and communication units that implement algorithms tailored towards fleet operations in joint manned-unmanned airspace. HIL integration of this unit to the in-house joint flight network simulator provides an extensive testing capability of these units and algorithms before the risky and costly real flight testing that involves manned and unmanned vehicles.

Acknowledgements The authors would like to thank ASELSAN for providing hardware support with regards to the manned flight simulator. Serdar Ates and Ismail Bayezit would like to thank TUBITAK (Turkey Science and Technology Research Foundation) for supporting them financially during their graduate education. Besides, the authors would like to thank Bahadir Armagan for his engineering support in every section of this study, Fatih Erdem Gunduz for re-designing Ground Station GUI and SmartCAN codes, Mirac Aksugur for mechanical support, Oktay Arslan for his work at joint flight network simulator integration, Captain Selim Etger for being the test pilot and Melih Fidanoglu for kindly providing the logistics support.

References

1. Aksugur, M., Inalhan, G., Beard, R.: Hybrid propulsion system design of a vtol tailsitter UAV. In: Wichita Aviation Technology Conference, Wichita, August 2008
2. Arslan, O., Armagan, B., Inalhan, G.: Development of a mission simulator for design and testing of C2 algorithms and HMI concepts across real and virtual manned-unmanned fleets. In: Hirsch, M.J, Commander, C.W., Pardalos, P.M., Murphey, R (eds.) Optimization and Cooperative Control Strategies. Lecture Notes in Computer Science, Springer (2008, in press)
3. Cetinkaya, A., Karaman, S., Arslan, O., Aksugur, M., Inalhan, G.: Design of a distributed c2 architecture for interoperable manned/unmanned fleets. In: 7th International Conference on Cooperative Control and Optimization, Gainesville, February 2007
4. Christiansen, R.S.: Design of an Autopilot for Small Unmanned Aerial Vehicles. Msc, Brigham Young University (2004)

5. Cummings, M.L., Guerlain, S.: An tnteractive decision support tool for real-time in-flight re-planning of autonomous vehicles. In: AIAA 3rd "Unmanned Unlimited" Technical Conference, Workshop and Exhibit, pp. 1–8, Chicago, 20–23 September 2004
6. Elston, B.J., Frew, E.: A distributed avionics package for small uavs. In: AIAA Infotech at Aerospace. Arlington, September 2005
7. Evans, J., Inalhan, G., Jang, J.S., Teo, R., Tomlin, C.J.: Dragonfly: A versatile uav platform for the advancment of aircraft navigation and control. In: Proceedings of the 20th Digital Avionics System Conference, Daytona Beach, October 2001
8. Gavrilets, V., Shterenberg, A., Martinos, I., Sprague, K., Dahleh, M.A., Feron, E.: Avionics system for aggressive maneuvers. IEEE Aerosp. Electron. Syst. Mag. **16**, 38–43 (2001)
9. Hollan, J., Hutchins, E., Kirsh, D.: Distributed cognition: Toward a new foundation for human-computer interaction research. ACM Trans. Comput.-Hum. Interact. **7**, 174–196 (2000)
10. How, J., Kuwata, Y., King, E.: Flight demonstrations of cooperative control for uav teams. In: 3rd Conference Unmanned Unlimited Technical Conference, Chicago, September 2004
11. Inalhan, G., Stipanovic, D.M., Tomlin, C.: Decentralized optimization, with application to multiple aircraft coordination. In: IEEE Conference on Decision and Control, Las Vegas, December 2002
12. Jang, J.S.: Nonlinear Control Using Discrete-Time Dynamic Inversion Under Input Saturation: Theory and Experiment on the Stanford Dragonfly UAVs. PhD, Stanford University, November (2003)
13. Johnson, E.N., Schrage, D.P.: The Georgia Tech unmanned aerial research vehicle: Gtmax. In: Proceedings of the AIAA Guidance, Navigation, and Control Conference, Austin, 11–14 August 2003
14. Karaman, S., Aksugur, M., Baltaci, T., Bronz, M., Kurtulus, C., Inalhan, G., Altug, E., Guvenc, L.: Aricopter : aerobotic platform for advances in flight, vision controls and distributed autonomy. In: IEEE Intelligent Vehicles Symposium, Istanbul, June 2007
15. Karaman, S., Inalhan, G.: Large-scale task/target assignment for uav fleets using a distributed branch and price optimization scheme. In: Int. Federation of Automatic Control World Congress (IFAC WC'08), Seoul, June 2008
16. Koyuncu, E., Ure, N.K., Inalhan, G.: A probabilistic algorithm for mode based motion planning of agile air vehicles in complex environments. In: International Federation of Automatic Control World Congress, Seul, July 2008
17. Mutlu, T., Comak, S., Bayezit, I., Inalhan, G., Guvenc, L.: Development of a cross-compatible micro-avionics system for aerorobotics. In: IEEE Intelligent Vehicles Symposium, Istanbul, June 2007
18. Bear: Berkeley aerobot team homepage. University of California: Berkeley Robotics Laboratory. http://robotics.eecs.berkeley.edu/bear/ (2008)
19. GPL OpenSource: The flightgear flight simulator (1996)
20. Shim, D.H., Sastry, S.: A situation-aware flight control system design using real-time model predictive control for unmanned autonomous helicopters. In: AIAA Guidance, Navigation, and Control Conference, vol. 16, pp. 38–43, Keystone, 21–24 August 2006
21. Ure, N.K., Koyuncu, E.: A mode-based hybrid controller design for agile maneuvering uavs. In: International Conference on Cooperative Control and Optimization, Gainesville, 20–21 February 2008
22. Valenti, M., Schouwenaars, T., Kuwata, Y., Feron, E., How, J.: Implementation of a manned vehicle - UAV mission system. In: AIAA Guidance, Navigation, and Control Conference and Exhibit, Providence, 16–19 August 2004

A Hardware Platform for Research in Helicopter UAV Control

Emanuel Stingu · Frank L. Lewis

Originally published in the Journal of Intelligent and Robotic Systems, Volume 54, Nos 1–3, 387–406.
© Springer Science + Business Media B.V. 2008

Abstract The topic discussed in this paper addresses the current research being conducted at the Automation and Robotics Research Institute in the areas of UAV helicopter instrumentation targeted towards aerobatic flight and heterogenous multi-vehicle cooperation. A modular electronic system was designed which includes sensors communicating over a CAN bus and an on-board computer that runs a special real-time Linux kernel. The same system is reusable on other types of vehicles and on the ground stations, facilitating the development of multi-vehicle control algorithms.

Keywords UAV helicopter · Real-time control · CAN bus · Multi-vehicle cooperation

1 Introduction

During the recent years, the contribution of unmanned systems to the military operations continued to increase while reducing the risk to human life. Unmanned Aircraft Systems (UASs) can play an important role in missions such as reconnaissance and surveillance, precision target location, signals intelligence, digital mapping. Because of the miniaturization of sensors and communication equipment, smaller aircraft can now perform the same mission that would have required a bigger aircraft a few years ago. This has obvious advantages in costs and logistics, but also enables completely new missions, such as operation at street level in urban environments.

E. Stingu (✉) · F. L. Lewis
Automation and Robotics Research Institute, University
of Texas at Arlington, Arlington, TX, USA
e-mail: pestingu@arri.uta.edu
URL: http://arri.uta.edu/acs/

F. L. Lewis
e-mail: lewis@uta.edu

UAV helicopters are particularly useful because of their ability to hover and to take off vertically.

Great research efforts were made towards the control of helicopter UAVs [1–4]. Most of them use linear models for the helicopter dynamics and are concerned mostly about hovering, forward flight or simple maneuvers. The exceptional capabilities of these small-scale helicopters to do aerobatic flight maneuvers [5] due to the high thrust to weight ratio (that can reach 3) are not completely explored. Even more, most of the platforms used to implement the algorithms were modified in order to add instrumentation and on-board computers. The common approach was to include most of the additional components into a heavy box under the helicopter body, by replacing the original skids with taller skids that could accommodate the equipment height. The extra weight placed well below the initial center of gravity enhances the stability of the helicopter, but severely limits the aerobatic flight capabilities.

This paper addresses the current research being conducted at the Automation and Robotics Research Institute in the areas of UAV helicopter instrumentation targeted towards aerobatic flight and heterogenous multi-vehicle cooperation. We offer a slightly different approach to what has already been done in this field, by proposing a modularized system that can be configured to work on different types of vehicles with minimal effort.

2 Problem Description

There are a few main ideas that have guided the design of the control system. *Reusability* is important in order to reduce the development time and costs. The same electronic modules and the same software can be used on the helicopters, on the ground vehicles and on the ground monitoring stations. The system is *modular*, consisting of discrete electronic modules that communicate over a CAN bus. This architecture allows the system to be *distributed* across the whole helicopter body (Fig. 1), having a small effect on the center of gravity and allowing aerobatic maneuvers.

Safety is a big concern because helicopters of this size are expensive and present the potential of being very dangerous to the operator. In order to increase the robustness of the electronic control system, the computing functions are handled by both the onboard Pentium M processor, and by a microcontroller on the Real-Time Module that can take control when the other fails. The microcontroller has access to an additional set of local (low precision) inertial sensors so it can safely land the helicopter when something goes wrong with the software or with the electronics. The CAN protocol is not sensitive to electromagnetic interference and assures a reliable communication link between the modules. The power supplies for different functions are separated and short-circuit protected. In case of a wiring failure, most of the system will still be operational. Various current, voltage, temperature and speed sensors allow prediction of failures before they get critical.

Low power consumption and small weight are very important for an UAV, because they determine its flight time and the maneuverability. The specially de-signed electronics is smaller and it weights less than the corresponding off the shelf components. All the power conversion from the battery to the various modules is done using high-efficiency switched-mode power supplies, thus optimizing the use of electric energy and allowing for longer flight times.

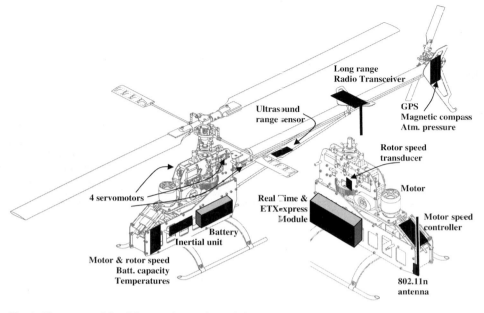

Fig. 1 Placement of the different electronic modules on the helicopter body

3 The Helicopter Platform

The mechanical platform is a Joker 2 fully aerobatic electric helicopter (Fig. 2), manufactured by Minicopter Germany. The main technical details are the following:

- Main rotor diameter = 1,560 mm
- Tail rotor diameter = 290 mm
- Weight = 4.5–5.0 kg (with batteries)

Fig. 2 The Joker 2 helicopter platform including the electronic control system

Fig. 3 The helicopter stand
with a Raptor 30V2 helicopter

- Maximum total weight = 6.8 kg
- Flight time = 8–20 min depending on flight style and load
- Batteries: two Lithium-Polymer, 8,000 mAh, 18.5 V
- Motor: Plettenberg 370/40/A2/Heli, three-phase brushless motor, 2.5 kW, low vibrations.

The main rotor has a Bell–Hiller stabilizer bar, but it can be replaced with a rigid rotor head with no stabilizer bar at all, for improved 3D aerobatic maneuverability. The choice of an electric helicopter instead of one powered by a combustion engine was due to system identification issues because of the fuel consumption and the change of the helicopter mass, which makes it impossible to obtain consistent measurements at every experiment. It is also easier to work with electric motors instead of combustion engines that have to be periodically tuned and have to be started manually.

Some of the experiments are done with the helicopter tied to a specially designed helicopter stand (Fig. 3) that allows the helicopter to tilt in any direction (within some limits), to rotate around the vertical axis, to go up and down and to rotate in a circle in the horizontal plane. The weight of the stand is cancelled using an air cylinder. The stand is most useful to trim the helicopter and to obtain the linear model for hovering.

4 Organization of the Helicopter Control System

The electronic control system is distributed in three locations: on the helicopter body, on the ground (base station) and at the human pilot for emergency situations (remote control). Because the onboard computer is sufficient for control, the ground base station is optional. The radio communication is realized using two distinct transmit/receive links. The first one, in the 900 MHz band, uses high-power transceivers to cover a radius of about 14 mi with simple omnidirectional antennas. The shortcoming of this connection is the fact that it only works at low speeds, having a maximum bit rate of 115.2 kbps. At this speed, the helicopters can transmit to the ground monitoring station only the most important flight variables and can receive commands from the ground station and from remote controls in case of emergency. The second radio link is a high-speed 802.11n wireless network, having a maximum throughput of 300 Mbps. The range is about 300 m. The network uses UDP packets and allows each device to communicate with any other. This allows the implementation of formation flight algorithms, because the helicopters, the ground vehicles and the ground stations can easily share information in real-time. If the high-speed connection is lost, the long-range link is used to maintain control and make the vehicles return inside the area covered by the high-speed radio link.

4.1 The On-board System

The main idea that has driven the design of the on-board electronic system was the ability to spread the components across the helicopter body in such a way that the center of gravity does not shift too much compared with the original platform. The approach of having a large box containing all the electronic system under the helicopter body was inacceptable because the center of gravity would have moved down too much and would have made the helicopter more stable, preventing aerobatic maneuvers. This issue only exists for small-scale helicopters.

Because the system components are spread apart, a reliable communication channel had to be adopted. A factor that puts more pressure on the choice of communication protocol is the fact that the helicopter is electric. The motor can have current spikes of 100 A. The average current consumption is of 25 A for hovering. The speed controller for the brushless motor has to switch the power to different windings synchronously to the rotor movement and controls the power delivered to the motor by PWM. All this high-current and high-voltage switching causes additional electromagnetic interference that can potentially disrupt the communication between the modules of the control system. The long-range radio transceiver has a maximum transmitting power of 1 W. Some of the energy transmitted can be captured by the wiring on the helicopter body and rectified inside the semiconductor components, causing additional communication errors. The solution to all these problems was to choose CAN as the communication standard. The Controller Area Network (CAN) is a high-integrity serial data communications bus for real-time control applications originally developed by Bosch for use in cars. It has very good error detection and confinement capabilities, which coupled with the differential signaling ensure a high degree of robustness. It allows real-time communication due to the following characteristics: relatively high speed (1 Mbps), small latency due to short packets, and node addressing with bitwise arbitration that prevents collisions and allows prioritization of messages.

The block diagram of the on-board electronic system is shown in Fig. 4.

The most important component is the Real-Time Module (RTM). It acts as a communication bridge between the CAN bus and the USB bus, provides power to all the modules and implements some simple control algorithms that are activated in emergency situations. It is also a carrier board for the Pentium M ETXexpress module that actually runs the complex control algorithms (in normal conditions). The ETXexpress module runs Red Hat Enterprise Linux with a real-time kernel. It connects through the PCI express bus to a 802.11n wireless network adapter installed

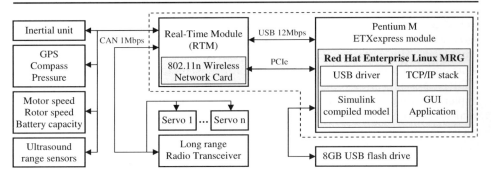

Fig. 4 Organization of the on-board helicopter control system

on the RTM and also communicates with the RTM using a 12 Mbps USB link. Each sensor or group of sensors is contained in a separate plastic box and communicates with the Real-Time Module via a CAN bus at a data rate of 1 Mbps. In addition to receiving commands, the servomotors can also send their position back to the RTM.

4.1.1 The ETXexpress Module

There are various possible choices for the architecture of an on-board computer. DSP and FPGA platforms are attractive due to the high degree of parallelism, the low power consumption or the good floating-point performance. Unfortunately, all the benefits vanish when it comes to actually implement a complete system. The software components that allow interfacing to USB, Ethernet and PCI express as well as the drivers or libraries necessary to communicate with peripherals like USB video cameras and wireless network cards are not open-source or free (in general). The goal was to build a cheap system where one can have complete control over the software and also be able to easily extend the functionality when needed. Besides that, the ease of implementation of the control algorithms was also considered. Because the helicopter is a research platform, most of the algorithms will not be optimized for speed or for parallelism. They will need strong floating-point performance and might not run optimally on fixed-point DSPs. FPGA implementations require the conversion of the C or Matlab code into VHDL or Verilog code, then testing and synthesizing, a process that requires specialized knowledge in FPGA design.

Due to the reasons stated above, the choice for the architecture of the on-board computer was to use an x86 general-purpose processor that has the largest amount of software support and adapts reasonably well to any type of control algorithm. The best performance per watt ratio (especially for floating-point tasks) at the time of system design was reached by the Intel Pentium M processors (Dothan) and by the Intel Core 2 Duo processors (for parallel algorithms).

The smallest form factor for the CPU board that contained the desired processors and all the necessary peripherals was the COM Express computer-on-module (also named ETXexpress). It is a highly integrated off-the-shelf building block based on a PCI Express bus architecture that plugs into custom made, application-specific carrier boards (in our case, the Real-Time Module). COM Express modules measure just 95 × 125 mm and include generic functions such as video, audio, Ethernet, storage interfaces and USB ports that are needed for most applications. The components present on the modules are the processor, the chipset (with on-board graphics), the DDR memory, one Gigabit Ethernet controller and the necessary switched-mode power supplies that convert the input 12 V voltage to the necessary levels needed by the different circuits. Due to their small size, the modules do not have specialized connectors for the various peripherals. Instead, they connect through a pair of high-density connectors to the carrier board, which routes the necessary signals to their connectors if needed. The RTM makes available five USB ports, 1 Gb Ethernet port and a combined analog video and TV output. It also routes one PCI express lane to a mini PCI express wireless adapter and one USB channel to the RTM microcontroller. There is a small number of other signals used, mainly to allow the RTM microcontroller to do power management and supervise the behavior of the ETXexpress module. Most of the signals available from the ETXexpress module are not used. Especially important are the unused PCI express lanes that can be routed to a FPGA on the RTM PCB in a future design.

The hardware configuration of the ETXexpress modules is compatible with Microsoft Windows and Linux, so they can be used the same way as normal computers, without the need of customized software.

4.1.2 The Real-Time Module

The Real-Time Module is built around a 32-bit ARM microcontroller running at 48 MHz. The main role of the microcontroller is to support the communication on the CAN bus and on the USB bus. It has a lightweight free real-time operating system (called exactly like that: FreeRTOS) that provides the basic functionality expected from a RTOS: tasks, preemptive and cooperative scheduling, semaphores, mutexes, queues, critical sections and many others. Special drivers were written to support low-latency communication for the USB bus, the CAN bus and the RS-232 serial ports. They use circular buffers and take advantage of the DMA transfer capabilities of the microcontroller, allowing the hardware to do most of the work and guaranteeing the fastest possible response to communication events without requiring software intervention.

The microcontroller connects via USB with the ETXexpress module or to an external computer. The channel is used twice during a sample period: at the beginning, the microcontroller sends the sensor data to the ETXexpress module, and at the end of the sample period it receives the commands to be sent to the actuators. The time taken by the USB communication has to be minimized and that is why all efforts were done to optimize the drivers for low latency in order to use the USB bus at its full capacity. The USB connection is also used by a bootloader that allows the firmware to be updated easily.

The RTM provides two CAN channels for connecting external sensors, actuators and special function devices. The communication uses CANopen [6], a CAN-based higher layer protocol. It provides standardized communication objects for real-time data, configuration data as well as network management data. Each device on the CAN bus has its own identifier and an associated Object Dictionary (OD) that provides a complete description of the device and its network behavior. All messages are broadcast on the CAN bus. Based on the entries in the OD, each device implements a filter that allows the acceptance of only the messages of interest. The following predefined types of messages exist on the network:

- Administrative messages

  They are based on a master–slave concept. On a CAN bus there is only one master (for each specific service) and one or multiple slaves. The following services are implemented: Synchronization—used to synchronize all the devices to the sample time. The RTM is the master. The CAN identifier associated to this message has the maximum priority and always wins the bitwise arbitration with all the other messages that are ready to be sent at the same time. Because each CAN node has a hardware timer that timestamps the CAN messages, the synchronization is precise to 1 μs. Boot-up—the slaves have four possible states: Initializing, Pre-operational, Operational and Stopped. The slaves send this message to indicate to the master that they have transitioned from state Initializing to state Pre-operational. The RTM sends this message to start, stop or reset one or all the slaves.

- Service Data Objects

  These are messages that access the Object Dictionary of the slave devices. The master (the RTM) can thus interrogate each slave and determine the supported network services and the available real-time variables and can update its configuration. While the slaves are in the Pre-operational state, the master configures the sample rate, the time offset of different actions relative to the SYNC message and other specific parameters. When the configuration is done for all the devices on the CAN bus, the master transitions them to the Operational state. In order to allow updating of the firmware for the CAN devices without removing them from the system, a CAN bootloader was implemented. It is active in the Pre-operational state and accessible using SDO messages. The RTM receives the CAN identifier for the device that has to be programmed and the HEX file over the USB bus from the ETXexpress module or from an external computer and sends it via the CAN bus to the device to be programmed.

- Process Data Objects

  These are messages that encapsulate the contents of real-time variables that are sent or received by the CAN devices during each sample period. They are based on a producer–consumer concept. Data is sent from one (and only one) producer to one or multiple consumers. In normal conditions, the RTM is the consumer for the data produced by the sensors, and the actuators are the consumers for the commands sent by the RTM. In the case where the RTM stops responding, the CAN devices change behavior and implement a distributed control algorithm that tries to stabilize the helicopter and land. In order to accomplish this, the actuators become consumers for the sensor data.

  The USB interface implements the Communication Device Class (CDC) that appears as a virtual COM port to the USB host, which greatly simplifies the application programming. It uses three endpoints. The default control endpoint 0 is used for device management. The other two are a pair of IN and OUT endpoints that can use Bulk or Isochronous transfers. In the current implementation, Bulk transfers are used because they guarantee the correct delivery of data and no significant latency was observed when only one device is connected to the USB root hub. In the future, to improve the real-time behavior, isochronous transfers will be used because they guarantee a minimum latency. The retransmission of data in case of errors will be handled in software. The USB communication uses two buffers for each data endpoint. While the software handles the contents of one buffer, the hardware works with the other. This way, the hardware interface does not generally have to wait for the software to fill the buffers, allowing the full bandwidth of USB to be used.

The Real-Time Module also contains 10 switched-mode power supplies to power all the different modules installed on the helicopter (see Fig. 5). All are current limited and short-circuit protected and automatically recover from an over-current condition. All the power supplies can be individually enabled by the microcontroller and all report over-current conditions for easy localization of defects in the system. A high-current switched-mode Li-Po battery charger was included to eliminate the need of removing the battery for external charging. It also supplies the internal power bus when the DC adapter is connected. It gives priority to the internal devices and only the remaining current is used to charge the battery. The cell balancer is needed to guarantee that all cells charge equally even after a high number of charge/discharge cycles. The charge and discharge current is permanently monitored by the microcontroller in order to create the function of a "fuel gauge" that will inform the pilot and the control algorithm of the available flight time for the on-board

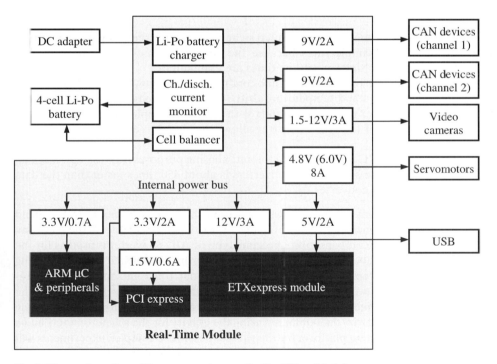

Fig. 5 The configuration of the power system on the Real-Time Module

electronics. Each CAN channel is powered by its own power supply for increased reliability. A short-circuit on the power lines of one channel will not disable all the functionality of the external system. The 1.5–12 V programmable power supply was included to allow powering video cameras for stereo vision and any other device that might need to be added in the future.

4.1.3 The Inertial Measurement Unit

The Inertial Measurement Unit (IMU) is one of the most important components of the control system. Its measurements are critical for estimating the full state of the helicopter during flight when the other measurements for absolute position and orientation are absent or imprecise. All efforts were made to reach the maximum possible accuracy of this instrument using relatively inexpensive sensors. It uses MEMS accelerometers and gyroscopes with high internal resonant frequency in order to insure reliable vibration rejection over a wide frequency range and in order to have a higher bandwidth.

Most of the commercial solutions use a single A/D converter with a multiplexer to convert all the six analog channels to digital representation. Each sensor is sampled at a different instant and the data set provided for all the sensors is not consistent if the sample rate is low. Another issue is the bias and sensitivity calibration with temperature, where the usual three-point calibration technique does not offer sufficient precision.

Because the cost of analog to digital converters is much lower than the cost of the inertial sensors, the solution to all of the above problems was to use a dedicated sigma–delta A/D converter for each channel. This approach has the following advantages:

- Provides consistency of the sampled signals (all are sampled simultaneously).
- Allows faster sample rates because the converter time is fully used by one channel and multiplexer settling times are eliminated.
- Avoids the use of complex signal conditioning circuitry that can cause bias, sensitivity errors and temperature drift. This can be achieved because there is no need to buffer the sensor outputs as there is no multiplexer present any more and the input of the A/D converter allows the signal to completely settle after each sample.
- Only a simple RC filter is needed for anti-aliasing purposes because the sampling frequency of the sigma–delta converters is about 450 times faster than the data rate of the A/D converter.

Even more, a custom solution allows for setting the sample points at well-defined moments in time, synchronized with the rest of the sensors available on the platform.

For temperature compensation, the same type of A/D converter is used as for the inertial sensors to avoid introducing additional errors to the final results by using imprecise or noisy temperature data. There are about 100 points used for temperature calibration. For all these temperature points the raw measurements for all the sensors were collected. They include the bias values obtained with the IMU being stationary and sensitivity data obtained using the gravity for the accelerometers and a constant-speed rotating platform for the gyroscopes. More complex experiments will be designed in order to enhance the calibration performance and take into account the cross-axis effects and the alignment errors.

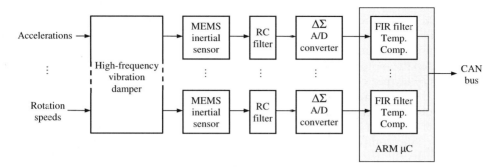

Fig. 6 The signal paths inside the IMU

Another aspect that had to be addressed especially in the case of the helicopter platform was to find a way to eliminate the vibration effects from the output of the IMU. Using electric helicopters instead of gas-powered ensures that there are less vibrations, but this is not enough. Using mechanical vibration dampers between the helicopter body and the IMU deteriorates the precision of the measurements by adding biases, phase delay and some nonlinearities, all of these being temperature dependent. Using analog filters at each sensor output complicates the design too much and doesn't allow for adaptability. The problem was solved using a combination of solutions (Fig. 6). First of all, the vibrations with a frequency higher or close to the internal resonant frequency of the sensor were filtered using a mechanical damper with very good characteristics but also very simple: dual-sided adhesive tape of a certain height. A RC filter was used to lower the bandwidth of the sensors to about 2 kHz. This limits the demodulation noise inherent to this kind of resonant sensors. The sigma–delta A/D converters sample the signal at 1.8 MHz and provide output data at a rate of 4 kHz. The IMU is driven by a 32-bit ARM microcontroller that applies a FIR filter that down-samples the data to a rate programmed on the CAN bus (currently 100 Hz). The filter coefficients are also programmed over the CAN bus during the initialization of the sensors.

4.1.4 The System Monitor Module (SMM)

The safety of the helicopter and of the human operators has to be enforced using all possible means. The operation of an electric helicopter of this size requires the strict observation of some specific parameters during flight. The System Monitor Module is the interface between the low-power electronic system (that contains the on-board computer and most of the sensors, implementing the control algorithm) and the high-power electro-mechanical system that actually makes the helicopter fly. The two electronic/electric systems have separate batteries in order to prevent interferences from the high-current and high-voltage system going into the sensitive low-voltage system. Because the SMM monitors most of the parameters in the power supply domain of the helicopter motor, it receives its power from the main Li-Po batteries. The CAN bus is placed in the power domain of the on-board computer. To allow communication over the CAN bus, the SMM uses optical isolation between its microcontroller and the CAN transceivers.

The SMM measures the following parameters: the voltage for each of the two main batteries, the current through the main batteries, their temperature, the motor temperature, the motor speed and the main rotor speed. It also provides the command to the motor speed controller (as a servo-type PWM signal).

Lithium-Polymer cells are not allowed to reach a voltage lower than 3.0 V. If they do, irreversible reactions take place that can severely limit the battery capacity for subsequent charge cycles or even worse they can cause fire. Therefore a critical function of the SMM is to continuously monitor the voltage for each battery. In the case where the control system does not react and land the helicopter, the SMM slowly lowers the RPM of the motor to do a forced landing by itself. The current monitoring function, together with the voltage monitoring, allows measurement of the instantaneous power required by the helicopter. This is used to implement a "fuel gauge" that indicates the remaining battery time, but can also be used in the control algorithms to optimize for power consumption.

The motor is a three-phase brushless motor with no sensors to measure the shaft angle or speed. The Electronic Speed Controller uses PWM power signals to control the speed. It connects two phases of the motor at one time to the battery and measures the third to get the back-EMF voltage generated by the motor while it rotates. The zero-crossings of this measured voltage are used to detect the angle of the rotor and to switch the power to the next pair of phases. Because installing speed sensors on the motor is not a viable solution, the measurement is done on the command given by the speed controller to the low-side FET transistors that switch one of the phases. The phase-switching is synchronous with the rotor movement therefore the exact motor speed can be detected. The speed of the main rotor of the helicopter is sensed using an integrated Hall sensor. The two speeds are not necessarily an expression of the gear ratio. The motor drives the main rotor through a timing belt and a gearbox. Further, there is a one-way bearing installed that allows the main rotor to freewheel. One more case when the two speeds do not follow the gear ratio is when there is a mechanical failure and the timing belt skips teeth. In this situation, the main rotor speed is smaller than the minimum expected from the gear ratio and the motor speed. This case is continuously monitored and signaled to the control algorithm and to the human operator in order to land the helicopter before the problem gets worse.

4.1.5 The Servomotors

The nonlinear model of the helicopter can be written in the most general form

$$\dot{x} = f(x, u).$$

The u vector can have multiple components, but four of them are the different displacements of the servomotors that control the angles of the blades:

$\delta_{front}, \delta_{left}, \delta_{right}$ – determine the swashplate position

δ_{tail} – determines the angle of the tail rotor blades

Usually servomotors receive the reference angle as an input and an internal feedback loop rotates the servomotor disc to that position. In steady-state, the true position of the servomotor will be equal to the reference command. Because the reference is known, it is used in many similar projects to estimate the real position of the servomotor (which is not available for measurement) by approximating the dynamics of the servomotor with a time delay or a better model (Fig. 7).

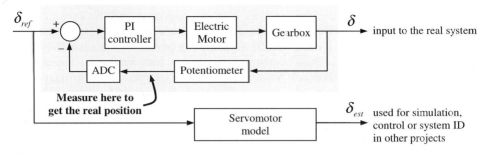

Fig. 7 Servomotor block diagram

The problem with this approach is the fact that in most situations the servomotor model is not known, and even if it is known, the dynamic depends on a few parameters like external forces or the supply voltage. The solution is to measure the voltage on the potentiometer installed inside the servomotor that provides the position feedback to the internal control loop. For each servomotor, an extra PCB was added to the existing electronics in order to interface it to the CAN bus (Fig. 8). The new electronic module reads the potentiometer voltage and generates the PWM signal that is used by the servomotor to calculate the reference position. This way, the true position of the servomotors is available to the parameter estimation algorithms in order to achieve better precision.

4.1.6 The Radio Transceivers

The helicopter communicates using two radio interfaces. For safety reasons it uses a long-range (14 mi) radio link on 900 MHz such that it is always able to receive commands from the remote control and can communicate with the optional base station. The high-speed link is a 802.11n wireless network connection, used to communicate with all the other devices (helicopters, ground vehicles, base stations).

The long-range radio transceiver is a module available on the CAN bus. It includes an XTend RF module manufactured by Digi International. The RF data rate is 125 kbps. At this relatively low speed, only the most important states of the helicopter can be sent in real-time at a rate of 100 Hz. The complete information is sent using the 802.11n wireless network. There are three devices present on the 900 MHz network: the helicopter, the remote control and the base station on the ground. To ensure the

Fig. 8 Extra PCB used to measure the position of the servomotor

Servomotor

Extra PCB

CAN bus & power

deterministic delivery of the data packets, transmit collisions must be avoided. This is achieved by having the devices synchronize with the helicopter using the packets sent by it as a reference. During the 10 ms sampling period of the control system, each of the three devices present on the network has its own window assigned when it sends data. They keep their own internal timing so they remain synchronized and are able to send data at the right time even if some radio packets transmitted by the helicopter are not received due to radio interference.

The 802.11n network connection is made using an Intel 4965AGN mini PCI express wireless card installed on the RTM PCB but controlled by the ETXexpress module running Red Hat Enterprise Linux. It offers support for three antennas. Currently only one vertical antenna is installed on the helicopter body, but in the future two more will be installed orthogonal to the existing one to allow reliable data transfer during aerobatic maneuvers when the orientation of the helicopter changes.

4.1.7 The Vision System

Video cameras, especially when installed in a configuration for stereo vision, can be cheap sensors that provide a great amount of information useful for control and navigation. Currently our interest is focused on system identification and nonlinear control of the helicopter, where video cameras are not absolutely required. However the electronic system was designed with cameras in mind. Prosilica GC650C cameras were considered for this application. They communicate using the GigE Vision Standard on Gigabit Ethernet. The video data is provided directly in digital format and uncompressed. The processor overhead for capturing the data using the network adapter is about 3% when jumbo frames are used. This is much better than using USB where due to the small data packets a large amount of CPU time is taken just to transfer data to memory. Currently the ETXexpress module has only one Gigabit Ethernet connection. A small external Ethernet switch is necessary to connect both cameras simultaneously. In the future designs, a dual-port PCI express Gigabit Ethernet controller will be implemented on the RTM PCB to avoid using the switch, or an extra DSP board with two Ethernet connections will be added.

Stereo vision requires very precise synchronization of the trigger of the two cameras. Most if not all of the USB cameras don't provide such a feature. The GC650C cameras allow for an external trigger input and can also generate trigger signals. This way, one camera can generate the trigger and the second can use it for synchronization, with no additional logic needed. Other parameters such as the exposure time, white balance and gain can be configured simultaneously over the Gigabit Ethernet interface by the software running on the ETXexpress module. Another important feature needed for fast-moving vehicles is to have progressive (non-interlaced) image capture, otherwise the images would be almost unusable.

A third small USB camera will be used for landing on a moving platform installed on a ground vehicle. The camera will point downwards to get the pattern drawn on the platform, which will be used by the vision processing algorithms to extract the orientation and the distance of the helicopter relative to the ground vehicle.

To maintain the center of gravity high enough so aerobatic maneuvers are not affected, the cameras are not installed under the helicopter body using elongated skids, but on the sides (Fig. 9). This approach limits the movements the cameras can do without being obstructed by the helicopter body, but it can be used especially if the cameras are fixed.

Fig. 9 Placement of the video
cameras on the helicopter
body

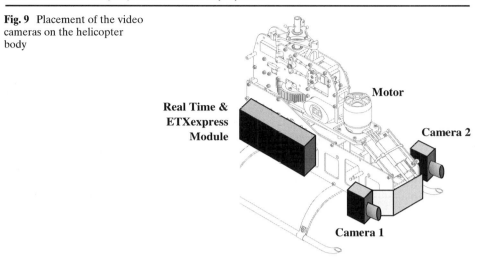

Real Time &
ETXexpress
Module

Motor

Camera 2

Camera 1

4.2 The Remote Control

The experiments involving the helicopter need the intervention of a human pilot. It
could be just the need to do simple testing of the basic functionality of the helicopter
platform, to activate the automatic control functions or to take over in case of various
failures and land safely. The use of remote controls from the RC helicopter hobby
industry adds an unnecessary risk factor into the system. It is well-known that many
of them are sensitive to radio interference. Some newer models that use the 2.4 GHz
frequency range solve this problem, but have limited range.

A completely new electronic board was designed for the remote control. A Spek-
trum DX6i model was used as the platform that had to be customized. It contains the
enclosure, buttons, joysticks and the graphical LCD screen. The original electronic
board was replaced with the new design. The low-capacity Ni-MH batteries that
were provided can not be charged unless they are completely discharged, limiting the
usage flexibility. They were replaced by a higher capacity Li-Po battery that allows
the remote control to work for several experiments without the need to be recharged.
The electronic board implements a Li-Po battery charger, so the battery doesn't have
to be removed from the enclosure each time it is charged.

The potentiometers of the joysticks, the buttons and the LCD connect to the new
board. An XTend RF transceiver operating on 900 MHz allows commands to be
sent to the helicopter and important parameters to be displayed to the pilot on the
LCD screen. The voltage and the current of the battery are continuously sampled
to estimate the remaining operating time and to give warning to the pilot before the
battery runs out.

4.3 The Base Station

The Base Station (Fig. 10) is an optional component of the control system. The on-
board computer of the helicopter can run the control algorithm by itself and can store
the history of all the signals in the system. The main function of the base station is to
display detailed information to the operators on the ground and to provide reference
ground pressure and DGPS corrections to the helicopter.

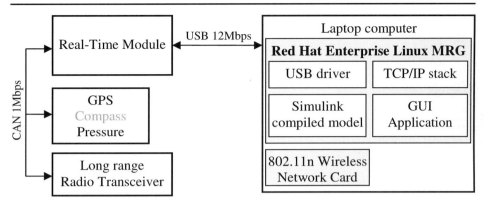

Fig. 10 Organization of the ground monitoring station

The electronic system of the helicopter was designed to be modular and reusable. The base station does not need additional hardware. It can use a set of the modules already available for the helicopter. There is no need to use an ETXexpress module on the ground. A laptop computer is more functional. The RTM can interface via the USB bus with the laptop computer the same way it does with the ETXexpress module.

5 Heterogenous Multi-vehicle Control

The helicopter platform will be included in an "ecosystem" of different types of vehicles. Currently in the development phase there are a relatively large size ground vehicle and a quad-rotor helicopter (Fig. 11).

Fig. 11 Three types of vehicles for multi-vehicle control

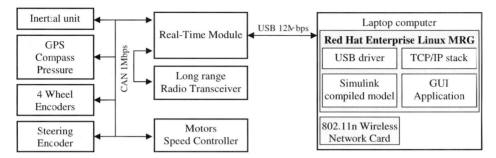

Fig. 12 The electronic system on the ground vehicle

The ground vehicle is built using a 4 × 4 all-terrain wheelchair base (the X4 Extreme model, manufactured by Innovation in Motion). It has four independent motors that drive each wheel separately and a passive steering system for the front wheels. The driving range is 19 mi and the maximum speed is 5.5 mph. It will be used as a dynamic landing platform for the helicopters.

In Fig. 12 it can be seen that most of the electronic components are those already designed for the helicopter. The ground vehicle can carry a heavy load, so there is no reason to optimize the system for low weight by using an ETXexpress module. Instead, a laptop computer is installed as in the case of the base station. The long-range radio transceiver, the inertial unit and the GPS-compass sensor module are still necessary for the control algorithm. The movement of the wheels and the steering direction are measured using custom-made incremental magnetic encoders. They are much more robust to vibrations, mechanical displacements and dust compared to the optical encoders and can achieve the same precision for this application. The motors are controlled by a high-power speed controller that is installed by default on the platform. The original control system used the CAN bus to communicate with the speed controller, so no changes were necessary to interface it to the RTM.

The design of the electronic system for the quad-rotor helicopter is different be-cause it does not have the possibility to carry a heavy payload. Therefore a powerful on-board computer can not be implemented. The control algorithm is executed on a ground base station. The on-board electronic system measures different system states, sends them to ground using a radio transceiver and, after the control algorithm processes the information, receives the commands that have to be given to the motors. The block diagram can be seen in Fig. 13.

The design of the helicopter electronic system was done using a highly modular-ized hierarchical electronic schematic. It was very easy to select the parts needed from different modules and put them together on a single PCB in order to create the Quad-rotor Real-Time Module (Q-RTM). The inertial sensors are less precise than in the case of the helicopter, but much easier to interface.

The radio transceiver is an XBee-PRO module manufactured by Digi Interna-tional. It operates on 2.4 GHz and has a RF data rate of 250 kbps. The range is about 1.6 mi. This model was preferred to the XTend RF module used on the helicopter due to the higher data rate and the smaller size, the range being of secondary importance. The hardware and software interfacing is practically the same. The ARM processor is again the same as on the helicopter RTM. This allows most of the software to be

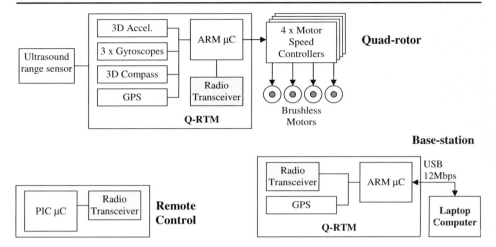

Fig. 13 The electronic system for the quad-rotor helicopter

shared on the two platforms. The remote control uses an identical electronic board as the helicopter, but instead of the XTend module it has an XBee-PRO module installed. This module is designed to communicate using the ZigBee protocol, but in this mode it can not achieve real-time performance. Instead, the lower-level 802.15.4 protocol (on which ZigBee is based) is used. To avoid latencies and possibly long dropouts in communication, the transmission mode is set not to wait for any acknowledge from the destination. The three devices present in the network use the same synchronization mechanism as in the case of the helicopter long-range transceiver to avoid collisions while transmitting.

The multi-vehicle control is realized using the 802.11n wireless network to transmit the necessary data between any two entities through UDP packets. Each helicopter, ground vehicle, their base-stations and the quad-rotor base-stations behave the same relative to this network, allowing for a unified framework where the vehicle type is not important.

6 Real-Time Control

Hard real-time systems consider the late delivery of correct results as a system failure. The control system of the UAV helicopter is such an example, where both the correctness of the results and their timely delivery must occur for the aircraft to operate safely. The hardware and the software must guarantee the deterministic execution of the control loop. The main operations that have to take place during one sample period are shown in Fig. 14. For slow systems where the sample frequency is low, the entire control process can be implemented on an easy to use operating system, like Microsoft Windows or Linux. For high sample frequencies, a real-time operating system is mandatory for deterministic execution. Usual choices are QNX Neutrino or VxWorks, the latter being used on the Spirit and Opportunity Mars Exploration Rovers. They provide very good support for real-time applications and enhanced reliability compared to the mainstream operating systems. A shortcoming

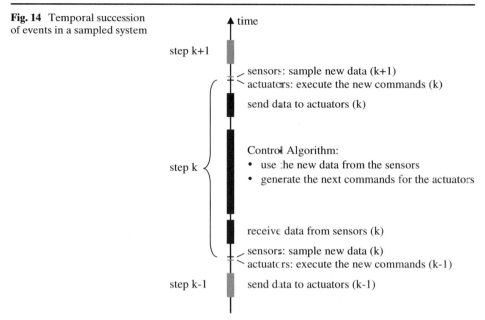

Fig. 14 Temporal succession of events in a sampled system

is the lack of support for hardware. Although they come with drivers for an entire range of peripherals, it is difficult to extend the functionality to devices that are not already supported.

Because of the ease of use and the broad support for hardware, real-time capabilities were added to Windows and Linux. The main approach was to add a second real-time kernel that can preempt the main kernel. This allows Windows or Linux to run as a task in the new real-time kernel. The determinism is very good, but the initial problem related to the hardware support is only partially solved. A real-time task is not allowed to make Windows or Linux system calls. In order to make use of the hardware support, like network devices, USB peripherals and storage devices, the real-time task has to send the request to a Windows or Linux task that in turn forwards it to the appropriate driver. Anything running in the Windows or Linux context is not deterministic. A control loop can be implemented in the real-time kernel, but it still can not make use of the Windows or Linux services in real-time. This might be fine with some applications that don't require deterministic performance for the network or USB communications, but is a major limiting factor for the current project.

In the last few years there were efforts to make the 2.6 Linux kernel real-time. Ingo Molnar (now an employee of Red Hat) provides a patch that improves the responsiveness of the kernel dramatically. In December 2007 Red Hat has released a beta version of Red Hat Enterprise MRG (Messaging Real-Time Grid) that includes the optimized kernel.

Each computer (on the vehicles and on the base stations) runs Red Hat Enterprise Linux MRG. The control is provided by a multi-threaded process that handles USB communications with the Real-Time module, network communications with the other base station(s) and/or the other vehicles and executes a compiled Simulink model with the control algorithm. On the base stations, a separate process with a lower priority is used to display the necessary information to the pilot and to accept high level commands.

The control algorithms are implemented in Simulink. Special blocks were created for the sensors and actuators. There are also blocks that represent an entire helicopter, a ground vehicle, a base station or a pilot remote control, such that formation control algorithms can be implemented very easily. The Simulink model is converted to C code using the Real-Time Workshop and then compiled into the real-time Linux application, which runs on the ETXexpress module or on the laptop computer. The application provides the Simulink blocks with all the necessary data obtained from the Real-Time Controller or from the wireless network.

7 Conclusion

Research topics in UAV control become more and more complex. The industry and the Army are reluctant to accept complicated theoretical solutions without a practical implementation to prove their applicability. This paper presented a hardware platform that can be used to implement aerobatic UAV control, formation flight and aerial–ground vehicle coordination in order to help validating the theoretical results in practice.

References

1. Gavrilets, V., Mettler, B., Feron, E.: Nonlinear model for a small-size aerobatic helicopter. In: Proceedings of the AIAA Guidance, Navigation and Control Conference. Montreal, Quebec, Canada (2001)
2. Castillo, C.L., Alvis, W., Castillo-Effen, M., Moreno, W., Valavanis, K.: Small unmanned helicopter simplified and decentralized optimization-based controller design for non-aggressive flights. Int. J. Syst. Sci. Appl. **1**, 303–315 (2006)
3. Vachtsevanos, G., Tang, L., Drozeski, G., Gutierrez, L.: From mission planning to flight control of unmanned aerial vehicles: strategies and implementation tools. Annu. Rev. Control **29**, 101–115 (2004)
4. Dong, M., Chen, B.M., Cai, G., Peng, K.: Development of a real-time onboard and ground station software system for a UAV helicopter. AIAA J. Aerospace Comput. Inform. Commun. **4**(8), 933–955 (2007)
5. Gavrilets, V., Mettler, B., Feron, E.: Dynamic model for a miniature aerobatic helicopter. MIT-LIDS report, no. LIDS-P-2580 (2003)
6. CAN in Automation Group. CANopen. Application layer and communication profile. Draft Standard 301 Revision 4.2 (2002)

A Development of Unmanned Helicopters for Industrial Applications

David Hyunchul Shim · Jae-Sup Han · Hong-Tae Yeo

Originally published in the Journal of Intelligent and Robotic Systems, Volume 54, Nos 1–3, 407–421.
© Springer Science + Business Media B.V. 2008

Abstract In contrast to the wide adoption of unmanned aerial vehicles (UAVs) in the military sector, a far less number of successful applications have been reported on civilian side due to high cost, complexities in operation, and conflicts with existing infrastructure, notably the access of civilian airspace. However, there exist a few cases where UAVs have been successfully adopted to delegate dull, dirty, dangerous tasks in civilian sectors. In this paper, we consider one of such cases, the agricultural application of UAVs and present our work to develop an industrial UAV platform. The developed system is capable of fully autonomous flights with a payload of 30 kg for more than an hour. We present the overview of our program as well as detailed description of the development processes. We believe, when combined with highly reliable hardware and adequate adaptation to the civilian demands, in a not too distant future more UAVs will successfully serve the needs of public.

Keywords Unmanned aerial vehicle · UAV design · Flight control system

D. H. Shim (✉)
Department of Aerospace Engineering, KAIST, Daejeon, South Korea
e-mail: hcshim@kaist.ac.kr
URL: http://fdcl.kaist.ac.kr/~hcshim

J.-S. Han · H.-T. Yeo
Oneseen Skytech, Co. Ltd., Busan, South Korea

J.-S. Han
e-mail: osst@oneseen.net

H.-T. Yeo
e-mail: htyeo@oneseen.net
URL: http://www.oneseen.com

K. P. Valavanis et al. (eds.), *Unmanned Aircraft Systems*. DOI: 10.1007/978-1-4020-9136-0_22   407

1 Introduction

Unmanned aerial vehicles (UAVs) have been proven as an effective and affordable solution that complements military operations traditionally performed by human pilots. Many missions that are unacceptably strenuous or dangerous have been successfully delegated by UAVs as witnessed in recent military campaigns. In civilian sectors, UAVs can be used for many applications such as law enforcement, traffic report, disaster monitoring, aerial photography and more. However, civilian UAVs are yet to see its prime time due to several reasons. First, they are still too expensive to be justified in many civilian applications. Safe and sane operation of UAVs requires well-trained crews who are familiar with all aspects of the system, including general principles of flight dynamics, electronics, communication, navigation systems and more. Social and legal barriers such as airspace conflict must be resolved before fully embracing the concept of civilian UAVs. Rigorous certification process for airworthiness should be enacted, but such processes are still in preparation as of this writing.

There exist a few applications where UAVs have been proven to have substantial advantages over conventional approaches in civilian sectors. For example, in many parts of Asia including Korea, the rice farming constitutes the largest portion of a country's agriculture and it is very important to improve the productivity by proper fertilization and pest control. Whereas full-size airplanes are used for large-scale farming in the US or some other countries, many Asian countries have smaller and highly fragmented rice patties. In Japan, since 1980s, Yamaha Corporation, renowned for their lineup of motor bicycles and recreational vehicles, has pioneered the concept of unmanned helicopters for agricultural applications. Yamaha has developed a lineup of highly successful remotely controlled helicopters, R-50 and RMAX (and its variants). These were initially scaled-up radio-controlled helicopters with spraying device attached underneath the fuselage, flown by the pilot on the ground. One of the key advances Yamaha has contributed is the Yamaha Attitude Control System, which helps to improve the handling quality of inherently unstable helicopter platforms. Originally being an add-on option for now phased-out R-50 and standard equipment for RMAX, this device performs attitude stabilization using high-quality accelerometers and rate gyros. The most recent version, RMAX Type IIG, is capable of autonomous navigation aided by the onboard GPS and heading sensors [1]. Yamaha helicopters are now used by many farmers with government's subsidiaries in Japan and have been adopted as a research platform by many agencies in the US and worldwide until the export was banned in 2007 presumably to protect the sensitive technology from leaking to foreign countries.

As seen from the example of Yamaha helicopters, however, UAVs are a viable option for civilian applications if the system is highly reliable, easy to operate, and performs superior to conventional method so that the extra costs and efforts can be justified. Motivated by such examples, the authors have formed a joint research to develop a fully autonomous industrial helicopter platform that can carry 30 kg payload for an hour. A range of autonomy will be provided varying from simple stability augmentation to a fully automated flight sequence from take-off to landing. The developed system is anticipated to meet not only the agricultural needs but also many other applications, both civilian and military (Fig. 1).

Fig. 1 Oneseen Skytech's industrial helicopter X-Copter1 (*left*) and its test flight (*right*)

2 System Description

2.1 System Requirements

Our goal is to create an unmanned helicopter platform with a sufficient payload-lifting capability and operation time so that it can perform a variety of aerial missions in a highly autonomous manner. Coincidently, in spring 2007, the Korean government announced a funding opportunity for the development of an unmanned agricultural helicopter system that meets the following requirements:

– An unmanned helicopter system that has 80 + kg maximum take-off weight (MTOW), payload of 30 kg or more, and an operation time of an hour or more.
– Development of onboard flight control system and remote ground control station

- Onboard flight control system for attitude control
- Fully automatic crop dusting sequence
- Design and fabrication of crop dusting device
- Construction of a ground control station
- Emergency landing system when loss of communication occurs

In order to improve the productivity, a rice paddy has to be sprayed several times a year with fertilizers and pesticides. When the spraying is done manually using high-pressure pumps and nozzles as commonly done in Korea, the chemical is most likely spread unevenly and excessively, causing environmental issues. When the chemical is inhaled by the operators, he or she may suffer lethal poisoning. Therefore, it has been suggested to spray them from an airplane, but not full-size aircraft as done in countries with larger farming areas, but smaller robotic helicopters. Aerial spraying can cover a much wider area easily and quickly than the conventional method and the amount can be precisely controlled to match the flight speed. The downwash of the rotor is also known to help the substances to distribute over a wider area and reach deeper parts of the plant.

The conceptual design process of civilian UAVs is quite different from that of military UAVs, which should be robust in a harsh environment. Military UAVs have justifications to employ advanced (and expensive) technologies such as satellite communication, military-grade IMUs and GPS and more. In contrast, commercial UAVs have to meet various constraints from the cost structure, public safety and operator training. The overall cost to acquire, operate, and maintain a commercial

UAV should be kept as low as possible so that average people, farmers in this case, can afford them. As a product to make profit in the emerging yet highly competitive civilian market, it is very important to select the right components, which would fit into the overall cost structure while delivering the right performance. The civilian UAVs also require a very high level of reliability for the sake of public safety. Unlike the military applications where certain level of mishaps may be accepted in the name of national defense, damages to civilians and properties by commercial UAVs can cause huge complications. The vehicle should be also simple and straightforward enough to be operated by average people without in-depth knowledge on the aircraft, navigation, control, etc.

2.2 Initial Sizing

In the conceptual design process, we begin with the initial sizing of the important geometries and engine ratings. We chose the conventional main-tail rotor configuration (as opposed to coaxial or tandem configurations) since it is relatively simple and well-understood. For initial sizing, we referred to the work in [2], which suggests empirical rules for main/tail rotor size and RPM, tail rotor lever arm, fuselage size, engine power and more. With MTOW of 100 kg, according to the rules, the main rotor diameter should be about 4 m, as shown in the following table (Table 1).

As the suggested value of 4 m for main rotor was deemed excessively large for transportation and safe operation in confined rice patties, it was reduced to 3.1 m and instead the chord length was increased to compensate for the lift loss.

Table 1 Sizing of main rotor diameter and chord length

The "square-cube" scaling law: $D \propto W_0^{1/3}$

$D = 0.980\, W_0^{0.308}$, where D is in [m] and W_0 is in [kg] ($\varepsilon^{AVER}=7\%$, $\varepsilon^{MAX}=30\%$, $R = .9606$)

W_0 [kg]	D [m]
100	4.048

$D = 9.133\, W_0^{0.380} / V_m^{0.515}$, V_m is in [km/hr] S/L ($\varepsilon^{AVER}=6\%$, $\varepsilon^{MAX}=21\%$, $R = .9744$).

W_0 [kg]	V_m [km/hr]	D [m]
100	120	4.465
100	238	3.138

$c = .0108\, W_0^{0.540} / N_b^{0.714}$, where c is in [m] and W_0 is in [kg] ($\varepsilon^{AVER}=10\%$, $\varepsilon^{MAX}=41\%$, $R = .9535$).

Wo [kg]	nb	c [m]
100	2	0.0792

Chart (left top): Main Rotor Diameter (m) vs Main Rotor Diameter Estimation (m); legend: $D = f(W_0)$, $D = f(W_0, V_m)$, Estimation.

Chart (left bottom): Main Rotor Blade Chord (m) vs Main Rotor Blade Chord Estimation (m); legend: Database configurations, Estimation.

Table 2 Sizing of engine using MTOW (=100 kg)

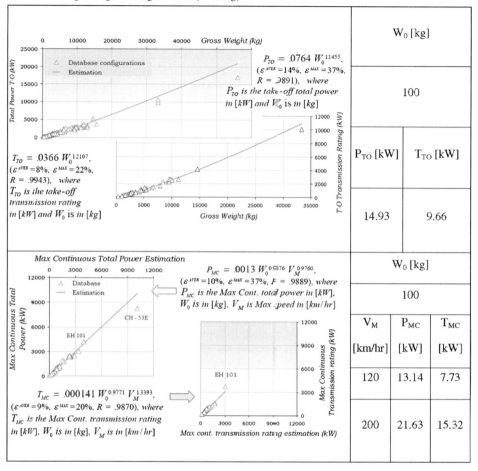

Another important design parameter is the engine size. According to the guideline summarized in Table 2, the take-off power should be about 15 kW and the maximum continuous power is to be no more than 20 kW in the extreme flight condition 120 km/h. Therefore, in order to meet the requirement of the max power of 15–20 kW (20–27 horsepower) and the given total weight limit, a lightweight gasoline engine is the natural choice for their power rating, operational time and cost. We proceed with the rest of important sizing factors based on the guideline and the result is summarized in Table 3 (Fig. 2).

3 System Design

A development of a helicopter UAV from scratch requires the in-depth study in the following fields:

Structure: frame, landing gear, tail boom
Powerplant: engine, transmission, fuel tank/pump

Table 3 Specification of X-Copter 1 (Fig. 2)

Model	Oneseen Skytech X-Copter1
Dimensions	Width: 0.74 m
	Length: 2.680m/3.645m (main rotor included)
	Height:1.18m
	Main rotor diameter: 3.135 m
	Tail rotor diameter: 0.544 m
Weight	Empty weight: 83 kg
	Payload: 30kg
Engine	Rotary engine, 4-cycle
	Cylinder volume: 294 cc
	Power: 38HP/49HP Max
	Water-cooled with radiator
	Fuel: gasoline:lubricant mixture (50:1 ratio)
	Fuel tank volume : 10 liters
	Electric starter
	Onboard generator (12V, 8A)
Flying time	60~90 minutes depending on the load and flight conditions
Avionics	Navigation: DGPS-aided INS
	GPS: NovAtel OEM V-2
	IMU: Inertial Science ISIS IMU
	Flight Computer: PC104 Pentium III 400MHz
	Communication: Microhard radio modem (900 MHz)
	Onboard status indicator light and control panel
Autonomy	Attitude stabilization
	Waypoint navigation with automatic take-off and landing
	Airspeed-sensitive spraying
	Automatic collision-free path generation and tracking

Aerodynamic components:	main and tail rotors, blades, fuselage,
Control mechanism:	swash plate, control linkages, servo actuators, stabilizer linkages
Avionics:	flight control, sensors, control panel, generators, status lights
Payload device:	spraying device, tanks

The entire vehicle is built around the engine, transmission, and main rotor mast assembly, to which the landing gear, the avionics enclosure, and the tail boom unit are attached. For weight reduction, lightweight aluminum alloy and carbon-fiber composites are used throughout the vehicle. Three-dimensional design software SolidWorks™ and CosmosWorks™ are used throughout the design and manufacturing stage for faster and efficient prototyping. The system is designed to allow easy disassembly for transportation. The major components are readily accessed by opening the cowls for easy maintenance (Fig. 3).

For main rotor components that should be resilient against high stress loading and fatigue, not forging process is favored over CNC machining to achieve a significantly higher level of strength at a less weight and a shorter manufacturing time. The transmission case is built with a sand aluminum casting process aided by 3D modeling and rapid prototyping process. The transmission is made of cast aluminum

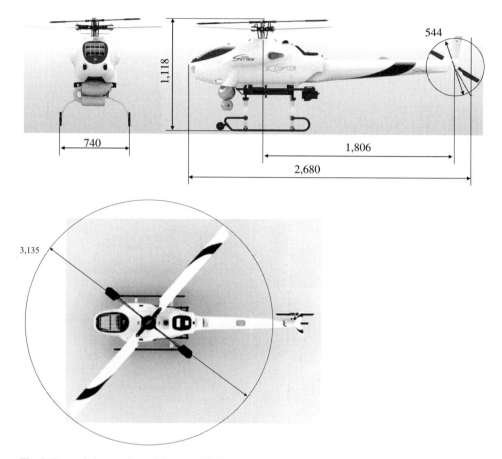

Fig. 2 Front, left, top view of Oneseen X-Copter 1 with key dimensions

AC4C, which is heat treated to enhance the strength. Some other prototype parts such as the stabilizer paddles are built by silicon molding and rapid prototyping. The landing gear is built by fiber reinforced plastic for light and strong structure (Figs. 4 and 5).

Based on the sizing results presented in Section 2.2, a number of commercially available engines are evaluated for our platform. Initially, we chose Rotax FR125Max engine, and another type of engine is currently evaluated to achieve a higher level of payload. The gear ratio for the main rotor is determined by the engine's torque/power characteristics and the recommended main/tail rotor speed found in Section 2.2.

The main rotor is chosen to have a Bell-Hiller type stabilizer, which delays the flapping response time, increases the damping of roll and pitch responses, and lessens the load of the servo actuators for swash plate [3]. For larger helicopters, it may be acceptable not to use a stabilizer mechanism especially when onboard flight

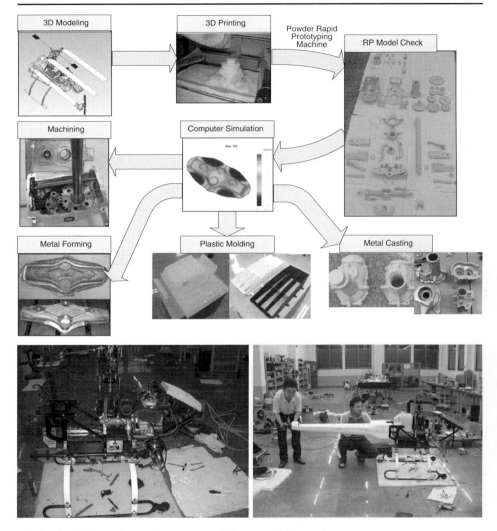

Fig. 3 The design and assembly processes of Oneseen X-Copter 1

control is available to improve the flight characteristics in a more flexible manner than the preconfigured mechanical link. Therefore, we are considering a design without a stabilizer to lower the overall cost and complexity while restoring the maneuverability in near future.

In order to meet the autonomy required, the vehicle is equipped with an onboard navigation and control system. Here, the component selection process for flight control system requires a trade-off study between the performance and the cost. Unlike fixed-wing airplanes, helicopters need accurate velocity feedback for stabilization, which requires a precision navigation solution using a pair of high-accuracy IMU and GPS. Some of those sensors are not only prohibitively expensive (>US$10,000) but also restricted from civilian access. With the accuracy level typically achieved by non-military grade IMUs, mating with a differential GPS is highly desired for

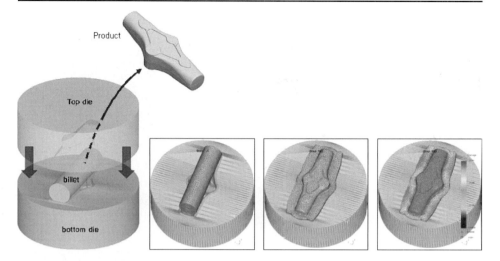

Fig. 4 Hot forging process for main rotor head

improving the overall navigational accuracy. However, DGPS are also expensive, complicated to set up and operate, and vulnerable to the loss of communication link or the degradation of GPS signal quality. Well aware of such trade-offs, we begin with a differential GPS and a high-end commercial IMU for the prototyping, but they will be eventually replaced with lower-cost IMUs and GPS.

The CPU for the flight controller is initially based on PC104, which offers a great flexibility and efficiency needed in the development stage. The Intel CPUs on PC104 are much faster than most embedded processors and there are many hardware and operating systems to run on this machine. However, as PC104 is somewhat off from the desirable price range especially for volume production, we plan to replace it with lower-cost embedded processors such as StrongARM processors. In this research, we use QNX real-time operating system for their real-time capability and efficient development environment as proven in many previous research programs [4, 5]. In Fig. 6, the hardware and the software architectures are presented.

For controller design, we follow the author's previous work in [3] and [4]. Helicopters are inherently unstable in near-hover condition and only stabilized with proper horizontal linear velocity feedback. The attitude (angle and rate) feedback damps the attitude response further without stabilizing the entire vehicle dynamics, and it gives the operator a sensation that the helicopter is less sensitive to the disturbance and more sluggish to stick commands. The attitude estimation can be done just with three orthogonally positioned rate gyros if they have relatively small drift over average mission time. In the fully autonomous crop dusting mode, the vehicle should be guided along a conflict-free path while maintaining a constant amount of spraying. Automatic flight needs precision navigation sensors consisting of three orthogonal accelerometers and a quality GPS in addition to the rate gyros. Using GPS for civilian helicopter UAVs is a tricky choice as any sudden degradation of GPS solution due to signal occlusion or jamming may lead to unexpected motion of the helicopter. However, without any immediate alternative to GPS-aided INS correction, it is considered as an inevitable choice for the time being.

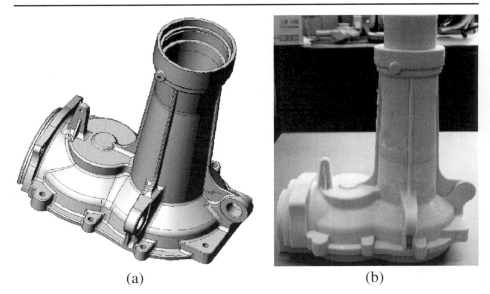

(a)   (b)

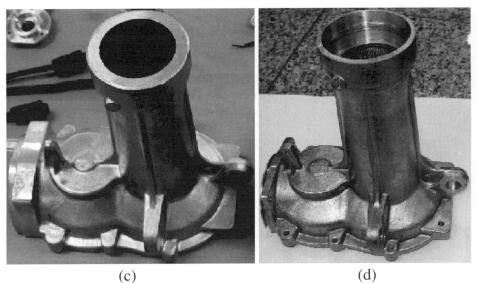

(c)   (d)

Fig. 5 Aluminum casting process for a transmission case. **a** Computer model. **b** Rapid prototyped insert. **c** Aluminum cast without finish. **d** Final product

For safe and reliable operation, the system should have self-diagnostic function and a clear user interface to inform of the current status. X-Copter 1 has numerous sensors embedded to many safety-critical components such as engine temperature, fuel level, battery voltage, etc and the flight computer gives appropriate visual signals according to the current situation. The green light gives the visual assurance that everything is okay while the yellow and the red lights indicates there are degradation which requires immediate attention or emergency procedure, respectively (Fig. 7).

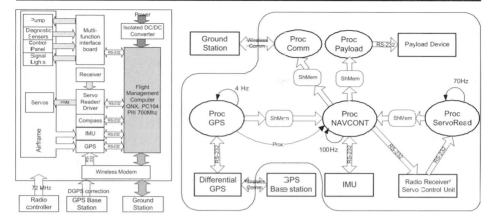

Fig. 6 The hardware and software architectures of X-Copter 1

For agricultural application, it is rather inconvenient to operate a dedicated ground station in the field of activity although running one would greatly help the operators to understand the current status of the vehicle. The entry model with attitude stabilization only may be fully operational without any special ground station. For higher-end models, a PDA or a notebook-based ground station are provided to specify the waypoints or crop-dusting areas. Snapshots of proposed ground stations are shown in Fig. 8.

For agricultural missions, an add-on pump system is designed and built as shown below. The pump can be turned on and off by the flight computer or directly by the operator remotely. The spraying amount can be regulated to match the vehicle's speed so that field can be evenly sprayed (Fig. 9).

4 Flight Tests

As pointed out, the reliability is one of the key concerns for civilian UAVs and therefore the prototypes are put into a series of rigorous tests before release. After basic function check upon finishing the assembly, the vehicle is put into a rig test where the vehicle is tied down to a counterbalanced teetering platform that offers

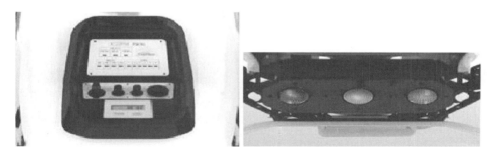

Fig. 7 The onboard control panel (*left*), the tail status light (*right*)

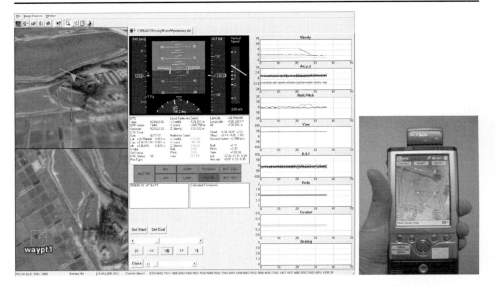

Fig. 8 Screenshots of the full-screen version (*left*) and a PDA version (*right*) ground stations

three degrees of freedom: rotation about the center, up-down motion, and the rotation about the platform attached at the end of teetering arm (Fig. 10). For safety, the test area is fenced around with heavy-duty steel frame and wire mesh to keep any flying objects from hitting operators. The rig test is a very important step to check various parts of the system in terms of function, endurance, and overall vibration level at a minimal risk before the free flight.

After a thorough validation on the rig, the vehicle is taken out to a field for free-flight test to check the vehicle's overall status in full detail. As shown in Fig. 10, the vehicle showed a satisfactory performance in terms of flight characteristics and handling.

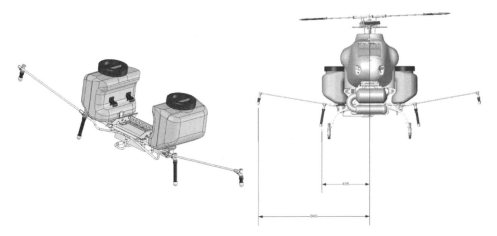

Fig. 9 The pump assembly (*left*) and the dimensions of pump system (*right*)

Fig. 10 Rig test (*left*) and field test (*right*)

In parallel with the airframe development, the proposed flight control system is developed and validated using a smaller, more manageable platform as a surrogate. By this approach, the navigation sensor, hardware, software and control algorithms can be validated without risking the expensive prototype. In Fig. 11, the surrogate vehicle based on a commercially available electric helicopter is shown. The flight control system is based on multi-loop PID system, which is a proven industrial approach. The waypoint navigation algorithm [3] is implemented on top of the control system in the hierarchical system. Using the proposed navigation and control system, the surrogate vehicle is put to many test flights including a long and a relatively high-altitude flight, which is a sequence of five waypoints that is 2,300 m long at 180 m above ground level. The test results from this flight are given in Fig. 12.

With the successful test of the airframe and the flight control system, we are currently working towards the full integration of these two major components. The vehicle will be put to a series of test flights to evaluate all aspects of the system in terms of performance, reliability and ease of operation.

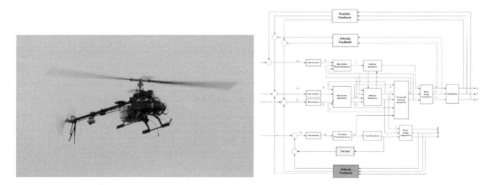

Fig. 11 The surrogate vehicle in automatic flight (*left*) and the control system architecture (*right*)

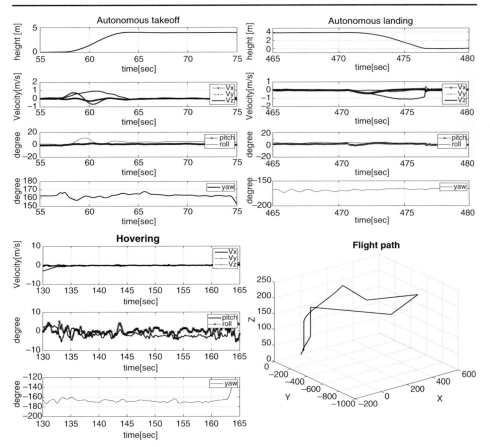

Fig. 12 Test flight results of the surrogate vehicle flying a waypoint sequence 2,300 m long and 180 m high AGL

5 Conclusion

In this paper, we have introduced the design goal, initial sizing, and the detailed descriptions on the various parts of our industrial helicopter design. With a partial support from Korean Government, the authors are currently working as a team to develop unmanned helicopters that offer 30 kg in payload and the flight time about an hour or more for agricultural applications. The helicopter platform has passed the rig test and field flight test so far and will be put into more rigorous tests in the coming months.

Acknowledgements The authors gratefully acknowledge for financial support by Korea Ministry of Knowledge and Economy.

References

1. Yamaha Corp.: Yamaha Corp. Website, https://www.yamaha-motor.co.jp/global/industrial/sky/history/rmax-iig (2005)
2. Omri, R., Vladimir, K.: Helicopter Sizing by Statistics. AHS Forum 58, June (2002)
3. Shim, D.H.: Hierarchical control system synthesis for rotorcraft-based unmanned aerial vehicles. Ph. D. thesis, University of California, Berkeley (2000)
4. Shim, D.H., Chung, H., Sastry, S.: Autonomous exploration in unknown urban environments for unmanned aerial vehicles. IEEE Robot. Autom. Mag. 13, 27–33, September (2006)
5. Shim, D.H., Kim, H.J.: A flight control system for aerial robots: algorithms and experiments. IFAC Control Engineering Practice, October (2003)

The Implementation of an Autonomous Helicopter Testbed

R. D. Garcia · K. P. Valavanis

Originally published in the Journal of Intelligent and Robotic Systems, Volume 54, Nos 1–3, 423–454.
© Springer Science + Business Media B.V. 2008

Abstract Miniature Unmanned Aerial Vehicles (UAVs) are currently being re-searched for a wide range of tasks, including search and rescue, surveillance, reconnaissance, traffic monitoring, fire detection, pipe and electrical line inspec-tion, and border patrol to name only a few of the application domains. Although small/miniature UAVs, including both Vertical Takeoff and Landing (VTOL) vehi-cles and small helicopters, have shown great potential in both civilian and military domains, including research and development, integration, prototyping, and field testing, these unmanned systems/vehicles are limited to only a handful of labora-tories. This lack of development is due to both the extensive time and cost required to design, integrate and test a fully operational prototype as well as the shortcomings of published materials to fully describe how to design and build a "complete" and "operational" prototype system. This work attempts to overcome existing barriers and limitations by detailing the technical aspects of a small UAV helicopter designed specifically as a testbed vehicle. This design aims to provide a general framework that will not only allow researchers the ability to supplement the system with new technologies but will also allow researchers to add innovation to the vehicle itself.

Keywords Mobile robots · Helicopter control · Helicopter reliability ·
Modular computer systems

R. D. Garcia (✉)
Army Research Laboratory, Aberdeen Proving Ground, Aberdeen, MD 21005, USA
e-mail: duster3@gmail.com

K. P. Valavanis
Department of Electrical and Computer Engineering,
School of Engineering and Computer Science,
University of Denver, Denver, CO 80208, USA
e-mail: kvalavan@du.edu, kimon.valavanis@du.edu

1 Introduction

The field of robotics, for the past few decades, has not only allowed humans to expand their abilities, it has allowed them to surpass them. This is possible through the shift from teleoperated robotics to fully autonomous robotics which attempts to remove or reduce the need for the human component. The ability of an autonomous system to surpass the capabilities of a human operator are obvious when one considers the limitations of the human body in locomotive speed and decision making as compared to the operating speed of the newest computer systems and the speed and precision of direct current motors.

More recently, humans have seen the capabilities of robotics expand to the flight control of both commercial and military vehicles. This shift has even begun to include highly agile and low cost Miniature Unmanned Aerial Vehicles (MUAVs). These MUAVs make it feasible for low budget research and testing of control, fusion and vision algorithms, and system integration.

There is currently great interest in the area of UAV research including search and rescue [1–4], surveillance [5–10], traffic monitoring [11–14], fire detection [15–18], pipe and electrical line inspection [19, 20], border patrol [21, 22], and failure tolerance [23] to name only a few. Although UAVs have shown great potential in many of these areas their development as a whole has been somewhat slow. This can be attributed to the difficulty in purchasing, replicating, or developing an autonomous vehicle to validate new ideas.

To date there currently exist several development oriented, RC based, autonomous helicopters mainly owned and operated by academic societies or government sponsored organizations and most notably include the Massachusetts Institute of Technology (MIT), Carnegie Mellon University (CMU), Stanford, Georgia Tech, University of Southern California, Berkeley, and CSIRO (see Table 1). Although these vehicles are development oriented they are generally designed for in-house development and do not supply sufficient information to fully replicate their vehicles, control systems, and algorithms.

There also currently exist several commercial vendors of MUAV helicopters. These most notably include Rotomotion and Neural Robotics Inc. (NRI). Although both of these companies produced vehicles that have shown great abilities to maneuver autonomously, all of their vehicles operate on strictly proprietary hardware and software. The inability to manipulate or change either the software or hardware severely limits the use of these vehicles as development machines.

Although prior and ongoing research has shown the enormous benefits and interest in the development of MUAVs, the migration of these ideas from paper to realization has been greatly limited. To realize the full benefit of these vehicles and to alleviate the gap between innovative ideas and innovative technologies there must be a continuing medium for development and testing.

2 Platform and Hardware

Hardware is the building block of all unmanned vehicles and a great source of difficulty when designing a testbed vehicle. Decisions made about hardware can significantly decrease or increase the complexity and functionality of an unmanned

Table 1 Laboratories with development UAV helicopters

University	Hardware and sensors	Software

Massachusetts Institute of Technology [24–26]	X-Cell 60 Helicopter
	ISIS-IMU (100 Hz &
	  0.02 deg/min drift)
	Honeywell HPB200A Altimeter
	  (2 ft accuracy)
	Superstar GPS (1 Hz)
		QNX Operating System
		13-state extended Kalman
		  filter (state estimation)
		LQR based control

Carnegie Mellon University [26–29]	Yamaha R-Max Helicopter
	Litton LN-200 IMU (400 Hz)
	Novatel RT-2 DGPS
	  (2 cm accuracy)
	KVH-100 flux-gate compass
	  (5 Hz)
	Yamaha laser altimeter
		VxWorks Operating System
		13-state extended Kalman
		  filter (state estimation)
		Control based on PD and
		  H_∞ control

Stanford University [30, 31]	XCell Tempest 91 Helicopter
	Microstrain 3DM-GX1 (100 Hz)
	Novatel RT-2 DGPS
	  (2 cm accuracy)
	DragonFly2 cameras
	  (position est.)
		Undisclosed Operating System
		12-state extended Kalman
		  filter (state estimation)
		Differential Dynamic Programming
		Extension of Linear
		  Quadratic Regulator

Georgia Institute of Technology [32–34]	Yamaha R-50 Helicopter
	ISIS-IMU (100 Hz &
	  0.02 deg/min drift)
	Novatel RT-2 DGPS with
	  2 cm accuracy
	Radar and Sonar Altimeters
	HMR-2300 triaxial
	  magnetometers
		OS: QNX, VxWorks, Linux
		Real-time CORBA
		Object Request Broker Arch.
		17-state extended Kalman filter
		Neural networks control
		  (Feedback linearization)

University of California Berkeley [35, 36]	Yamaha R-Max & Maxi Joker
	Boeing DQI-NP INS/GPS system
	Novatel Millen RT-2 DGPS
	  (2 cm accuracy)
		VxWorks Operating System
		No state estimation
		  (provided by sensor)
		Reinforcement Learning control

University of Southern California [37, 38]	Bergen twin Industrial Helicopter
	ISIS IMU (100 Hz &
	  0.02 deg/min drift)
	Novatel RT-2 DGPS
	  (2 cm accuracy)
	TCM2–50 triaxial magnetometer
	Laser altimeter (10 cm
	  accuracy, 10 Hz)
		Linux Operating System
		16-state Kalman filter
		  (state estimation)
		Decoupled PID based control

CSIRO [26, 39]	X-Cell 60 Helicopter
	Custom embedded IMU
	  with compass (76 Hz)
	Ublox GPS with WAAS
	  (2 m accuracy)
	Stereo vision for
	  height estimation
		LynxOS Operating System
		Velocity estimation using vision
		2×7-state extended
		  Kalman filters
		Complimentary filters
		PID based control

JPL [40, 41]	Bergen Industrial Helicopter NovAtel OEM4 DGPS ISIS IMU (2 cm accuracy) MDL ILM200A laser altimeter TCM2 compass	QNX real-time OS Behavior based and H_∞ control Extended Kalman filter (state estimation) Image-based motion estimates

system. For these reasons great effort was taken in selection all hardware, interconnections, and mounts utilized on the autonomous helicopter testbed. Note that detailed information about individual hardware components and justifications as well as detail schematics are available in [42].

2.1 Platform

The platform chosen for the autonomous helicopter testbed is the electric Maxi-Joker II. The Maxi-Joker II helicopter has the following characteristics:

- Manufacturer: Joker
- Main Rotor Diameter: 1.8 meter
- Dry Weight: 4.22 kg (w/o batteries)
- Dimensions: 56 × 10.25 × 16.5 in (w/o Blades)
- Payload Capacity: 4.5 kg (after batteries)
- Endurance: 15–20 min
- Motor Battery: 37 V (10 A) Lithium Polymer
- Servo Battery: 4.8 V (2.0 A) NiMh
- Engine: Plettenberg HP 370/40/A2 Heli
- Speed Controller: Schulze future 40/160H

This platform was chosen due to its cost, approximately $3000 USD ready-to-fly, desire to avoid carrying and storing explosive fuel, reduced vibrations, relatively small size, and ability to handle wind gust exceeding 20 mph.

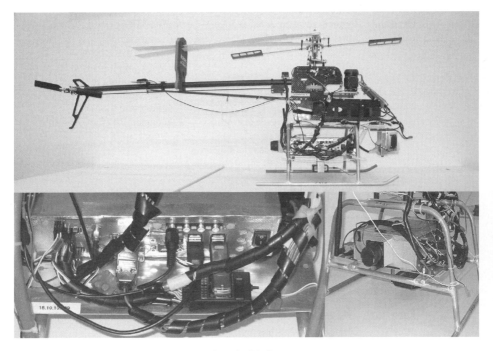

Fig. 1 Mosaic of the completely assembled testbed helicopter

Note that no modifications were made to the Maxi Joker II kit or any of the above mentioned equipment. The kit, motor, speed controller as well as all support equipment were assembled and setup as instructed in the manufacturer supplied manuals.

2.2 Hardware

The main hardware components of the UAV system consist of:

- Pentium M 755 2.0 GHz Processor
- G5M100-N mini-ITX motherboard
- Microstrain 3DMG-X1 IMU
- 2 Gigs of Crucial 333 MHz RAM

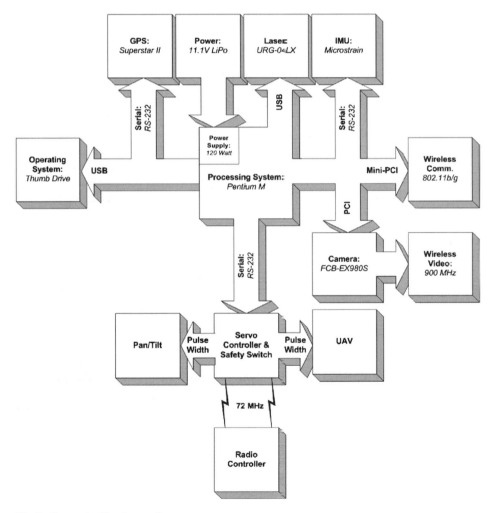

Fig. 2 Conceptual hardware diagram

- Novatel Superstar 2 GPS receiver (5 Hz, 5 V model)
- Microbotics Servo/Safety Controller (SSC)
- Thunderpower 11.1 V 4.2Ah LiPo Battery
- Intel Pro 2200 802.11B/G Mini-PCI wireless card
- URG-04LX Hokuyo laser range finder
- IVC-200G 4channel frame grabber
- 120 Watt picoPSU-120 Power supply
- Sony FCB-EX980S module camera

Figure 1 depicts the equipped Joker Maxi 2 with the testbed equipment.

This hardware configuration is used because of its high computational capabilities, various Input/Output (I/O) ports, small size, low heat emission, and cost. Figure 2 depicts the overall concept for the testbed.

3 Software

Although almost any modern Operating System (OS) could be utilized on this testbed, the helicopter has been strictly operated by Linux distributions. Linux was chosen due to stability, ease of modification, and heavy use in both the academic and industrial arenas. The utilized distributions have included multiple versions of Slackware, versions 10.0 through 12.0, and Gentoo. The currently utilized distribution is Slackware 12.0. Slackware was chosen due to ease of installation and ability to boot from a USB thumbdrive. Also note that for the testbed the OS operates entirely out of a RAM drive. Due to the physical limitations of RAM, this requires that the OS be of minimal size to allow for sufficient RAM for operating processes.

From conception the software for the testbed was designed to be highly modular and to support an operating structure that could be dynamically modified. Modularity in the design allows for code reuse, quick integration, and ease of understanding. Dynamic modification simply means that the system is able to remove or add functionality during operation. To support these design requirements, software is developed as a set of processes that run concurrently and pass information through shared memory structures. By utilizing this method of development a monitoring process can start and stop processes as needed by the testbed.

3.1 Servo Cyclic and Collective Pitch Mixing

The Maxi Joker II platform utilizes a three point swashplate and controls the roll, pitch, and collective through Servo Cyclic and Collective Pitch Mixing (CCPM), referred to as servo mixing. Servo mixing provides a mechanically simpler design without sacrificing accuracy. The testbed converts traditional roll, pitch, and collective commands, ranging from $[-1,1]$, into Pulse Width (PW) values which are then converted to CCPM and passed to the vehicle's servos. Specific roll, pitch, and collective commands correspond to a percentage of distance from neutral PW to the maximum or minimum PW. Equation 1 represents the calculation used to determine the PW value corresponding to a command. For Eq. 1, Max_{P_i} is the maximum PW value for servo 'i', Min_{P_i} is its minimum PW value, N_{P_i} is its neutral PW value, O_{P_i}

is the calculated PW for servo 'i', and α is the control command with a value ranging from $[-1,1]$.

$$O_{P_i} = \begin{cases} (\text{Max}_{P_i} - N_{P_i}) * \alpha & \text{for } \alpha \geq 0 \\ (N_{P_i} - \text{Min}_{P_i}) * \alpha & \text{for } \alpha < 0 \end{cases} \tag{1}$$

CCPM mixing first mixes the lateral, or roll, command. This is done because roll commands do not require modification to the front servo's position. The algorithm first determines the direction of servo movement from neutral by checking to see if the command is negative or positive. Once this is determined the PW deviation is calculated using Eq. 1 for the left servo. This value is then added to both the right and left servo neutral values. This will lower the left servo the exact same amount that it will raise the right servo hence preserving the current collective value. Note that the deviation is added to both servos because the right servo is mounted inversely to the left and thus an addition to both will lower one and raise the other.

Next the CCPM mixing algorithm mixes longitudinal, or pitch, commands. This is also done using Eq. 1 for the front servo. Once the deviation is calculated, its value is added to the front servo's neutral position. Due to the configuration of the swashplate there is a 2:1 ratio for front to left and right servo commands. This is due to an unequal ratio of distance between the front control arm and the two side control arms. Thus, a single PW change in the front servo only corresponds to a 0.5 PW changes in the left and right servos. Keeping this in mind, the deviation calculated for the front servo is then divided by two and added to the left servo and subtracted from the right servo. Note that alterations to both the left and right servo values are changes made to the values already calculated in the lateral, or roll, part of the mixing algorithm.

The last command that must be mixed is the collective command. Collective's min, neutral, and max values are not directly connected to a particular servo. These values describe the amount of change in the right, left, and front servos that are allowed for by the collective command and for the testbed ranges from 0 to 160 μs. Mixing first calculates the PW deviation using Eq. 1. This value is then added to the negative of the collective's neutral. This value represents the PW change for the collective input and is added to the front and right servo's value and subtracted from the left servo. It should be mentioned that the Maxi Joker-2 has a "leading edge" rotor. Thus, the swashplate must be lowered to increase the collective. This is why the PW deviation calculated for the collective is added to the negative value of the collective's neutral.

Although PW commands for yaw are not calculated using CCPM it is noteworthy to mention that it is calculated using Eq. 1. As yaw commands correspond to a single servo, the value calculated in Eq. 1 can be added directly to the yaw's neutral value to calculate the desired PW.

3.2 Positional Error Calculations

The calculations described in this section represent the algorithms used to calculate the positional error, or offset, of the testbed. Specifically, position error represents the distance, in feet, from the current position to a desired position.

Positional error is typically calculated by determining the distance between the current GPS coordinate and a desired position on a calculated flight path. Flight paths are represented by straight lines between latitude and longitude waypoints.

The desired position on the flight path is determined by finding intersections between a fixed circle around the vehicle and the current flight path. To determine these intersections the software must first calculate the distance from the vehicle to both the previous and current waypoints.

The testbed utilizes the Earth model for calculating lateral, E_X, and longitudinal, E_Y, distances, in feet, between GPS coordinates, which is well defined in [43]. Note that since only lateral and longitudinal distances are calculated using the equations in [43], the altitude variable is always held constant using the current GPS altitude. Given the short range and low altitudes of flights this will not adversely affect distance calculations.

Once E_X and E_Y are calculated and converted to feet, the software can now determine the vehicle's relative position to its flight path. This can be done by determining if a secant or tangent line exists. This equation is well defined in [44] and [45] and will not be repeated here. If intersections exist, the testbed will then determine which intersection is closest to its next waypoint, using direct line distance, and then use that point as an intermediary goal. The x and y distances are then calculated to this intermediary goal, also defined in [44] and [45]. If the intersection does not exist, then the vehicle is too far off course to use intermediary goals. In this case, positional x and y error will be calculated by determining the direct line distance from the testbed to the next waypoint.

The last step in determining the positional error is to transform the x and y distances from the world coordinate frame to the local coordinate frame. The local coordinate frame utilizes the heading of the vehicle as the Y axis. These rotated values are provided to the controllers as error in the pitching direction and error in the rolling direction. Figure 3 depicts an example of determining positional error.

It should be noted that altitude, or Z axis, positional error is strictly calculated by determining the difference, in feet, between the altitude set point and GPS supplied altitude. This value is supplied to the collective controller as collective positional error.

Heading error is the deviation of the current heading from the desired heading. Calculating heading error is done by determining the shortest distance from the

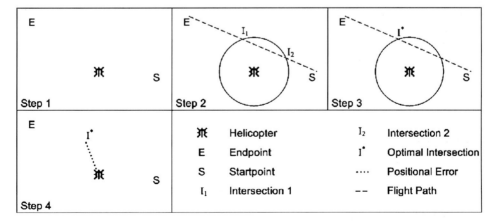

Fig. 3 Demonstration of calculating positional error

current heading to the desired heading. This is typically done by subtracting the current heading from the desired heading. Due to the heading's range of -180 to 180 degrees a check must be performed to assure that movement in the other direction would not be optimal. This is simply done by determining if the difference between the current heading and the desired heading is greater than 180 or less than -180. Once the optimal error is determined it is provided to the yaw controller as heading error.

Although velocity is typically one of the key elements in controlling a helicopter, it is by far the most difficult to accurately obtain. On the testbed, velocity calculations are performed by integrating the accelerations provided by the IMU. The testbed attempts to compensate for drift, bias, and noise in the velocity calculations using GPS calculated velocities and first order Kalman filters.

Integration of accelerations to acquire velocity first requires the removal of gravity which is calculated during initialization of the vehicle. Thus this procedure must first rotate the gravity vector to the local coordinate frame. This is performed using

$$
\begin{bmatrix} g'_y \\ g'_z \\ g'_x \end{bmatrix} = \begin{bmatrix} 1 & 0 & 0 \\ 0 & \cos(-C_{D2R}(\psi)) & -\sin(-C_{D2R}(\psi)) \\ 0 & \sin(-C_{D2R}(\psi)) & \cos(-C_{D2R}(\psi)) \end{bmatrix} * \begin{bmatrix} 0 \\ g_z \\ 0 \end{bmatrix} \tag{2}
$$

to rotate the gravity vector g about the X axis and then

$$
\begin{bmatrix} g''_y \\ g''_z \\ g''_x \end{bmatrix} = \begin{bmatrix} \cos(-C_{D2R}(\theta)) & -\sin(-C_{D2R}(\theta)) & 0 \\ \sin(-C_{D2R}(\theta)) & \cos(-C_{D2R}(\theta)) & 0 \\ 0 & 0 & 1 \end{bmatrix} * \begin{bmatrix} g'_y \\ g'_z \\ g'_x \end{bmatrix} \tag{3}
$$

to rotate vector g' about the Y axis. ψ and θ represent the Euler angles roll and pitch respectively. Make note that in Eq. 2, only gravity readings on the Z axis of the gravity vector are rotated. Since the vehicle is stable when the gravity vector is calculated all accelerations are deemed to be on the Z axis. Gravity accelerations recorded on the X and Y axis are assumed to be erroneous and will be systematically filtered out in the drift calculation. Due to this fact it is unnecessary to rotate the gravity vector about the Z axis.

The rotated gravity vector, g'', can now be subtracted from the IMU provided accelerations. The new acceleration vector, referred to as a, is then partially rotated to the world coordinate frame where an approximated drift, discussed later in this section, is subtracted. For the purpose of this paper a partial rotation simply refers to a rotation procedure that does not rotate about all three axes. The rotation procedure for a is performed using

$$
\begin{bmatrix} a'_y \\ a'_z \\ a'_x \end{bmatrix} = \begin{bmatrix} \cos(C_{D2R}(\theta)) & -\sin(C_{D2R}(\theta)) & 0 \\ \sin(C_{D2R}(\theta)) & \cos(C_{D2R}(\theta)) & 0 \\ 0 & 0 & 1 \end{bmatrix} * \begin{bmatrix} a_y \\ a_z \\ a_x \end{bmatrix} \tag{4}
$$

to rotate the vector a about the Y axis and then

$$
\begin{bmatrix} a''_y \\ a''_z \\ a''_x \end{bmatrix} = \begin{bmatrix} 1 & 0 & 0 \\ 0 & \cos(C_{D2R}(\psi)) & -\sin(C_{D2R}(\psi)) \\ 0 & \sin(C_{D2R}(\psi)) & \cos(C_{D2R}(\psi)) \end{bmatrix} * \begin{bmatrix} a'_y \\ a'_z \\ a'_x \end{bmatrix} \tag{5}
$$

to rotate vector a' about the X axis. This partial rotation is performed as drift calculations are stored in this coordinate system. Once the drift vector has been subtracted from a' the rotation to the world coordinate frame is completed using

$$\begin{bmatrix} a_y''' \\ a_z''' \\ a_x''' \end{bmatrix} = \begin{bmatrix} \cos\left(-C_{D2R}\left(\phi\right)\right) & 0 & \sin\left(-C_{D2R}\left(\phi\right)\right) \\ 0 & 1 & 0 \\ -\sin\left(-C_{D2R}\left(\phi\right)\right) & 0 & \cos\left(-C_{D2R}\left(\phi\right)\right) \end{bmatrix} * \begin{bmatrix} a_y'' \\ a_z'' \\ a_x'' \end{bmatrix} \qquad (6)$$

where ϕ is the yaw Euler angle.

The next step performed in calculating velocity is to average a''', with the $a_{\tau-1}'''$, defined as the previous control loops a'''. This is done to account for the loss of information due to the use of discrete sensor readings. It is assumed that variations in acceleration are linear for the short period of time between IMU readings. Thus an average of the prior and current accelerations should provide a more accurate value for integration.

The averaged acceleration vector, $\overline{a'''}$, is then integrated using IMU data timestamps. This calculated velocity vector is then added to two currently stored velocity vectors, V_I and V_F. The vectors V_I and V_F represent velocities calculated for the entire operation of the vehicle using only integrated accelerations and integrated accelerations corrected by GPS respectively. These vectors will be discussed later in this section. Note that all velocity readings are stored in the world coordinate frame and are only temporarily transformed to the local coordinate frame to supply roll, pitch, and collective velocities to the controllers.

When new GPS positions are available, corrections to the current velocity and velocity calculations are performed. This includes calculating an approximation of bias and fusing the GPS and IMU calculated velocities.

Although GPS velocity can simply be calculated by determining the distance traveled divided by the amount of time that has passed, this method typically performs poorly. One reason is that calculating velocity from discrete readings, such as GPS coordinates, introduces an error caused by the resolution of the readings. The value of this error increases significantly as the amount time used for the calculation decreases. This GPS velocity error will now be defined as

$$G_{E_A}\left(\Delta\tau\right) = \left|\frac{M_A}{\Delta\tau}\right| \qquad (7)$$

where M is the resolution of the sensor readings, in feet, on axis A and $\Delta\tau$ is the amount of time that has passed, in seconds, between distances readings. This is significant when considering calculating GPS velocity from two consecutive readings. At its lowest resolution the NMEA GPGGA message can only provide position at approximately a six inch resolution. If the GPS is operating at 5 Hz and velocity is calculated using consecutive readings the GPS velocity error could be as large as 2.5 ft/sec. Note that to reduce this error, $\Delta\tau$ must be increased. Simply increasing $\Delta\tau$ assumes that the system is stagnate during that particular time interval. Although this assumption may be valid periodically, it is not valid for every possible time interval. A more optimal algorithm should dynamically increase $\Delta\tau$, as appropriate.

To assure that the testbed utilizes a more accurate GPS velocity a new method which dynamically updates $\Delta\tau$ was developed. This method attempts to approximate the current GPS velocity by utilizing past GPS velocities that are deemed plausible.

The algorithm first calculates multiple velocity vectors from GPS movements that occurred over the last one second. These vectors are calculated using

$$
\begin{bmatrix} T_{i_Y} \\ T_{i_Z} \\ T_{i_X} \end{bmatrix} = \begin{bmatrix} (F\left(E_y\left(\tau, \tau - i\right)\right)) * \dfrac{Hz_G}{i} \\ (F\left(E_z\left(\tau, \tau - i\right)\right)) * \dfrac{Hz_G}{i} \\ (F\left(E_x\left(\tau, \tau - i\right)\right)) * \dfrac{Hz_G}{i} \end{bmatrix},
\tag{8}
$$

where T_i represents the temporary GPS velocity vector calculated using 'i + 1' GPS readings, F is a function for converting from meters to feet, E is a function for calculating the distance vector between the current GPS position, τ, and the $\tau - i$ GPS position using the World Model equation, and is the operating frequency of the GPS. As these calculations are only performed for the past one second's worth of data, 'i' will range from $[1, Hz_G]$ and Hz_G vectors will be calculated.

Initializing the GPS velocity vector V_G to T_1 and beginning with $i = 2$, the elements of vector T_i are compared to calculated error thresholds. This is done using

$$
V_{G_A} = \begin{cases} T_{i_A} & \text{for } V_{G_A} - G_{E_A}\left(\dfrac{i-1}{Hz_G}\right) < T_{i_A} < V_{G_A} + G_{E_A}\left(\dfrac{i-1}{Hz_G}\right) \\[3mm] V_{G_A} - G_{E_A}\left(\dfrac{i-1}{Hz_G}\right) & \text{for } T_{i_A} \le V_{G_A} - G_{E_A}\left(\dfrac{i-1}{Hz_G}\right) \\[3mm] V_{G_A} + G_{E_A}\left(\dfrac{i-1}{Hz_G}\right) & \text{for } T_{i_A} \ge V_{G_A} + G_{E_A}\left(\dfrac{i-1}{Hz_G}\right) \quad \text{or} \quad T_{i_A} * T_{i-1_A} < 0 \end{cases}
\tag{9}
$$

if T_{i_A} is positive or

$$
V_{G_A} = \begin{cases} T_{i_A} & \text{for } V_{G_A} + G_{E_A}\left(\dfrac{i-1}{Hz_G}\right) \le T_{i_A} \le V_{G_A} - G_{E_A}\left(\dfrac{i-1}{Hz_G}\right) \\[3mm] V_{G_A} + G_{E_A}\left(\dfrac{i-1}{Hz_G}\right) & \text{for } T_{i_A} \le V_{G_A} + G_{E_A}\left(\dfrac{i-1}{Hz_G}\right) \quad \text{or} \quad T_{i_A} * T_{i-1_A} < 0 \\[3mm] V_{G_A} - G_{E_A}\left(\dfrac{i-1}{Hz_G}\right) & \text{for } T_{i_A} \ge V_{G_A} - G_{E_A}\left(\dfrac{i-1}{Hz_G}\right) \end{cases}
\tag{10}
$$

if T_{i_A} is negative. Note that subscript A represent a particular axis (i.e. lateral, longitudinal, or vertical). Updates to V_G continue until either all T_i vectors have been exhausted or until T_i fails to adhere to the boundaries set by V_{G_A} and G_E. Boundary failures are determined by the inequality

$$
V_{G_A} - G_{E_A}\left(\frac{i-1}{Hz_G}\right) \le T_{i_A} \le V_{G_A} + G_{E_A}\left(\frac{i-1}{Hz_G}\right)
\tag{11}
$$

in Eq. 9 and

$$
V_{G_A} + G_{E_A}\left(\frac{i-1}{Hz_G}\right) \le T_{i_A} \le V_{G_A} - G_{E_A}\left(\frac{i-1}{Hz_G}\right)
\tag{12}
$$

Fig. 4 Example of six GPS
readings along a single axis

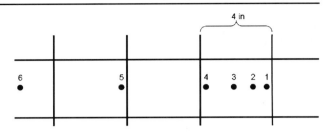

in Eq. 10. If the threshold is broken for a particular element in vector T_i then that axis has used all the data deemed valid for its computation. The failed axis is then finalized by setting the corresponding axis in V_G with the closest valid value in inequality Eq. 11 or 12 to the failed T_i element. Although other elements in vector V_G may continue to be updated, axes whose thresholds have been broken cease to be updated. Note that T_i vectors can be individually calculated and compared rather than calculating all possible vectors as described in Eq. 8. Calculating vectors as needed will decrease the computational complexity of the algorithm. The algorithm was described as it is for clarity and ease of understanding.

For clarity, Fig. 4 details an example of GPS data from a 5 Hz receiver along a single axis. Note that the position data is labeled 1–6 where 1 is the most current data and 6 is the oldest and the grid represents the lowest resolution of the GPS on that particular axis. Using the above method to calculate the current velocity the algorithm would first calculate the base velocity T_1 using positions 1 and 2. This would provide a velocity of zero ft/sec. Next the algorithm would calculate the velocity using positions 1, 2, and 3. This would also provide a velocity of zero ft/sec which does not violate the inequality Eq. 11, i.e. the range $(-1.7,1.7)$. Using positions 1, 2, 3, and 4 again provides a velocity of zero which also does not violate Eq. 11, i.e. the range $(-0.8,0.8)$. Using positions 1, 2, 3, 4, and 5 provides a velocity of 0.8 ft/sec. This velocity violates Eq. 11, i.e. the range $(-0.6,0.6)$, and thus stops further integration of GPS data on that particular axis. Since the temporary velocity is violated on the positive side of the threshold the final velocity along that axis is set to zero, the last valid velocity, plus the threshold, 0.6 ft/sec.

The method described above allows $\Delta\tau$ in Eq. 7 to be dynamically increased by utilizing past data that appears to be valid and thus decreasing the velocity error due to resolution. Figure 5 depicts velocity calculations for a single run of the testbed using the standard velocity calculation algorithm and the one used by the testbed.

Velocities calculated from IMU integrated accelerations are typically subject to some level of drift. This drift is typically constant for very short periods of time. Drift

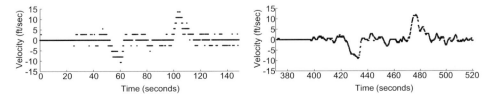

Fig. 5 Flight velocities calculated using the standard method (*left*) and method (*right*)

Fig. 6 Flight velocities
calculated using integrated
accelerations without drift
corrections

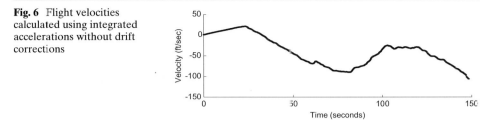

can heavily influence velocity calculations and will typically render the velocities useless within a matter of seconds. Figure 6 depicts velocities calculated using IMU supplied accelerations with only the gravity vector removed.

During operation the velocity algorithm continually attempts to calculate drift and remove it from the accelerations readings. This is first done by calculating the difference between a strictly IMU calculated velocity, V_I, and the velocity calculated by GPS, V_G. This is performed using

$$S = \left(V_{I_\tau} - V_{G_\tau}\right) - \left(V_{I_{\tau-H}} - V_{G_{\tau-H}}\right) \tag{13}$$

where S is the slope vector of the offset, τ is the current time step, and H is the number of time steps in one second. This vector is then rotated about the Z axis using

$$\begin{bmatrix} S'_y \\ S'_z \\ S'_x \end{bmatrix} = \begin{bmatrix} \cos\left(C_{D2R}\left(\phi\right)\right) & 0 & \sin\left(C_{D2R}\left(\phi\right)\right) \\ 0 & 1 & 0 \\ -\sin\left(C_{D2R}\left(\phi\right)\right) & 0 & \cos\left(C_{D2R}\left(\phi\right)\right) \end{bmatrix} * \begin{bmatrix} S_y \\ S_z \\ S_x \end{bmatrix}. \tag{14}$$

S' is the added to any previously stored bias represented by vector B. Bias, B, is stored in this coordinate frame for two reasons. First, bias is being calculated for specific axis of sensors within the IMU. Thus the bias should be stored utilizing the orientation of the IMU. Second, complete rotation to the local coordinate frame would be redundant. B is only used to subtract bias from the acceleration vector, a''. As this vector is already in the correct coordinated system it is unnecessary to fully rotate B. Figure 7 depicts flight velocities calculated by integrating the accelerations with and without bias removal.

It should be noted that during testing it was determined that an offset between the GPS calculated velocities and the IMU calculated velocities existed. The IMU velocities where preceding the GPS velocities by approximately one second. To account for this the slope calculation described in Eq. 13 is offset by one second. Instead of calculating the difference between the current GPS velocities and the

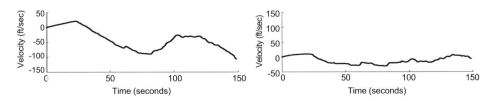

Fig. 7 Flight velocities calculated with (*right*) and without (*left*) drift corrections

current IMU velocity, the slope is calculated by comparing the current GPS velocity with the IMU velocity from one second prior, i.e.

$$S = \left(V_{I_{\tau-H}} - V_{G_\tau}\right) - \left(V_{I_{\tau-2H}} - V_{G_{\tau-H}}\right). \tag{15}$$

Once drift has been approximated the GPS velocity vector V_G is fused with $V_{F_{\tau-H}}$ using

$$\begin{bmatrix} V_{Fy_\tau} \\ V_{Fz_\tau} \\ V_{Fx_\tau} \end{bmatrix} = \begin{bmatrix} V_{Fy_\tau} \\ V_{Fz_\tau} \\ V_{Fx_\tau} \end{bmatrix} + \begin{bmatrix} K_y \\ K_z \\ K_x \end{bmatrix} * \left(\begin{bmatrix} V_{Gy} \\ V_{Gz} \\ V_{Gx} \end{bmatrix} - \begin{bmatrix} V_{Fy_{\tau-H}} \\ V_{Fz_{\tau-H}} \\ V_{Fx_{\tau-H}} \end{bmatrix} \right), \tag{16}$$

where the vector K represents the standard first order Kalman gain. Equation 16 is performed to remove offset from the velocity calculation that were not previously accounted for by drift. Figure 8 details velocities calculated using only GPS, only bias corrected IMU, and fused IMU and GPS.

3.3 Acceleration Variant Calculation

Controllers for the testbed, described later in this section, utilize accelerations to help determine the correct control response. Although accelerations are provided directly by the IMU they are unfiltered and typically do not correspond well with the velocity calculations above. To account for this the accelerations are recalculated, now referred to as the acceleration variant, using the filtered and fused velocity vectors.

First, the difference between the current velocity, V_{F_Ω}, and the last seven velocities, $V_{F_{\Omega-1}}, V_{F_{\Omega-2}}, \ldots, V_{F_{\Omega-7}}$, is calculated. The use of seven velocities provides adequate smoothing without a large delay in acceleration calculation. This was determined through experimentation with various size sets of velocities. These seven

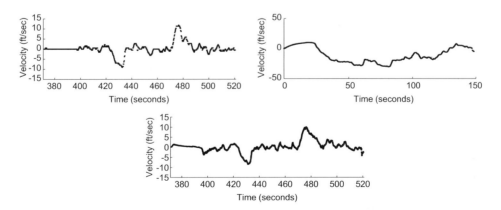

Fig. 8 Flight velocities calculated using GPS (*top left*), bias corrected IMU (*top right*), and fused GPS/IMU (*bottom*)

values are then averaged to produce the change in velocity, $\overline{\Delta F_V}$. The acceleration variant, A_V, is then calculated using

$$A_V = \frac{\overline{\Delta V_F}}{\Omega - (\Omega - 7)},\qquad(17)$$

where Ω represents the timestamp for V_{F_Ω} and $(\Omega - 7)$ represents the timestamp for $V_{F_{\Omega-7}}$. The acceleration variant is then passed through a first order Kalman filter and provided to the pitch and roll controllers as accelerations. Figure 9 depicts a comparison of the acceleration variant and raw accelerations provided by the IMU.

3.4 Trim Integrators

To assure that progress is being made by the helicopter the positional error, velocity, and acceleration of the vehicle are constantly monitored. During navigation the flight algorithm is constantly calculating three integrators. These integrators are used as trim values for the roll, pitch and collective and are added to the PW outputs of these axes. Calculations are performed by incrementing or decrementing the integrator by a small constant amount. These calculations are only performed when the vehicle is outside of a positional threshold of its goal and no progress, be it though acceleration or velocity, is being made. Direction of the integration is based on the direction of progress that needs to be made.

The roll and pitch integrators are also used to alter the neutral orientation of the helicopter. The neutral orientation is considered to be the orientation at which the vehicle will hover. Further discussion of this aspect is detailed in next section.

3.5 Antenna Translations

The GPS antenna for the testbed is mounted towards the end of the tail boom on top of the horizontal fin. This location is utilized to assure that the GPS antenna is not located near any of the vehicle's structures or components that might interfere with its satellite reception. This method of installation induces a constant error into the positional data of the vehicle. Due to the size of the vehicle the offset is fairly small (<2.5 feet). Although the error is small, it does create issues during heading rotations of the vehicle.

Heading rotations cause the vehicle to rotate about the main shaft of the vehicle. As the GPS antenna is mounted on the tail of the vehicle, a rotation about the main shaft appears as a positional movement around a circle. This positional movement is also calculated as a GPS velocity. To insure that the testbed can safely and efficiently

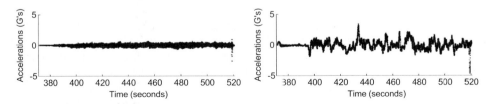

Fig. 9 Flight accelerations from IMU (*left*) and variant calculated (*right*)

change headings during flight two translations are performed to remove both the positional error and velocities caused by the mounting location of the GPS antenna.

Positional translations are performed to remove the offset between the GPS antenna and the rotational axis of the helicopter. The positional translation algorithm first determines the antenna's positional error by rotating its constant offset in the local coordinate frame to the world coordinate frame. This is done using

$$
\begin{bmatrix} P'_y \\ P'_z \\ P'_x \end{bmatrix} = \begin{bmatrix} \cos\left(-C_{D2R}(\phi)\right) & 0 & \sin\left(-C_{D2R}(\phi)\right) \\ 0 & 1 & 0 \\ -\sin\left(-C_{D2R}(\phi)\right) & 0 & \cos\left(-C_{D2R}(\phi)\right) \end{bmatrix} * \begin{bmatrix} P_y \\ P_z \\ P_x \end{bmatrix} \tag{18}
$$

where P is the positional offset of the GPS antenna in the world coordinate frame. Note that the x component of the positional offset should always be zero in the local coordinate frame. This is because the antenna is located on the boom of the helicopter and thus located on the y axis of the local coordinate frame.

P' is then converted to offsets in decimal hours format using

$$
L_X = \left(P'_X \times M_X\right)/1,000,000 \tag{19}
$$

to calculate the longitudinal offset and

$$
L_Y = \left(P'_Y \times M_Y\right)/1,000,000 \tag{20}
$$

to calculate the lateral offset. Note that the product of the offset, P', and the GPS resolution, M, is divided by one million to adhere to NMEA positional units.

Although the offsets are now in a form that can be directly removed from the GPS provided positions, to do so may cause unexpected results. As previously mention there is a noticeable time offset between IMU data and GPS data. This offset was approximated to be a one second delay. To account for this, changes to the positional translation, Tp, are delayed for up to one second. This is done by calculating and storing the difference between the corrections calculated in Eqs. 19 and 20 for the current GPS time step and the previous GPS time step using

$$
\begin{bmatrix} Ts_{X_\tau} \\ Ts_{Y_\tau} \end{bmatrix} = \begin{bmatrix} L_{X_\tau} \\ L_{Y_\tau} \end{bmatrix} - \begin{bmatrix} L_{X_{\tau-1}} \\ L_{Y_{\tau-2}} \end{bmatrix}. \tag{21}
$$

Note that Ts only stores data for the last one second's worth of GPS data and its size is equal to the frequency of the GPS data. As such, when a new set of data is added to Ts the oldest set of data must be removed. Any non-zero value removed for Ts is added to Tp. This ensures that all translations will be incorporated into the position calculation within one second.

It should be noted that GPS positional data may not be delayed for a full second. Thus this algorithm monitors changes in the positional data and determines if these changes correspond with changes caused by a rotation. This is done by comparing the difference between the current GPS position and the previous GPS position, Ld. If a difference exists on the lateral, Ld_Y, or longitudinal, Ld_X, axis it is compared with that axes corresponding values stored in T_S. These comparisons are used to make corrections to the position translation and are calculated using

$$T p_A = \begin{cases} T p_A + T s_{A_{\tau-i}} & \text{for } L d_A = T s_{A_{\tau-i}} \\ T p_A + T s_{A_{\tau-i}} & \text{for } L d_A > T s_{A_{\tau-i}} > 0 \text{ or } L d_A < T s_{A_{\tau-i}} < 0 \\ T p_A + L d_A & \text{for } T s_{A_{\tau-i}} > L d_A > 0 \text{ or } T s_{A_{\tau-i}} < L d_A < 0 \\ T p_A & \text{otherwise} \end{cases} \quad (22)$$

$$T s_{A_{\tau-i}} = \begin{cases} 0 & \text{for } L d_A \geq T s_{A_{\tau-i}} > 0 \text{ or } L d_A \leq T s_{A_{\tau-i}} < 0 \\ T s_{A_{\tau-i}} = T s_{A_{\tau-i}} - L d_A & \text{for } T s_{A_{\tau-i}} > L d_A > 0 \text{ or } T s_{A_{\tau-i}} < L d_A < 0 \\ T s_{A_{\tau-i}} & \text{otherwise} \end{cases} \quad (23)$$

and

$$L d_A = \begin{cases} L d_A = L d_A - T s_{A_{\tau-i}} & \text{for } L d_A > T s_{A_{\tau-i}} > 0 \text{ or } L d_A < T s_{A_{\tau-i}} < 0 \\ L d_A = 0 & \text{for } T s_{A_{\tau-i}} \geq L d_A > 0 \text{ or } T s_{A_{\tau-i}} \leq L d_A < 0 \\ L d_A & \text{otherwise} \end{cases} \quad (24)$$

where A represents the axis, either longitudinal or lateral, and i represents a variable used to traverse the data stored in Ts. Note that i is initialized to be equal to the frequency of the GPS and is reduced by one after Eqs. 22, 23, and 24 have been performed on both the lateral and longitudinal axis. The comparison calculations are concluded after the $i = 0$ calculations are performed. It should be mentioned that values stored in Ts are modified in Eq. 23. This done to ensure that corrections made during these calculations are not reused in later calculations.

Values Tp_y, the latitude offset, and Tp_x, the longitude offset, are always removed from the GPS supplied latitude and longitude before positional error calculations are performed. This assures that the positional error represents the error from the rotational axis, or main shaft, of the helicopter and not the location of the GPS antenna.

Although the positional translation algorithm can effectively translate the GPS location, its translations cannot be directly used by the velocity calculation. The positional translation algorithm will only delay a known offset for up to one second. It is feasible that the GPS position take longer than one second to recognize the movement. Thus, velocities calculated using the translated positions would calculate a velocity from the offset induced by the position translation algorithm and an equal but opposite velocity from the delayed GPS position.

To account for these issues the velocity translation algorithm was developed. This algorithm is designed to remove velocities that would be calculated due to rotations but will never increase or create a velocity. Unlike the positional translation algorithm which attempts to match the GPS antenna's position with the vehicles position, this algorithm only attempts to remove differences between consecutive GPS positions that may have been caused by heading rotations.

The velocity translation algorithm, like the positional translation algorithm, first determines the antenna's positional error using Eq. 18 and then calculates the lateral and longitudinal offsets using Eqs. 19 and 20. Ts is then calculated using Eq. 21. Note that any unused data that is removed from Ts due to time is simply discarded and is in no way utilized in this algorithm. This method prevents the algorithm from ever creating or increasing a velocity. Unlike the position translation algorithm, the velocity translation algorithm does not use Eqs. 22, 23, and 24 to calculate modifications to the translation. Instead this algorithm uses

$$
Tv_A = \begin{cases}
Tv_A + Ts_{A_{\tau-i}} & \text{for } Ld_A = Ts_{A_{\tau-i}} \\
Tv_A + Ts_{A_{\tau-i}} & \text{for } Ld_A > Ts_{A_{\tau-i}} > 0 \text{ or } Ld_A < Ts_{A_{\tau-i}} < 0 \\
Tv_A + Ld_A & \text{for } Ts_{A_{\tau-i}} > Ld_A > 0 \text{ or } Ts_{A_{\tau-i}} < Ld_A < 0 \\
Tv_A & \text{otherwise}
\end{cases}
\qquad (25)
$$

Eqs. 23 and 24 to directly calculate the velocity translation, Tv, which is always initialized to zeros. As previously described, i represents a variable used to traverse the data stored in Ts and is initialed to equal to the frequency of the GPS. The variable is reduced by one after Eqs. 23, 24, and 25 have been performed on both the lateral and longitudinal axis. The comparison calculations are concluded after the $i = 0$ calculations are performed.

Values Tv_Y, the latitude offset, and Tv_X, the longitude offset, are always removed from the GPS supplied latitude and longitude before the GPS velocity calculations are performed. Note that modifications to the GPS position data are not carried outside of the velocity calculation and do not affect any other algorithm or calculation.

4 Controller

Fuzzy Logic was chosen as the methodology for developing the controllers for the helicopter testbed. Fuzzy logic provides multiple advantages to control development including the use of linguistic variables, functionality using imprecise or contradictory input, ease of modification, and the ability to directly implement knowledge from multiple experts.

The helicopter controllers were developed with several key aspects in mind. First, the controllers needed to be as robust as physically possible. This was required as the project was originated with the desire to operate a small fleet of autonomous helicopters. Since the dynamics of small RC based UAVs varies greatly even between replicas, any controllers developed for use on multiple UAV's had to be robust. The design of robust controllers coupled with modular software and modular hardware would allow for the fastest implementation of an autonomous UAV fleet.

One important aspect in the design of the controllers is the creation of fuzzy rules based on general RC helicopter flight and not on the Joker Maxi 2 helicopter flight. This allowed focus to be placed on providing control based on the concepts of flight rather than the specifics of one particular helicopter. This is done by utilizing an expert in RC helicopter flight that had little experience with the Joker Maxi 2. This assured that decisions were not based on a single platform.

Although helicopter dynamics are heavily coupled, the degree to which they are coupled is heavily influenced by the types of flight being performed. Aggressive maneuvers such as stall and knife edge turns require heavy compensation from all aspects of control. Non-aggressive maneuvers such as hovering and simple waypoint flight can be virtually decoupled. Decoupling of the control is further assisted by the rate and resolution at which newer technology sensors operate. Although the controllers may not immediately compensate for coupling the input from sensors allows compensation to be performed before a noticeable degradation occurs. The work presented here decouples control into four categories: collective control, yawing control, roll control, and pitch control. Note that throttle control is simply held constant during flight.

Although the controllers described by this work were designed without a highly developed model they were designed for controlling a specific class of vehicles, i.e. helicopters. This class of vehicle has several fundamental properties that classify it as a helicopter. Thus, these fundamental properties correspond to a generic model for this class of vehicle.

First, only two controllable surfaces exist on a helicopter: the main rotor and the tail rotor. In the local coordinate frame, the main rotor can directly control a vertical force and two angular forces (latitudinal and longitudinal). The tail rotor can directly control an angular force about the local vertical axis. Thus, there is only a single controllable aspect of the vehicle that causes a non-angular velocity. This non-angular velocity, now referred to as lift, can only create a vertical force in the local coordinate frame. Thus, velocities within the world frame are directly connected to the orientation of the vehicle and can be calculated using

$$V_{\text{TOT}} = \int F_{\text{MR}} + {}^{\iota}F_{\text{G}} + {}^{\iota}F_{\text{D}} + {}^{\iota}F_{\text{E}}, \tag{26}$$

where ${}^{\iota}F_{\text{G}}$ is the force vector produced by gravity in the local frame, F_{MR} is the force vector produced by the main rotor in the local frame, ${}^{\iota}F_{\text{D}}$ is force vector produced by drag in the local frame, and ${}^{\iota}F_{\text{E}}$ is the force vector produced by all other miscellaneous forces in the local frame. Miscellaneous forces encompass all other forces acting on the helicopter including wind, temperature, weight distribution, etc. Note that ${}^{\iota}F_{\text{E}}$ is assumed to be far smaller than ${}^{\iota}F_{\text{G}}$ or F_{MR}. As ${}^{\iota}F_{\text{E}}$ begins to approach either of these forces the vehicle will most likely become uncontrollable.

It should now be mentioned that V_{TOT} has a natural threshold which prevents the velocity vector from growing without bound. As V_{TOT} increases the drag, ${}^{\iota}F_{\text{D}}$, will increase as an opposing force. Since, ${}^{\iota}F_{\text{G}}$ has a constant strength, F_{MR} is bound by the mechanics of the vehicle, and ${}^{\iota}F_{\text{E}}$ is considered to be small, V_{TOT} will ultimately be constrained by ${}^{\iota}F_{\text{D}}$.

Second, the bulk weight for helicopters is, and should be, centrally located under the main rotor. Thus the center of gravity falls below the vehicle's natural rotational axis located at the center of the main rotor. This design causes a simulated pendulum effect where the bulk weight of the vehicle will naturally swing below the main rotor. This pendulum effect is then dampened by external forces such as drag. Thus, in the absence of a large angular force the vehicle will naturally stabilize in a near horizontal fashion. This fact allows the controllers to prevent excessive angles with very little control effort.

Using this generic and heavily generalized information a controller was developed to calculate a desired velocity and then achieve that velocity. The controller attempts to maneuver and stabilize the vehicle simply by controlling the velocities of the vehicle.

Position error is the driving force behind maneuvering the vehicle. The controllers described in this work are designed to utilize the position error to determine a desired velocity. Note that desired velocity is strictly a fuzzy variable which describes the desired direction and strength for velocity. This desired velocity is then used by the controllers, along with the state of the vehicle, to determine control output. Desired velocity is calculated for the lateral, longitudinal, and vertical axes as well as the heading orientation. The heading velocity for the testbed is controlled strictly by the heading hold gyro. Thus, the desired velocity calculated for the heading is sufficient for controlling the tail rotor.

It should be noted that the desired velocity is never actually calculated. It is described here to show the decision processes that the fuzzy controllers attempt to mimic. This type of description is used for the remainder of this section.

As control is based on the desired velocity the first input evaluated by the fuzzy controllers is the velocity of the vehicle. From Eq. 26, velocities are directly proportional to the lift and orientation of the helicopter. Assuming that the lift is essentially stagnant, the lateral and longitudinal velocities for a rigid main rotor helicopter can be controlled by the orientation of the vehicle. Thus the lateral and longitudinal controllers calculate a desired orientation by comparing the desired velocity with the current velocity.

Vertical velocities are controlled by calculating a desired change in lift, or collective. This desired change is based on the difference between the current velocity and the desired velocity. Note that during flight the natural lift, or neutral collective, is assumed to be enough lift to counteract gravity and thus create vertical stability. Desired velocities create the need to reduce or increase the natural lift.

Although the controller does its best to determine orientations and lift that will provide the desired velocities, the specific velocity obtained given an orientation or lift is heavily dependent on the type and configuration of the vehicle. In an attempt to compensate for this type of variation, the desired orientations are modified according to the acceleration variant input.

The acceleration variant input determines if the rate at which the velocity is changing is appropriate. For the lateral and longitudinal axes this input attempts to modify the desired orientation to fit the vehicle being used. If the desired angles are producing a velocity at too great of a rate the desired angles are reduced. If the desired angles are producing a velocity at too slow a rate the desired angles are increased. This concept is also used to modify the change in lift. Once the change in lift has been corrected by the acceleration variant the desired collective is output from the controller. Note that these modifications do not carry from operation to operation and are only calculated using the most current state data.

Now that the desired angles have been calculated the controller must determine an output capable of achieving these angles. This is performed by calculating a desired angular rate which identifies the direction and speed at which the vehicle will rotate. This value is based on the difference between the current and desired angles.

To assure that the vehicle does not obtain an orientation that is difficult to control limitations were designed into the fuzzy rules. These rules allow angles to be achieved within a constrained threshold. The values of these constraints are designed directly into the fuzzy controllers via the two extreme Membership Functions (MFs) of the orientation inputs. As the vehicle begins to approach the assigned thresholds for orientation the desired angular rates become skewed. This skew reduces any desired angular rate that would increase the orientation of the vehicle. Once the vehicle has surpassed the threshold for orientation the controller will only calculate desired angular rates that decrease the angle of the vehicle.

The four fuzzy controllers for the testbed were developed in Matlab utilizing Sugeno constant fuzzy logic and a weighted average defuzzification method. All rules for the controllers are based on the 'and' method and use membership products to

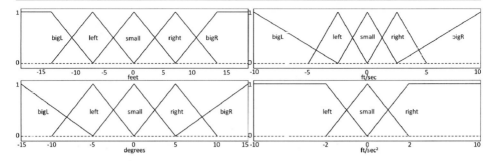

Fig. 10 Membership functions for positional error (*top left*), velocity (*top right*), angle (*bottom left*), and acceleration variant (*bottom right*) for the roll controller

determine the strength of each rule. Each controller has a single output which ranges from [−1,1] corresponding to the minimum and maximum PW for that particular control respectively. It should be noted that all of the inputs, with the exception of orientation (angle), for the controllers are in various units of feet (i.e. feet, feet/second, feet/second^2). Orientation inputs are based on Euler angles and are in units of degrees.

Several assumptions were made during the design of the controllers. First, control for both the roll and pitch is based on the assumption that a level vehicle will create minimal velocity and accelerations. Although this statement is not typically valid, there is an orientation that is optimal for creating minimal velocities. This optimal orientation may or may not be perfectly level and is periodically dynamic. The optimal orientation is based on multiple variables including weight distribution, mechanical setup, and wind to name only a few. To compensate for this the flight algorithm implements three integrators, discussed in the previous section. These integrators are not only used to adjust the vehicles trims but also to modify its internal definition of level. If the vehicle drifts off course or cannot reach a desired position the integrator slowly increases the vehicles trim in that direction. In parallel the vehicles definition of level is rotated in the direction of the integrator. This allows the vehicle to continuously attempt to find the optimal definition of level which will

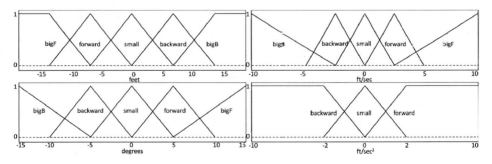

Fig. 11 Membership functions for positional error (*top left*), velocity (*top right*), angle (*bottom left*), and acceleration variant (*bottom right*) for the pitch controller

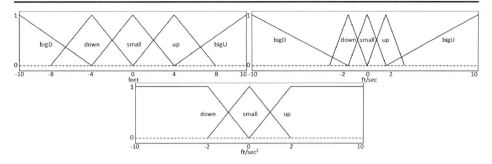

Fig. 12 Membership functions for vertical error (*top left*), velocity (*top right*), and variant acceleration (*bottom*) for the collective controller

ultimately increase the accuracy of the controller and validate the afore mentioned assumption.

Second, the collective controller assumes that a neutral, or zero, command will not create any vertical velocities. This statement is also typically untrue. The hovering collective varies greatly based on the wind, throttle, and battery charge, to name only a few. Thus the collective integrator continuously updates the trim value of the collective. This dynamically increases or decreases the neutral value of the controller to its optimal location thus validating the assumption.

The decisions for both the roll and pitch controllers are based on four inputs: positional error, velocity, orientation, and acceleration variant (discussed earlier in this section). Note that all inputs are in the local coordinate frame of the helicopter. Figures 10 and 11 detail the exact makeup of the roll and pitch controllers. It should be noted that exterior membership functions extend to both $-\infty$ and ∞. This is to assure that any possible value for these inputs can be appropriately handled by the controller.

It should be noted that rules for both the roll and pitch controllers are identical. This is deemed valid as the tail of an RC helicopter has minimal effect on these two axes. The effect of the tail is further minimized by the heading hold gyro which is responsible for keeping the heading stable.

The collective controller is responsible for all vertical movement of the helicopter. Decisions for control are based on three inputs: vertical position error, vertical velocity, and vertical acceleration variant. Note that all inputs are in the local coordinate frame of the helicopter. Figure 12 details the exact makeup of the collective controller. It should be noted that exterior membership functions extend to both $-\infty$ and ∞. This is to assure that any possible value for these inputs can be appropriately handled by the controller.

Fig. 13 Membership function for heading error for the yaw controller

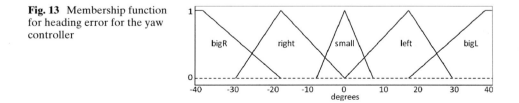

Table 2 Operation times of batteries

Battery	Maximum endurance	Laboratory set maximums
11.1 V 4.2Ah LiPo	45–90 min	40 min
11.1 V 0.5Ah LiPo	18 h	10 h
4.8 V 2.0Ah NiMh	45 min	30 min
37 V 10Ah LiPo	16–17 min	12 min

The yaw controller is responsible for all heading changes of the helicopter. Decisions for control are based on a single input heading error. Only a single input is required to successfully stabilize and control the heading of the helicopter due to the use of the heading hold gyro. Utilizing of a heading hold gyro converts all control inputs into rate commands. If the control input is zero then the gyro attempts to keep the angular rate at zero. If the control input is not zero then the gyro attempts to create an angular rate proportional to the control. Figure 13 details the exact makeup of the yaw controller. It should be noted that exterior membership functions extend to both $-\infty$ and ∞. This is to assure that any possible value for these inputs can be appropriately handled by the controller.

To assure that every possible combination of inputs is accounted for a rule is developed for each possible combination of input in each controller. This is foremost done to assure that the fuzzy rule set is complete. This method also assures that every combination of inputs is distinctly evaluated by the expert. This assures that input combinations would not be incorrectly grouped together and assigned a single rule. This method creates 375 distinct rules for both the roll and pitch controllers, 75 rules for the collective controller, and 5 rules for the yaw controller. Although the specific rules are not presented here, they are available in [42].

At this time there is no widely accepted mathematical method for proving stability of fuzzy controllers on a non-linear system without an accurate vehicle model. Although this work cannot prove stability of the described controllers, this work attempts to supply a level of reliability through extensive testing discussed in the following section and described in detail in [42].

Although the testbed is developed for field testing of UAV technology it is not designed to replace simulation testing. First, field testing requires a large amount of overhead time. The time spent simply moving the equipment to the flight location and prepping the vehicles can take hours. Second, field testing has an inherent level of danger. Any testing performed in the field, no matter what safety precautions are taken, inherently places the vehicle, onlookers, and surrounding environment in a potentially hazardous situation. For these reasons initial testing of controllers and most software algorithms are tested in simulation.

The research lab utilizes the X-Plane flight simulator for testing and the initial development of software for the testbed helicopter. X-Plane is an FAA certified

Table 3 Flight envelope tested in simulation

*Total wind speed did not exceed 23 mph

Variable	Range
Flight speed	0–10 mph
Wind	0–23 mph
Wind shear	0–20 degrees
Wind gusts	0–23 mph*

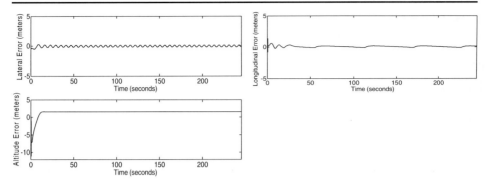

Fig. 14 Positional error for zero wind hovering

commercial flight simulator that supports both fixed wing and VTOL vehicles. Besides the standard vehicle models, X-Plane supports customized vehicles that can be modeled directly in X-Plane. Specific details of the simulator, simulation model, and Simulink control model can be found in [42] and [46].

5 Experiments and Results

Due to the shear breadth of this research the experimentation had to be performed on multiple levels. This included testing and validation of individual sections of the work and testing throughout the integration phase. Experimentations were performed in lab settings, simulations, and outdoors depending on the specific type of experiment. Types of experimentation included endurance, heat, shock, vibration, payload limitations, CPU utilization, and controller validation.

Payload experimentations were performed to determine the absolute maximum weight that could be carried by the vehicle. The payload limitation was determined by increasing the weight of the vehicle using half pound weights until the vehicle was deemed unsafe by the RC pilot or was unable to lift weight out of ground effects.

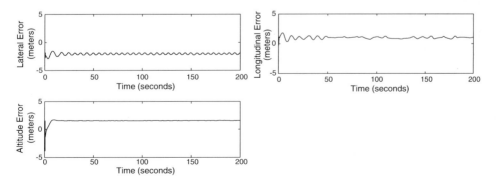

Fig. 15 Positional error for dynamic wind hovering

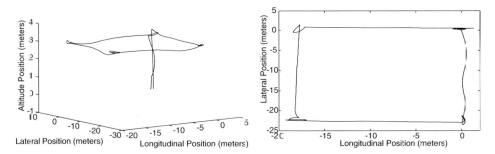

Fig. 16 3D (*left*) and birds-eye-view (*right*) of a square flight path with no wind effects and stopping at each waypoint

Experimentation showed a maximum payload of approximately 5.5 kg above the stock vehicle's flight weight.

Endurance experiments were performed on all batteries utilized on the testbed. Endurance results for each individual battery are detailed in Table 2. It should be noted that the experiments used to test the 11.1 V 4.2Ah LiPo included various powering scenarios. This was done by varying the CPU utilization, sensor output rates, and communication rates. It should also be mentioned that the endurance of individual batteries is a function of their age and usage. As such the endurance of the equipment they power is a function of the batteries age and usage.

Due to the design of the processing hardware enclosure there was some concern for heat dissipation. Thus multiple experiments were performed to determine if any failures would result from internal heat. These experiments varied from only a few hours up to three days. Even after three days of continuous operation the enclosed processing system operated without failure. Internal temperatures during experimentation reached 168 degrees for the system and 150 degrees for the CPU.

As this work describes a helicopter testbed capable of multiple areas of development, the processing system must have free clock cycles available during operation. To determine the amount of CPU available during flight the testbed was initialized and operated as for normal flight. CPU utilization during flight varied between zero and two percent and averaged approximately 0.8%.

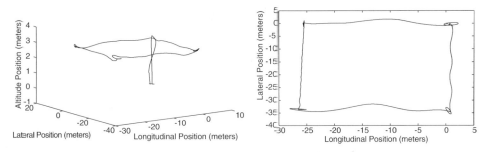

Fig. 17 3D (*left*) and birds-eye-view (*right*) of a square flight path with a 5 kt wind, 20 degree wind shear, and stopping at each waypoint

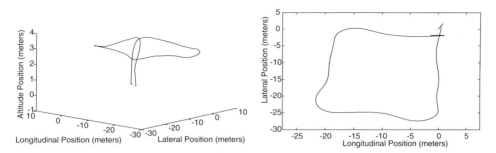

Fig. 18 3D (*left*) and birds-eye-view (*right*) of a square flight path without stopping at waypoints

To validate the controllers developed for the testbed multiple experiments were performed. These experiments included multiple flights using multiple flight patterns under varying environmental conditions. Experiments were performed in both simulation and on the actual hardware in an outdoor environment.

5.1 Simulation Experiments

Simulation experiments were performed utilizing the X-Plane simulator and Matlab 2006a, detailed in [42]. Experiments were performed with and without wind. These experiments included hovering and waypoint navigation. Two forms of waypoint navigation were performed including navigation that required the vehicle stop and stabilize at each individual waypoint and navigation that did not allow the vehicle to stabilize at any waypoint. The overall flight envelope tested in simulation is available in Table 3.

Hovering test varied in length of time from several minutes to several hours. These tests were first performed using no wind, constant wind (~5 kt), and wind with dynamically changing speeds and direction (0–5 kt and 0–20 degrees shear). Figures 14 and 15 detail the results of selected hovering experiments in simulation. It should be noted that constant offsets do exist in the results due to the lack of an integration term in the simulation controller.

Flight test were designed to test the vehicles ability to maneuver to waypoints under varying conditions including wind variation and point stabilization. This was

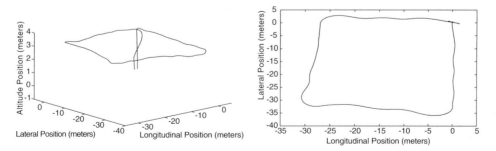

Fig. 19 3D (*left*) and birds-eye-view (*right*) of a square flight path with a 5 kt wind and 20 degree wind shear

Table 4 Flight envelope tested in the field	Variable	Range
	Flight speed	0–5 mph
	Time	Day & night
	Wind	0–15 mph
	Wind shear	0–5 degrees
	Wind gusts	0–5 mph
	Cloud cover	Clear-Overcast
	Rain	Clear-Misting

done by first allowing the vehicle to stabilize at individual waypoints before being allowed to continue, Figs. 16 and 17. The second set of experiments tested the vehicles ability to maneuver between waypoints without being allowed to stabilize at waypoints, Figs. 18 and 19. Make note that at each waypoint the vehicle was command to rotate in the direction of the new flight path. Thus, the vehicle is always moving in the forward direction.

5.2 Field Experiments

Although simulation experiments provide a level of assurance in the ability of the controllers they are far from conclusive. The controllers were designed to operate on a helicopter testbed and therefore must be validated on the actual hardware in realistic environments.

The helicopter was tested in a non-optimal outdoor environment in an urban setting. The test area is approximately 70 meters wide and 100 meters long. The test area is surrounded by multiple buildings including a four story parking garage to the north west, a three story office building to the north, and a four story office building to the south. This environment created varying wind effects and less than optimal GPS reception.

Field experiments included waypoint navigation, hovering, takeoff, and landing. These experiments were first performed individually and then as an integrated set. The overall flight envelope tested in the field is available in Table 4.

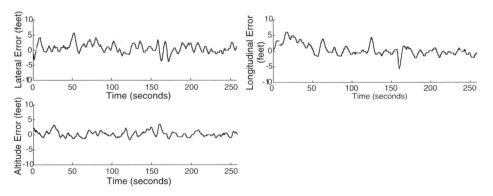

Fig. 20 Positional error during hover

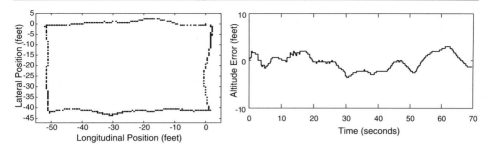

Fig. 21 Birds-eye-view of square flight path (*top*) and altitude error (*bottom*)

Hovering was naturally the first outdoor experiment to be performed on the testbed. These experiments were performed by having the safety pilot takeoff the vehicle and position it in a hover approximately 50 ft off of the ground. Once the vehicle was stable the safety pilot relinquished control to the testbed. The testbed was then responsible for holding the position where it was located when control was relinquished. Figure 20 details the positional error over time that occurs during a continuous hover in a very light wind with very moderate gusts. Note that the first 35 to 40 seconds of flight contained a strictly positive longitudinal error. This error was partially due to the longitudinal wind and vehicle setup. It should be mentioned that position data used to plot all of the figures in this section were gathered directly from the GPS unit. As such, all of gathered position data are subject to the errors and noise associated with GPS. To compensate, various videos of autonomous flight are available with the original research, on file with the USF library in Tampa, Florida, or available at www.uavrobotics.com.

Once hovering had been sufficiently tested experiments were performed to validate the controller's abilities to maneuver between points. These experiments consisted of the safety pilot taking off the vehicle and positioning it in a hover approximately 50 ft above the ground. The safety pilot would then relinquish control and the vehicle would begin transitioning hard coded waypoints. Once the vehicle had transitioned the entire set of waypoints it would maintain a hover at the last

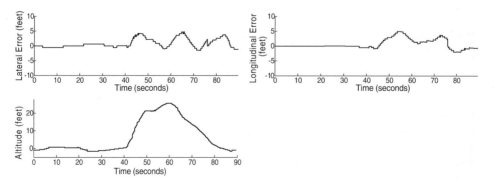

Fig. 22 Lateral error (*top*), longitudinal error (*middle*), and altitude (*bottom*) for a typical takeoff and landing experiment

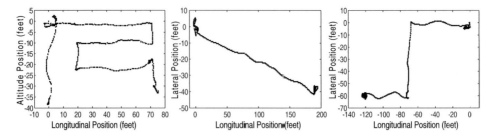

Fig. 23 Field experiments of the vertical step (*top left*), straight line (*top right*), and square-s (*bottom*) flight paths

waypoint. Figure 21 details a typical path followed by the testbed while performing a square pattern of four waypoints.

Once the vehicle had shown a consistent ability to successfully perform flight maneuvers, experiments were performed to validate the controller's ability to take off and land successfully. These experiments required the safety pilot to relinquish control before the vehicle was powered up. Once control was relinquished the vehicle would power up the main rotor and lift off. After a predetermined altitude had been reached, ~30 ft, the controller began the landing phase of the experiment. Once the vehicle had successfully landed the main rotor was powered down. Figure 22 details the vertical error as well as the lateral and longitudinal positions during one of these experiments. Note that lateral and longitudinal errors during takeoff are due to the extreme shifts in both the direction and magnitude of accelerometer drift. These errors will eventually be removed by the drift calculation described in the previous section.

The last step in testing the controllers was to perform flights that were integrated sets of all the abilities programmed into the testbed. This included takeoff, landing, hovering, and navigation. These experiments included several flight patterns including the square-s, straight line, and vertical steps. Note that vertical steps are transitions in the longitudinal and vertical axes Figure 23 detail the flight paths of the testbed for all three flight patterns.

6 Conclusions and Future Work

This research presented the technical aspects of an unmanned helicopter testbed. This vehicle was specifically designed to support removal, modification, and replacement of both hardware and software as deemed necessary by the end user. This is made possible through the design and implementation of modular hardware and software as well as the use of Commercial off the Shelf (COTS) components. This vehicle has performed over three hundred fully autonomous flights and represents one of only a handful of fully autonomous helicopters in the world. It is distinct due to its ground up design as a testbed vehicle. It is this researcher's hope that this work will provide the opportunity and motivation for others to join and help further this technology.

The helicopter testbed has shown the ability to autonomously navigate waypoints, hover, takeoff, and land as well as the ability to filter and fuse data without relying on a vehicle specific model. The possible future work that can be accomplished from this point is almost limitless. The vehicle is designed for modification and testing and as such lends itself to a multitude of areas. This includes vision processing, controller design, software architecture design, hardware design, filtering and fusion, and mechanical design to name only a few.

Foreseeable work in the immediate future would most likely include controller updates that attempt to separate uncontrollable forces into external and internal categories and correctly compensate for these forces. Controllers could also be updated or developed to utilize state feedback to permanently modify the controller's outputs. One specific feedback update would be to allow the acceleration variant to permanently update the desired angle calculated by the controller.

Another foreseeable update includes mounting and testing the chassis and processing system on other platforms. This could include variations of types and sizes of helicopters as well as other types of unmanned vehicles including fixed wing aircraft and ground vehicles. Note that ground vehicle usage has been tested to a limited degree in [47] and [48] where the processing system, sensors, and software described in this work were ported to a four wheel drive RC truck. This UGV has been utilized in UAV/UGV coordination, swarm control, and pattern formation.

Vision, although only briefly mention in this work, could also provide a large amount of information about the state of the vehicle as well as provide the vehicle the ability to accomplish many more tasks. Types of information could include state data regarding the velocity, relative position to an object, or even failure detection for the vehicle. Vision processing algorithms could also provide the vehicle the ability to identify and track objects, perform statistical calculations, and be readily fault tolerant.

Although it is not detailed in this work the use of a Graphical User Interface (GUI) would greatly improve the appeal of this testbed. This interface could easily be designed to allow users to dynamically change flight paths and gather information in a more user friendly environment.

Acknowledgements   This research has been partially supported by a fellowship through the Oak Ridge Institute for Science and Education (ORISE), ONR Grants N00014-03-01-786 and N00014-04-10-487, SPAWAR, and the DOT through the USF CUTR Grant 2117-1054-02.

References

1. Hrabar, S., et al.: Combined optic-flow and stereo-based navigation of urban canyons for a UAV. In: Proceedings of the 2005 IEEE Intenational Conference on Intelligent Robots and Systems (2005)
2. Quigley, M., Goodrich, M.A., Beard, R.W.: Semi-autonomous human-UAV interfaces for fixed-wing mini-UAVs. In: Proceedings of the 2004 IEEE/RSJ International Conference on Intelligent Robots and Systems (2004)
3. Ryan, A., Hedrick, J.K.: A mode-switching path planner for UAV-assisted search and rescue. In: 44th IEEE Conference on Decision and Control, 2005 European Control Conference. CDC-ECC'05, pp. 1471–1476 (2005)
4. Ryan, A., et al.: An overview of emerging results in cooperative UAV control. In: 43rd IEEE Conference on Decision and Control, 2004 European Control Conference. CDC–ECC'04 (2004)
5. Freed, M., Harris, R., Shafto, M.: Human-interaction challenges in UAV-based autonomous surveillance. In: Proceedings of the 2004 Spring Symposium on Interactions Between Humans and Autonomous Systems Over Extended Operations (2004)

6. Freed, M., Harris, R., Shafto, M.G.: Human vs. autonomous control of UAV surveillance. In: AIAA 1st Intelligent Systems Technical Conference, pp. 1–7 (2004)
7. Kingston, D., Beard, R., Casbeer, D.: Decentralized perimeter surveillance using a team of UAVs. In: Proceedings of the AIAA Conference on Guidance, Navigation, and Control (2005)
8. Loyall, J., et al.: Model-based design of end-to-end quality of service in a multi-UAV surveillance and target tracking application. In: 2nd RTAS Workshop on Model-Driven Embedded Systems (MoDES). Toronto, Canada, May 2004
9. Nygaards, J., et al.: Navigation aided image processing in UAV surveillance: preliminary results and design of an airborne experimental system. J. Robot. Syst. **21**(2), 63–72 (2004)
10. Quigley, M., et al.: Target acquisition, localization, and surveillance using a fixed-wing mini-UAV and gimbaled camera. In: Proceedings of the 2005 IEEE International Conference on Robotics and Automation, pp. 2600–2605 (2005)
11. Coifman, B., et al.: Surface transportation surveillance from unmanned aerial vehicles. In: Procs. of the 83rd Annual Meeting of the Transportation Research Board, pp. 11–20 (2004)
12. Coifman, B., et al.: Roadway traffic monitoring from an unmanned aerial vehicle. IEE Proc. Intell. Transp. Syst. **153**(1), 11 (2006)
13. Dovis, F., et al.: HeliNet: a traffic monitoring cost-effective solution integrated with the UMTS system. In: 2000 IEEE 51st Vehicular Technology Conference Proceedings, 2000. VTC 2000-Spring Tokyo (2000)
14. Srinivasan, S., et al.: Airborne traffic surveillance systems: video surveillance of highway traffic. In: Proceedings of the ACM 2nd International Workshop on Video Surveillance & Sensor Networks, pp. 131–135 (2004)
15. Casbeer, D.W., et al.: Cooperative forest fire surveillance using a team of small unmanned air vehicles. Int. J. Syst. Sci. **37**(6), 351–360 (2006)
16. Martínez-de Dios, J.R., Merino, L., Ollero, A.: Fire detection using autonomous aerial vehicles with infrared and visual cameras. In: Proceedings of the 16th IFAC World Congress (2005)
17. Ollero, A., et al.: Motion compensation and object detection for autonomous helicopter visual navigation in the COMETS system. In: Proceedings of ICRA 20004 IEEE International Conference on Robotics and Automation, vol. 1, pp. 19–24 (2004)
18. Rufino, G., Moccia, A.: Integrated VIS-NIR hyperspectral/thermal-IR electro-optical payload system for a mini-UAV. Infotech@ Aerospace, pp. 1–9 (2005)
19. Hausamann, D., et al.: Monitoring of gas pipelines – a civil UAV application. Aircr. Eng. Aerosp. Technol. **77**(5), 352–360 (2005)
20. Ostendorp, M.: Innovative airborne inventory and inspection technology for electric power line condition assessments and defect reporting. In: 2000 IEEE ESMO-2000 IEEE 9th International Conference on Transmission and Distribution Construction, Operation and Live-Line Maintenance Proceedings, pp. 123–128 (2000)
21. Girard, A.R., Howell, A.S., Hedrick, J.K.: Border patrol and surveillance missions using multiple unmanned air vehicles. In: 43rd IEEE Conference on Decision and Control. CDC (2004)
22. Morris, J.: DHS using Northrop Grumman UAV in Arizona border patrol flights. 2004 [cited 2008 July 18]; Available from: www.aviationweek.com/aw/generic/story_generic.jsp?channel=hsd&id=news/HSD_WH1_11104.xm
23. Garcia, R.D., Valavanis, K.P., Kandel, A.: Fuzzy logic based autonomous unmanned helicopter navigation with a tail rotor failure. In: 15th Mediterranean Conference on Control and Automation. Athens, Greece (2007)
24. Gavrilets, V., et al.: Avionics system for a small unmanned helicopter performing aggressive maneuvers. In: The 19th Digital Avionics Systems Conferences, 2000. Proceedings. DASC (2000)
25. Gavrilets, V., et al.: Avionics system for aggressive maneuvers. Aerosp. Electron. Syst. Mag., IEEE **16**(9), 38–43 (2001)
26. Saripalli, S., et al.: A tale of two helicopters. IEEE/RSJ Int. Conf. Intell. Robots Syst. **1**(1), 805–810 (2003)
27. Kanade, T., Amidi, O., Ke, Q.: Real-time and 3D vision for autonomous small and micro air vehicles. In: 43rd IEEE Conference on Decision and Control. CDC (2004)
28. Mettler, B., et al.: Attitude control optimization for a small-scale unmanned helicopter. In: AIAA Guidance, Navigation and Control Conference, pp. 40–59 (2000)
29. Mettler, B., Tischler, M.B., Kanade, T.: System identification of small-size unmanned helicopter dynamics. Annu. Forum Proc. Am. Helicopter Soc. **2**, 1706–1717 (1999)
30. Abbeel, P., et al.: An application of reinforcement learning to aerobatic helicopter flight. In: Proceedings of Neural Information Processing Systems (NIPS) Conference, Vancouver, B.C., Canada (2007)

31. Abbeel, P., Ganapathi, V., Ng, A.Y.: Learning vehicular dynamics, with application to modeling helicopters. In: Proceedings of Neural Information Processing Systems (NIPS) (2006)
32. Johnson, E.N.: UAV Research at Georgia Tech. Presentation at TU Delft (2002)
33. Johnson, E.N., Mishra, S.: Flight simulation for the development of an experimental UAV. In: Proceedings of the AIAA Modeling and Simulation Technologies Conference (2002)
34. Johnson, E.N., Schrage, D.P.: The Georgia Tech unmanned aerial research vehicle: GTMax. In: Proceedings of the AIAA Guidance, Navigation, and Control Conference (2003)
35. Meingast, M., Geyer, C., Sastry, S.: Vision based terrain recovery for landing unmanned aerial vehicles. In: 43rd IEEE Conference on Decision and Control. CDC. vol. 2, pp. 1670–1675 (2004)
36. Ng, A.Y., et al.: Autonomous helicopter flight via reinforcement learning. In: Proceedings of Advances in Neural Information Processing Systems (NIPS) (2004)
37. Kelly, J., Saripalli, S., Sukhatme, G.S.: Combined visual and inertial navigation for an unmanned aerial vehicle. In: International Conference on Field and Service Robotics (2007)
38. Mejias, L., et al.: Visual servoing of an autonomous helicopter in urban areas using feature tracking. J. Field Robot. **23**(3), 185–199 (2006)
39. Roberts, J.M., Corke, P.I., Buskey, G.: Low-cost flight control system for a small autonomous helicopter. Proc. IEEE Int. Conf. Robot. Autom. **1**, 546–551 (2003)
40. Montgomery, J.F., et al.: Autonomous helicopter testbed: a platform for planetary exploration technology research and development. J. Field Robot. **23**(3/4), 245–267 (2006)
41. Unknown: The Autonomous Helicopter Testbed. 2004 [cited 2008 January 28]; Available from: http://www-robotics.jpl.nasa.gov/systems/system.cfm?System=13
42. Garcia, R.D.: Designing an autonomous helicopter testbed: from conception through implementation, in computer science engineering, p. 305. University of South Florida, Tampa (2008)
43. Carlson, C.G., Clay, D.E.: The earth model – calculating field size and distances between points using GPS coordinates. In: Site Specific Management Guidelines, p. 4. Potash & Phosphate Institute (1999)
44. Weisstein, E.W.: Circle-line intersection. [cited 2007 October 19]; Available from: http://mathworld.wolfram.com/Circle-LineIntersection.html
45. Rhoad, R., Milauskas, G., Whipple, R.: Geometry for Enjoyment and Challenge, rev. ed. McDougal, Littell & Company, Evanston, IL (1984)
46. Ernst, D., et al.: Unmanned vehicle controller design, evaluation and implementation: from MATLAB to printed circuit board. J. Intell. Robot. Syst. **49**, 23 (2007)
47. Barnes, L., Alvis, W.: Heterogeneous swarm formation control using bivariate normal functions to generate potential fields. In: Proceedings of the IEEE Workshop on Distributed Intelligent Systems: Collective Intelligence and Its Applications (DIS'06), vol. 00, p. 85–94 (2006)
48. Barnes, L., Fields, M., Valavanis, K.: Unmanned ground vehicle swarm formation control using potential fields. In: Mediterranean Conference on Control & Automation, Athens, Greece, 2007

Modeling and Real-Time Stabilization of an Aircraft Having Eight Rotors

S. Salazar · H. Romero · R. Lozano · P. Castillo

Originally published in the Journal of Intelligent and Robotic Systems, Volume 54, Nos 1–3, 455–470.
© Springer Science + Business Media B.V. 2008

Abstract We introduce an original configuration of a multi rotor helicopter composed of eight rotors. Four rotors, also called main rotors, are used to stabilize the attitude of the helicopter, while the four extra rotors (lateral rotors) are used to perform the lateral movements of the unmanned aerial vehicle (UAV). The main characteristic of this configuration is that the attitude and translation dynamics are decoupled. The dynamic model is obtained using the well known Euler–Lagrange approach. To validate the model, we performed real-time experiments using a simple nonlinear control law using optical flow and image processing.

Keywords Modeling · Control · UAV · Multirotors

Partially supported by SNI-CONACyT México.

S. Salazar (✉)
Instituto de Investigaciones Eléctricas
Reforma 113 Col. Palmira, CP 62490 Cuernavaca, Mor., Mexico
e-mail: ssalazar@iie.org.mx

H. Romero · R. Lozano · P. Castillo
Heudiasyc UMR 6599 CNRS-UTC, BP 20529 Compiègne, France

H. Romero · R. Lozano · P. Castillo
LAFMIA UMR 3175 CNRS-CINVESTAV, Av. IPN 2508, AP 14-740 Mexico D.F., Mexico

H. Romero
e-mail: hromerot@hds.utc.fr

R. Lozano
e-mail: rlozano@hds.utc.fr

P. Castillo
e-mail: castillo@hds.utc.fr

1 Introduction

An unmanned aerial vehicle (UAV) is a powered aircraft that does not carry a human operator, uses aerodynamic forces to provide lift, can fly autonomously (by an onboard computer) and/or be piloted remotely (by radio control), and can be recovered for repeated flights. UAVs can perform those missions considered dull, dirty, or dangerous for their manned counterparts. Examples include orbiting endlessly over a point for communications relay or jamming (dull), collecting air samples to measure pollution or CW/BW toxins (dirty), and flying reconnaissance over hostile air defenses (dangerous). And while some will still contest it, it is increasingly accepted that UAVs cost less to build (two to five times the weight of the pilot in specialized equipment is needed just to support him, not to mention double, triple, even quadruple redundant systems to ensure his return, not just the aircraft's) and to operate (pilot proficiency flying is eliminated or maintained on cheap semi-scale UAVs, oxygen servicing is eliminated) [2].

Future UAVs will evolve from being robots operated at a distance to independent robots, able to self-actualize to perform a given task. This ability, autonomy, has many levels emerging by which it is defined, but ultimate autonomy will require capabilities analogous to those of the human brain by future UAV mission management computers.

The push of unmanned flight has been the driving or contributing motivation behind many of the key advancements in aviation: the autopilot, the inertial navigation system, and data links, to name a few. Although UAV development was hobbled by technology insufficiencies through most of the 20th century, focused efforts in small, discrete military projects overcame the problems of automatic stabilization, remote control, and autonomous navigation [2]. The last several decades have been spent improving the technologies supporting these capabilities largely through the integration of increasingly capable microprocessors in the flight and mission management computers flown on UAVs [4, 12, 13]. The early part of the 21st century will see even more enhancements in UAVs as they continue their growth.

UAVs are categorized by size (from a couple centimeters to several dozen meters long), and mission performance (endurance, operating altitude, etc). Of course, the whole purpose of the UAV is to carry a payload. For the moment, the primary mission assigned to UAVs is aerial observation and surveillance. UAVs are obviously an advantageous solution for missions where a live crew offers no real benefits, or where the risk are too high. They offer a number of benefits that manned vehicles can't match, including a continuous presence, endurance, responsiveness, low-observability and operational versatility.

In general, for UAVs to carry out the missions under consideration, all aspects of the system will have to be improved: the vehicle itself, payloads and especially sensors, transmission systems, onboard intelligence, sharing authority with the human operator and data processing (3D optical or radar imaging, etc.) and sensor fusion [11–13].

We introduce, in this paper, an original configuration of a multi rotor helicopter composed of eight rotors. Four rotors, also called main rotors, are used to stabilize the attitude of the helicopter, while the four extra rotors (lateral rotors) are used to perform the lateral movements of the UAV (see Fig. 1). The main characteristic of this configuration is that the attitude and translation dynamics are almost totally

Fig. 1 The eight-rotors helicopter

decoupled. In our knowledge it is the first helicopter having this configuration. In this paper, we extend our results and present the nonlinear model taking into account aerodynamics effects. To validate the model we propose a nonlinear control strategy and we apply it in real-time experiences.

The outline of this paper is as follows: In Section 2 we develop the mathematical nonlinear model of the eight rotors helicopter. In Section 3 we present the control strategy. Section 4 presents the platform description, some experiences are presented in Section 5 and finally conclusions are given in Section 6.

2 Mathematical Model

The aim of this section is to develop a nonlinear dynamic model for the eight-rotor rotorcraft using the Euler–Lagrange approach. The equations describing the attitude and position of the UAV are basically those of a rotating rigid body with six degrees of freedom. Those equations can be separated into kinematic and dynamic equations.

2.1 Characteristics of the Helicopter

The aerial robot under consideration consists of a rigid cross frame equipped with eight rotors as shown in Fig. 2. In order to avoid the yaw drift due to the reactive torques, the main rotors are configured such that the right and left rotors rotate clockwise and the front and rear rotors rotate counterclockwise. Using the same philosophy, the external lateral motors located over the same axis rotate in the same direction to avoid the roll and pitch drift phenomena.

The up-down motion is achieved by increasing or decreasing the total thrust of the main rotors. The yaw motions are achieved through a differential speed generated by each couple of rotors. A movement in the yaw direction is produced by increasing (decreasing) the speed of the front and rear motors while decreasing (increasing) the speed of the right and left motors [6, 7]. Forward/backward, left/right are performed through a differential control strategy of the thrust generated by each lateral rotor.

2.2 Euler–Lagrange Equations

The dynamical model of the aircraft will be obtained using the Lagrangian method. This approach is based in the two basic forms of energies contained in dynamical systems: the kinetic energy and the potential energy. Let $\mathcal{I} = \{\vec{i}, \vec{j}, \vec{k}\}$ be an external reference set of axes, and let $\mathcal{B} = \left\{\vec{i}, \vec{j}, \vec{k}\right\}$ denote a set of coordinates fixed in the rigid aircraft as is shown in Fig. 2.

Let $q = (\xi, \eta) \in \mathbb{R}^6$ be the generalized coordinates vector for the helicopter, where $\xi = (x, y, z) \in \mathbb{R}^3$ denote the position of the center of mass of the rotorcraft relative to the frame $\mathcal{I}$, and $\eta = (\psi, \theta, \phi) \in \mathbb{R}^3$ are the Euler angles (yaw, pitch and roll) that describe the orientation of the aircraft.

The translational and rotational kinetic energy of the rotorcraft are expressed as

$$T_{\text{trans}} \triangleq \frac{m}{2}\dot{\xi}^T\dot{\xi}$$

$$T_{\text{rot}} \triangleq \frac{1}{2}\dot{\eta}^T J\dot{\eta}$$

where m denotes the mass of the rotorcraft and $J = W_\eta^T I W_\eta$ is the inertia matrix, more details see [14].

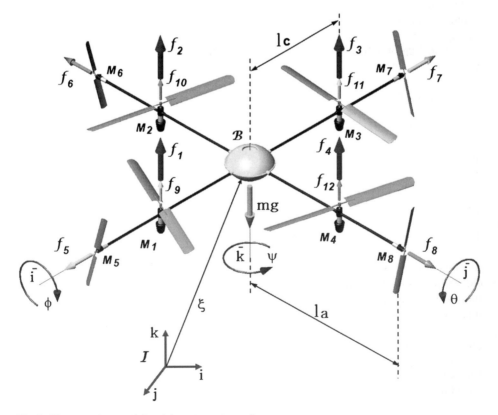

Fig. 2 Vectors scheme of the eight-rotor rotorcraft

The only potential energy which needs to be considered is due to the standard gravitational force g. Therefore, potential energy is expressed as

$$U = mgz$$

Following the approach, the Lagrangian is given by

$$\mathcal{L}(q, \dot{q}) = T_{\text{trans}} + T_{\text{rot}} - U$$

$$= \frac{m}{2}\dot{\xi}^T\dot{\xi} + \frac{1}{2}\dot{\eta}^T J\dot{\eta} - mgz$$

The model for the full eight-rotors aircraft dynamics is obtained from the Euler–Lagrange equations with external generalized force, F,

$$\frac{d}{dt}\frac{\partial\mathcal{L}}{\partial\dot{q}} - \frac{\mathcal{L}}{\partial q} = F$$

where $F = (F_\xi, \tau)$. F_ξ defines the translational force applied to the aerial robot due to the control inputs and relative to the frame $\mathcal{I}$ and τ is the generalized moments vector. The small body forces are ignored, and we only consider the principal control inputs u, u_x, u_y and τ. They represent the total thrust, the control input for lateral motors located in x-axis, the control input for lateral motors located in y-axis and the generalized moments respectively.

2.2.1 Forces

Note from Fig. 2 that, the force applied to the mini-helicopter relative to the frame $\mathcal{B}$ could be defined as

$$\bar{F} = u_x\bar{\mathbf{i}} + u_y\bar{\mathbf{j}} + u_z\bar{\mathbf{k}}$$

where

$$u_x = f_6 - f_8$$
$$u_y = f_5 - f_7$$
$$u_z = u + f_9 + f_{10} + f_{11} + f_{12}$$

with

$$u = f_1 + f_2 + f_3 + f_4; \quad f_i = k\omega_i^2, \quad i = 1, \ldots, 4$$

where $k > 0$ is a parameter depending on the density of air, the radius, the shape, the pitch angle of the blade and ω_i is the angular speed of the motor "i" ($M_i, i = 1, \ldots, 4$). There exist additional forces $f_j, \forall\, j = 9, \ldots, 12$, acting over each main rotor, see Fig. 2. These forces are due to the airflow generated by the lateral rotors, $\tilde{f}_i, i = 1, \ldots, 4$. It means that the magnitude of vectors f_j, is a function of the incoming lateral air flow coming from the corresponding lateral rotor, see Fig. 3.

In order to clarify the notation we will use the subscripts p for the main rotor and subscript s for the lateral rotor. The thrust $f_k = f_i + f_j$, $k, i = 1, \ldots, 4$, $j = 9, \ldots, 12$, could be expressed as (see Fig. 3)

$$f_k = 2\rho A_p \hat{V} V_\upsilon \tag{1}$$

Fig. 3 Analysis of the main
and lateral thrusts

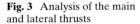

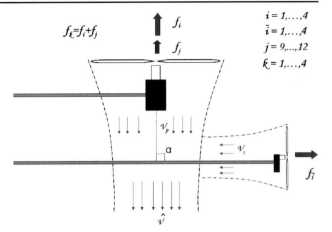

where $V_p{}^1$ is the induced wind speed in the main propeller and $\hat{V}$ is the total induced wind speed by the set of rotors (main and lateral rotors), this is given by

$$\hat{V} = \left[\left(V_s \cos\alpha + V_p \right)^2 + \left(V_s \sin\alpha \right)^2 \right]^{\frac{1}{2}} \tag{2}$$

where α is the angle between the main rotor axis and the lateral rotor axis[2].

For simplicity and obvious reasons for the translational dynamics, we choose $\alpha = 90°$, thus, introducing Eq. 2 into Eq. 1 we have

$$f_k = 2\rho A_p V_p^2 \left(1 + \frac{V_s^2}{V_p^2} \right)^{\frac{1}{2}} . \tag{3}$$

Note that, the additional nonlinear term $\left(V_s^2 / V_p^2 \right)$ appearing in this equation is related with the airflow produced by the corresponding lateral rotor. Nevertheless, this extra term has an almost linear behavior mainly for large values of V_s. From the design of the helicopter we have that $V_p > V_s$, so, $(V_s / V_p) < 1$. Thus, it follows that $\left(1 + V_s^2 / V_p^2 \right)^{\frac{1}{2}} < \sqrt{2}$. The term (V_s^2 / V_p^2) in Eq. 3 will need to be compensated to effectively decouple the translational and rotational displacements. In practice, the dynamics in the mini-helicopter due to this term have a slow behavior.

The force, f_i, in each one of the four main rotors is affected by the lateral thrust $f_{\tilde{i}}$, of the corresponding lateral rotor. The lateral thrust in turn depends on the control actions u_x and u_y applied to lateral motors. Considering identical lateral motors, the force f_j, $j = 9, ..., 12$ can be expressed as follows

$$f_9 = b u_{x_r}, \quad f_{10} = b u_{y_r}, \quad f_{11} = b u_{x_l}, \text{ and } f_{12} = b u_{y_l}$$

---

[1]The induced wind speed in a propeller is defined as follows $V = \left(\frac{f}{2\rho A} \right)^{\frac{1}{2}}$; where f is the thrust generated by the propeller, ρ is the air density and A is the propeller area [9].

[2]Notice that without extra lateral rotor $V_s = 0$, this implies that $\hat{V} = V_p$, $f_j = 0$, and Eq. 1 becomes $f_k = f_i = 2\rho A_p V_p^2$; $\forall k, i = 1, ..., 4$

where $b \geq 0$ is a constant. Then, the control input u_z can be rewritten as follows

$$u_z = u + b\bar{u} \tag{4}$$

where $\bar{u} = u_{x_r} + u_{y_r} + u_{x_l} + u_{y_l}$. Consequently

$$F_\xi = R\bar{F}$$

where R is the transformation matrix representing the orientation of the rotorcraft from $\mathcal{B}$ to $\mathcal{I}$. We use c_θ to denote $\cos\theta$ and s_θ for $\sin\theta$.

$$R = \begin{bmatrix} c_\psi c_\theta & c_\psi s_\theta s_\phi - s_\psi c_\phi & c_\psi s_\theta c_\phi + s_\psi s_\phi \\ s_\psi c_\theta & s_\psi s_\theta s_\phi + c_\psi c_\phi & s_\psi s_\theta c_\phi - c_\psi s_\phi \\ -s_\theta & c_\theta s_\phi & c_\theta c_\phi \end{bmatrix}$$

2.2.2 Moments

The generalized moments on the η variables are

$$\tau \triangleq \begin{bmatrix} \tau_\psi \\ \tau_\theta \\ \tau_\phi \end{bmatrix} = \begin{bmatrix} \sum_{j=}^{4} \tau_{M_j} \\ (f_2 - f_4)l_c + b u_x \\ (f_3 - f_1)l_c + b u_y \end{bmatrix}$$

where l_c is the distance from the center of gravity to any internal rotor and τ_{M_j} is the couple produced by motor M_j.

The Euler–Lagrange equation is expressed as

$$m\ddot{\xi} + \begin{bmatrix} 0 & 0 & mg \end{bmatrix}^T = F_\xi$$
$$J\ddot{\eta} + C(\eta, \dot{\eta})\dot{\eta} = \tau \tag{5}$$

where $C(\eta, \dot{\eta})$ is the Coriolis matrix. In order to further simplify the analysis, let us propose a change of the input variables

$$\tau = C(\eta, \dot{\eta})\dot{\eta} + J\tilde{\tau}$$

where $\tilde{\tau} = \begin{bmatrix} \tilde{\tau}_\psi & \tilde{\tau}_\theta & \tilde{\tau}_\phi \end{bmatrix}^T$ are the new inputs. Then

$$\ddot{\eta} = \tilde{\tau} \tag{6}$$

Rewriting Eqs. 5, 6

$$m\ddot{x} = u_x c_\theta c_\psi - u_y \left(c_\phi s_\psi - c_\psi s_\theta s_\phi\right) + \left(s_\phi s_\psi + c_\phi c_\psi s_\theta\right) u_z \tag{7}$$

$$m\ddot{y} = u_x c_\theta s_\psi + u_y \left(c_\phi c_\psi + s_\theta s_\phi s_\psi\right) - \left(c_\psi s_\phi - c_\phi s_\theta s_\psi\right) u_z \tag{8}$$

$$m\ddot{z} = -u_x s_\theta + u_y c_\theta s_\phi - mg + c_\theta c_\phi u_z \tag{9}$$

$$\ddot{\psi} = \tilde{\tau}_\psi \tag{10}$$

$$\ddot{\theta} = \tilde{\tau}_\theta \tag{11}$$

$$\ddot{\phi} = \tilde{\tau}_\phi \tag{12}$$

where the horizontal plane coordinates are represented by x and y, and z is the vertical position. ψ is the yaw angle around the z axis, θ is the pitch angle around the (new) y axis, and ϕ is the roll angle around the (new) x axis. The control inputs u is the main thrust or main collective input (directed out the bottom of the aircraft), and $\tilde{\tau}_\psi$, $\tilde{\tau}_\theta$ and $\tilde{\tau}_\phi$ are the new angular moments (yawing moment, pitching moment and rolling moment). The control inputs to develop lateral displacements are denoted by u_x and u_y.

3 Control Strategy

In this section, we propose a simple nonlinear control law to stabilize the eight rotors helicopter at hover.

Define

$$\tilde{\tau} = \begin{bmatrix} -\sigma_{g_1}(k_{\psi_1}(\psi - \psi_d)) - \sigma_{g_2}(k_{\psi_2}\dot{\psi}) \\ -\sigma_{g_3}(k_{\theta_1}(\theta - \theta_d)) - \sigma_{g_4}(k_{\theta_2}\dot{\theta}) \\ -\sigma_{g_5}(k_{\phi_1}(\phi - \phi_d)) - \sigma_{g_6}(k_{\phi_2}\dot{\phi}) \end{bmatrix} \tag{13}$$

where $\sigma_{g_i}(\cdot)$ is a saturation function, i.e., $|\sigma_{g_i}(\cdot)| \leq g_i$, $\forall g_i > 0$, $i = 1, ..., 6$, and $k_{l,\ell} > 0$, $\forall l = \psi, \theta, \phi$; $\ell = 1, 2$, are constant. Then

$$\ddot{\psi} = -\sigma_{g_1}(k_{\psi_1}(\psi - \psi_d)) - \sigma_{g_2}(k_{\psi_2}\dot{\psi})$$

$$\ddot{\theta} = -\sigma_{g_3}(k_{\theta_1}(\theta - \theta_d)) - \sigma_{g_4}(k_{\theta_2}\dot{\theta})$$

$$\ddot{\phi} = -\sigma_{g_5}(k_{\phi_1}(\phi - \phi_d)) - \sigma_{g_6}(k_{\phi_2}\dot{\phi})$$

3.1 Stability Analysis

We will prove the stability of the closed-loop system 6–13 using the pitch dynamics, the proof for the others attitude dynamics are similar. Then, let us consider the system

$$\dot{\theta}_1 = \theta_2$$

$$\dot{\theta}_2 = \tilde{\tau}_\theta \tag{14}$$

Propose the following control input

$$\tilde{\tau}_\theta = -\sigma_{g_3}(k_{\theta_2}\theta_2) - \sigma_{g_4}(k_{\theta_1}\theta_1)$$

Define the following positive function $V_1 = \frac{1}{2}\theta_2^2$, then

$$\dot{V}_1 = \theta_2\dot{\theta}_2 = -\theta_2\left[\sigma_{g_3}(k_{\theta_2}\theta_2) + \sigma_{g_4}(k_{\theta_1}\theta_1)\right]$$

Note that if $|k_{\theta_2}\theta_2| > g_4$ then $\dot{V}_1 < 0$. This implies that $\exists T_1$ such that $|\theta_2(t)| \leq g_4/k_{\theta_2}$ $\forall t > T_1$, choosing $g_3 > g_4$, this yields

$$\tilde{\tau}_\theta = -k_{\theta_2}\theta_2 - \sigma_{g_4}(k_{\theta_1}\theta_1) \quad \forall t > T_1 \tag{15}$$

Define

$$v = k_{\theta_2}\theta_1 + \theta_2$$

differentiating the above and using Eq. 15, it yields

$$\dot{v} = k_{\theta_2}\dot{\theta}_1 + \dot{\theta}_2 = k_{\theta_2}\theta_2 + \tilde{\tau}_\theta$$
$$= -\sigma_{g_4}\left[(k_{\theta_1}/k_{\theta_2})\,(v - \theta_2)\right] \quad \forall\, t > T_1$$

Define

$$V_2 = \frac{1}{2}v^2$$

this yields

$$\dot{V}_2 = v\dot{v} = -v\sigma_{g_4}\left[(k_{\theta_1}/k_{\theta_2})\,(v - \theta_2)\right]$$

Note that, if $|v| > g_4/k_{\theta_2}$, then $\dot{V}_2 < 0$. This implies that $\exists\, T_2 > T_1$ such that

$$|v| \le g_4/k_{\theta_2} \tag{16}$$

If

$$k_{\theta_2}^2 \ge 2k_{\theta_1} \tag{17}$$

then, $(k_{\theta_1}/k_{\theta_2})\,|v - \theta_2| \le g_4$. Thus, from Eqs. 15, 16 and 17 we obtain for $t > T_2$

$$\tilde{\tau}_\theta = -k_{\theta_1}\theta_1 - k_{\theta_2}\theta_2 = -K^T\bar{\theta} \tag{18}$$

where

$$K = \begin{pmatrix} k_{\theta_1} \\ k_{\theta_2} \end{pmatrix} \text{ and } \bar{\theta} = \begin{pmatrix} \theta_1 \\ \theta_2 \end{pmatrix}$$

Rewriting system 14, we have

$$\dot{\bar{\theta}} = A\bar{\theta} + Bu$$

where

$$A = \begin{pmatrix} 0 & 1 \\ 0 & 0 \end{pmatrix},\ B = \begin{pmatrix} 0 \\ 1 \end{pmatrix}$$

Using Eq. 18, we have

$$\dot{\bar{\theta}} = (A - BK^T)\bar{\theta}.$$

Then, we need to choose k_{θ_1} and k_{θ_2} such that the matrix $(A - BK^T)$ is stable and Eq. 17 is valid.

3.2 Translational Subsystem

In view of the above it's follows that $\psi, \theta, \phi \to 0$ as $t \to \infty$. Then, for a time T_3 large enough ψ, θ and ϕ are arbitrarily small, therefore, Eq. 5 becomes

$$m\ddot{x} = u_x$$

$$m\ddot{y} = u_y$$

$$m\ddot{z} = -mg + u + b\bar{u} \tag{19}$$

To stabilize the position of the aircraft, we propose

$$u = mg - b\bar{u} - m\sigma_{g_7}(k_{z_1}(z - z_d)) - m\sigma_{g_8}(k_{z_1}\dot{z})$$

$$u_x = -m\sigma_{g_9}(k_{x_1}(x - x_d)) - m\sigma_{g_{10}}(k_{x_1}\dot{x})$$

$$u_y = -m\sigma_{g_{11}}(k_{y_1}(y - y_d)) - m\sigma_{g_{12}}(k_{y_1}\dot{y}) \tag{20}$$

Introducing Eq. 20 into Eq. 19, we have

$$\ddot{z} = -\sigma_{g_7}(k_{z_1}(z - z_d)) - \sigma_{g_8}(k_{z_1}\dot{z})$$

$$\ddot{x} = -\sigma_{g_9}(k_{x_1}(x - x_d)) - \sigma_{g_{10}}(k_{x_1}\dot{x})$$

$$\ddot{y} = -\sigma_{g_{11}}(k_{y_1}(y - y_d)) - \sigma_{g_{12}}(k_{y_1}\dot{y})$$

Then, this implies (see method to θ dynamics) that $\ddot{z}, \ddot{x}, \ddot{y}, \dot{z}, \dot{x}, \dot{y} \to 0$ and $z \to z_d$, $x \to x_d$ and $y \to y_d$.

4 Platform Architecture

The platform is composed of an eight-rotor aircraft which has two RABBIT microprocessors RCM3400 on board. This microprocessor has the following main features: module running at 29.4 MHz, 512K flash memory, four PWM outputs, six serial ports, two input capture channels, more details see [3, 5, 15] . The microprocessor one runs the control algorithm in real-time to stabilize the eight-rotor rotorcraft, therefore it reads the information provided by the INS (Inertial Navigation System). The second microprocessor is used to compute the PWM level output to control the lateral rotors, using one camera that provides the velocity using the optical flow measurements and another camera for obtained the x and y position [1, 10].

The closed loop algorithm for the altitude used an infrared sensor, which works from 0.20 to 1.5 m. The inertial measurement unit is composed of three-axis gyro with range of ±300°/s, three-axis accelerometer with range of ±6 g and three-axis magnetometer. The INS provides the angular rate, the acceleration, and the direction of the earth magnetic field with a sample rate of up to 50 Hz. We have also tested

Fig. 4 ξ-Position of the rotorcraft

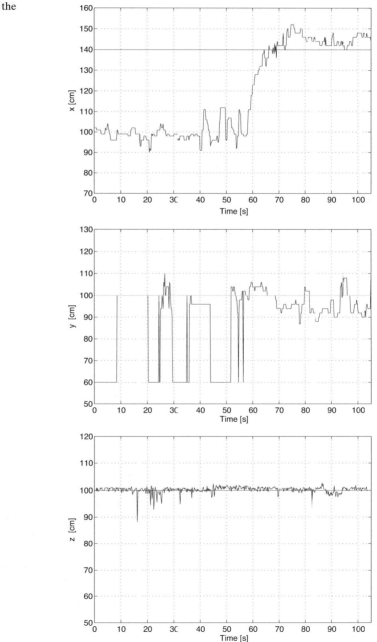

our platform using the measurements given by two cameras [8]. The position and velocity are available directly from the optical flow and video processing provides by the cameras at sample rate of 10 Hz [1, 10].

Fig. 5 η-Angles the rotorcraft

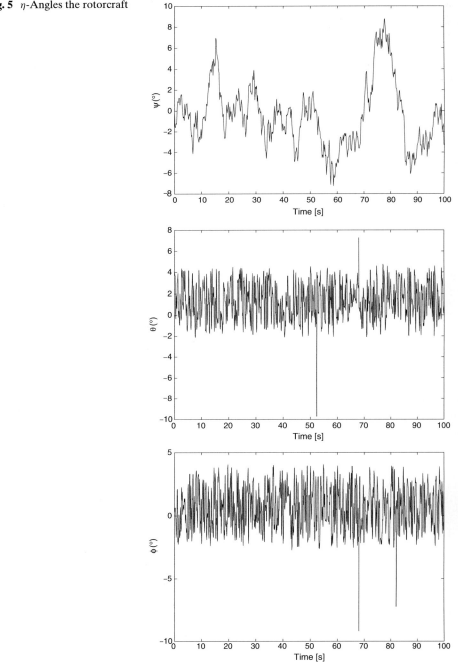

5 Experimental Results

Real-time experiment results are presented in this section to validate the performance of the eight-rotor rotorcraft during autonomous hover flight. The control
gains of equations were adjusted in practice to obtain a fast aircraft response but
avoiding mechanical oscillations as much as possible. The parameters were also
chosen in such a way that the aircraft attitude remains very close to a desired point.

Fig. 6 Translational control
inputs

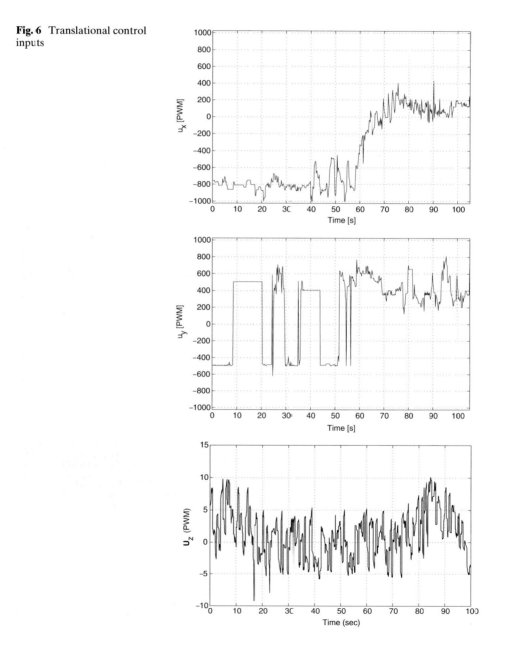

In the experiment, the mini-helicopter was stabilized at hover applying the proposed control strategy, we have obtained an acceptable behavior which is shown in Figs. 4 and 5. The control objective is to make the rotorcraft hover while $\xi = (140, 100, 100)$ cm and $\eta = (0, 0, 0)°$ (Figs. 4 and 5). Figures 6 and 7 show the control signals applied to the aircraft.

Fig. 7 Altitude control inputs

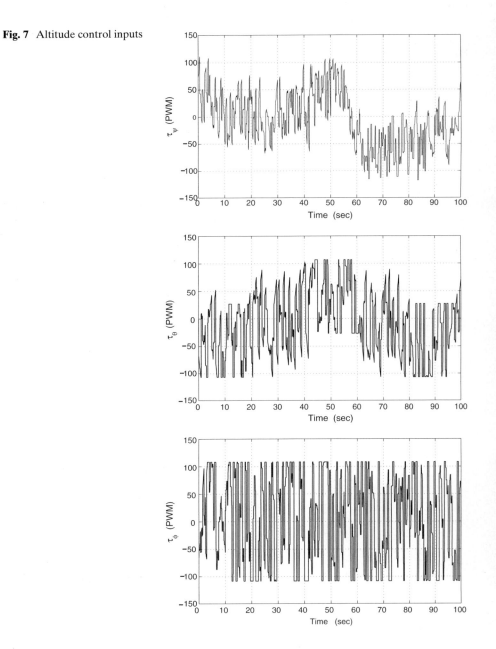

Those inputs control are added to offset set points $\sigma\_(\cdot)$ in the following structure

$$u_{xl} = \sigma\_x + u_x$$

$$u_{xr} = \sigma\_x - u_x$$

$$u_{yl} = \sigma\_y + u_y$$

$$u_{yr} = \sigma\_y - u_y$$

$$\tau_{\theta_2} = \sigma\_\theta + \tilde{\tau}_\theta + \tilde{\tau}_\psi + u_z + k_1 u_{yr}$$

$$\tau_{\theta_4} = \sigma\_\theta - \tilde{\tau}_\theta + \tilde{\tau}_\psi + u_z + k_1 u_{yl}$$

$$\tau_{\phi_1} = \sigma\_\phi + \tilde{\tau}_\phi - \tilde{\tau}_\psi + u_z + k_2 u_{xr}$$

$$\tau_{\phi_3} = \sigma\_\phi - \tilde{\tau}_\phi - \tilde{\tau}_\psi + u_z + k_2 u_{xl}$$

where the offsets are chosen when the vehicle begins the take-off and k_1 k_2 are the constants values.

6 Conclusions

In this paper we have presented the stabilization of a UAV having eight–rotors. One of the main characteristics of these configurations is that the translational and rotational dynamics are almost decoupled. Therefore we were able to apply a simple nonlinear control strategy to compensate the remaining dynamical coupling and to achieve the aircraft stabilization and location. The control algorithm considers the system constraints and is affordable for implementation in real-time applications. The real-time experiments have shown an acceptable performance of the helicopter applying the proposed control law.

References

1. Bouguet, J.Y.: Pyramidal implementation of the Lucas Kanade feature tracker. In: Technical report Intel Corporation (1999)
2. Lyon, D.: A military perspective on small unmanned aerial vehicles. Instrumentation and Measurement Magazine, IEEE, 7(3), 27–31 (2004)
3. Salazar-Cruz, S., Lozano, R.: Stabilization and nonlinear control for a novel tri-rotor mini-aircraft. In: Proc. of IEEE International Conference on Robotics and Automation, pp. 2924–2929 (2005)
4. Green, W.E., Oh, P.Y., Barrows, G.L.: Flying insect inspired vision for autonomous aerial robot maneuvers in near-earth environments. In: Proc. of IEEE International Conference on Robotics and Automation (2004)
5. Romero, H., Benosman, R., Lozano, R.: Stabilization and location of a four rotors helicopter applying vision. In Proc. American Control Conference ACC. pp. 3931–3936 (2006)
6. Castillo, P., Lozano, R., Dzul, A.: Stabilization of a mini rotorcraft with four rotors. Control Systems Magazine, IEEE 25, 45–55 (2005)
7. Tayebi, A., McGilvray, S.: Attitude stabilization of a four-rotor aerial robot. In: Proc. of Conference on Decision and Control, IEEE CDC., vol. 2, pp. 1216–1221 (2004)
8. Hartley, R., Zisserman, A.: Multiple View Geometry in Computer Vision, 2nd Edn. In: Cambridge University Press, ISBN 0521540518 (2004)
9. McCormick Jr., B.W.: Aerodynamics of V/STOL Flight. Dover Publication Inc. (1999)

10. Beauchemin, S.S., Barron, J.L.: The computation of optical flow. In: ACM Computing Surveys, **27**, 433–467 (1995)
11. King, C.Y.: Virtual instrumentation-based system in a real-time applications of GPS/GIS. In: Proc. Conference on Recent Advances in Space Technologies, pp. 403–408 (2003)
12. Sasiadek, J.Z., Hartana, P.: Sensor Fusion for Navigation of an Autonomous Unmanned Aerial Vehicle. In: Proc. International Conference on Robotics and Automation, vol. 4, pp. 429–434 (2004)
13. Yoo, C.-S., Ahn, I.-K.: Low cost GPS/INS sensor fusion system for UAV navigation. In: Proc. Digital Avionics Systems Conference, vol. 2, pp. 8.A.1-1–8.A.1-9 (2003)
14. Goldstein, H.: Classical Mechanics, 2nd Edn. Addison Wesley Series in Physics, Adison-Wesley, U.S.A. (1980)
15. Castillo, P., Lozano, R., Dzul, A.: Modelling and Control of Mini-Flying Machines. Springer-Verlag in Advances in Industrial Control. ISBN: 1-85233-957-8 (2005) July

An Overview of the "Volcan Project": An UAS for Exploration of Volcanic Environments

G. Astuti · G. Giudice · D. Longo · C. D. Melita ·
G. Muscato · A. Orlando

Originally published in the Journal of Intelligent and Robotic Systems, Volume 54, Nos 1–3, 471–494.
© Springer Science + Business Media B.V. 2008

Abstract This paper presents an overview of the *Volcan Project*, whose goal is the realization of an autonomous aerial system able to perform aerial surveillance of volcanic areas and to analyze the composition of gases inside volcanic plumes. There are increasing experimental evidences that measuring the chemical composition of volcanic gases can contribute to forecast volcanic eruptions. However, in situ gas sampling is a difficult operation and often exposes scientists to significant risks. At this aim, an Unmanned Aircraft System equipped with remote sensing technologies, able to sense the plume in the proximity of the crater, has been developed. In this paper, the aerial platform will be presented, together with the problems related to the flight in a hard scenario like the volcanic one and the tests performed with the aim of finding the right configuration for the vehicle. The developed autonomous navigation system and the sensors unit for gas analysis will be introduced; at the end, several experimental results will be described.

Keywords Unmanned aircraft system · Volcanic plume gas sampling · Autonomous navigation system · Hardware in the loop

Abbreviations

ADAHRS	air data attitude and heading reference system
COTS	commercial off-the-shelf
DIEES	Dipartimento di Ingegneria Elettrica, Elettronica e dei Sistemi
EKF	extended Kalman Filter
FCCS	flight computer control system
GBS	ground base station

G. Astuti · G. Giudice · D. Longo · C. D. Melita · G. Muscato (✉) · A. Orlando
Dipartimento di Ingegneria Elettrica, Elettronica e dei Sistemi (DIEES),
Università degli Studi di Catania, Viale A. Doria 6, Catania, Italy
e-mail: gmuscato@diees.unict.it
URL: http://www.robotic.diees.unict.it

K. P. Valavanis et al. (eds.), *Unmanned Aircraft Systems*. DOI: 10.1007/978-1-4020-9136-0_25   471

GCI	ground control interface
GDLS	ground data link system
GUI	graphical user interface
HIL	hardware in loop
HOTAS	hands on throttle-and-stick
INGV	Istituto Nazionale di Geofisica e Vulcanologia
PIC	pilot-in-command
PIL	pilot-in-loop
R/C	remotely controlled
SACS	servo actuator control system
UAS	unmanned aircraft system
UAV	unmanned aerial vehicle
VHUD	virtual head up display
WP	waypoint

1 Introduction

Nowadays Unmanned Aerial Vehicles (UAVs) are very often adopted as a powerful instrument for environmental monitoring [1–5] and several NASA projects are involved in the study of aerial platforms for earth science [6, 7]. The study of volcanoes represents one of the most interesting research activity in the field of robotics applied to earth science; even if several aerial platforms have been developed to be adopted in volcanic scenarios [8, 9], the collection and analysis of gas inside volcanic plumes is still an open issue.

The "Dipartimento di Ingegneria Elettrica, Elettronica e dei Sistemi" (DIEES) at the University of Catania, Italy, is involved in several research projects concerning volcanoes and related problems. Some of these activities are within the Robovolc project [10–13], whose aim is to develop robotic systems for the exploration and analysis of volcanic phenomena.

In situ gas sampling is one of the more reliable techniques to obtain information about the concentration of the main components of the fumes; the composition analysis of gasses emitted by volcanoes is crucial for volcanologists to understand volcanoes behaviors. Close field data collected during eruptions can be used as input data for computer simulation of volcanic activity, to improve forecasts for long-lived volcanic phenomena, such as lava flow eruptions [14–16].

For example measurements on the concentration ratios between different acidic gas species during the recent Mt. Etna volcano eruptions provided insights onto the depth of volatile exsolution and the evolution of volcanic activity. Similar measurements performed on Stromboli volcano during the 2002–2003 eruption evidenced very high S/Cl ratios just before paroxysm activities and a gradual decrease parallel to the exhausting of the eruption and the resuming Strombolian activity.

In situ gas sampling often exposes scientists to significant risks, these however still being unable to measure CO_2, the most important volatile in volcanic fluids. As a consequence the adoption of a flying machine seems to be the right choice: all the required instrumentation can be carried directly into the plume to obtain measures collected before contamination due to the interaction with the atmosphere.

The measurements of CO_2 in volcanic gas plume can be achieved by flying inside the plume an UAV mounting onboard ad-hoc sensors: CO_2/SO_2 ratios in Etna's plume will be measured by using a field-portable multi-sensor gas analyzer. The aim is to sample the dense "near vent" plume: in-situ measurements will be concentrated on the plumes of *Voragine* and *North-East* craters, the two currently most active degassing vents on the Mount Etna volcano. This operation will be devoted to the UAV that will be driven inside the plume by receiving GPS waypoint data from a base station. During the measurements in both crater's plumes, air will be actively pumped into infrared and electrochemical cells (working in series), allowing the real-time measurement of CO_2 and SO_2 plume concentrations. Finally, the correlation of this data with the GPS positions will give the needed map of the gases concentration.

This paper represents an overview of the *Volcan Project* [17–20], concerning the development of an autonomous aerial system, able to collect and analyze the gases into volcanic plumes. The partners involved in the Volcan Project are the DIEES, the "Istituto Nazionale di Geofisica e Vulcanologia"(INGV) and a local company, the OTe Systems. After the description of a typical mission, the entire system will be presented. The aerial platform will be described: several tests and experimental flights have been executed to choose the right configuration for the vehicle and to ensure the desired performance at a high altitude and in the hard volcanic environment.

A navigation system has been realized with the aim of making the UAV autonomous and able to perform a complete gas sampling mission without human intervention. The architecture and the sub-modules of the system will be presented, together with the achieved experimental results.

2 Planned Mission

Since most of the areas of Mt. Etna close to the main craters are very rough, it was decided to locate the base camp into the *"Piano delle Concazze"* ($37°45'55.59''$ N, $15°00'50.35''$ E), placed in the proximity of the foot of the *North-East Crater*. This large plateau allows take-off and landing thanks to its almost level ground in a zone very close to the "Valle del Bove" (*Valley of the Ox*) that is a large depression in the south-east side of the volcano where the volcanic plume, emitted by the *Voragine* and *North-East* craters, is usually driven by the wind. As a matter of fact, data recorded by the weather stations placed in the neighbourhood of the volcano usually report a wind blowing from North to North West in the direction of the *Valle del Bove*.

Moreover, *Piano delle Concazze* height is 2,800 m above sea level, an altitude close to that of the plume (3,000–3,500 m ASL), and at a distance of about 2,000 m from the zone of interest, where the greatest concentrations are expected for the chemical gases.

The volcanic cloud is usually about 500 m wide, so the cruise speed inside the plume must be lower than 17 m/s (60 km/h) in order to collect about 30 samples, with a sampling time of 1 s, for each passage into the plume.

Figure 1 shows a view of the zone of interest, as seen from the *Valle del Bove*, while Fig. 2 offers an aerial view of the Mount Etna volcano. The orange dashed line represents the planned mission. Figure 3 is a picture taken from the south side of the *Valle del Bove* depression showing the target plume emitted by the *Voragine* crater.

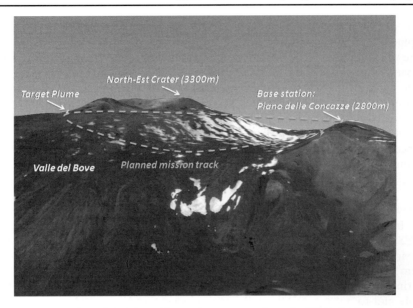

Fig. 1 A perspective view of the Mount Etna volcano as seen from the "Valle del Bove" valley. The *orange dashed line* represents the planned mission track: after manual take-off from the base camp placed in "Piano delle Concazze," the plane will autonomously execute the mission going inside the plume, sampling the gases of interest and coming back to the landing strip

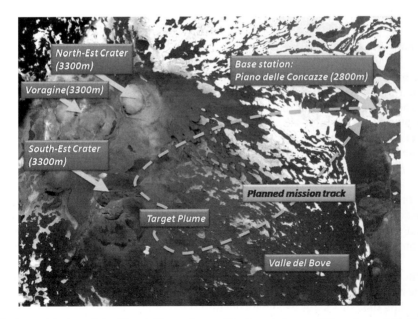

Fig. 2 An aerial view of the Mount Etna volcano. The target plume of *Voragine* and *North-East* craters is visible even in this picture taken from the satellite

Fig. 3 The Mount Etna volcano as seen from the south side of the Valle del Bove depression. The target plume is clearly visible in the photo

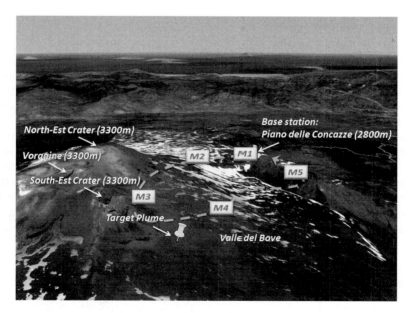

Fig. 4 An aerial view of a typical mission (*dashed line*; phases M2 to M4 will be autonomously performed by the UAV following a pre-planned trajectory, while take-off (M1) and landing (M5) will be pilot-in-command operations

Fig. 5 June 2007 mission executed at an altitude of 2,800 m with a little trainer model. *Blue line* is part of the track extracted from the mission log. The on-board autopilot was in automatic navigation through waypoints; the *yellow stars* represent the two assigned waypoints. The series of turns was caused by the strong North–North West wind (80 km/h) that hindered the course of the plane

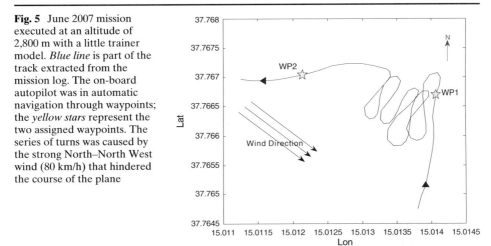

Figure 4 shows the main phases of a typical mission:

-   M1: remotely operated (pilot in command) take-off
-   M2: initial climb out
-   M3: waypoints based autonomous flight (measurement phase)
-   M4: base camp approach and descent phase
-   M5: remotely operated (pilot in command) landing

One of the main problem related to this kind of mission is the presence of very high wind and strong turbulences: during the past missions and on-the-spot investigations executed at a high altitude on the top of the Mount Etna volcano, very high winds have been noticed. In a mission of June 2007 executed at an altitude of 2,800 m with a little trainer model, the strength of north-west wind reached a speed of more than 80 km/h.

Figure 5 shows part of the track held by the plane during this mission; two waypoints were assigned (yellow stars) and the *Mission Mode* was activated in the on-board autopilot (see Section 4), enabling the automatic navigation through waypoints. As can be observed, once the first waypoint WP1 was reached, the plane turned left pointing to the second waypoint, but a strong wind gust hindered the course of the aircraft. A series of turns was then needed to reach the second waypoint.

3 The Aerial System

The choice and the design of an aerial system intended for operations in hard environments and conditions, involve taking into account several constraints: among the others, major significance have the problems related to high altitude flight and gas sampling technique. More details about the Volcan UAV design can be found in [17–20].

The main project specifications are:

- The adoption of an electric engine for propulsion. This constraint is mainly imposed by the volcanologists: gas emitted by an internal combustion engine could alter the chemical composition of the volcanic cloud and invalidate the collected measures
- Autonomy of about 30 min and a working range of 3 km: the volcanic plume is at a distance of about 2,000 m from the base camp
- A payload of about 5 kg: a high payload is required in order to carry the onboard avionics (autonomous navigation system, radio link, INGV sensors, data logger)
- A minimum cruise speed of about 40 km/h to collect a satisfactory number of samples into the plume, as described in the section dedicated to the planned mission
- A maximum cruise altitude of 4,000 m

Several test flights and laboratory experiments have been performed during the last three years to evaluate the materials, the engine and the avionics in a hard environment like the volcanic scenario and to establish the quality, performance and reliability of the entire system at a high altitude, where some problems, like a reduction of thrust and efficiency of the engine or a drop in the lift, can occur.

The main difference between flying at sea level and flying on the top of the volcano is the air density, which decreases as the height increases. Moreover, it depends on temperature, pressure and humidity. Plane wings and propeller behavior are strongly related to the air density; moreover also gasoline engine behavior depends on all these parameters. As general rule, all these components reduce their performances as the air density decreases, so a suitable design of the wing profile and a suitable choice of the propeller must be done. It must be taken into account that an electrical engine does not suffer from air density reduction.

In this section the aerial platform, chosen taking into account the results of the conducted experiments, is presented.

3.1 The Aerial Platforms

The aircraft was selected taking into account the problems related to the flight at an altitude above 3,000 m (9,840 ft) and in presence of strong turbulences with wind speed greater than 80 km/h (43.2 kn).

Considering the project constraints and following several design trials, the Volcan UAV was designed. Inspired by the configuration of the more famous Aerosonde [21], the realized plane has a fuselage made in carbon fiber and fibreglass and wooden wing and V-tail, a wing span of 3 m, a total weight of 13 kg, a 2,000 W brushless motor, Li-Po battery packs and a maximum cruise speed of 150 km/h. It meets all the specifications described above and can be seen in Fig. 6. Nowadays, the Volcan is launched from a car, but a fixed launching system is under study in order to allow the take-off even in locations where a suitable flat area for a car is not available or in presence of snow.

In [17–20] further information about the Volcan UAV can be found.

Several missions have been conducted to test the behavior of the plane in volcanic environment: a good maneuverability and a high stability were reached even in harsh conditions.

Fig. 6 The Volcan UAV

To make the development easier and faster and to minimize risks for the Volcan UAV, a commercial low cost model airplane was used in the initial phase. The plane was chosen taking into account the same considerations done for the realization of the Volcan UAV, but with the aim of making simpler take-off and landing operations, as well as the assembling and the transportation. After the examination of several kinds of plane models, the Graupner Taxi2400 [22] was chosen as the *DPVolcan* (Development Phase Volcan); with a wing-span of 2.39 m and a wing area of 84 dm^2, this model is a powerful plane for aero-towing of large scale sailplanes (Fig. 7). Its robustness and stability makes this aircraft well suited to fly in volcanic environments; a large fuselage and a high payload allow carrying all the on-board electronic and sensor modules.

3.2 The Engine

Even if the final configuration will require an electric motor, both electric and internal combustion engines have been tested to compare their behaviors at a high altitude. An experimental workbench was used to measure by means of an electronic dynamometer the static thrust of several motors with different types of propellers; although the static thrust is quite different from the dynamic thrust furnished by the engine during the real flight, this tests allowed to understand:

- The loss of efficiency at a high altitude for several types of engines
- The right type of propeller in terms of length and pitch for the engines under test

Moreover, the carried out experiments had the purpose of finding the best engine-propeller combination for the Graupner Taxi2400 plane, so the experiments were focused on two engines suggested for this kind of aircraft.

In Table 1 the main features of the tested engines are reported.

Fig. 7 The *DPVolcan*. Designed for aero-towing, this big-scale Graupner Taxi2400 model combines a good maneuverability with an excellent robustness, permitting the use in volcanic environment

Table 1 The main features of the engines tested at a low altitude (100 m ASL) and at a high altitude (2,585 m ASL)

Engine name	Engine type	Type of feeding
Graupner Compact650	Electric	6 Li-Po cells (22 V 4 A 30 C)
Roto 35Vi	Two-stroke 34.8 cm^3	Petrol

Table 2 reports the results of trials executed at the DIEES laboratory at an altitude of 100 m above sea level together with the data obtained during the tests performed at an altitude of 2,585 m ASL. To check the measurement procedure for accuracy, power absorption and RPM have been monitored respectively for the electric motor and the internal combustion engine, comparing the measured values with the expected ones.

The absolute and percentage reductions of the produced static thrusts, due to the change in the altitude, are also shown in Table 2; as it was expected, there is a remarkable loss in terms of thrust at a high altitude. The best performance was obtained by the electric engine with an 18\12 propeller and by the internal combustion engine with a 21\10 propeller.

However, several considerations suggested the adoption of an internal combustion engine for the Graupner Taxi2400 model:

- Because of the frequent presence of high wind, an engine with better performances than the one suggested by Graupner was chosen.
- During the development phase, the adoption of an electric engine is wasteful in terms of costs and time. As a matter of fact, a powerful electric motor needs very expensive high capacity Li-Po batteries and several spare batteries must be available. Moreover, batteries have to be replaced very often causing a significant waste of time due to the take-off and landing procedures.

Therefore, a ZDZ 80 cm^3 two-stroke petrol engine was adopted for the DPVolcan plane; this twin-cam motor assures a static thrust of 15 kg at a low altitude and has a nominal power of 5.88 kW (7.89 HP). Several tests are under examination in order to verify the correlation between the measures of interest (CO_2, SO_2... gas plume) and the emissions produced by the petrol combustion. If these tests will show that the measures are not corrupted by the exhausts, than the internal combustion engine might be adopted on the final plane configuration.

Table 2 This table reports the static thrust produced by the engine under test with several types of propellers during the trials executed at an altitude of 100 m above sea level and 2,585 m ASL

Engine	Propeller length\ pitch [in]	Static thrust (100 m ASL) [kg]	Static thrust (2585 m ASL) [kg]	Static thrust absolute reduction [kg]	Static thrust percentage reduction [%]
Compact650	20\10	3.50	3.00	0.50	14.28
Compact650	18\12	2.50	1.70	0.80	32.00
Roto 35Vi	22\12	6.00	4.80	1.20	20.00
Roto 35Vi	22\10	6.50	5.60	1.10	13.84
Roto 35Vi	21\12	6.00	5.20	0.80	13.33
Roto 35Vi	21\10	7.00	6.00	1.00	14.28

Absolute and percentage reduction caused by the change in altitude are presented in the last two columns.

4 The Vehicle Control and Mission Management System

The environmental and working conditions in which the plane has to carry out the mission, suggested the adoption of an autonomous navigation system for the plane. There are two main problems that do not allow using a classic remote controlled UAV.

The first one is the long distance between a safe place and the volcanic plume. The volcanic scenario is very rugged and often it is not possible to find a runway in proximity of the crater; take-off is usually executed in safe areas from where the R/C teleoperations are very difficult due to the distance between the take-off area and the target plume.

The second problem is related to the loss of the line of sight when the plane is flying inside the fumes: the gas within the plume is usually very dense and does not allow a visual recognition of the vehicle.

The main project specifications for the vehicle control and mission management system, are then:

– An autonomous navigation system (except for take-off and landing)
– Path planning through way points
– Local and remote data log of the measures
– Real-time visualization on a user-friendly PC interface

A navigation system was then developed with the aim of making the Volcan UAV autonomous: a set of avionic modules that allow configuring the plane in every state of the flight and make the UAV capable of autonomous flight and payload control, was realized.

The block diagram of the developed Unmanned Aircraft System (UAS) is shown in Fig. 8; blue blocks represent the boards/devices developed by DIEES and OTe Systems, green blocks are the equipments developed by INGV, while orange blocks stand for Commercial Off-the-Shelf (COTS) devices.

The red square in the *On-Board Avionics* group, holds the devices that compose the core of the developed autopilot; the implemented navigation and stability algorithms allow obtaining a fully autonomous flying platform able to perform a complete preplanned mission.

Several features make the implemented autopilot versatile and flexible, allowing to reconfigure the system and to adapt its behavior to the mission requirements; the main peculiarities are:

– The type of control can be selected between a full waypoints based autonomous navigation and a pilot in the loop (PIL) mode. In the first case the autopilot takes care of both navigation and stability of the plane: the mission is planned via waypoints, placing on a georeferred map the position of each waypoint (WP) at the beginning of the mission. The mission can be easily modified during its execution by adding/changing/removing waypoints in the map. In PIL mode, the plane can be teleoperated from the control station by using a Hands on Throttle-and-Stick (HOTAS) while the on-board autopilot attends to the stability of the aircraft.
– A user-friendly interface and a radio link allow a continuous exchange of data between the plane and the control station; the graphical user interface (GUI) is used to display the plane position in real time on a map during the mission,

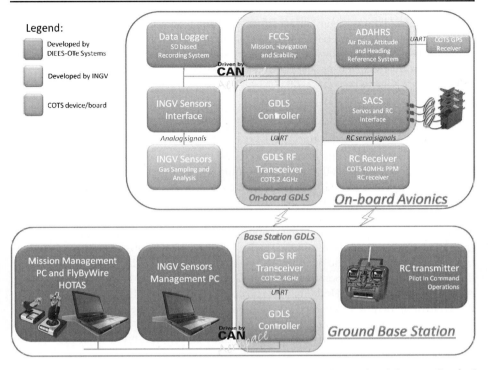

Fig. 8 The Volcan Unmanned Aircraft System: the on-board avionics consist of the autopilot (*red square*), the data-logger and the boards dedicated to the INGV sensors. The Ground Base Station allows both to plan and manage the mission and to display the INGV sensors data. A COTS 2.4 GHz radio link system is used for the data exchange between the plane and the ground station

to monitor some UAV parameters such as battery levels, speed, position and orientation, sensors measurements.

- On line setting of navigation and stability algorithms parameters allows to reconfigure the autopilot, adapting its response to new supervened conditions. This important feature is fundamental in volcanic environments, where changes in the weather are very frequent.

In the following, the modules constituting the developed UAS will be described into detail.

4.1 INGV Sensors and Data Logger

The gas analyzer integrates an infrared spectrometer for CO_2 determination (model Gascard II, calibration range 0–3,000 ppmv; accuracy ±2%; resolution 0.8 ppmv) and an electrochemical sensor specific to SO_2 (SO2-S-100 model Membrapor, calibration range 0–200 ppmv; accuracy 2%; resolution 0.5 ppmv).

The conditioning circuits of the *INGV Sensors Interface* board allow to deal with the analog signals coming from gas sensors; a Microchip microcontroller provided with a 10 bit AD converter acquires sensors signals and sends the converted data to the data-logger board and the radio link interface through the CAN bus.

The on-board data-logger has a double function: it acts as a black-box recording all the flying parameters and data and, at the same time, stores all the measures coming from the INGV sensors together with the plane position derived from the on-board GPS receiver. In such a way, it is possible to know the exact position of each collected sample.

4.2 FCCS – Flight Computer Control System

The Flight Computer Control System (FCCS) provides for autonomous navigation and stability control, as well as acts as a command and stability augmentation system. This device works as a real autopilot able to perform a complete planned mission through waypoints navigation; moreover, it performs pilot-aiding functionalities to implement Pilot-In-Loop operations.

The FCCS processor accomplishes both flight-control and mission completion tasks. The control thread, running in the FCCS processor, implements three PID control loops that compute the positions of the aircraft control surfaces, required to bring the plane to the desired position, heading, speed and altitude. Data coming from the Air Data Attitude and Heading Reference System (ADAHRS) are processed by the control algorithms to provide output commands to the Servo Actuator Control System (SACS) via CAN bus.

The implemented operative modes, selectable during the flight from the *Mission Management PC* GUI, are:

Mission Mode

This is the modality that allows performing a complete autonomous mission: the path is assigned to the autopilot by fixing a series of geo-referred waypoints on a map in the *Mission Management PC*.

Stability Augmented Fly-by-Wire

This mode provides for the main Pilot-In-Loop operations; the pilot acts on the HOTAS of the Mission Management PC to fly the aircraft. Moving the control stick, the pilot sets the attitude angles rates and the throttle position; the maneuver parameters are sent to the autopilot through the *Ground Data Link System* (GDLS). The FCCS then acts on the control surfaces as required to satisfy the pilot's demand and to ensure the stability of the aircraft.

Flight Hold

Three buttons in the GUI of the *Mission Management PC* allows engaging *Airspeed Holding, Heading Holding* and *Altitude Holding* modes.

When *Airspeed Holding* mode is activated, the FCCS works to maintain the pre-selected airspeed by controlling the engine thrust. In *Heading Holding* mode, the roll angle is regulated to follow a pre-selected heading angle, while in *Altitude Holding* altitude by pitch regulation is activated.

Flight Hold may be activated in the *Mission Mode*, as well as in the *Stability Augmented Fly-by-Wire*.

Flight Director

The *Flight Director* mode may be used to offer visual cues to the pilot; the FCCS computes the maneuvers, in terms of roll and pitch angles, needed to reach a programmed destination (*Mission Mode*) or to maintain a selected flight state (*Flight Hold*) and sends them to the ground station through the radio link.

In the *Mission Management PC* GUI, a dedicated *Flight Director Visual Interface* has been integrated in both pitch and roll angles panel instruments; maneuver angles computed by the FCCS algorithms are displayed together with the actual angles.

In Fig. 9 the two panel instruments are shown: on the left the roll angle indicator, on the right the pitch angle one. The white Volcan pictures represent the actual attitude angles, while the colored ones are the maneuvers suggested by the FCCS to reach the desired flight condition: the pilot must turn right and pitch up until the white Volcans overlap colored ones.

Attitude Control Mode

This modality is helpful in the tuning of navigation and stability control algorithms. In the *Attitude Control Mode*, the FCCS performs as a stability controller, providing for attitude (Roll and Pitch) regulation.

To tune the autopilot, a reference attitude angle is assigned and the response of the plane, in terms of rising time, overshoot, settling time and steady-state error, is used to find the appropriate values for the PID control loops.

Configuration Mode

Direct access to control algorithm parameters is implemented in order to allow fine tuning. *Configuration Mode*, combined with *Assisted Mode*, permits to easily and quickly perform the on line setting of navigation and stability algorithms parameters, allowing observing the real response of the aircraft during the flight.

The FCCS processor, a 8 bit Microchip PIC, is mounted on the same electronic board of the Air Data Attitude and Heading Reference System as is shown in Fig. 10.

During bench tests, the use of a Hardware In Loop (HIL) architecture based on X-Plane simulator has been an ideal platform to observe the behavior of our own customized plane. In [17] the FCCS architecture is described, together with the

Fig. 9 The attitude panel instruments of the Mission Management PC. When the FCCS is in Flight Director mode, the maneuvers suggested by the autopilot (*colored pictures*) are shown together with the actual attitude angels (*white pictures*): the pilot can follow the FCC's advices by moving the HOTAS so as to overlap the pictures

Fig. 10 The developed Air Data Attitude and Heading Reference System. The sensors set is shown, together with the ADAHRS processor. On the *top-left*, the FCCS processor

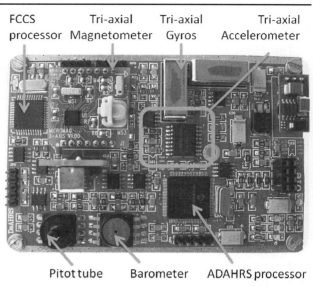

Hardware in the Loop technique, based on the adoption of X-Plane flight simulator by Laminar Research [23], used in the development phase of the FCCS board.

As described in [17], control architecture is based on the adoption of two cascade PID controllers for the attitude regulation:

1. *Heading-by-Roll* regulation: the desired course is the reference value for the outer loop that computes the roll angle set point for the inner control loop. The last calculates the desired positions for the ailerons and rudder servomotors.
2. *Altitude-by-Pitch* regulation: the desired altitude is given as set-point for the outer control loop, whose output represents the pitch angle reference for the inner loop. The output of the inner control loop is the desired position for the elevator servomotors.

A simple PID controller is used for the airspeed regulation; the reference value is the assigned altitude, while the output is the desired position for the throttle servomotor.

Figure 11 shows a block diagram depicting the interaction of the FCCS with the other autopilot modules. The set-points values depends on the FCCS operating mode; in *Mission Mode* and *Fly Hold* mode, reference values are given as input to the outer control loops, that compute the set-point for the inner loops. In *Stability Augmented Fly-by-Wire* and *Attitude Control* modes, the outer loops are disengaged and reference values are directly given as input to the inner control loops.

4.3 ADAHRS – Air Data Attitude and Heading Reference System

Information coming from MEMS tri-axial accelerometers, tri-axial piezo gyros, tri-axial magnetometers, temperature sensor, barometric altimeter, airspeed pressure sensor and GPS are integrated by a sensor fusion algorithm, based on Kalman Filtering techniques, running on a high performance high-speed digital signal processor to provide a complete 6DoF attitude and position solution.

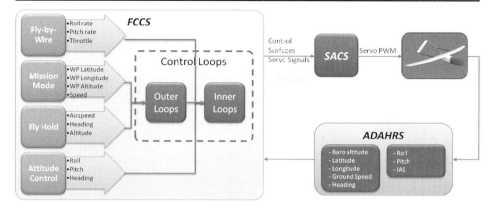

Fig. 11 A block diagram depicting the interaction between the FCCS, the Servo Actuator Control System and the Air Data Attitude and Heading Reference System

Figure 12 depicts the architecture of the developed board while in Fig. 10 the board is shown. The sensor suite of this strap-down low cost inertial navigation unit consists of:

- Freescale MMA 1220D accelerometer for the Z-axis
- Freescale MMA 2260 accelerometers for X- and Y-axes
- Tokin tri-axial piezo gyros
- Freescale MPX 5010 differential pressure sensor for air speed measurement (Pitot tube)
- Freescale MPXAZ4115A absolute pressure sensor for altitude determination
- Second order hardware filters for analog sensors filtering
- COTS GPS receiver
- Dallas DS18S20 temperature sensor
- Pni MicroMag3 three-axis magnetic sensor module
- Microchip dsPIC33FJ256GP710 general purpose digital signal controller

Fig. 12 A block diagram showing the architecture of the ADAHRS; analog and digital information coming from the sensors are processed by a sensor fusion algorithm based on an Extended Kalman Filter to reconstruct the pose and the navigation data of the plane. Filtered data are sent over CAN bus by using the CANaerospace protocol

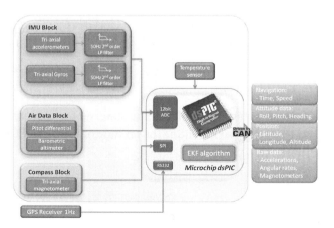

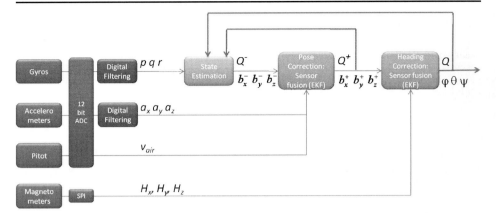

Fig. 13 A block diagram of the two-stage Extended Kalman Filter implemented in the dsPIC digital signal controller of the ADAHRS board. Quaternion algebra is used to make the algorithm faster and more accurate

This unit represents the second generation of developed ADAHRS; the first board [18] was based on a 8 bit Microchip PIC and the heading angle was computed by using the course angle furnished by the plane positions coming from the GPS receiver. The first board was successfully used during the development phase and in the first campaign of flights; even if no problem was noticed during the executed missions, this kind of approach could suffer in presence of high and/or lateral wind: if the plane is flying crabwise, the course obtained by the GPS is different from the real heading. The second generation ADAHRS solves this problem thanks to the tri-axial magnetic sensor module.

A two-stage Extended Kalman Filter (EKF) [24] computes plane attitude by combining sensors measures: a block diagram of the sensor fusion algorithm implemented in the dsPIC is shown in Fig. 13. Quaternion algebra makes developed algorithms faster and more accurate.

Accelerometers and gyros analog signals are acquired by the dsPIC internal 12bit ADC. Several tests have been executed to study sensors behavior: vibration dependences, static and dynamic responses have been analyzed to implement ad-hoc digital filters able to guarantee high performance in all conditions and environments. Moreover, sensors drifts and noises have been examined to better identify the process and measurement error covariance matrices of the implemented EKF.

4.3.1 Attitude Estimation

In the first stage of the filter, the state vector consists of the unit quaternion $Q = \begin{bmatrix} q_0 & q_1 & q_2 & q_3 \end{bmatrix}^{\mathrm{T}}$ representing the attitude of the plane [25] and the low frequency gyro drifts $\mathbf{b} = \begin{bmatrix} b_x & b_y & b_z \end{bmatrix}^{\mathrm{T}}$:

$$\mathbf{x}_1 = \begin{bmatrix} q_0 & q_1 & q_2 & q_3 & b_x & b_y & b_z \end{bmatrix}^{\mathrm{T}}$$

In the *State Estimation* block, the unit quaternion $\mathbf{Q}^-$ is computed by using the well known set of differential equation relating the unit quaternion and the angular rates, taking into account the gyro biases [26–29]:

$$\omega_x = p - b_x$$
$$\omega_y = q - b_y$$
$$\omega_z = r - b_z$$

where p, q, r are the measures provided by the tri-axial gyros; gyro drifts are modeled as a first order system with a time constant tuned according to the biases behavior registered during the sensors characterization phase.

The linear dynamic accelerations a_x, a_y, a_z, coming from the tri-axial accelerometers, and the air speed V_{air}, provided by the Pitot tube, are used to compute static gravitational components [30]. These are used in the first EKF (*Pose Correction* block) to correct attitude computation and to calculate the corrected unit quaternion $\mathbf{Q}^+$ and gyro biases.

4.3.2 Heading Estimation

In the second stage of the EKF (*Heading Correction* block), magnetic field components $\mathbf{H} = \begin{bmatrix} H_x & H_y & H_z \end{bmatrix}^T$ measured by the magnetometers along the three axes of the plane allow to correct the heading angle.

The implemented filter computes the error between:

– the heading angle estimated through the attitude unit quaternion $\mathbf{Q}^+$ coming from the first stage;
– the heading angle calculated by using magnetometers measures.

The error is then used by the EKF to correct the attitude quaternion $\mathbf{Q}^+$.

The unit quaternion $\mathbf{Q}$ and the attitude angles φ, θ, ψ, together with sensors' raw data, are sent over CAN bus by using the CANaerospace protocol.

The main ADAHRS frequencies are:

– Anti-aliasing filters: second order 50 Hz low-pass filters
– ADC sampling: 150 Hz
– Digital filtering: 20 Hz FIR/IIR
– Sensor fusion algorithm:
    State estimation: 80 Hz
    Pose correction: 80 Hz
    Heading/yaw correction: 20 Hz
– Transmission over CAN bus: 50 Hz

Figure 14 shows a comparison between the developed ADAHRS and a commercial inertial measurement unit, the MTi by XSens; during the test, the two inertial measurement unit where fixed to the same swinging plane and attitude angles were measured.

In [18] the Hardware in the Loop architecture based on X-Plane flight simulator by Laminar Research used in the development phase of the ADAHRS board was described, together with the obtained experimental results.

Fig. 14 MTi roll angle (*blue line*) vs. ADAHRS estimated roll angle (*red line*). Sampling time is 20 ms

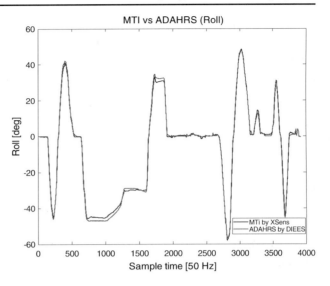

4.4 SACS – Servo Actuator Control System

The SACS receives the servo commands via CANaerospace and simultaneously controls the attached RC servos using PWM outputs. Every SACS can control up to eight servos or other peripherals and allows to activate manual override, through a normal R/C Radio command.

The SACS has two modes of operation:

–   Pilot-In-Command (PIC) mode: servo signals from R/C receiver are assigned directly to the servomotors. This mode is used during the take-off and landing procedures, when the plane is operated by the pilot through the R/C transmitter.
–   Autopilot (UAV) mode: servo output signals are driven by CANaerospace frame coming from the FCCS. This is the mode assigned when the autopilot is in command.

Upon system startup the PIC mode shall be active; switching to UAV mode is possible through a control signal coming from R/C receiver.

The SACS also supplies electrical power to the servos which it derives from the onboard batteries; opto-coupled servo signals increase power supply noise immunity. It also monitors voltage and current absorption of the batteries and transmits this information via CANaerospace.

4.5 GDLS – Ground Data Link System

These modules provide for the radio link between on board electronics devices and the ground station; they are based on the COTS MHX2400 transceiver by Microhard System:

–   2.4 GHz ISM band
–   maximum allowable transmit power = 1 W

- transparent, low-latency communications providing true 115 kbaud operation
- 60 mi, line of sight, with gain antenna

GDLS implements a CANaerospace bus in a transparent mode. Data coming from the CAN bus are compressed by the *GDLS Controller* (Fig. 8) to increase the throughput, then are sent to the transmitting *RF Transceiver* by using the UART serial bus. Data received by the *RF Transceiver* over the RF link, are sent via UART to the correspondent *GDLS Controller*, decompressed and delivered via CAN bus.

4.6 GBS – Ground Base Station

The *Ground Base Station* (GBS) allows setting and modifying the plane configuration, to plan and monitor the mission, to manage payload and sensors.

The two notebooks of the base station run the developed application programs:

- *INGV Sensors Management PC*: a dedicated software allows managing and displaying data coming from the on-board gas sensors together with the sampling position provided by the GPS. Real-time visualization as well as statistical analysis are possible.
- *Mission Management PC*: a *Ground Control Interface* (GCI) allows easy "click and fly" operations, facilitating powerful mission planning, monitoring and in-flight adjustment.

The GCI is divided in several sections:

Plane configuration

This section is a useful tool used to configure specific UAV platform parameters e.g. propulsion system, special control surfaces (VTail, XTail, flaperons and so on), payload control functions.

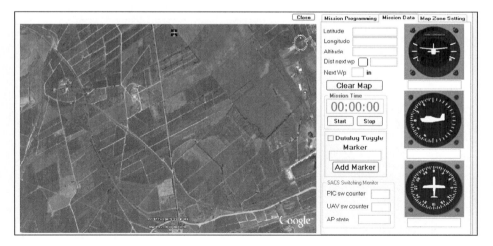

Fig. 15 The *Mission Management* section of the *Ground Control Interface*. On the *right*, the panel instruments are shown. The georeferred map allows to plane the mission via simple point n' click waypoints placing

Mission Management

This section, shown in Fig. 15, allows configuring and managing the mission; the main features are:

– Mission planning via point'n click waypoints placing on a geographically referred map
– In-flight mission reprogramming and WPs uploading
– Flight plan save and retrieve capabilities
– Plane position visualization in a georeferred map. The same map is also used to display the planned waypoints and the executed mission path
– Attitude angles and flight parameters (airspeed, ground speed, altitude, batteries level, system alerts, faults and status) visualization on the panel instruments

This section permits also to engage and set the *Airspeed Holding, Heading Holding* and *Altitude Holding* when the FCCS is in the *Flight Hold* mode.

FCCS Configuration

This section allows to set-up the FCCS. To make easier and faster the integration process of FCCS with a specific airplane, GCI offers a dedicated configuration tool for:

– Real time setting of the control loops parameters during the flight
– Real time plotting of the most significant data involved in the control loops tuning

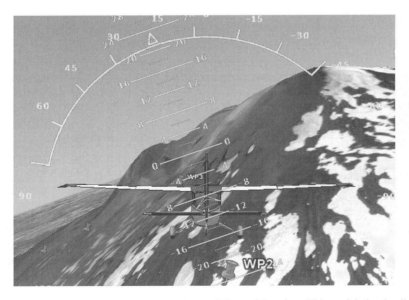

Fig. 16 The *Virtual Head Up Display* of the *Ground Control Interface*. This useful plug-in allows to know the plane attitude through the virtual instruments without obstructing the pilot's view when Fly-by-Wire functionalities are engaged

Fly-by-Wire Management

This section is used for the main Pilot-In-Loop operations; the pilot acts on the HOTAS to fly the aircraft. Moving the control stick, the pilot sets the attitude angles rates and the throttle position; the maneuver parameters are sent to the autopilot through the Ground Data Link System (see *Stability Augmented Fly-by-Wire* mode in the section dedicated to the FCCS).

To facilitate Fly-by-Wire operations, a special plug-in has been developed to display the position and the attitude of the plane through Google Earth, in a 3D virtual environment; moreover, a *Virtual Head Up Display* (VHUD) allows to presents navigation data without obstructing the pilot's view (see Fig. 16).

In addition, a *Flight Director Visual Interface* has been integrated in both pitch and roll angles panel instruments, in order to display the visual cues sent by the FCCS when the *Flight Director* mode is engaged (see the section dedicated to the FCCS).

The result is a virtual cockpit able to help Pilot-in-Loop operations and giving the pilot all the required information.

5 Experimental Results

Several flight campaigns have been executed during the last year to test the FCCS and ADAHRS board and to tune the autopilot parameters; the localization, navigation and stability algorithms have been tested and calibrated with several types of plane and propulsion systems to be assured of the best performances with all of the possible aircraft configurations. Many missions at high altitude have been performed, in order to test the behavior of the developed UAS in a hard environment.

The validity of the UAS was confirmed by several flight tests.

Figure 17 shows the roll angle when a wide change in roll angle was required in presence of a high wind; while the plane was flying with a roll angle of $-30°$, a $30°$ roll angle reference was assigned to the autopilot in Attitude Control Mode. The blue line is the measured roll angle, the red one is the reference angle. The sampling time was 100 ms.

Fig. 17 Measured roll angle (*blue line*) when aircraft was flying with a roll angle of $-30°$ and a roll reference value of $30°$ (*red line*) was assigned to the autopilot. The sampling time was 100 ms

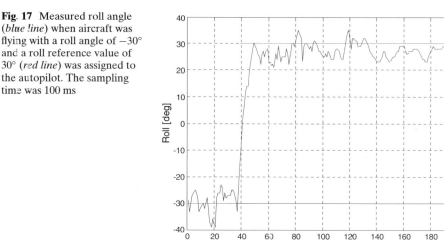

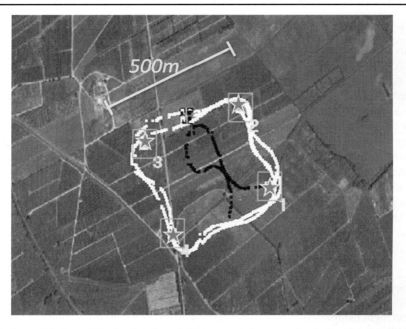

Fig. 18 An aerial view of the track followed during a mission executed in the Summer 2007. The *red stars* represent the four assigned waypoints; after the take-off (*black part* of the path), mission mode was activated. The *white trajectory* represents the mission course executed several times before landings

Nowadays, the system is extremely reliable and stable and allows to execute a complete autonomous mission.

Figure 18 shows an aerial view of one of the missions executed in the summer 2007; a squared path with a perimeter of 2,000 m was assigned by fixing four WPs (red stars). After the take-off and climbing (black track), the mission mode was activated, enabling the navigation through WPs; the white pattern represents the path followed by the DPVolcan with the trajectory executed several consecutive times.

6 Conclusions

In this paper an overview of the state of the Volcan project, concerning the development of a UAS for gas analysis and exploration in volcanic environment, was described. The system allows to perform autonomously a complete measurement mission with the exception of take-off and landing phases. The system allows the users easily to change mission parameters and to monitor all flight data and gas analysis sensors during the flight. Moreover several functionalities have been introduced to facilitate the mission executions, such as *Stability Augmented Fly-by-Wire* and *Flight Hold* mode.

Experimental tests have been performed confirming the validity and reliability of the designed system. Extensive measurement campaigns in Mt. Etna volcanic area are planned for the summer 2008.

Acknowledgements This work was carried out in cooperation with OTe Systems, Catania, Italy
(http://www.otesystems.com).

References

1. Ramanathan, V., et al.: Warming trends in Asia amplified by brown cloud solar absorption. Nature **448**, 575–579 (2007)
2. Ramanathan, V., Roberts, G., Corrigan, C., Ramana, M. V., Nguyen, H.: Maldives AUAV Campaign (MAC): observing aerosol–cloud–radiation interactions simultaneously from three stacked autonomous unmanned aerial vehicles (AUAVs). Available at: http://www-abc-asia.ucsd.edu/MAC/MAC_proposal_FINAL_2005July05.pdf (2005)
3. Holland, G.H., et al.: The aerosonde robotic aircraft: a new paradigm for environmental observations. Bull. Am. Meteorol. Soc. **82**, 889–901 (2001)
4. Valero, F.P.J., Pope, S.K., Ellingson, R.G., Strawa, A W., Vitko, J. Jr.: Determination of clear-sky radiative flux profiles, heating rates, and optical depths using unmanned aerospace vehicles as a platform. J. Atmos. Ocean. Technol. **13**, 1024–1030 (1996)
5. Bland, G., Coronado, P., Miles, T., Bretthauer, J.P.: The AEROS Project experiments with small electric powered UAVs for earth science. In: Proceedings of Infotech@Aerospace. American Institute of Aeronautics and Astronautics, 26–29 Sep. 2005, Arlington, VA, USA
6. NASA—Ames Research Center, Earth Science Division. Available at: http://geo.arc.nasa.gov/
7. NASA—Dryden Flight Research Center. Available at: http://www.nasa.gov/centers/dryden/research/ESCD/index.html
8. Saggiani, G., et al.: A UAV system for observing volcanoes and natural hazards. AGU Fall Meeting Abstracts (2007)
9. Patterson, M.C.L., et al. Volcano surveillance by ACR Silver Fox. Infotech@Aerospace, 26–29 September 2005, Arlington, VA
10. The ROBOVOLC project homepage. Available at: http://www.robovolc.diees.unict.it
11. Muscato, G., Caltabiano, D., Guccione, S., Longo, D., Coltelli, M., Cristaldi, A., Pecora, E., Sacco, V., Sim, P., Virk, G.S., Briole, P., Semerano, A., White, T.: ROBOVOLC: a robot for volcano exploration – result of first test campaign. Ind Robot Int. **30**(3), 231–242 (2003)
12. Caltabiano, D., Muscato, G.: A robotic system for volcano exploration. In: Kordic, V., Lazinica, A., Merdan, M. (eds.) Cutting Edge Robotics. Advanced Robotic Systems Scientific Book, pp. 499–519. Pro Literatur, Germany, ISBN:3-86611-038-3 (2005)
13. Service Robots Group – Università di Catania. Available at: http://www.robotic.diees.unict.it/
14. Aiuppa, A., Federico, C., Paonita, A., Pecoraino, G., Valenza, M.: S, Cl and F degassing as an indicator of volcanic dynamics: the 2001 eruption of Mount Etna. Geophys. Res. Lett. **29**(11), 1559 (2002). doi:10.1029/2002GL015032
15. Symonds, R., Rose, W.I., Bluth, G.J.S., Gerlach, T.M.: Volcanic-gas studies: methods, results and applications. In: Carroll, M.R., Halloway, J.R. (eds.) Volatiles in Magmas. Reviews in Mineralogy, vol. 30, pp. 1–66. Mineralogical Society of America, Chantilly, VA (1994)
16. Stix, J., Gaonac'h, H.: Gas, plume and thermal monitoring. In: Sigurdsson, H. (ed.) Encyclopaedia of Volcanoes, pp. 1141–1164. Academic, New York (2000)
17. Astuti, G., Longo, D., Melita, D., Muscato, G., Orlando, A.: Hardware in the loop tuning for a volcanic gas sampling UAV. In: Valavanis, K.P. (ed.) Advances in Unmanned Aerial Vehicles, State of the Art and the Road to Autonomy. Intelligent Systems, Control and Automation: Science and Engineering, vol. 33. Springer, New York ISBN:978-1-4020-6113-4 (2007)
18. Astuti, G., Longo, D., Melita, C.D., Muscato, G., Orlando, A.: HIL tuning of UAV for exploration of risky environments. In: Proceedings of the IARP Workshop HUDEM'08 Robotics for Risky Environments and Humanitarian De-Mining. Il Cairo, Egypt, 28–30 March (2008)
19. Longo, D., Melita, D., Muscato, G., Sessa, S.: A mixed terrestrial aerial robotic platform for volcanic and industrial surveillance. In: Proceedings of the IEEE International Conference on Safety, Security and Rescue Robotics 2007. Rome (Italy), 27–29 September (2007)
20. Caltabiano, D., Muscato, G., Orlando, A., Federico, C., Giudice, G., Guerrieri, S.: Architecture of a UAV for volcanic gas sampling. In: Proceedings of the ETFA2005 10th IEEE International Conference on Emerging Technologies and Factory Automation. Catania, Italy, 19–22 September (2005)
21. The Aerosonde Robotic Aircraft homepage. Available at: http://www.aerosonde.com/
22. Graupner homepage. Available at: http://www.graupner.de/

23. X-Plane simulator by Laminar Research homepage. Available at: http://www.x-plane.com/
24. Brown, R.G., Hwang, P.Y.C.: Introduction to Random Signals and Applied Kalman Filtering. Wiley, New York (1992)
25. Diebel, J.: Representing Attitude: Euler Angles, Unit Quaternions, and Rotation Vectors. Stanford University, Stanford, CA
26. Kim, A., Golnaraghi, M.F.: A quaternion-based orientation estimation algorithm using an inertial measurement unit. In: Proceedings of the IEEE Position Location and Navigation Symposium, 2004 (2004)
27. Gebre-Egziabher, D., et al.: A gyro-free quaternion-based attitude determination system suitablefor implementation using low cost sensors. In: Proceedings of the IEEE Position Location and Navigation Symposium 2000 (2000)
28. Jang, J.S., Liccardo, D.: Small UAV Automation Using MEMS. IEEE A&E Systems Magazine, pp. 30–34. May (2007)
29. Jang, J.S., Liccardo, D.: Automation of small UAVs using a low cost MEMS sensor and embedded computing platform. In: 25th Digital Avionics Systems Conference. 15 October (2006)
30. Eldredge, A.M.: Improved state estimation for miniature air vehicles. Master's degree thesis, Department of Mechanical Engineering, Brigham Young University, December (2006)

FPGA Implementation of Genetic Algorithm for UAV Real-Time Path Planning

François C. J. Allaire · Mohamed Tarbouchi ·
Gilles Labonté · Giovanni Fusina

Originally published in the Journal of Intelligent and Robotic Systems, Volume 54, Nos 1–3, 495–510.
© Springer Science + Business Media B.V. 2008

Abstract The main objective of an Unmanned-Aerial-Vehicle (UAV) is to provide an operator with services from its payload. Currently, to get these UAV services, one extra human operator is required to navigate the UAV. Many techniques have been investigated to increase UAV navigation autonomy at the Path Planning level. The most challenging aspect of this task is the re-planning requirement, which comes from the fact that UAVs are called upon to fly in unknown environments. One technique that out performs the others in path planning is the Genetic Algorithm (GA) method because of its capacity to explore the solution space while preserving the best solutions already found. However, because the GA tends to be slow due to its iterative process that involves many candidate solutions, the approach has not been actively pursued for real time systems. This paper presents the research that we have done to improve the GA computation time in order to obtain a path planning generator that can recompile a path in real-time, as unforeseen events are met by the UAV. The paper details how we achieved parallelism with a Field Programmable Gate Array (FPGA) implementation of the GA. Our FPGA implementation not

F. C. J. Allaire (✉) · M. Tarbouchi
Electrical and Computer Engineering Department, Royal Military College of Canada,
Kingston, ON K7K7L6, Canada
e-mail: francois.allaire@rmc.ca

M. Tarbouchi
e-mail: tarbouchi-m@rmc.ca

G. Labonté
Mathematics and Computer Science Department, Royal Military College of Canada,
Kingston, ON K7K7L6, Canada
e-mail: labonte-g@rmc.ca

G. Fusina
Defence R&D of Canada–Ottawa, Ottawa, ON K1A0Z4, Canada
e-mail: Giovanni.Fusina@drdc-rddc.gc.ca

only results in an excellent autonomous path planner, but it also provides the design foundations of a hardware chip that could be added to an UAV platform.

Keywords UAV · Path planning · Genetic algorithms · FPGA · Real-time

1 Introduction

UAVs are used for many different services. In the military context, research has been performed to improve the autonomy of the UAV with different types of missions in view:

- Suppression of enemy air defence [1–3];
- Air-to-ground targeting scenario [4];
- Surveillance and reconnaissance [5];
- Avoidance of danger zones [6–15]; and
- Command, control, communication, and intelligence [16]

In the civilian context, UAVs could also be used for purposes such as:

- Weather forecast/ hurricane detection;
- Urbane police surveillance;
- Farm field seeding; and
- Border surveillance

All these scenarios require a common task: Path Planning. This task is crucial to the well being of the mission and to ensure the safety of both the mission and the UAV. Facing the reality of a dynamic world, UAVs need to re-plan their path in real-time. Currently, there is a human operator, dedicated to the UAV navigation, who is responsible for that function. This paper explains a solution that provides UAVs with an autonomous real-time path planning capability based on the Genetic Algorithms (GA) method implemented on a FPGA circuit board. This solution not only meets the real-time requirements of UAVs, but, with some additional works, would provide a hardware design ready to be inserted into UAV military and/or civilian platforms.

Firstly, the paper presents different existing path planning algorithms; it explains why the GA was selected. Secondly, the specific GA method used is detailed. Thirdly, the paper presents details of the GA design implementation on the FPGA circuit board. Finally, the results obtained will be discussed.

2 Path Planning Techniques

There are many path planning techniques used within the mobile robot world. Each one has its strengths and its weaknesses. Some techniques better fit an indoor environment, while others an outdoor environment. Some techniques better fit a fixed environment and others a dynamic environment. Some are better suited to rover robots, other ones to flying robots. This section gives an overview of these existing techniques, which can be divided in three groups:

- deterministic techniques;
- probabilistic techniques; and
- heuristics techniques

This section's aim is to highlight why the GA method, from the heuristics techniques, is an excellent candidate for the path planning of the UAV, which is a flying robot within a dynamic environment.

2.1 Deterministic Algorithms

The deterministic group covers the techniques that are using fixed cost equations. These always provide the same results for a given scenario. They tend to strive to get the shortest path, which may not be the optimal solution.

2.1.1 "A-Star" Family

The best known member of this family is Disjkstra's algorithm, which computes the cost of going through a cell, and every cell surrounding the current position, until it reaches the goal position. A second member, A-Star, improved this intensive search by adding a heuristic constant that directs the search only towards cells that are in the direction of the goal position. A third member, D-Star, further improved the search process by adding a second heuristic to minimize the recalculation in case of meeting unforeseen obstacles. A fourth member, AD-Star uses an inflation factor to get to a sub-optimal solution quickly, meeting real-time requirements. This paper will not go deeper in details for these techniques; [17] provides a good comparison of these four different techniques. However, from Fig. 1 [18], we can extrapolate how the search could get mathematically heavier as we move from a 2D-rover environment to a 3D-UAV world. Moreover, UAVs are traveling through large areas;

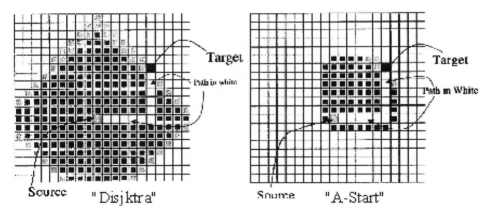

Fig. 1 Disjktra's and A-Star algorithm computation [18]. The Source is in an obstacle-free environment and tries to find the shortest path to go to the Target. *Black cells* represent those in which the algorithm had to evaluate the cost of passing through to find the best path

hence the algorithms have a larger number of cells to evaluate. As the evolution of the A-Star family shows, one needs heuristics to get away from this computational cumbersomeness.

2.1.2 Potential Field

The Potential Field concept involves representing obstacles by repulsive forces and the target by an attractive force. The robot needs to vector-sum the forces, and uses the resulting vector as its velocity direction command. The weakness of this concept is that there can be local minima of the potential field close to obstacles, in particular concave ones. Moreover, the computation complexity increases as we move to the 3D-UAV world, as vector summation has an additional component to deal with. Reference [19] elaborates some techniques (e.g., "the Mass–Spring–Damper System") that overcome the minima problem but introduce new problems (e.g., "high-resolution far from obstacles, but low-resolution near to obstacles").

2.1.3 Refined Techniques

Deterministic approaches tend to be computationally demanding because they perform many calculations to get exact predictable results with known situations. References [19, 20] developed refined techniques and demonstrated that a good way to get around the computational problem is to split the problem into two levels of path-definition: high path-definition for the neighbouring environment, within a specific radius from the UAV, and a coarse path-definition for environment outside that radius. These techniques aim at reducing the computation time as much as possible and to be able to re-compute the path in real-time to overcome unforeseen events.

2.2 Probabilistic Algorithms

The probabilistic group includes techniques that are using probabilities to model the uncertainties of the UAV world. Markov Decision Process (MDP) is one of these techniques. Reference [20] comments on the weakness of this method. What needs to be understood is that even if MDP addresses the uncertainties of the UAV world, MDP is a complex modeling process that does not take into consideration important constraints such as the time constraint, which is important when dealing with a team of UAVs. This research did not consider the probabilistic group because of the complexity of developing these models. This complexity will, most likely, prevent them from meeting the real-time requirement.

2.3 Heuristic Algorithms

The heuristic group covers the techniques that mix concepts from the deterministic and the probabilitic groups by using some heuristic representation of the problem. Two of these techniques used in the path planning are:

- Artificial Neural Networks, and
- Genetic Algorithm (GA)

2.3.1 Artificial Neural Networks

Artificial neural networks are based on the corcept of the human cerebral cells, which can be conditioned or trained to behave as desired. The difficulty with this method is to get the proper training data to let the neural network learn the proper behaviour. One could want to train a neural network to be expert in path planning, but he would face the incapability, time wise or data wise, to train the system to overcome all possible unforeseen events.

2.3.2 Genetic Algorithm

The GA method is based on Darwin's theory of evolution where crossovers and mutations can generate better populations. GA uses randomness to cover the whole search area of potential solutions for the UAV path planning optimization problem and it uses determinism in keeping the best solutions. GA thus takes advantage of both deterministic and probabilistic approaches, while using simple operators to improve solutions: crossovers and mutations. GA provides high quality solutions [19, 21, 22]. Moreover, GA creates many possible solutions at a time and works on plain data without any a priori information [23]. However, because of the GA's slow iterative process (see the While-loop of Fig. 2), active research on this technique have not been pursued for the development of a real-time path planner.

Our research recognises the GA's advantages; and with its FPGA implementation, it overcomes the GA's computational disadvantage by considerably improving the processing time. This permits a GA-based real-time path planner for UAVs.

3 The Genetic Algorithm Used

An application of the Genetic Algorithm to path-planning, developed by Cocaud [24],—sponsored by the Defence R&D Canada (DRDC) in Ottawa—was used as an initial starting point for this research. This section describes this implementation by giving details on the following elements:

- Initial population characteristics,
- Selection technique,
- Crossover operator,
- Mutation operators,

Fig. 2 Pseudo-code of the genetic algorithm

```
GENETIC ALGORITHM;
    Initialize the population;
    Evaluate initial population;
    WHILE convergence criteria is NOT satisfied, DO
        Perform competitive Selection;
        Crossover solutions to generate new solutions;
        Mutate new solutions;
        Evaluate new solutions;
        Update old Population with new solutions;
    END-WHILE.
```

- Population update technique,
- Evaluation technique, and
- Convergence criteria

All details were set according to Cocaud's [24] experimentations and recommendations.

3.1 Population Characteristics

A population is composed of a certain number of solutions that are called chromosomes. Each chromosome represents a path-solution. It is composed of a certain number of genes that represent the transitional waypoints. It was found through experimentation that a population of path-solutions with around 30 to 40 different paths provided good solutions. The initial formulation of the Genetic Algorithm application also ensured that each potential path-solution does not exceed 20 transitional waypoints, but has at least 2 transitional waypoints. Each transitional waypoint is characterized by its three spatial coordinates. Starting waypoint and ending waypoint are fixed for all path-solutions, hence these are not considered as genes. Size of the population is fixed during the whole GA process.

The initial population is composed of a fixed number of path-solutions, each one having initially two randomly generated transitional waypoints. Thus, each initial path-solution has initially exactly four waypoints, starting and ending waypoints included.

3.2 Selection

The initial GA implementation of Cocaud [24] used the standard roulette-wheel method to select the parent path-solutions for the crossover phase. This selection method computes a selection probability for each path-solution. The probability is proportional to the path-solution fitness: the better is the path, the higher the probability is; the worse is the path, the smaller the probability is. Once the size of the probabilities has been set, a random–number–generator system provides a number that selects one of the path-solutions based on these probabilities.

3.3 Crossover

The crossover operation affects two selected path-solutions and permutes random sub-paths between the two solutions. A specific region is usually selected by one or two cutting-point(s) and only sub-paths within that region are permuted. The initial GA implementation always selects the middle of the path for the cutting-point, while cutting randomly before or after the middle waypoint for path with odd number of waypoints. Hence, each parent path is exchanging its transitional waypoints that are after the middle waypoint. Crossover is performed on 70% of the population, creating "children".

3.4 Mutation

Five different mutation methods were used by the initial GA implementation to speed-up the convergence: the perturb mutation, the insert mutation, the delete mutation, the smooth-turn mutation, and the swap mutation. Mutations are done over 5% of the child population. Children are randomly selected with probabilities following a normal distribution.

3.4.1 Perturb Mutation

The perturb mutation selects randomly a waypoint of the selected child-path and modifies the values of its three coordinates randomly, while keeping them within an area of the size equal to 15% of the size of the full map (see Fig. 3). Before being accepted, the new waypoint is always checked to ensure it is not obstructed.

3.4.2 Insert Mutation

The insert mutation performs the same operation as the perturb mutation, except that it keeps the initial waypoint and adds a new waypoint after it in the path, while respecting the maximum number of waypoints per path (see Fig. 4).

3.4.3 Delete Mutation

The delete mutation randomly selects a waypoint and removes it from the current path, while respecting the minimum number of waypoints per path (see Fig. 5).

3.4.4 Smooth-Turn Mutation

The smooth-turn mutation evaluates the sharpness of each turn angle within a path. If it finds an angle that exceeds a specific threshold, a waypoint is added in such a way that the new angle is larger than the original one.

3.4.5 Swap Mutation

This mutation operator was used in [24] only when missions with multiple-objectives were considered. Because path planning usually addresses one objective at a time, we shall not consider this mutation hereafter.

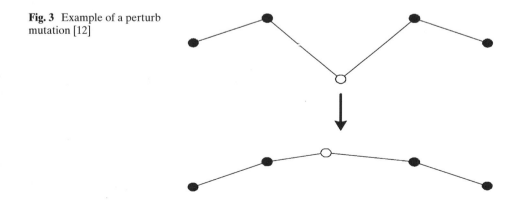

Fig. 3 Example of a perturb mutation [12]

Fig. 4 Example of an insert
mutation [12]

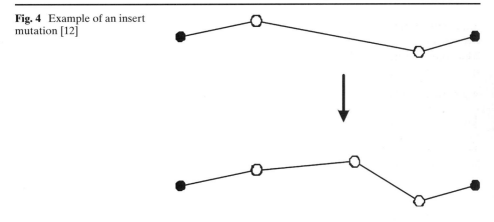

3.5 Population Update

This step consists in building a new population with the right number of path-
solutions, using the children population and the old population. The new population
used by the initial GA implementation is composed of: the entire children population; 5% of the best path-solutions from the old population; and it completes the
population count by randomly selecting enough paths from the 95% leftover from
the old population. This last selection uses the roulette-wheel technique. Each time
one of the leftover solutions is selected, the roulette-wheel is recomputed without
that selected solution.

3.6 Evaluation

Once the new population is created, each of its path-solutions has its fitness
(goodness) reassessed. This evaluation phase was not modified from the original
implementation of [24] in our research because the latter approach requires some
improvements that will not be covered in this publication.

Fig. 5 Example of a delete
mutation [12]

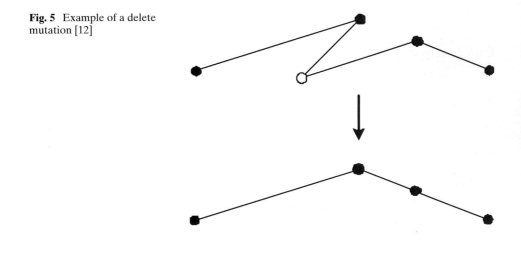

3.7 Convergence Criteria

The convergence criterion tells the GA when to stop its iterative process. More than one criterion could be used. The following three are commonly used:

- Fixed number of generations (iterations);
- Fixed level of fitness (goodness), calculated by the evaluation; and/or
- Fixed amount of time

The initial GA implementation stops the iterative process after 50 to 70 generations to have enough time for the GA to converge without wasting time with an already converged GA.

4 FPGA Design

A Field Programmable Gate Array (FPGA) consists of a programmable chip where the logic gates can be rearranged as needed by its user. This technology provides the flexibility to develop a processor dedicated to a specific operation instead of using a general purpose processor as used within personal computers. The chip area used by the processor can then be optimized by incorporating parallel processing. This is why some research [25–36] has already attempted different implementations of GA on FPGAs. However, none of them was applied to the UAV path-planning, where genes have a third dimension, namely the altitude or the Z axis, to take into consideration. None of them considered implementing elaborate mutation operators like the ones present in this research design. Nor did any of them use heavy parallelism while keeping the standard GA structure. On these three different points our research brings some new perspectives.

Section 4 will describe the details of the hardware design implementing the GA described in Section 3 on a FPGA circuit board. Figure 6 shows a block diagram of the design. The whole algorithm is running on the FPGA except for the evaluation phase. Evaluation runs on the PC, as does the simulation environment. This section will detail the design by covering the following points:

- Some structural modifications to the initial GA implementation;
- Control Unit;
- Receiving Unit;
- Selection Unit and Population Update;
- Crossover Unit;
- Mutation Unit; and
- Transmitting Unit

4.1 Structural Modification to the Initial GA Implementation

To integrate the FPGA hardware design shown in Fig. 6 to the initial PC-based GA executable, some modifications were required:

4.1.1 Fixed Population Characteristics

The first modification was to fix the population size to a predefined value of 32 path-solutions, in order to control the size of the hardware used from the FPGA board.

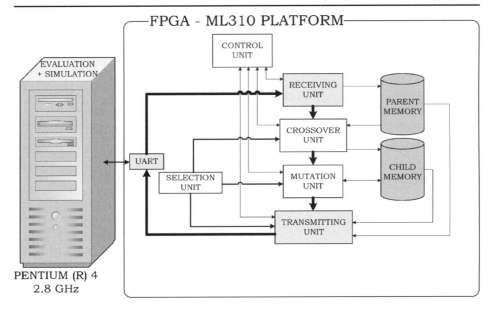

Fig. 6 This design overview shows the main hardware modules on the FPGA. The *dark thick arrows* provide an idea of the process flow

Fixing the size of the population also permitted fixing the number of the children generated at each iteration to 22 (69% of the old population). In the same line, the number of the best path-solutions kept from the old population, in the new population, was fixed to 2 (6% of the old population). Having these numbers fixed procured a time saving over having to dynamically calculate these values.

4.1.2 Sorting Algorithm

The second modification consisted in inserting a sorting algorithm after the evaluation phase. Having the path-solutions sorted based on their fitness score permits a reduction in the time required to find the best solutions within the population. This modification also permits fixing once and for all the roulette-wheel probabilities proportional to the rank of the solution; instead of re-computing the probabilities each time the fitness values were modified during the iterations.

After studying a number of sorting algorithms [37], the QuickSort algorithm was seen to be the fastest for the size of the population with which we dealt. Therefore, the QuickSort algorithm was implemented to minimize the amount of computations added to the evaluation phase.

4.2 Control Unit

The Control Unit turns "ON" one specific unit at a time. Once the specific unit completes its function, it sends back to the Control Unit an "Over" signal for the Control Unit to turn "OFF" that specific unit and turn "ON" the next unit in the GA process.

The Control Unit follows this process:

1. Waits on a "Listen" state to see if it will find a byte command on the UART that contain the number of chromosomes to be received;
2. When that command is received, it turns "ON" the Receiving Unit to receive the parent population from the PC;
3. The Crossover Unit is turned "ON" to crossover the parent population, hence creating the child population;
4. The Mutation Unit is turned "ON" to mutate 5 children from the child population; and
5. The Controller Unit turns "ON" the Transmit Unit to send the updated new population

Once the transmission is over, the PC computes the value of the individuals in the population; sorts the population based on the new fitness scores; and sends back the new population to the FPGA.

4.3 Receiving Unit

The Receiving Unit, once turned "ON" by the Control Unit, directly communicates with the UART to receive each path-solution of the parent population. The Receiving Unit saves the received population into the Parent Memory.

4.4 Selection Unit and Population Update

The Selection Unit is the only unit that is always "ON" independently from the Control Unit. The Selection Unit continuously provides 24 different random address numbers (11 for the crossover, 5 for the mutation and 8 for the Population Update), which change each clock cycle.

The Population Update consists of selecting the two best parents (predetermined addresses), the entire children population (predetermined addresses), and 8 random path-solutions (dynamic addresses) from the worst old parents to create the new parent population. Hence, the Selection Unit's random address numbers provide the only 8 dynamic addresses required for the Population Update to select the 32 path-solutions of the new parent population within one clock cycle.

4.5 Crossover Unit

The Crossover Unit is composed of 11 identical modules, which permute the way-points past the parent-paths middle-point (middle-point is rounded down for odd number of points) between the two parent-paths to generate two child-paths. The crossover module uses the same crossover technique as [24]; however, instead of performing the crossover sequentially, it is done entirely in parallel. In one clock cycle, the Crossover Unit uses the 11 outputs from the Selection Unit twice, and combines them differently. It results in having 22 parents that are available for the crossover within one clock cycle.

4.6 Mutation Unit

The Mutation Unit is composed of one Perturb Mutation module, two Inserter Mutation modules, and two Delete Mutation modules; all of these modules are running in parallel. The smooth mutation was not included because the insert mutation could produce a similar effect. The different mutation modules use the respective techniques from [24].

4.7 Transmitting Unit

The Transmitting unit (see the Transmitting Unit on Fig. 6) directly communicates with the UART to transmit all the 32 chromosomes of the new updated population.

5 Results Discussion

5.1 Testing Environment

To compare computation time between the initial GA implementation (described in Section 3) and this research's final FPGA implementation design (descried in Section 4), this research used a simulation environment of a canyon (see Fig. 7). The simulation environment provided a coarse atmospheric model with a constant unidirectional wind. The starting point was fixed at one extremity of the map and the

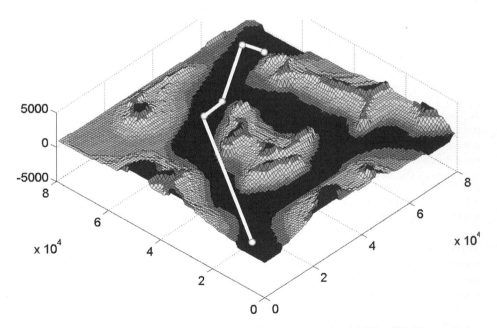

Fig. 7 Testing environment map (80 × 80 × 3.5 km), with a resolution of 500 × 500. The resolution of the Z axis was determined by the 16-bit integer used to define the height of each cell of the X and Y plan

Table 1 Timing results

GA phase	Initial PC-based GA	Modified GA with FPGA	Speed improvement
Selection and crossover	94 ms	8.85 µs	10,000×
Mutation	2 ms	18 µs[a]	111×
Population update	30 ms	600 ns[b]	50,000×
Evaluation	60 s/50 s	N/A[c]	1.2×[d]

The time values are averages captured after multiple trials. The PC used was a Pentium (R) 4 with a CPU of 2.8 GHz. The FPGA used was a Virtex-II Pro (2vp30ff896) with a system clock running at 100 MHz

[a]The Mutation was only tested within the simulation environment provided by Active-HDL version 7.1 (ALDEC, Inc's developing software). This restriction came from the fact that the current FPGA platform (the ML310) didn't have enough resources to host the full design; a FPGA ML410 (with twice the resources of the ML310) would have been required. However, before testing the Selection/Crossover system on hardware, testing was also performed with Active-HDL. Simulation results predicted properly the results obtained with the FPGA board connected to the C-code executable (running on the PC). We could therefore extrapolate this rectitude to the Mutation simulation prediction

[b]We can have access within on clock cycle to all the updated population (within 10 ns). Hence after 60 generations, we spent 600 ns to update the population. Some may want to multiply by the average number of waypoints-by-chromosome (5) this number since we can read only one waypoint by clock cycles (this is true only if we want to copy these information within a new memory space). Even with this multiplication, we still have a speed improvement of ' 10,000×"

[c]The evaluation phase was not implemented on FPGA in our research, because the original implementation approach [24] requires some improvements that will not be covered in this publication

[d]Sorting the population improved by 1.2 times the original computation time. This is due to the fact that a sorted population requires less search time to find the fitness statistic used to compute the resulting fitness score of each population member

ending point was fixed at the other extremity of the map. The scenario environment was fixed during the testing trials.

Multiple trials were first run with the C-code executable [24], as a stand-alone system. Average timing was extracted from these trials to provide the results exhibited in Table 1. Trials were run on a Pentium (R) 4 CPU 2.8 GHz.

The second set of trials incorporated the QuickSort algorithm within the C-code executable (running on the same PC); while the system, with the Selection and the Crossover Unit, was running on the FPGA (ML310 Xilink platform).

5.2 Timing Improvement Explanation

For both the crossover and the population update phases, the improvement reached a factor of 10,000. This is mainly due to the fact that the initial GA implementation had to sequentially select one solution at a time after generating a new random number for each selection, while this process is done in parallel in the FPGA. The recompilation of the selection-wheel was also a cumbersome task in the initial GA implementation. In the FPGA, these tasks are performed within one clock cycle (10 ns). For the crossover unit, the full parallelism used by duplicating the same crossover module (11 times) resulted in such a good improvement.

The mutation time improvement reaches a factor of 100. This is because the original GA implementation performed fewer mutations than the hardware design does. The difference in number of mutations comes from the fact that the initial GA used probabilities of a mutation to be applied to a child chromosome, while the hardware design uses a fixed percentage of the child population to be mutated. However, being able to perform the 5 mutations simultaneously still provides an improvement by a factor of 100 in computation time over the initial GA implementation.

5.3 Future Work

The first step should be to optimize the evaluation phase for the members of the population. Secondly, a FPGA of the same calibre as the Virtex-4 LX200, which has 10 times more resources than the ML310 platform used here, should be acquired. If a hardware design could evaluate simultaneously the fitness of the 22 children solutions, we could theoretically improve the evaluation process time by a factor of at least 1,000 times. The whole GA path planning algorithm would then be totally running on the FPGA and would have the independence required to be connected to another UAV hardware platform. With that improvement, we would have a re-planning capability that provides a near optimal solution that, we estimate, would run in approximately 100 ms. Hence, a Sperwer (length of 4.2 m), flying at 150 km/h, would only cover a distance of 4.17 m before having its new directed path to follow. This distance is within the safety zone of the size of the aircraft considered by most of the path planning algorithms. We would then have a real-time path planner.

The third step would be to increase the confidence in the safety of the path planner by testing our system with a realistic tracker algorithm within a more sophisticated simulation environment. Then a live test with a real UAV would be realistic.

There is also room to optimize the current hardware design to save some area and some power dissipation, since our present design was oriented solely towards speed acquisition.

6 Conclusion

This research has successfully improved the processing time of a Genetic-Algorithm-based path planning algorithm. The FPGA implementation provides speed-ups of about 10,000 over the software executable. This improvement permits the UAV community to envisage research to incorporate a real-time path re-planning system, which can handle unforeseen events, into future UAV systems.

Acknowledgements F. C. J. Allaire thanks C. Cocaud, MEng from University of Ottawa, for his C code. It provided the foundation of this research project. F. C. J. Allaire thanks Dr. G. Labonté and Dr. M. Tarbouchi for their guidance throughout the research. F. C. J. Allaire also thanks J. Dunfield, MEng from Royal Military College of Canada, for the great support provided throughout the research. F. C. J. Allaire thanks M. Fricker, MEng from Royal Military College of Canada, for his support with paper redaction. Finally, F. C. J. Allaire thanks Dr. D. Al Khalili and J.-L. Derome for the technical support provided regarding FPGA technology.

References

1. Orgen, P., Winstrand, M.: Combining path planning and target assignment to minimize risk in a SEAD mission. In: AIAA Guidance, Navigation, and Control Conf., pp. 566–572. San Francisco, USA (2005)
2. Ye, Y.-Y., Min, C.-P., Shen, L.-C., Chang, W.-S.: VORONOI diagram based spatial mission planning for UAVs. Journal of System Simulation, China **17**(6), 1353–1355, 1359 (2005)

3. Yu, Z., Zhou, R., Chen, Z.: A mission planning algorithm for autonomous control system of unmanned air vehicle. In: 5th International Symposium on Instrumentation and Control Technology, pp. 572–576. Beijing, China (2003)
4. Vaidyanathan, R., Hocaoglu, C., Prince, T.S., Quirn, R.D.: Evolutionary path planning for autonomous air vehicles using multiresolution path representation. In: IEEE International Conf. of Intelligent Robots and Systems, vol. 1, pp. 69–76. Maui, USA (2001)
5. Rubio, J.C., Vagners, J., Rysdyk, R.: Adaptive path planning for autonomous UAV oceanic search missions. AIAA 1st Intelligent Systems Technical Conf., vol. 1, pp. 131–140. Chicago, USA (2004)
6. Shim, D.H., Chung, H., Kim, H.J., Sastry, S.: Autonomous exploration in unknown environments for unmanned aerial vehicles. In: AIAA Guidance, Navigation, and Control Conf., vol. 8, pp. 6381–6388. San Francisco, USA (2005)
7. Chaudhry, A., Misovec, K., D'Andrea, R.: Low observability path planning for an unmanned air vehicle using mixed integer linear programming. In: 43rd IEEE Conf. on Decision and Control. Paradise Island, USA (2004)
8. Theunissen, E., Bolderheij, F., Koeners, G.J.M.: Integration of threat information into the route (re-) planning task. In: 24th Digital Avionics Systems Conf., vol. 2, pp. 14. Washington, DC, USA (2005)
9. Rathinam, S., Sengupta, R.: Safe UAV navigation with sensor processing delays in an unknown environment. In: 43rd IEEE Conf. on Decision and Control, vol. 1, pp. 1081–1086. Nassau, USA (2004)
10. AIAA: Collection of technical papers. In: AIAA 3rd "Unmanned-Unlimited" Technical Conf., Workshop, and Exhibit, vol. 1, p. 586. Chicago, USA (2004)
11. Boskovic, J.D., Prasanth, R., Mehra, R.K.: A multi-layer autonomous intelligent control architecture for unmanned aerial vehicles. J. Aerosp. Comput. Inf. Commun., pp. 605–628. Woburn, USA (2004)
12. Zheng, C., Ding, M., Zhou, C.: Real-time route planning for unmanned air vehicle with an evolutionary algorithm. Int. J. Pattern Recogn. Artif. Intell. Wuhan, China 17(1), 63–81 (2003)
13. Boskovic, J.D., Prasanth, R., Mehra, R.K.: A multi-layer control architecture for unmanned aerial vehicles. In: American Control Conference, vol. 3, pp. 1825–1830. Anchorage, USA (2002)
14. Rathbun, D., Kragelund, S., Pongpunwattana. A., Capozzi, B.: An evolution based path planning algorithm for autonomous motion of a UAV through uncertain environments. In: AIAA/IEEE Digital Avionics Systems Conf., vol. 2, pp. 8D21–8D212. Irvine, USA (2002)
15. Shiller, I., Draper, J.S.: Mission adaptable autonomous vehicles. Neural Netw. Ocean Eng., pp. 143–150. Washignton DC, USA (1991)
16. Fletcher, B.: Autonomous vehicles and the net-centric battlespace. In: International Unmanned Undersea Vehicle Symposium, San Diego, USA (2000)
17. Ferguson, D., Likhachev, M., Stentz, A.: A guide to heuristic-based path planning. In: International Conf. on Automated Planning & Scheduling (ICAPS), vol. 6, pp. 9–18. Monterey, USA, Work Shop (2005)
18. McKeever, S.D.: Path Planning for an Autonomous Vehicle. Massachusetts Institute of Technology, Massachusetts, USA (2000)
19. Judd, K.B.: Trajectory Planning Strategies for Unmanned Air Vehicles. Dept. of Mech. Eng., Brigham Young Univ., Provo, USA (2001)
20. Chanthery, E.: Planification de Mission pour un Véhicule Aérien Autonome, pp. 15. École Nationale Supérieur de l'Aéronautique et de l'Espace, Toulouse, France (2002)
21. Sugihara, K., Smith, J.: Genetic algorithms for adaptive planning of path and trajectory of a mobile robot in 2d terrains. IEICE Trans. Inf. Syst. E82-D(1), 309–317 (1999)
22. Pellazar, M.B.: Vehicle route planning with constraints using genetic algorithms. In: IEEE National Aerospace and Electronics Conf., pp. 111–119 (1998)
23. Mostafa, H.E., Khadragi, A.I., Hanafi, Y.Y.: Hardware implementation of genetic algorithm on FPGA. In: Proc. of the 21st National Radio Science Conf., pp. 367–375. Cairo, Egypt (2004)
24. Cocaud, C.: Autonomous Tasks Allocation and Path Generation of UAV's. Dept. of Mech. Eng., Univ. of Ottawa, Ontario, Canada (2006)
25. Tachibana, T., Murata, Y., Shibata, N., Yasumoto, K., Ito, M.: A hardware implementation method of multi-objective genetic algorithms. In: IEEE Congress on Evolutionary Computation, pp. 3153–3160. Vancouver, Canada (2006)
26. Tachibana, T., Murata, Y., Shibata, N., Yasumoto, K., Ito, M.: General architecture for hardware implementation of genetic algorithm. In: 14th Annual IEEE Symposium on Field Programmable Custom Computing Machine, pp. 2–3. Napa, USA (2006)

27. Tang, W., Yip, L.: Hardware implementation of genetic algorithms using FPGA. In: 47th Midwest Symposium on Circuits and Systems, vol. 1, pp. 549–552. Hiroshima, Japan (2004)
28. Hamid, M.S., Marshall, S.: FPGA realisation of the genetic algorithm for the design of greyscale soft morphological filters. In: International Conf. on Visual Information Eng., pp. 141–144. Guildford, UK (2003)
29. Karnik, G., Reformat, M., Pedrycz, W.: Autonomous genetic machine. In: Canadian Conf. on Elect. and Comp. Eng., vol. 2, pp. 828–833. Winnipeg, Canada (2002)
30. Skliarova, I., Ferrari, A.B.: FPGA-based implementation of genetic algorithm for the traveling salesman problem and its industrial application. In: 15th International Conf. on Industrial and Eng. Application of Artificial Intelligence and Expert Syst. IEA/AIE, pp. 77–87. Cairn, Australia (2002)
31. Lei, T., Ming-cheng, Z., Jing-xia, W.: The hardware implementation of a genetic algorithm model with FPGA. In: Proc. 2002 IEEE International Conf. on Field Programmable, pp. 374–377. Hong Kong, China (2002)
32. Hailim, J., Dongming, J.: Self-adaptation of fuzzy controller optimized by hardware-based GA. In: Proc. of 6th International Conf. on Solid-State and IC Technology, vol. 2, pp. 1147–1150. Shanghai, China (2001)
33. Aporntewan, C., Chongstitvatana, P.: A hardware implementation of the compact genetic algorithm. In: Proc. of the IEEE Conf. on Evolutionary Computation, vol. 1, pp. 624–625. Soul (2001)
34. Shackleford, B., Snider, G., Carter, R.J., Okushi, E., Yasuda, M., Seo, K., Yasuura, H.: A high-performance, pipelined, FPGA-based genetic algorithm machine. Genetic Programming and Evolvable Machine **2**, 33–60 (2001)
35. Scott, S.S., Samal, A., Seth, S.: HGA: a hardware-based genetic algorithm. In: ACM 3rd International Symp. on FPGA, pp. 53–59. Monterey, USA (1995)
36. Emam, H., Ashour, M.A., Fekry, H., Wahdan, A.M.: Introducing an FPGA based genetic algorithms in the applications of blind signals separation. In: Proc. 3rd IEEE International Workshop on Syst.-on-Chip for Real-Time Appl., pp. 123–127. Calgary, Canada (2003)
37. Thiébaut, D.: Sorting algorithms. Dept. of Comp. Science, Smith College, Northampton, MA, USA (1997) Available: http://maven.smith.edu/~thiebaut/java/sort/demo.html (1997). Accessed 23 April 2008

Concepts and Validation of a Small-Scale Rotorcraft Proportional Integral Derivative (PID) Controller in a Unique Simulation Environment

Ainsmar Brown · Richard Garcia

Originally published in the Journal of Intelligent and Robotic Systems, Volume 54, Nos 1–3, 511–532.
© Springer Science + Business Media B.V. 2008

Abstract At the current time, the U.S. Army Research Laboratory's (ARL's) Vehicle Technology Directorate is interested in expanding its Unmanned Vehicles Division to include rotary wing and micro-systems control. The intent is to research unmanned aircraft systems not only for reconnaissance missions but also for targeting and lethal attacks. This project documents ongoing work expanding ARL's program in research and simulation of autonomous control systems. A proportional integral derivative control algorithm was modeled in Simulink (Simulink is a trademark of The MathWorks) and communicates to a flight simulator modeling a physical radio-controlled helicopter. Waypoint navigation and flight envelope testing were then systematically evaluated to the final goal of a feasible autopilot design. Conclusions are made on how to perhaps make this environment more dynamic in the future.

Keywords Simulink · MATLAB · UAV · UAS · Simulation · X-Plane · PID

International Symposium on Unmanned Aerial Vehicles, Orlando, FL, June 23–24, 2008.

A. Brown (✉)
National Institute of Aerospace,
Army Research Lab—Vehicle Technology Directorate,
Aberdeen Proving Grounds, MD 21005-5066, USA
e-mail: ainsmar.brown@nianet.org

R. Garcia
Motile Robotics Inc., Army Research Lab—Vehicle Technology Directorate,
Aberdeen Proving Grounds, MD 21005-5066, USA

K. P. Valavanis et al. (eds.), *Unmanned Aircraft Systems.* DOI: 10.1007/978-1-4020-9136-0_27   511

Nomenclature

θ	pitch attitude
ψ	yaw attitude
ϕ	roll attitude
ρ	density of fluid
λ	inflow coefficient
α	aerodynamic angle of attack
A	specified area
AR	Aspect ratio of lifting surface
C	Airfoil chord length
C_L	lift coefficient
C_D	drag coefficient
I	identity matrix
k	current discrete time step
K_p	proportional gain
K_d	derivative gain
K_i	integral gain
T	Thrust
V_E	velocity of element
X	body x coordinate
x	vehicle state
Y	body y coordinate
Z	body z coordinate

1 Introduction

Because of recent advancements in sensor and power plant technologies, the need for aircraft in the large size category has significantly diminished. Along with the development of modern autopilots, such as those designed by companies like Micropilot, Athena, and Procerus or in many cases universities with expansive computer engineering departments, the development of smaller vehicles similar to those shown in Fig. 1 has become more feasible and desirable.

Although the aircraft shown in Fig. 1 are smaller, lighter, quieter, and cheaper than many of those of the past, they are all products designed for the general consumer and as such, do not meet general requirements for survivability, safety, etc. The current trend is for research groups to use these R/C vehicles as test beds for their own autopilot software and hardware designs. This allows widespread research to be conducted without the high cost and typical security-restrictions of aircraft currently in operation with the armed forces. With only a few minor adjustments, these standard R/C aircraft become serious components of UAS research and development. One current issue affecting this type of development is the use of simulation and the role it should play throughout the development process. Current simulation software such as The MathWorks' Simulink and Laminar Research's X-Plane can provide highly accurate and influential test data. Although these data can simplify and expedite the development process, their role as a development tool has been minimal.

Fig. 1 Smaller scale radio-controlled (R/C) vehicles. From *top–down*, *left* to *right*: align T-Rex, WowWee Flytech Dragonfly, Minicopter Maxi Joker 2, and E-Flite Blade CX 2. Image reproduced from [7–10]

2 Background Concepts in Aeronautics

2.1 X-plane Model

The equations of motion for a 6°-of-freedom helicopter can be modeled a number of ways. Note that additional care must be taken with regard to angular orientation in a three-dimensional (3-D) space. The reason for this need of care is the ambiguity of a vehicle's orientation when the pitch attitude is equal to 90°. The traditional convention is the usage of Euler angles, and the second, less common method uses quaternion math to eliminate the ambiguity at a pitch attitude of 90°. The application of the quaternion method is described in the thesis by Green [2] and several navigation textbooks. For the purposes of this design, this helicopter will not operate during aggressive flight conditions. The main flight modes for the vehicle include hover and steady level forward flight. More specifically, the helicopter will never need to achieve a pitch attitude anywhere near 90°, which would cause a singularity and invalidate the Euler angle equations.

The simulation, as previously stated, depends on modeling and visualizations from Laminar Research's X-Plane. X-Plane models low-speed, high-speed, and to a lesser extent, transonic flight regimes with the use of a multiple-step process. A summary of information proved in [3], as well as the equations behind the theoretical model as explained in Nelson [4] and Prouty [5] follow. For reference, Prouty provides the background on the helicopter aerodynamics and flight modeling in this section, while Nelson explains the general concepts in automatic vehicle control used in Section 2.2.

The first step is to break the surfaces of the aircraft into smaller segments in order to use blade-element theory. This builds up the incremental forces on the entire vehicle by finding the forces on each "blade element," which can be thought of as a theoretical area over which a differential force is applied because of the aerodynamic properties of that segment. Although such elements can include a slice of a horizontal stabilizer or a slice of a main fuselage, they often are limited to certain bodies, depending on whether one is trying to determine lift or drag from a vehicle. A closed form expression for the general forces created by an aerodynamic surface is given in Eq. 2.1.1 in the form of a thrust expression:

$$dT = \frac{1}{2}\rho V_E^2 c \left[a\left(\beta - \phi - \alpha_i\right)dr\cos\left(\phi + \alpha_i\right) - C_d dr\left(\sin\phi + \alpha_i\right)\right] \bullet \text{\# of elements}$$

$$(2.1.1)$$

With an expression for the forces generated by the vehicle, linear and angular accelerations may be determined with the use of Newton's First Law.

The next step is to determine the velocities of each of the vehicle's components. Included in these equations are induced angle of attack, propwash, and downwash, all of which are relevant to rotorcraft aerodynamics. Induced angle of attack comes from the tip vortices that are generated as air wraps itself around the edges of an aerodynamic surface. As a result, a lower amount of lift is experienced by a wing or a tail section. Because helicopters have much smaller lifting surfaces (if any at all), as compared to fixed wing aircraft, this term can be considered very small or even negligible in many cases. The equations for induced angle of attack and its role in the effective angle of attack are given in Eqs. 2.1.2 and 2.1.3. The effective angle of attack is the net resultant angle of attack that a surface "feels" after dynamic losses and gains in lift attributable to various concepts in aerodynamics that are considered. In the case of a rotary wing, inflow is a major contributor.

$$\alpha_{effective} = \alpha_{geometric} - \alpha_i \qquad (2.1.2)$$

$$\alpha_i = \frac{C_L}{\pi \, AR} \qquad (2.1.3)$$

The equation for propwash, which is especially important in the case of a helicopter, comes from this idea of inflow and the disk method used in momentum theory. The prop, or rotor in this case, is treated as a continuous disk and is sliced into concentric rings. The following equation is solved in terms of power required:

$$P = \sqrt{\frac{T^3}{2\rho A}} \qquad (2.1.4)$$

All aerodynamic coefficients are taken from lift-curve slope plots for two-dimensional (2-D) airfoils. Prandtl–Glauert is used for compressible flow, and the diamond airfoil method is used for supersonic flow, which is not applicable to rotary wing flight. The Prandtl–Glauert equation, a simple correction factor for changes in fluid density at higher speeds, is given in Eq. 2.1.5. where "a" represents the lift-curve slope.

$$a_{comp} = \frac{a_{incompressible}}{\sqrt{1 - M^2}}$$

$$a = \frac{dC_L}{d\alpha} \qquad (2.1.5)$$

As of version 8.60, X-Plane had been using a rather unconventional reference system in relation to the accepted norm in aerospace engineering and navigation. In X-plane, V_x and V_z are velocities in reference to the surface of the earth, while rate of climb is given as V_y. In the context of this paper, the traditional aerospace convention will always be used, where V_x and V_y are in-plane velocities and V_z is the climb and descent rate.

2.2 PID Control Scheme

As is the case with most general control problems, there are a number of ways to implement a solution for acceptable controlability for a system. The only real requirement is to have a system that sends commands to each of the four main helicopter control output which include lateral cyclic, longitudinal cyclic, tail rotor collective, and main rotor collective. Different combinations of these controls can change the performance of an aircraft, depending on its mode of flight.

The PID used here includes a double loop system for hover control and a triple loop system for forward flight. The hover controller uses an angular orientation as the innermost loop and adds a body-frame velocity feedback loop to control the vehicle. The forward flight controller adds a global position controller as the outermost loop, similar to the approach described in Shim [6]. To attempt to design a collective controller with acceptable response, a complete PID controller with K_p, K_d, and K_i gains is used where needed. In other instances, PI or PD controllers may be used. The main rotor collective receives feedback from the vehicle's measured altitude, h, and the vehicle's measured climb speed, V_z.

$$Collective = K_p h + K_d dh + \int K_i h dh + K_p V_z + K_d dV_z + \int K_i V_z dV_z \qquad (2.2.1)$$

In principle, terms 2 and 5 in Eq. 2.2.1 overlap, and terms 1 and 6 overlap, but because these values will eventually come from different sensors, it is useful to track and respond to them separately.

Controls of the lateral and longitudinal cyclic pitch of the main rotor are given below.

$$Longitudinal\ cyclic = K_{p\theta}\theta + K_{d\theta}d\theta + \int K_{i\theta}\theta d\theta + K_{pV_x}V_x$$

$$+ K_{dV_x}dV_x - \int K_{iV_x}V_x dV_x$$

$$Lateral\ cyclic = K_{p\phi}\phi + K_{d\phi}d\phi + \int K_{i\phi}\phi d\phi + K_{pV_y}V_y$$

$$+ K_{dV_y}dV_y - \int K_{iV_y}V_y dV_y \qquad (2.2.2)$$

The final control output to be calculated is the tail rotor collective. Equation 2.2.3 shows the expression for the PID controller on the tail rotor. In forward flight, yaw attitude control is augmented with a feedback loop to the main rotor lateral control. This gives the system a faster overall response and limits the amount of work required by the tail rotor to maintain heading. Feeding yaw attitude to the main rotor lateral cyclic control also allows for more sophisticated maneuvers such as coordinated turns. The augmented lateral cyclic control of the main rotor is given in Eq. 2.2.4.

$$Tail\ rotor_{forward\ flight} = K_p \psi + K_d d\psi + \int K_i \psi d\psi \qquad (2.2.3)$$

$$Lateral\ Cyclic_{augmented} = K_p \phi + K_d d\phi + \int K_i \phi d\phi + K_{p\psi} \psi + K_{d\psi} d\psi + \int K_{i\psi} \psi d\psi$$
$$(2.2.4)$$

Through trial and error in simulation, it becomes evident that eliminating some feedback control improves the performance of the PID controller. In particular, setting the integral term for roll angle to zero in the case of hover improved the stability of the helicopter. The likely cause is that the helicopter becomes overly constrained with too many controls and needs to have the control algorithm relaxed to regain stability. Another instance that includes eliminating position feedback is hover. As long as the vehicle state is updated to the controller quickly enough, the vehicle is able to maintain sufficient hover status with drift bias. Considering how most GPS have a tendency to retain non-negligible random error in their output, even when tracking a single position, maintaining hover without GPS can be considered useful during short-term conditions, but of course would be impractical under most circumstances.

All PID controllers include low and high cut-offs for the maximum control output. When we normalize maximum control output to 1, cut-offs range from 0.1 to 0.65. The benefit of using cut-offs is to maintain the level of responsiveness generally given with larger gains, while avoiding long period over-compensation in the controls. Without the cut-offs, the controls will respond quickly with large gains, but they will have trouble maintaining stability within an appropriate time frame. Cut-offs are especially important for the integral feedback controls that employ an additive feedback that can grow to excessive magnitudes if no boundaries are placed on them.

3 Simulation

The basic layout of the simulation environment used in this work was designed by the second author [1] in previous work at the University of South Florida. With flight data from a commercial version of X-Plane simulator and feeding the data into the processing and control modeling environment in The MathWorks Simulink™, a fully functional simulation environment is realized. This allows the user to design any type of control algorithm necessary and easily integrate that algorithm into the Simulink model. This allows the user to experiment with anything from waypoint navigation to extreme 3-D competition maneuvers (Fig. 2).

Originally designed around a fuzzy logic controller, the simulator was converted to use PID control laws. Figure 3 shows a generalized view of the components

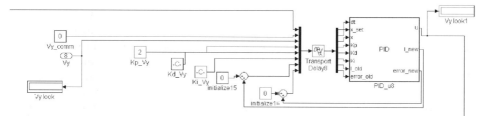

Fig. 2 Simulink PID controller block models

involved in the setup. The loop starts with the X-Plane data processing center. Data are first provided by X-Plane to the Simulink model via the User Datagram Protocol (UDP) communication. From there, current vehicle latitude, longitude, angular rates, velocities, and angular orientation are sent through processing blocks. The first of these blocks is responsible for filtering noise and/or drift from the sensors. The vehicle states are then sent to the second block responsible for transforming data from a fixed, earth-centered coordinate system to the body-fixed coordinate system of the helicopter.

Next, the waypoint processing station determines the error in position of the aircraft and uses this error to decide if the vehicle should move toward its next waypoint, land, or continue its current course of action.

After the waypoint station has decided what the helicopter's state should be, it passes its information to the logical controller selector. This block set is responsible for deciding which version of the PID controller should be activated to properly execute the specified flight mode. Thus, the PID controller that is active at a given time will change, depending on whether the vehicle is to hover, navigate directly to a waypoint, navigate with a user-specified orientation to the next waypoint, land, or end its mission. Generally speaking, the hover PID will be used when the current command is to sustain hover or transition to the next waypoint. Hover is used to change between two waypoints because heading changes occur more directly in the hover configuration. The hover PID is also used if the helicopter is on the ground and is told to take off or if it is in the air and is told to land.

Fig. 3 Generalized diagram of control system designed in Simulink

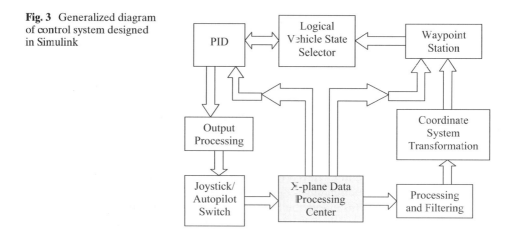

When the selector determines which PID to use, the data input is passed to all the controllers, and a switch on the other side of the PID block chooses only one set of controller output to pass into the joystick/autopilot switch. In this block, the user is given the choice of allowing the autopilot to run a mission or take control directly through a radio controller style computer joystick. This added dimension of controllability plays a major role in the flight-testing process. For instance, if the user sees that the autopilot is not functioning correctly and determines that the problem might be in the tail rotor controller, s/he can isolate that one particular controller and manually control it. If the vehicle's response to the autopilot is still problematic even in manual control of that specific control output, the user can deduce that his or her initial thought was wrong and look elsewhere for a solution.

4 Key Flight Modes

4.1 Trim State

Although PID controllers will eventually settle into a final value after receiving stable gains, errors may still be present even with a strong K_i term. As such, the helicopter must be trimmed out to satisfy the flight requirements. In the context of this simulation, this means adding constants to the set of control output after the PID controller has determined the best output command. In a real flight, this means steady state deflections to the control surfaces in addition to the commands being sent by the pilot or the autopilot. In the results section of this paper, one may notice some steady state errors in the vehicle orientations. It is accepted in helicopter dynamics that in the trim state, there will need to be a non-zero value for the roll orientation or the yaw orientation due to balancing of vehicle side forces in the absence of main rotor side cant. This model contains an orientation error in the roll direction. The result is that the helicopter will always fly with a non-zero roll attitude.

4.2 Waypoint Navigation

The waypoint data center is a Simulink S-function written in C that considers a number of variables to determine what would be the best next step for the UAS. In take-off, the waypoint data center considers how close the helicopter is to a commanded altitude. After the vehicle is within the desired altitude threshold, the system counts how many seconds the helicopter has maintained such an altitude. If the time requirement is met, the center will command the next waypoint. As the helicopter approaches each waypoint, the waypoint center continues to monitor heading, altitude, and the difference between desired and actual latitude and longitude coordinates. When all requirements are met for the current waypoint, the waypoint center commands the next location. When the vehicle has passed through all the waypoints, the center sends a command for the helicopter to land. After the vehicle lands, another command is sent to reverse the pitch in the main rotor to induce reverse thrust. This plants the helicopter to the ground and successfully ends the mission (Fig. 4).

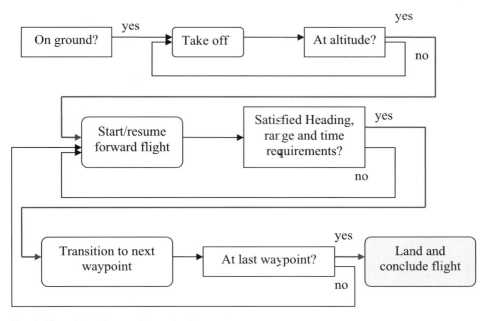

Fig. 4 Flow chart of waypoint navigation process

4.3 Elementary Pilot Control Augmentation System

Another commonly expected aspect of modern autopilots is the ability to augment pilot control input. A basic attempt to this feature was added to the simulation. The

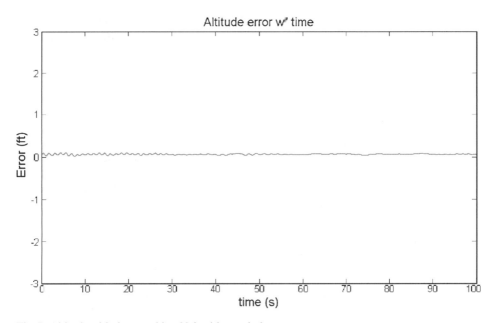

Fig. 5 Altitude with time tracking 20 ft with no winds

Fig. 6 Latitude and longitude
with time and no winds

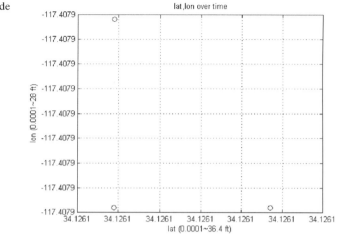

control gains were derived from a modified hover controller. The control scheme
consists of augmented climb and descent which is mutually exclusive to the rest of
the control. The other part of the control includes low speed forward, backward, left
and right flights, as well as vehicle yaw. The vehicle response is of the same quality
of the secondary forward flight autopilot (B) as given in Section 5.1 and Fig. 5. These
flight modes were augmented by the controller, based on the pilot's input. If the
right joystick is pointed exclusively in any of these four directions, the helicopter will
move in the given direction. If the pilot does not cleanly point the joystick in any one
direction, the vehicle will default into hover.

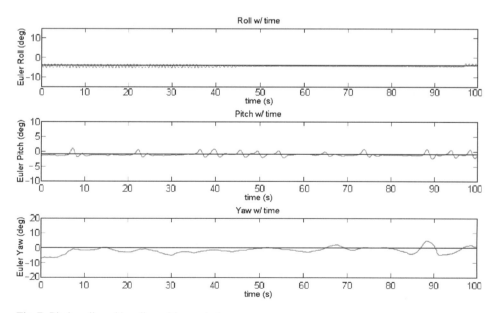

Fig. 7 Pitch, roll, and heading with no wind

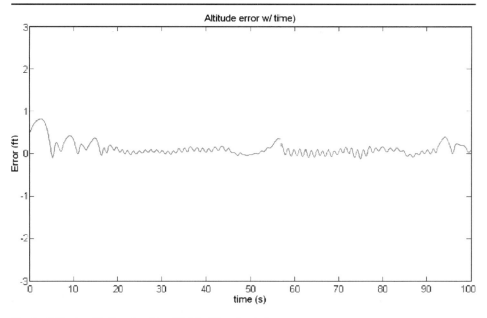

Fig. 8 Altitude with time tracking 20 ft in 20-mph winds

5 Results, Discussion, and Conclusion

The helicopter controller process continues to evolve over time. The following plots, discussion, and figures display the final results of the helicopter simulation and experimental setup and provide a perspective of the data evolution. This provides a better understanding of possible issues that may arise en route to the final result.

Fig. 9 Latitude and longitude with time in 20-mph winds

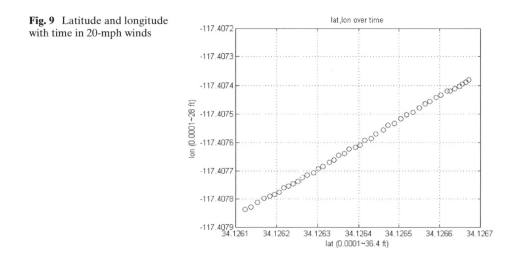

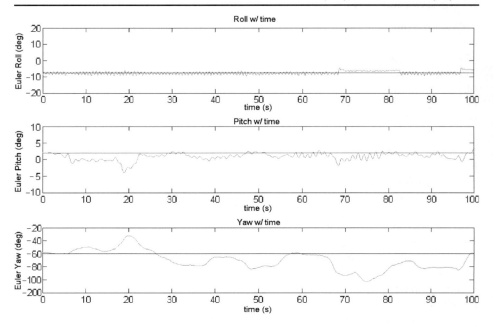

Fig. 10 Pitch, roll, and yaw in hover state against 20-mph winds

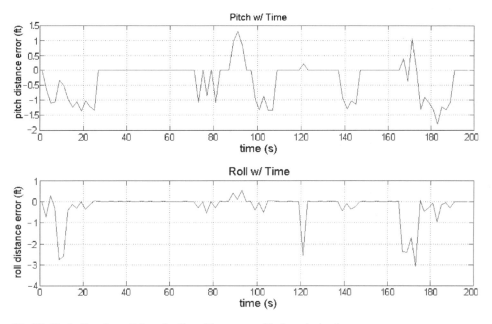

Fig. 11 Body direction pitch and roll position error with time during hover

5.1 Test Results

The set of controller parameters was acquired by a component build-up method. This method is similar to the recommended procedure given for tuning commercial off-the-shelf autopilots. The first step was to tune the tail collective controller while hovering the vehicle fractions of an inch on the ground. From there, the main rotor cyclic control was tuned at a slightly larger static main rotor collective. Finally, the main rotor collective was added to maintain the hover state of the rotorcraft about 5 ft off the ground. After satisfactory results were obtained with the hover controller, forward flight was tuned, making the considerations as noted in Section 2.2. All control gains in the preliminary test model and the validation model are given in Appendix (Fig. 6).

The hover state stands as a critically important flight mode for the VTOL, as one might expect. The experiment process started with tuning the PID controller for a hover condition. After this step was successfully completed, the process of achieving forward flight became somewhat simplified. Figures 7, 8, 9, 10, 11, and 12 demonstrate the hovering performance in different scenarios. A lot of the altitude tracking error results from a round off error of GPS position given in X-Plane of 30–60 in., but the results still show an average deviation of about 6 in. over 10 s in spite of round off.

The above figures show conditions for flight with and without wind. Note that the commanded pitch, roll, and yaw in Figs. 9 and 12 are shown as black lines where the results are shown as red lines. The results show that the responses without wind track very well, especially in the latitude and longitude position. Because of the expected noisy and inaccurate data that will come from an actual GPS, the hover controller

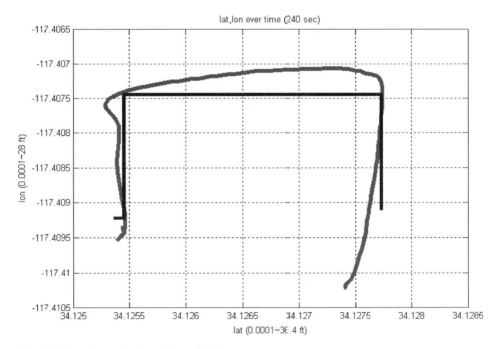

Fig. 12 Waypoint navigation with method A

does not use a position feedback. A controller such as this can quickly be integrated into the Simulink model and be considered acceptable over short periods of time. When we set the controller gains high but also limit their range of authority with high and low cut-offs, the helicopter is able to maintain nearly "dead-on" position for 90 s. Figure 8 shows three points which, when one looks at the two axes, are given to be virtually the same point. In fact, Fig. 8 has the same number of data collection points as Figs. 7 and 9, but their actual values are repeated as one of the three visible in Fig. 8. Figures 10, 11, and 12 show the same helicopter in 20-mile-per-hour winds. Though the vehicle does well at maintaining its desired heading, its position is much less stable than without the wind conditions. These winds are representative of a level 5 on the standard Beaufort scale.

Just as an exercise, since the proper implementation of position feedback for a PID controller in hover in an experimental setup has yet to be determined, a modest position feedback in hover for the simulation was implemented. We did this by only including proportional terms of −0.1 and −0.4 to both pitch and roll direction error, respectively. To attempt to make the control feedback a bit more akin to a physical autopilot controller, the position updates were reduced to 5 Hz to mimic the traditional update rate of GPS receivers. Figure 13 shows the results over a 200-s time span in X-plane.

In the case of waypoint navigation, multiple approaches can be considered valuable. In the case of getting from point to point quickly to deliver goods or follow faster ground and air vehicles, approach "A" is preferred. This approach is

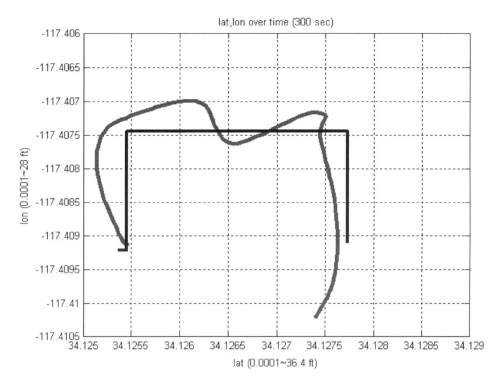

Fig. 13 Waypoint navigation as given in Fig. 13 but with method B

characterized by the helicopter pointing its nose in the direction of the waypoint and tilting its body forward to enter high-speed forward flight. As it approaches the current waypoint, its flight mode changes as it meets some boundary conditions. At this point, the vehicle picks up the next waypoint and adjusts its heading to continue. The following plot is shown in reference to this model.

The second approach is designed for use in close quarters and surroundings. This type of navigation is designated as approach "B". In this mode, the helicopter is allowed to travel to different waypoints at low speed and any given heading. This means the helicopter may have a steady state yaw attitude with respect to the next waypoint. This consequently results in much smaller but still non-zero roll and pitch attitudes to gain the desired velocity vector in addition to normal steady state trim conditions. The kink between the third and fourth waypoints in Fig. 5 is because the helicopter is attempting to correct for its desired heading. The vehicle was commanded to fly backwards and experienced some difficultly but was able to recover, as shown by the plot. Some of this instability is attributable to X-plane's poor treatment of the physics of small-scaled vehicles. In reality, tail control tends to be a bit better for R/C helicopters than is given in X-plane. The extension after the last waypoint in both plots is attributable to the landing sequence which is separate from the waypoint navigation.

6 Future Work

6.1 Validation and Long-Term Goals

The design of this simulation was based on having a tool to test control algorithms for actually UASs. Thus, as a means to validate the results of these simulations, building an R/C helicopter with autopilot hardware and software on board is a necessary step that will follow. Although one may argue that the Federal Aviation Administration-certified status of X-Plane would be enough, R/C helicopter setups can vary a great deal, even between two instances of the same aircraft design. This, coupled with typical simulation inaccuracies, makes hardware validation a logical step. Perhaps another method for validation would be a comparison to previous simulations of unmanned aircraft in the X-Plane and/or Simulink or some other comparable environment. Results from a literature search have given some indication as to what one should expect in this regard.

As a basic test for validation, the flight control er in this paper has been adjusted to control a Kyosho Concept 60SR II Graphite helicopter as described in the simulation and flight testing section of the dissertation by Shim. The actual vehicle was also modeled in X-Plane according to the estimated vehicle properties given in Tables 1 and 2 in "Appendix" of [6]. The graphic in Fig. 14 shows the model designed with consideration of all the large on-board avionics and autopilot equipment typical of a vehicle built in 1999. Results from emulating the flight plan in [6] are given in Figs. 15 and 16. The object of these figures is to show a similar flight plan as given in the paper by Shim. The relative precision of the flight plan in Shim's paper is about 1 m. Figures 15 and 16 show the autopilot's ability to navigate the modeled Kyosho helicopter to within 4 ft of the desired waypoint, which is on the same order of magnitude as 1 m (3.28 ft). The main conclusion is that the controller is adequate

Fig. 14 Berkeley URSA
minor 2 model in X-plane

to follow the waypoints given in the dissertation with the same vehicle parameters
provided in the dissertation. The controller was also able to do this with little or no
changes in the gains. Differences in the actual approaches can be attributed to things
such as the fact that the PID controller designed in Shim's paper is different than
the one designed in this paper. Additionally, Shim's paper takes advantage of the
limited flight envelope used for the waypoint navigation. As a result, he is able to

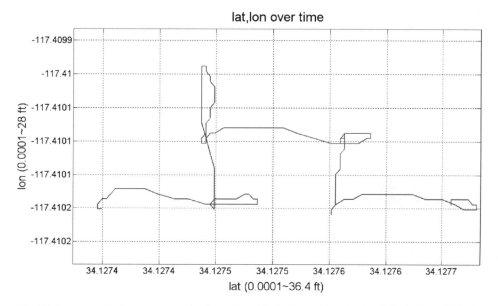

Fig. 15 Low speed, close range navigation plot of latitude and longitude of Berkeley vehicle in
X-plane/Simulink simulation environment

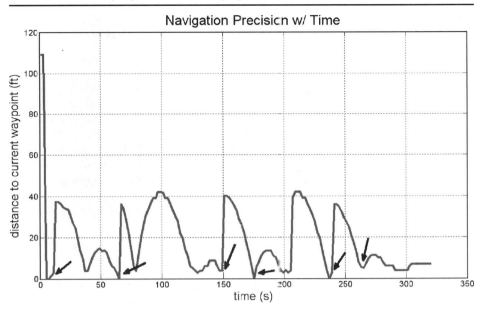

Fig. 16 Relative distance to current waypoint (*arrow*)

use linearized stability derivatives while the model in X-Plane uses an aerodynamic method with various aerodynamic losses as mentioned in Section 2.1 to generate the flight mechanics involved. The results of this validation thus point at the value of actually testing this controller on an experimental test bed such as an R/C helicopter.

In anticipation of the first suggestion for validation as a project, Section 6.2 discusses the equipment and expected procedure toward building such a mechanism.

Fig. 17 Pico-ITX board form-factor comparison. http://www.via.com.tw/en/ initiatives/spearhead/pico-itx/ Accessed 04, December 2007

6.2 Final Thoughts and Suggested Approach to Future Goals

The building of cheap UASs with commercially available equipment was once nearly impossible, but in the last 10 years, this impossibility has become more like a common practice to hobbyists and researchers alike. The general approach is to start with a moderately price R/C vehicle. In this case, some preferred platforms include Thunder Tiger's Raptor and Minicopter's Joker. These vehicular platforms would satisfy the payload requirement of approximately 5 lb and are readily accessible.

The autopilot will likely be programmed in C and placed on a Linux-based on-board PC with a Pico ITX board. Other possibilities include loading the autopilot on Qwerk, Gumstix, FPGA or Microchip[1] PIC platforms. All these computational platforms offer a small form factor as well as relatively advanced computational power output. Several papers have been published in recent years, documenting the use of such devices on unmanned vehicle test beds. The likelihood of these other platforms being used depends on the specific project and the requirements included with each project (Fig. 17).

Finally, to bring the UAS together, multiple sensors and mounting equipment will be required. This list includes an IMU/INS sensor, GPS, compass, altimeter laser and dampers (used to free the sensors of excessive vibrations). After all these items are fitted to the UAS platform and the proper ground test is completed, validation of the PID simulation will be possible.

Other likely research that will stem from this paper includes detailed analysis of sensor filtering techniques which may include Kalman filtering, some adaptive filtering or something like Bayesian filtering. The determination of this depends on what would best fit the data specific to this UAS project. To further improve the performance of the simulator and autopilot to follow, designing some type of intelligent control would be beneficial. This adaptive PID would have the ability to dynamically alter the gains independently without human input. Overall, the design of this test bed as a research vehicle will allow virtually unlimited possibilities in the future for research. Anything from formation flying, high level control augmentation, mission-specific scenario simulation, pursuit and evasion to machine learning techniques and vision-based navigation will become much more feasible after this step in further validation is taken. Also in the future, as more design parameters are available for rotorcraft in X-Plane, even if the system dynamics are still in a black box, simulated system identification may be a feasible option given the ability to record data out of X-Plane.

---

[1] PIC chips are a trademark of Microchip

Appendix

A1 Simulink Block Diagrams

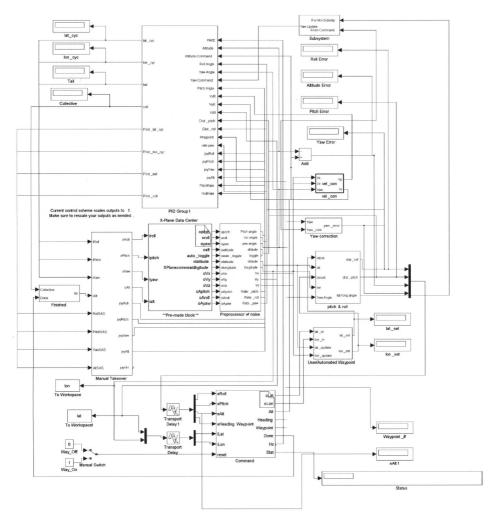

Complete Simulink block set (subsystems not shown)

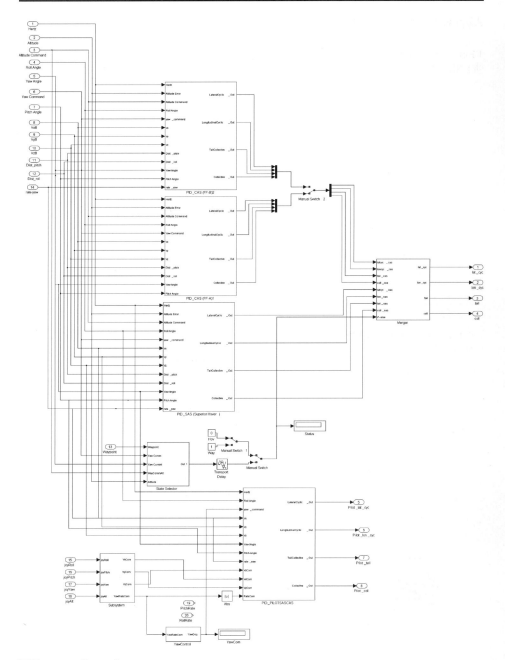

PID controller subsystems

A2 Basic Vehicle Specifications for Simulation

Table 1 Helicopter
characteristics

Vehicle property	Value
Main rotor radius	3 ft
Tail rotor radius	0.7 ft
Body max radius	0.5 ft
Max engine power	2.2 hp
Redline rpm (main)	1,250
Redline rpm (tail)	5,000
Vertical stabilizer length	1.4 ft
Horizontal stabilizer length	1.4 ft
Empty weight	13 lbs
Max take-off weight	16 lbs

Table 2 Control system gains
(USF model)

	Hover	FF-A	FF-B	Pilot CAS/SAS
$K_p\,\theta$	-1.75	$(-)0.1$	$(-)0.02$	-1.75
$K_d\,\theta$	-0.001	$(-)0.0005$	$(-)0.0005$	-0.001
$K_i\,\theta$	-0.5	$(-)0.0001$	$(-)0.0$	-0.5
$K_p\,\phi$	1.2	0.1	0.6	1.2
$K_d\,\phi$	0.0	0.025	0.04	0.0
$K_i\,\phi$	0.0	0.0	0.0	0.0
$K_p\,\psi$	0.004	0.0015	0.005	0.004
$K_d\,\psi$	0.4	0.0001	0.002	0.4
$K_i\,\psi$	0.001	0.08	0.001	0.001
$K_p V_x$	-0.5	0.0	-0.01	-0.5
$K_d V_x$	-0.0025	0.0	-0.0025	-0.0025
$K_i V_x$	-0.2	0.0	0.0	-0.2
$K_p V_y$	2	0.0	0.1	2
$K_d V_y$	0.001	0.0	0.1	0.001
$K_i V_y$	0.004	0.0	0.0	0.04
$K_p V_z$	0.05	0.005	0.05	2
$K_d V_z$	0.0015	0.0	0.0015	0.0015
$K_i V_z$	0.0	0.0	0.0	0.02
K_p h	1.5	0.05	1.5	0.0
K_d h	0.1	0.001	0.1	0.0
K_i h	2	0.0	2	0.0
K_p coupling	0.0	0.001	0.0	0.0

Table 3 Control system gains (Berkeley model)

	Hover	FF-A	FF-B	Pilot CAS/SAS
$K_p\,\theta$	−1.75	(−)0.2	−1.75	−1.75
$K_d\,\theta$	−0.001	(−)0	−0.001	−0.001
$K_i\,\theta$	−0.05	(−)0.01	−0.5	−0.5
$K_p\,\phi$	1.2	0.3	1.2	1.2
$K_d\,\phi$	0.0	0.035	0.0	0.4
$K_i\,\phi$	0.0	0.0	0.5	0.0
$K_p\,\psi$	0.004	0.0015	0.004	0.004
$K_d\,\psi$	0.4	0.0001	0.4	0.0004
$K_i\,\psi$	0.001	0.08	0.001	0.001
$K_p V_x$	−0.8	0.0	−0.08	−0.5
$K_d V_x$	−0.0025	0.0	−0.0025	−0.0025
$K_i V_x$	−0.2	0.0	−0.2	−0.2
$K_p V_y$	2	0.0	2	2
$K_d V_y$	0.001	0.0	0.001	0.001
$K_i V_y$	0.04	0.0	0.04	0.04
$K_p V_z$	0.05	0.005	0.05	4
$K_d V_z$	0.0015	0.0	0.0015	0.0015
$K_i V_z$	0.0	0.0	0.0	0.02
K_p h	1.5	0.05	1.5	0.0
K_d h	.1	0.001	.1	0.0
K_i h	2	0.0	2	0.0
K_p coupling	0.0	0.001	0.0	0.0

References

1. Garcia, R.D.: Designing an autonomous helicopter testbed: from platform selection to software design. Ph.D. Thesis, Department of CSEE, UVS. The University of South Florida (2008)
2. Green, W.E.: A multimodal micro air vehicle for autonomous flight in near-earth environments. Ph.D. Thesis, Drexel University (2007)
3. Meyer, A.: About X-plane. http://x-plane.com/about.html. Accessed 19, November 2007
4. Minicopter. www.minicopter-canada.com/. Accessed 20 November 2007
5. Nelson, R.C.: Flight Stability and Automatic Control. McGraw Hill, Boston (1998)
6. Prouty, R.W.: Helicopter Performance, Stability and Control. Kieger (printed 2002, errata 2005)
7. Shim, H.D.: Hierarchical flight control system synthesis for rotorcraft-based unmanned aerial vehicles. Ph.D. Thesis, The University of California: Berkeley (2000)
8. WowWee Products. Product Wiki. www.productwiki.com/wowwee-flytech-dragonfly/. Accessed 20 November 2007
9. www.commons.wikimedia.org/wiki/Image:Align_T-Rex_2386_mini.jpg. Accessed 20 November 2007
10. www.e-fliterc.com/Products/Default.aspx?ProdID=EFLH1250. Accessed 20 November 2007